Investigating the EARTH

Fourth Edition

Investigating the EARTH

Fourth Edition

WILLIAM H. MATTHEWS III
CHALMER J. ROY
ROBERT E. STEVENSON
MILES F. HARRIS
DALE T. HESSER
WILLIAM A. DEXTER

Sponsored by the American Geological Institute and based on the original Earth Science Curriculum Project.

Houghton Mifflin Company / Boston
Atlanta Dallas Geneva, Ill. Hopewell, N.J. Palo Alto Toronto

William H. Matthews III is Regents' Professor of Geology at Lamar University, Beaumont, Texas. In addition to articles in scientific and professional journals, Professor Matthews is the author of more than two dozen books. He is past president of the National Association of Geology Teachers and in 1965 received the NAGT Neil Miner Award. He is currently serving as Director of Education for the American Geological Institute and Chief Technical Consultant for the AGI-Encyclopaedia Britannica Educational Corporation Earth Science Film Series. Professor Matthews also serves as chairman of the Worldwide Commission on Geology Teaching of the International Union of Geological Sciences.

Chalmer J. Roy is Professor of Geology and Dean Emeritus of the College of Sciences and Humanities at Iowa State University. He is the author of over 50 scientific and educational publications. He has engaged in research in Alaska, served as a specialist in geological education in India, and acted as an educational consultant in the Philippines. Dean Roy was chairman of the ESCP steering committee.

Robert E. Stevenson is the Scientific Liaison Officer for the Office of Naval Research and is based at Scripps Institution of Oceanography. He has been a university professor and formerly directed oceanographic research for the U. S. Fish and Wildlife Service. Dr. Stevenson is the author of many scientific articles and has pioneered the new field of space oceanography.

Miles F. Harris is an occasional technical editor for the American Meterological Society. Formerly a research meteorologist for the U. S. Weather Bureau and Editor of the *Monthly Weather Review,* he is the author of a number of research articles appearing in the *Review* and other scientific and engineering journals. He has contributed to several encyclopedias, including *The Encyclopedia Americana* and *The Planet We Live On.* Mr. Harris is the author of two popular books for children and young adults, as well as a vocational guidance book, all in the field of meteorology, and has served as a consultant for the AGI-EBEC earth science film and filmstrip series.

Dale T. Hesser is Director of Science for the North Syracuse Schools, North Syracuse, New York. He has more than 20 years classroom teacher's experience, ranging from junior and senior high to NSF in-service, summer institutes, and teacher workshops. He holds both an M.S. and a Certificate of Advanced Studies in Science Education from Syracuse University. Mr. Hesser was one of the original ESCP test center teachers and in 1975 received the Outstanding Earth Science Teacher Award from the National Association of Geology Teachers.

William A. Dexter is a consulting geologist for Energy Operating Corporation in Dallas, in charge of research and development. He was an earth science instructor at St. Marks School of Texas for eighteen years and was director of the Planetarium and the William A. Dexter Observatory. Mr. Dexter has taught earth science, geology, and astronomy to science teachers at seven different colleges and universities. He is the author of several books and articles and was one of the original ESCP authors. In addition, he received the Outstanding Earth Science Teacher Award from the Texas Section of the National Association of Geology Teachers.

Printed in U.S.A.

Student's Edition ISBN: 0-395-32070-4

Teacher's Edition ISBN: 0-395-32071-2

Cover: The Alberta Rockies, Canada.

PREFACE TO THE FIRST EDITION

We would like to tell you how this book came about. Hundreds of people worked more than three years to prepare it. Why were so many people involved? Why did the work take so long?

The scientists and educators who planned *Investigating the Earth* wanted many different persons to be involved. They sought the help of scientists in many fields to make sure that the basic principles in all these fields formed an integrated and up-to-date story of planet Earth and its environment in space. They wanted advice from teachers using the book about how young people could best investigate and learn. Finally, they wanted the reactions and opinions of students like yourself—what was exciting for them and what helped them to learn.

At the beginning of this project, a planning group prepared an outline for a science book that would encompass the story of the planet Earth. They then invited 40 scientists and teachers to meet and write the first version of the book. Astronomers, geologists, geographers, geophysicists, meteorologists, oceanographers, soil scientists, science educators, and teachers came to Boulder, Colorado to prepare manuscript for the book.

The first version of *Investigating the Earth* was sent to 77 teachers in schools across the country. During that first year it was used by 7,500 students. Each week the teachers sent their comments and the comments of their students back to the ESCP staff. The following summer another group of writers assembled in Boulder to write a second version of the Text. Changes in that Text were based on the reactions of the teachers and students who had used the book. The second version was also evaluated in many schools and involved thousands of students. The comments of teachers and students were gathered each week, studied, and used to prepare the third and final version of the book during the spring and summer of 1966. This is the book you are now reading. The many people involved in its preparation hope that their efforts have produced a stimulating book, one that will make your investigation of the earth more interesting.

The contents of this book may raise many new questions in your mind. You will answer some of these questions yourself by observing and performing investigations. Some will be answered in the Text and others by your teacher. Many will remain unanswered. When you read newspapers and magazines you find that you are not alone in wondering about these unanswered questions. Thousands of people such as scientists, philosophers, and teachers are constantly inquiring into the unknown.

Although basic principles are modified slowly, many of the ideas presented in this book are changing rapidly as man expands his knowledge. You will find it interesting to understand and keep pace with these advances. The people who worked on *Investigating the Earth* have attempted to give you some of the exciting developments in earth science by letting you find answers for yourself. They hope that in this way you may better appreciate future discoveries and perhaps participate in them yourself.

Ramon E. Bisque
Robert L. Heller

PREFACE TO THE FOURTH EDITION

The authors of the Fourth Edition of *Investigating the Earth* were all part of the team of earth scientists who wrote the original ESCP text, and all but one were on the revision team that prepared the 1973 edition of *Investigating the Earth.* We have retained the original unified approach to earth science in this 1984 version of the book, while including the latest geoscience knowledge derived from global tectonics and the "new" geology. Additional information derived from recent developments in meteorology, oceanography, and astronomy has also been added. The Fourth Edition utilizes the investigative and inquiry-oriented approach of its predecessors, with numerous investigations and additional "hands on" activities to stimulate student involvement.

The material in this book was reviewed by many experienced earth science teachers, who have sent their comments to us. They contributed comments and suggestions to help make the book more useful. The authors acknowledge their help with thanks.

William H. Matthews III

NOTE: Any references to evolution in this volume are presented as theory rather than verified fact.

Key to Pronunciation

Some words in this book that may be unfamiliar to you or hard to pronounce are followed by a pronunciation guide in parentheses. The respelled words are divided into syllables and appear in *italic* letters. The guide syllable printed in capital letters should be given the greatest emphasis when you say the word. In the key below, the first column is a list of the letters or marks used in many dictionary pronunciation guides, with an example of the sound that each one stands for. The next column shows how the same sound appears in the guides in this book. The third and fourth columns list examples of words that contain that sound, and their pronunciation guides.

Letter or mark	*Appears in this book as:*	*Example*	*Guide*
ă (hat, map)	*a*	alphabet	*AL fuh beht*
ā (age, face)	*ay*	Asia	*AY zhuh*
â (care, air)	*ai*	share	*shair*
ä (father, far)	*ah*	farming	*FAHR mihng*
ch (child, much)	*ch*	China	*CHY nuh*
ĕ (let, best)	*eh*	test	*tehst*
ē (equal, see, machine, city)	*ee*	leaf	*leef*
		tangerine	*tan juh REEN*
ĭ (it, pin, hymn)	*ih*	system	*SIHS tuhm*
ī (five, ice)	*y*	alive	*uh LYV*
	eye	island	*EYE luhnd*
k (coat, look)	*k*	cake	*kayk*
ŏ (hot, rock)	*ah*	otter	*AHT ur*
ō (open, go, grow)	*oh*	solo	*SOH loh*
ô (order, all)	*aw*	normal	*NAWR muhl*
		always	*AWL wayz*
oi (oil, voice)	*oy*	boiling	*BOYL ihng*
		poison	*POY zuhn*
ou (house, out)	*ow*	fountain	*FOWN tuhn*
s (say, nice)	*s*	mice	*mys*
sh (she, revolution)	*sh*	ration	*RASH uhn*
ŭ (cup, butter, flood)	*uh*	study	*STUHD ee*
		blood	*bluhd*
ûr (term, learn, sir, work)	*ur*	earth	*urth*
u,o͝o (full, put, wood)	*u*	pull	*pul*
		wool	*wul*
ü,o͞o (rule, move, food)	*oo*	bruise	*brooz*
zh (pleasure)	*zh*	measure	*MEHZH ur*
ə (*a*bout)	*uh*	America	*uh MEHR uh kuh*
(tak*e*n, purpl*e*)	*uh*	middle	*MIHD uhl*
(penc*i*l)	*uh*	citizen	*SIHT uh zuhn*
(lem*o*n)	*uh*	lion	*LY uhn*
(circ*u*s)	*uh*	focus	*FOH kuhs*
(curt*ai*n)	*uh*	mountain	*MOWN tuhn*
(sect*io*n)	*uh*	digestion	*dy JEHS chuhn*
(fabul*ou*s)	*uh*	famous	*FAY muhs*

Table of Contents

UNIT THREE
THE ROCK CYCLE

UNIT FIVE PLATE TECTONICS

UNIT SIX
EXPLORING THE UNIVERSE

unit I
Introduction to the Earth

Prologue:

Investigating "Change"

CHANGES *are all around you. What does the word "change" mean? If you try to define change, you will see how difficult it is to give a meaning that is acceptable to everyone.*

Consider this definition. Change has taken place when anything that is observed is different from what it was when last observed. Is this definition accurate? Can you think of changes not described by this statement?

Changes are of two types: intentionally-caused changes and natural changes. If you perform an experiment you are trying to make something different in some way. You can make water salty by stirring some salt into it. Both the salt and the water have been changed. Change from day to night, spring to summer, or rainy to dry weather are changes we do not cause and cannot control.

Natural changes occur constantly, but you are seldom aware of any difference from one minute, or even one hour, to the next. Although a tree near your window may change with the seasons, the tree itself remains in place. If you were to glance out the window some morning and find that the tree had vanished, you would probably be surprised, perhaps even alarmed.

Sometimes things change fast. Suppose you heard that people had seen an island appear in one day *in the open ocean. One November day in 1963 in the North Atlantic Ocean near Iceland, a volcanic island suddenly emerged from beneath the surface of the ocean. The island, which is now called Surtsey, is shown in Figure 6-14.*

Sometimes things change without appearing to have changed at all. For example, a pebble lying on a beach is heated by the sun's rays. At night the pebble cools down. To anyone who did not touch the pebble at both times it would not seem to have changed. However, to an insect walking across the pebble during the day and again at night, the pebble would seem to have changed a great deal. So to some extent, recognizing change depends on point of view. What does this suggest about the definition of change given earlier?

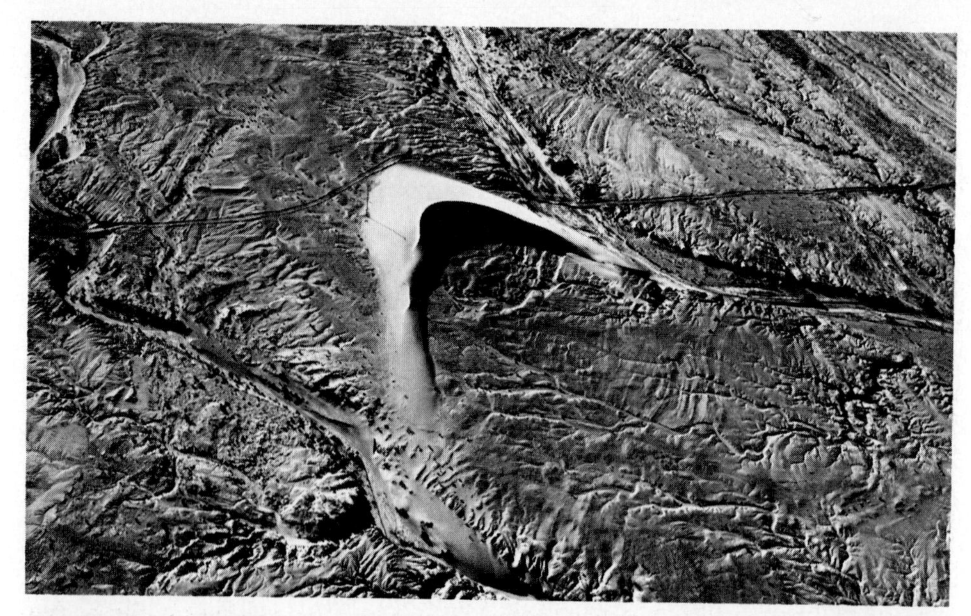

Figure P-1 Two photos of a sand dune, that were taken seven years apart. How can you prove which photo was taken earlier?

P-1 Observing

In determining whether changes have taken place, careful observations are necessary. Our ancestors made observations of the earth and of earthly objects for thousands of years. They had few of our instruments or systems of units but they were curious. It is remarkable how well they observed so many things. You have the most important equipment they had—the human mind.

Test your observational skills with this brief exercise. Figure P-1 consists of two photos of a sand dune in southern California. The photos were taken seven years apart. Observe the photos carefully and see if you can decide which photo is the earlier one. What can you tell, if anything, about the actual direction in which the dune is moving? At what time of day was each photo taken? What kinds of changes do you observe?

Frequently, direct observation does not give enough information to answer a question. Direct observations may be extended or tested by investigations. Investigations in earth science are nearly always based on questions left unanswered by direct observation. The number and importance of such questions is increasing rapidly.

As you perform the following investigation, consider what you are doing in each step. Is observation the only process you use?

Figure P-2 *Why does a piece of granite sink to the bottom of one beaker and float on the liquid in the other beaker?*

P-2 Investigating Mass, Volume, and Density

Look at the two beakers in Figure P-2. Each beaker contains a liquid and a solid. In the beaker on the left, the liquid is water and the solid is a piece of rock (granite). The beaker on the right also contains a piece of granite, but the liquid is mercury. What difference between the mercury and the water accounts for what you observe?

Before you can answer this question, you must learn about mass and density. The **mass** of a substance is the amount of matter it contains. The **density** is its mass per unit of volume. The density of a substance can be calculated by dividing its mass by its volume. The **volume** of a substance is the amount of space it occupies. Density of a solid is usually expressed in terms of grams (mass) per cubic centimeter (volume).

$$\text{Density} = \frac{\text{Mass (g)}}{\text{Volume (cm}^3\text{)}}$$

This means that you can obtain the density (D) by dividing the mass (M) by the volume (V), as follows:

$$D = \frac{M}{V}$$

Suppose that an object has a mass (M) of 100 grams and a volume (V) of 20 cubic centimeters. What is its density (D) in grams per cubic centimeter? The number you calculate is the mass (5 g) of one cubic centimeter, and thus the density, 5 g per cm^3 (5 g/cm^3), of the substance.

When you have completed this investigation, you will have found the densities of several objects and compared the density of ice with that of water.

PROCEDURE

MATERIALS
balance, objects for density determination, ruler (or other measuring instruments)

Calculate the density of each of the objects given to your group. To do this, you must know both the mass and the volume of the objects. Use a balance to determine the mass. Volume can be determined in many ways. After a class discussion, decide what method or methods you will use. Determine and record the mass and volume of each object. Make a table to help you record and organize your data. Then, calculate the density of each object.

DISCUSSION

1. What effect does the shape of a substance have on its density? Explain your answer.
2. What effect does the difference in the amount of the sample have on the density of the modeling clay? Explain your answer.
3. Arrange your materials in order of decreasing density.
4. What is your calculated value for the density of water?

MATERIALS
balance, ice cube, other equipment (depending on method used)

PROCEDURE

Now that you are familiar with density, you are ready for another problem. Observe the demonstration by your teacher. Using the materials at your station, determine the approximate density of an ice cube.

DISCUSSION

1. What is the approximate density of your ice cube?
2. Explain how you obtained this value.
3. Sometimes ice cubes have holes or air spaces in them. Would these spaces affect the density of the ice cube?

P-3 Hypothesis—Theory—Fact

Now you have had some experience with observing and investigating. You have also gained some experience in recording your observations and reaching conclusions. In your observations of density or your observation of the sand dune, you may have developed two or more possible answers to the same question. In science such a possible but not positive answer is called an **hypothesis** (*hy PAHTH uh sihs*). By further testing you might have gotten rid of all but one of your hypotheses. You still may not have been really sure of the answer. More testing even with additional equipment might neither prove nor deny your remaining hypothesis. When an hypothesis has been tested in every way possible but has not been proven, it can safely be called a *theory* but not a *fact*.

P-4 Causes of Change

You might agree that the kind of matter that changed most during Investigation P-2 was the ice. The explanation in this instance seems rather simple. The ice became warmer and melted at room temperature. The ice took in heat energy. From this observation you can conclude that change may result from the interaction of matter and energy. Water is a very unusual kind of matter as well as one of the most important substances on Earth. You know that water can exist in the solid, liquid, and gaseous (vapor) states at room temperature. If you take heat from water and cool it to a temperature below 0°C (Celsius) it will crystallize to form ice. At any temperature above 0°C the ice will melt. If water is heated to a temperature of 100°C it will boil away as a vapor. Can you think of an experiment to prove that there is nearly always some water vapor in the air? You will soon discover that many changes in the atmosphere, the oceans, and the solid earth are caused by interactions between matter and energy.

Matter also changes through interaction with force. One well known force is the **force of gravity.** The earth attracts all objects at or near its surface. Earth's gravity is one aspect of a broader theory of gravitational attraction. It seems that every bit of matter has this attraction for every other bit of matter. Testing the theory of gravitation has led to an assumption that all matter contains particles within the atoms that produce this attraction. Scientists are busy trying to prove the existence of this particle—the graviton. Watch for gravity and gravita-

Figure P-3 *This sign in Colorado reminds motorists of the force of gravity.*

tional attraction in the chapters ahead. It is an important force in many of the earth changes you will investigate (Figure P-3). You will also learn of other forces that interact with matter to produce and resist change.

Most changes in objects involve motion of some kind. Objects may grow, shrink, change position, come apart, or come together with other objects. These are obvious examples of motion. It has been said that every atom in your body will be replaced by another atom in seven years. Some additional atoms will be added as you grow. Atomic replacement can occur in any object, is a form of change, and requires the motion of atoms. Motions are produced by forces and usually involve energy, as you will discover.

Motion of an object takes time, and there must be space in which the motion can occur. You have heard much about "space exploration," "nearby space," and "outer space." Outer space is important because it is where the earth, the other planets, and the stars move and go through other changes. These motions and changes require periods of time ranging from minutes to many millions of years.

Most of the motions observed in the atmosphere, oceans, and solid earth also require space and time. A flash of lightning (Figure P-4) requires the space between a cloud and the earth (or another cloud) and about one fifty-thousandth of a second of time. Contrast the rate of a lightning flash with the rate of uplift in a mountainous area. Some areas of the Sierra Nevada in California have been measured carefully over a period of years. The land in these areas is rising at the rate of a few centimeters per century. You might say that lightning occurs instantly and the uplift of mountains takes forever (Figure P-5). The words "instantly" and "forever" are not very useful descriptions of time. You will find that most of the changes investigated in this course occur at rates expressed in the usual units of time—seconds, minutes, hours, days, years. Some earth changes, such as mountain uplift, may be de-

Figure P-4 *Lightning passes between clouds and the earth's surface in Wyoming.*

scribed as being "older" or "younger" than some other changes.

Consider time in a personal way. How old was your mother when you were born? There will be a time when you are half as old as your mother. Does this mean that you are aging faster than your mother? She will always be older than you but only once will she be twice as old. Actual and relative time are both important in earth science.

P-5 Models

Most scientists find it helpful to illustrate their hypotheses, theories, facts, and guesses. Such illustrations help scientists make sure that they understand their ideas. "If you understand it you can draw a diagram of it." The illustrations also help others understand the idea and look for ways to test it. For want of a better name these illustrations are called **models.** There are two kinds of models commonly used by scientists: scale or physical models, and mental models (Figure P-6).

You are familiar with models of people (dolls), airplanes, trains, and houses, among others. These are called **scale models** because they represent the real object at a smaller size. Small objects can be illustrated by models of a larger size.

Figure P-5 *The bench mark shown in the closeup and indicated by the arrow is in Cajon Pass, California. It rose 17 cm in 37 years.*

***Figure P-6** Models of an atom (left) and the earth (right). Which would you think is a mental model and which a physical model?*

The scale is expressed as a fixed ratio between the size of the model and the size of the real object. If a doll 25 cm (centimeters) tall is a scale model of a basketball player 200 cm tall, the ratio is 25:200 or 1:8 or 1/8. In other words one unit (centimeter, or width of one's finger) on the model equals eight units on the real thing.

The globe (30 or 40 cm in diameter) in your classroom is a scale model of the earth, which is 12,740 km (kilometers) in diameter. The globe is so small compared to the real thing that any statement of the ratio is almost meaningless. It can be done as follows: 30 cm: 1,274,000,000 cm or 1 cm: 42,500,000 cm or 1:42,500,000.

Earth scientists also use models in which both time and space are represented to scale. You may investigate some aspects of erosion by streams on a stream table, as the students are doing in Figure P-7. This allows you to observe in a few minutes and on an area of about half a square meter what might happen in years on an area of a few square kilometers. Accuracy in these experiments would require that everything be carefully scaled, which is impractical in your work.

Sometimes scientists find it helpful to have models of things they cannot see directly. Mental models are often based on behavior, rather than on observable physical properties. A good example of a mental model is the **atom.** An ancient Greek philosopher invented the word "atom" to refer to the smallest indivisible particle of matter. By the 16th century A.D. it was proposed that all solid matter consisted of tiny balls held together by hooks. This model helped in explaining the behavior of solids. No one has ever seen an atom but atoms have been measured. In our present model, the hooks have been replaced by electromagnetic forces. The atomic theory is far advanced but the best existing model will continue to be tested for many years. The present model of the atom is a mental model based more on the behavior than on the measurements of atoms.

Models of the atom illustrate the most important value of models—they can change. The real objects change either slower or faster than the models. Changes in models reflect the change in available knowledge about the object. Changes in knowledge come mainly as a result of testing models. Models provide guides to testing, the best way in which science can progress toward the truth.

This suggests two very interesting facts about science that you should keep in mind. First, science is an invention of the human mind motivated by curiosity and supported by imagination and logic.

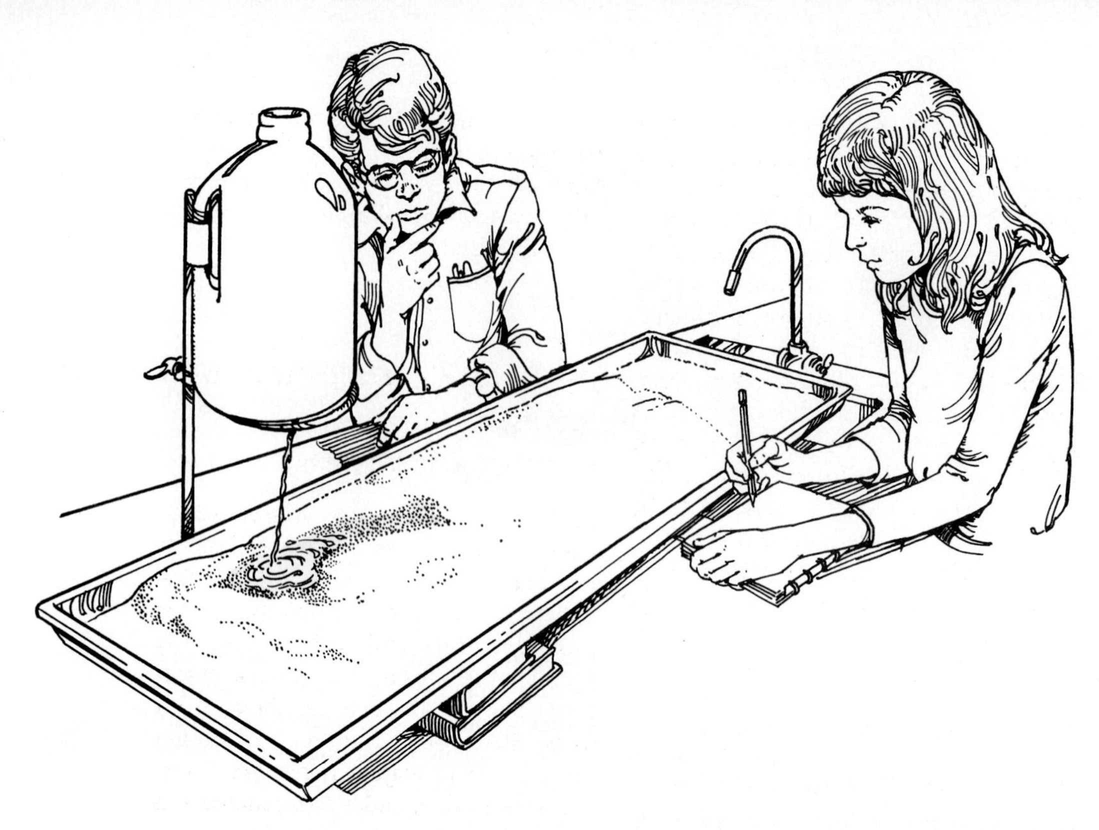

Figure P-7 *Water flowing down a stream table is a model of a real stream flowing across the land.*

Second, in spite of all we know, or think we know, much more remains to be learned. Science is an unfinished business.

P-6 Reporting Investigations

In this book there are many investigations. Ever since people have carried out investigations, they have felt the need to keep records or reports of them. Why is this practice useful? What should a report include?

There is no single best way to record and report an investigation. Generally a good report should include three main parts of an investigation: (1) why you did the investigation, (2) what you did, and (3) what you found out. Another way of expressing this would be (1) purpose, (2) procedure, and (3) results. In some of the investigations in this book the purpose and procedure will be stated for you. In this case, briefly write each of these in your report so that the investigation can be reviewed later. When you yourself design the procedure, it is particularly important that you describe exactly what you did. Then someone else reading the report could repeat your investigation in order to test your results.

When scientists perform investigations, they write reports similar to the ones you will write for your investigations. If the reports are published, the work becomes useful to other scientists. The results can be tested by others and used to discover more about the problem they are working on. Similarly, writing reports will help you to organize your information and allow you to share it with your classmates.

In the investigations that follow you will be observing changes that take place inside the earth and in the air. Many observations are required in each investigation. During the weeks ahead, try to organize the information you gather so that you can look for patterns of change. If you discover patterns, determine whether they are patterns in time or space or both. See if any of the patterns suggests a model to you. Try to discover whether or not the changes are predictable. If the changes do not seem to be predictable, try to determine what information might be needed to predict them. How could the needed information be gathered?

P-7 Investigating Patterns of Change—Weather Watch

Weather is the condition of the atmosphere at any one time and place.

When you have completed this investigation, you will have discovered weather patterns in your area. To discover what weather patterns are in your region, you will have to collect information for a long time.

PROCEDURE

The class should observe and measure the weather for the next two and one-half months. Figure P-8 is one possible way to organize your data.

During this time you will keep track of changes in the atmosphere from one day to the next. Use a wall chart to record these changes and see if you can discover any patterns to them. Your observations and records will involve time patterns (in the form of a graph) and space patterns (in the form of a map). See Appendix B to find out what to measure.

Discuss with your teacher how the chart can be made so that your observations and measurements will contribute most to the discovery of patterns. By the time you are studying Chapter 3 you may be ready to make some of your own weather predictions.

P-8 Investigating Patterns of Change—Earthquake Watch

In February 1971 the ground shook under Los Angeles, the most heavily populated county in America. The first upheaval collapsed hospitals, tossed around bridges and slabs of highway concrete, and tore up sewers, water pipes, and electric cables. In the next few days there were hundreds of minor quakes and 12 major aftershocks. Sixty-two persons died under collapsing beams and rubble. Many more would have died except that the earthquake took place early in the morning, before highways and streets were crowded.

Scientists estimate that as many as 1000 to 5000 measurable earthquakes take place around the world every day! Do earthquakes occur everywhere? How often do major quakes occur?

When you have completed this investigation, you will have made long-term observations that will give you the answers to these questions.

PROCEDURE

You will use the data from sensitive detecting instruments called **seismographs** (*SYZ muh grafs*). Seismographs detect the almost continuous trembling of the earth at many different points on the globe. Analysis of the data from three or more seismographs can pinpoint the time and location of an earthquake. It also indicates their depth and magnitude. Perhaps you can discover patterns in earthquake locations and frequency.

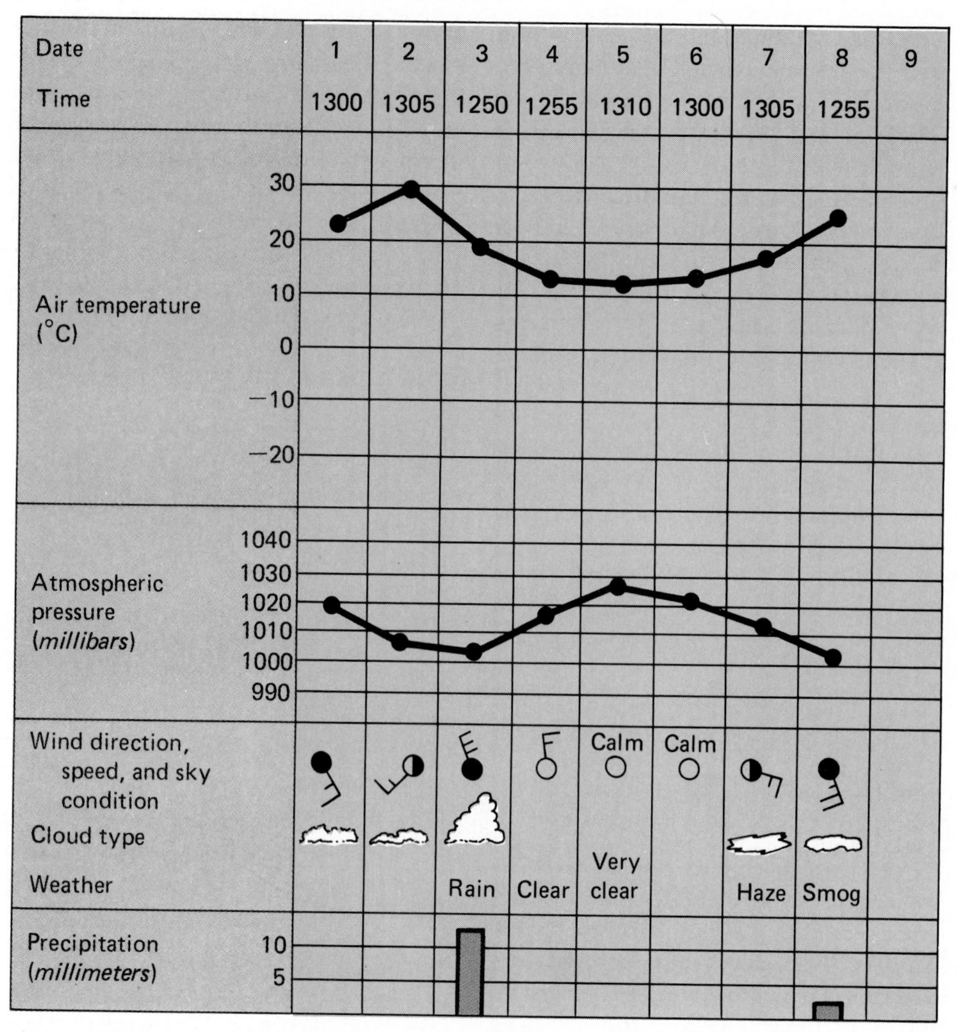

Figure P-8 *A sample weather watch chart.*

The information needed for this investigation comes from the United States Department of the Interior, Geological Survey.

The geographic location of an earthquake on the earth's surface is not the same as its point of origin. The actual place from which an earthquake originates is called the **focus.** It is beneath the surface. The geographic point on the earth's surface directly above the focus is called the **epicenter** (*EHP uh sehn tur*). Epicenter locations are given in latitude and longitude. The focus is given as shallow (less than 70 km), intermediate (70–300 km), or deep (300–700 km). Very few quakes occur below 700 km.

Use map pins to plot the positions of the epicenters. Pins of different colors should be used for shallow, intermediate, and deep earthquakes.

P-9 Looking Ahead

As you do more investigations, you will see that there is no set route for an investigation to follow. Observing and gathering information lead to questions and possible interpretations or hypotheses. These hypotheses suggest further observations, measurements, calculations, and testing.

Even with the many new instruments and equipment available today, methods of observation in the sciences are still limited. There are still places to which we cannot go. We cannot yet visit a star or go very far down into the earth. Instruments are still imperfect, too. However, instruments are constantly increasing our powers of observation. Investigations are becoming more complicated and the amount of information they provide is increasing from day to day. New information always leads to new questions. The more we investigate, the more we find to investigate.

1 The Earth in Space

STONEHENGE, pictured here, was completed about 1650 B.C., yet only recently have we discovered a probable reason for its construction. We now believe that the prehistoric people who built this ring of rocks used it as a highly accurate instrument for keeping track of the motions in the sky of the sun, moon, and stars. An electronic computer has shown that this stone observatory could have enabled the users to keep account of the passage of time, to follow the seasons, and probably even to predict the exact day of eclipses of the sun and moon.

Even though the early scientists at Stonehenge may have made all of these observations, it probably never occurred to them that the earth itself was moving, as well as the sun, moon, and stars. They believed that when the sun rose it really was coming up around the rim of the earth. They thought that the sun would make its journey through the skies, set in the west, and mysteriously find its way back to rise again the next day. On about the same day in early summer each year, the sun came up exactly over the pointed rock called the sighting stone, placed beyond the ring of rocks. From that day on for six months, the sun would rise each day a little more to the south and then it would move northward to again rise behind the stone the next summer.

There is nothing to suggest that these people had any explanation for what they observed. The monument they constructed is evidence enough that they considered their observation important.

AFTER COMPLETING THIS CHAPTER, YOU SHOULD BE ABLE TO:

1. **explain how ancient observers discovered the directions north, south, east, and west.**
2. **list the observations and problems which led to testing of models of the solar system from Ptolemy to Newton.**
3. **explain why someone placed a bouquet and note on the tomb of Newton when the astronauts landed on the moon.**
4. **list and describe each of the more important motions of the earth.**
5. **list the kinds of information that are shown on maps used for a variety of purposes.**
6. **explain the importance of thinking to scale and thinking spherical in earth science.**
7. **explain why a flat map cannot be accurate in shape, size, direction, and distance.**

Druids to Ptolemy

1-1 Ancient Observers

Ancient civilizations developed in the areas around the eastern end of the Mediterranean Sea. Much of this area is desert and the clear night air is excellent for observing the stars. People everywhere in this region must have observed the stars, the moon, and the sun. The earliest and most complete records of their observations have come to us from the ancient Greeks.

The ancient Greeks thought that the earth was fixed, and that the sun, moon, and stars revolved around it. They recognized one star in the northern sky that appeared to remain in the same position. Nearly all other stars rotated slowly about this star. There were five stars that seemed to wander across the background of the others. To these they gave the name *planets* (wanderers).

When the ancient Greeks traveled northward they noticed that the "fixed" star in the northern sky appeared higher and higher above the horizon. When they returned home the star returned to its familiar position. They also noticed that sailing ships putting out to sea slowly disappeared, but in a curious way. The hull of the ship disappeared first and finally the top of the sail and tip of the mast. The ships disappeared in exactly the same fashion regardless of the direction in which they were moving. These observations were interpreted to mean that the earth is spherical.

Much earlier (3000 B.C.) the Babylonians had divided the circle into 360 parts representing the daily steps of the sun in

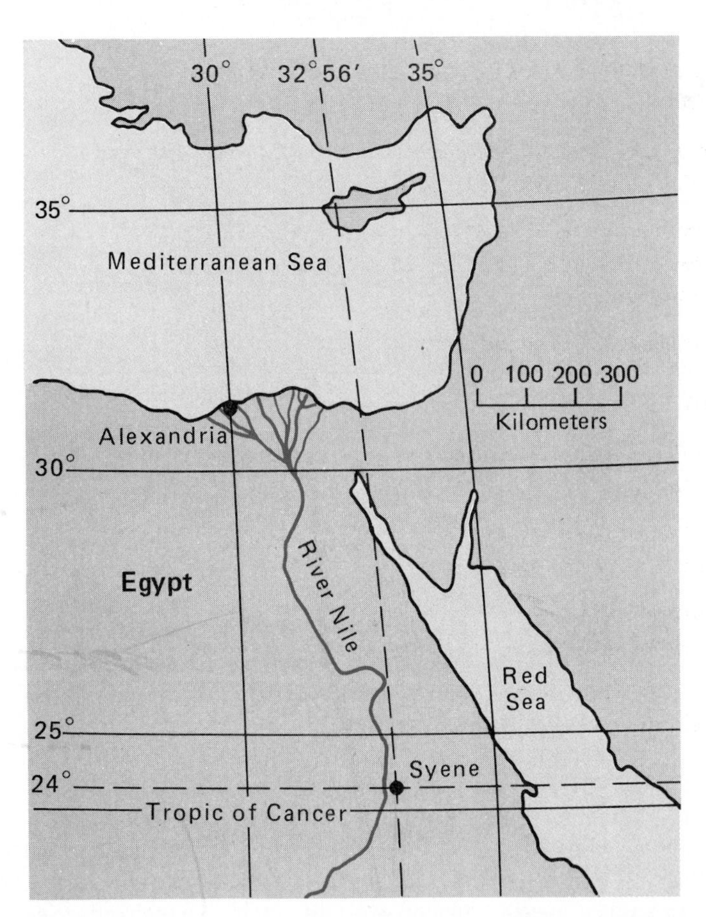

Figure 1-1 *The region involved in the calculations of Eratosthenes.*

its yearly journey across a background of stars. The Greeks used the same divisions. Today we know them as **degrees** (°). To provide greater precision in measuring circles we now divide degrees into 60 **minutes** (′), and minutes into 60 **seconds** (″). If the earth is spherical we can measure distances on its surface in degrees, minutes, and seconds.

A Greek scientist, Eratosthenes (*eh ruh TAHS thuh neez*), made a surprisingly accurate estimate of the earth's circumference about 230 B.C. In the great library in Alexandria he read about a deep vertical well near Syene (now Aswan) in southern Egypt. The well was entirely lit up by the sun at noon once a year. Eratosthenes reasoned that at this time the sun must be directly overhead, with its rays shining straight into the well. Alexandria is almost 900 km north of Syene. He knew that in Alexandria at noon on that same day a vertical object cast a shadow. Therefore, the sun was not directly overhead there.

Eratosthenes could now measure the circumference of the earth by making two assumptions—that the earth is spherical and that the sun's rays are essentially parallel. He set up a vertical post at Alexandria and measured the angle of its shadow (7° 12′) when the well at Syene was completely sunlit. Eratosthenes knew from geometry that the size of the measured angle equaled the size of the angle at the earth's center between Syene and Alexandria.

The angle was $\frac{1}{50}$ of a circle, and the distance between Syene and Alexandria was 5000 stadia (*STAY dee uh*). He multiplied 5000 by 50 to find the earth's circumference. His result of 250,000 stadia (about 39,250 km) is close to modern measurements.

Look at Figure 1-1, which shows the relationships of the areas involved in the

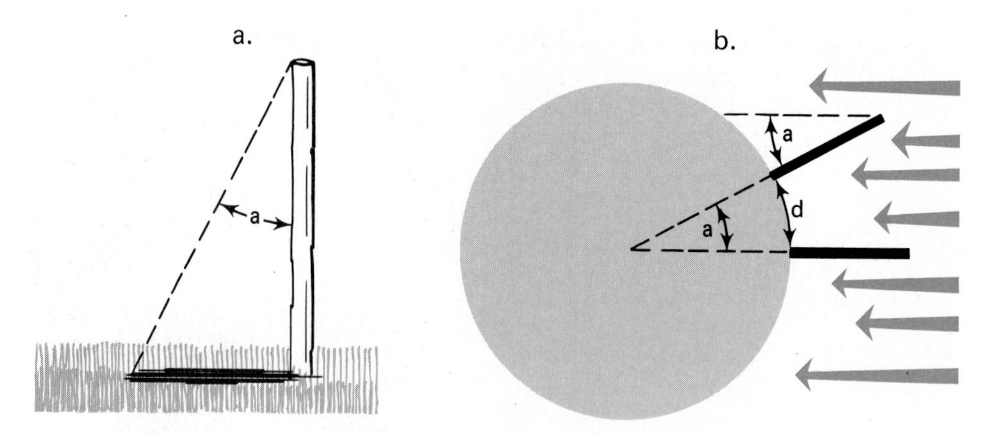

Figure 1-2 *Find the angle of a midday shadow in your school yard. Then copy the drawing in* **b.** *to use in calculating the earth's circumference.*

calculations by Eratosthenes. There may have been some errors in his knowledge and assumptions. If the sun shone down the well at noon on only one day each year the well must have been located on the Tropic of Cancer (23°30′). This is more than 60 km south of Syene. He may have assumed that Syene was due south of Alexandria, which it is not. Also, the stade was not a precise unit of measurement. There were at least three stadia in common use in his day. If he used the stade equal to 157 of our meters and the well was located at 24° due south of Alexandria, he would have gotten the result shown above. Other common stadia were equal to 185 and 210 m, respectively. We are not able to check his facts and assumptions in detail but we do know that his method was correct. He may have recognized some possible errors, for he is said to have adjusted his answer upward fom 250,000 to 252,000 stadia.

MATERIALS
globe and two small sticks with suction cups, protractor, large stick, string, thumbtack

1-2 Investigating the Size of the Earth

When you have completed this investigation, you will have tested the method used by Eratosthenes. Try the following procedure and see if you can improve or confirm his result.

PROCEDURE

Estimate the angle *a* of the shadow cast in your school yard at midday by a stick placed straight up in the ground. (See Figure 1-2a.) Copy Figure 1-2b using your own measured angle. Show on the diagram where you would place a vertical stick that would cast no shadow.

You can calculate the circumference of the earth if you know the distance, *d*, between the two sticks on the real world and the angle of the cast shadow *a*. You can find the distance on a globe.

Set up the globe and the sticks with suction cups to look like your diagram. Measure the distance between the sticks in millimeters, using a flexible rule. Using the given scale of the globe, translate the measured distance to kilometers. You can then calculate the circumference of the earth by the following formula.

$$\frac{\text{Distance around globe}}{\text{distance between sticks}} = \frac{360°}{\text{angle } a}$$

$$\text{or } \frac{D}{d} = \frac{360°}{a}$$

This ratio can be stated as follows: Part of the distance around the globe is to the entire distance as angle *a* is to the angle of the full circle.

DISCUSSION

1. How close is your calculated result to the accepted value of 40,075 km?
2. What was your percentage error?
3. Can you identify any errors or limitations in your procedure that might account for your error, if any?

Figure 1-3 *Crates' model of the earth (150 B.C.).*

1-3 Ancient Conclusions

Ancient observers and thinkers not only measured the earth but recognized the directions north, south, east, and west. They drew maps of the area known to them as "Oecumene" (inhabited world) and expressed real doubts about there being any other inhabited areas. In 150 B.C., however, Crates produced a globe, perhaps the first such model of the earth. This globe showed north-south and east-west lines much like those on a modern globe or map. It also showed a land area Oecumene, (*EH kyoo meen*), based on an earlier map by Eratosthenes, and three other hypothetical land areas as shown in Figure 1-3.

About 100 B.C. another Greek calculated the circumference of the earth to be equal to only 28,200 km. This reduced the calculated size of the earth by about one-fourth. There had been some real concern about how little was known about such a large earth. The smaller version was a welcome idea.

In 150 A.D. a Greek astronomer (and geographer) by the name of Ptolemy (*TAHL uh mee*) put it all together. He concluded that the earth was a stationary sphere and that the sun, moon, and planets revolved around it in circular orbits, as shown in Figure 1-4. He thought the earth was the center of the universe and he accepted the smaller size in preference to the nearly exact figure of Eratosthenes. His views were accepted until the sixteenth century. This was the first model of our planetary system that we know about.

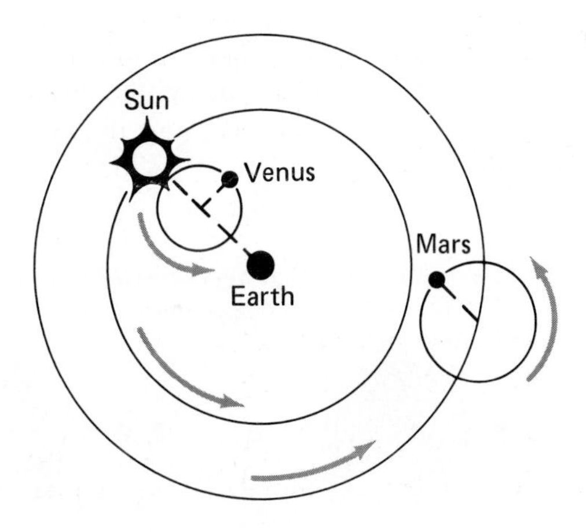

Figure 1-4 *Ptolemy's model of the solar system (150 A.D.).*

Columbus accepted the prevailing idea that the earth is spherical and the smaller size as approved by Ptolemy. His urge to sail west to Asia was practical only on the smaller earth. With his ships and crew it would have been impossible to sail to Asia on the larger earth. His venture was saved by the existence of America about where he thought Japan would be. He died in 1506 believing he had reached Asia (India) on a small earth. He did nothing to change the prevailing spherical model of the earth.

Thought and Discussion

1. Name the planets known to the ancient Greeks.
2. Name the planets discovered since ancient days.
3. How did Eratosthenes measure distance on the earth's surface in degrees and minutes?
4. What contribution did Columbus make to the testing of the prevailing model of the earth?

Copernicus to Newton

1-4 Toward Modern Ideas

A revival of interest in learning during the 16th and 17th centuries in Europe produced many changes in the models of the earth and of the planetary system. The major problem with the Ptolemaic earth-centered system was that it did not permit prediction of future positions of the planets. In 1543 a Polish astronomer, Nicolaus Copernicus (*koh PUR nuh kuhs*), announced a model in which the sun is at the center of the planetary system. The earth and the other planets revolve around the sun.

To explain the observed motion of the stars, Copernicus suggested that the earth rotates about a north-south axis. The earth, instead of being the fixed center of the universe, was now thought to be a revolving and rotating member of a family of planets. This model received wide support because it provided a more simple and workable explanation of planetary motions.

The telescope was invented around 1600 A.D. Galileo (*gal uh LAY oh*), an Italian astronomer, used a very simple telescope to observe the planets, moon, and stars. In 1610 he announced some of his results. The moon, which was known to shine by reflected sunlight, was revealed to have a rough rocky surface rather than being a smooth sphere. The big surprise, however, was the discovery of four new "stars" near the planet Jupiter. What Galileo saw on repeated observations of Jupiter was patterns such as those in Figure 1-5. From these patterns he reasoned that these new "stars" were

Figure 1-5 *Arrangements like these of the moons of Jupiter showed Galileo that the moons were orbiting the planet.*

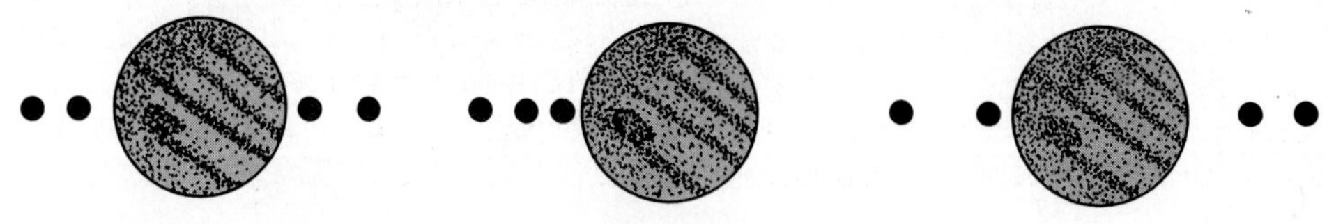

not part of the fixed starry background of the planets, but were four bodies moving along with Jupiter. Furthermore, these bodies were revolving around Jupiter as our moon revolves around the earth. The identification of other rotating and revolving bodies in the planetary system was evidence in favor of the Copernican model.

Observers were now convinced that the earth was one of six planets, each revolving around the sun and rotating about its own axis. There were also moons or natural satellites that revolved about some planets and rotated. The description of the system was so good that future positions of each of the bodies could be predicted. What was the cause of all this motion?

1-5 Newton's Contribution

For many centuries the prevailing idea had been that motion was an unusual state. Then in the early seventeenth century many lines of evidence indicated that motion seemed to be the rule and nonmotion to be unusual.

The person who did the most to explain motion was Isaac Newton. He began his studies before he was twenty years old and published his major results in the book *Mathematical Principles of Natural Philosophy* in 1687 at the age of forty-five. Newton had reached an understanding of his three laws of motion and of planetary motion some years earlier. He delayed publication because he had to create the mathematics (calculus) needed to state a quantitative proof.

Newton provided the laws of motion, and also the universal law of gravitation, which explains the motions of planets and of many other objects.

The first law of motion states that any object will remain at rest, or in uniform motion in a straight line, unless a force acts upon it. This stubborn tendency of an object to keep doing what it is already doing is called **inertia** (*ihn UR shuh*). The second law explains the relationship between force, the mass of an object, and changes in the speed or direction of motion. Newton's third law states that for every action there is an equal and opposite reaction. The force (thrust) of rocket engines may either lift the rocket or push the earth away, depending on which (the earth or the rocket) has greater inertia.

Once the laws of motion had been tested, Newton sought an explanation for the observed motion of the planets and their satellites. This led to his statement of the Law of Universal Gravitation: "Any two material objects attract each other with a force proportional to the product of their masses and inversely proportional to the square of the distance between them." In other words, the larger the masses the greater the attraction, but at twice the distance the attraction is only one-fourth as great. The cause of this attraction is still not fully known. But few ideas have survived as much testing as this one.

ACTION

MATERIALS
golf ball, ping-pong ball

Some of these ideas about motion are a little difficult to believe without some testing. Try a simple test to discover the rate of fall of two balls of different mass. First use a golf ball and a ping-pong ball that are about the same size but have different masses. Hold one ball in each hand (between the tips of thumb and index finger) and drop both from a height of 1 to 2 m at the same instant. Have someone watch from a few meters away to observe the start and finish of the fall. It may take a bit of practice before you can release both balls at the same time. Remember that the rate of fall is due to the attraction between the earth and each of the balls. What can you say about the rates of fall? How do you explain your observations?

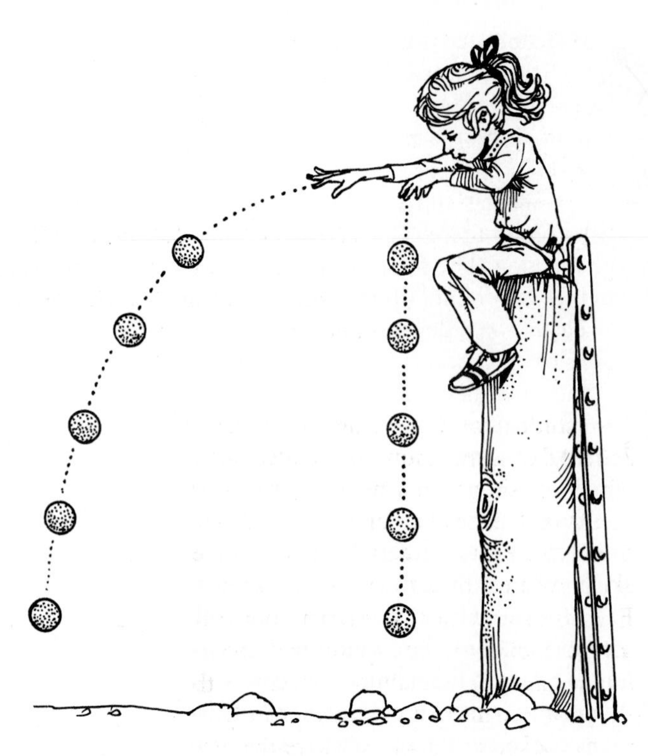

Figure 1-6 Drop one ball and throw a similar one horizontally. Will they hit the ground at the same time?

ACTION

Suppose you go to the top floor (or roof) of a building overlooking an open yard and try an experiment. Try to drop a golf ball or a stone vertically to the yard and at the same instant throw a similar (in mass and shape) object horizontally (no up or down direction) so that it will land in the yard (Figure 1-6). Which object traveled a greater distance? Which object reached the ground first in each of the best executed drops? Can you suggest a factor, other than gravity, that might influence the travel time of one object more than the other?

1-6 We Did It, Sir Isaac!

The Law of Universal Gravitation says simply that every mass of matter exerts a force of attraction on every other mass of matter. This force decreases as the square of the distance between any two bodies. The force is, however, directly proportional to the masses of the objects involved.

Newton concluded that the planets revolve around the sun because the gravitational attraction of the sun keeps them from traveling in straight lines. He assumed that the sun had a much greater mass than all the planets combined. We now know that 99.8 percent of the mass of the whole solar system is in the sun. The mass of the sun is adequate to hold a planet as far away as Pluto, which is 39.5 times farther away from the sun than is the earth.

Now you can imagine shooting an object horizontally somewhere above the atmosphere at such a speed that it would circle the earth forever but still be forever "falling" toward the earth instead of travelling in a straight line. This is what is done when we put a satellite in orbit. The satellite is boosted to some point high in the atmosphere where friction is very low. Then it is aimed in a direction parallel to the earth's surface at a carefully determined velocity—the orbital velocity. If the velocity is too great, the satellite will go off into space. If the velocity is too low, the satellite will fall back to earth. Vanguard I, the first satellite launched by the United States (March 1958), is still in orbit. It is making more than 3800 revolutions of the earth each year and may stay in orbit for two centuries or more.

From Vanguard I to a moon landing in 1969 required just over ten years of improvements in satellite technology. The most important improvement was more exact knowledge about gravitation so that astronauts could escape from the earth's gravity, enter the gravitational field of the moon, and later return to Earth. Is it any wonder that when the first astronauts landed on the moon

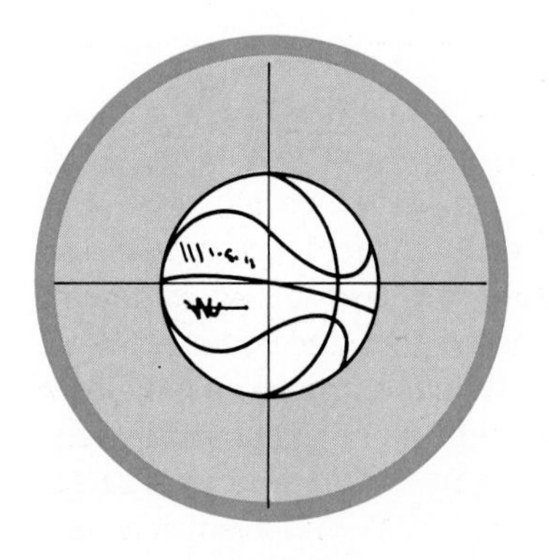
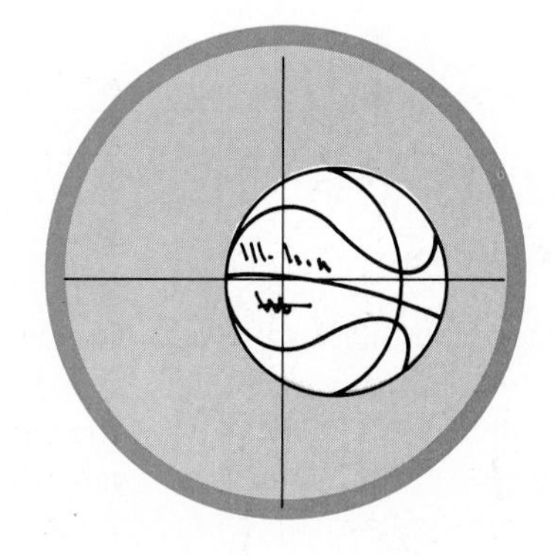

Figure 1-7 *If a basketball 150 m away were moved 10 cm, the change would look about like this through a telescope.*

someone placed a bouquet of flowers on the grave of Newton with a note reading "Sir Isaac, Eagle has landed!"

The law of gravitation permits the prediction of the future positions of the planets with great accuracy. The planet Uranus was discovered in 1781 as a result of careful observation with a telescope. Once its motion had been accurately plotted, theory said that its position at any future date could be predicted. For a few decades all seemed to be going according to theory. By 1840, however, either Uranus or the theory was in trouble. Neither you nor I would consider the facts any cause for alarm. Uranus was 1.5 minutes of arc from where it was supposed to be.

Suppose you use a telescope to describe the precise location of a basketball that is 150 m away. You center the cross hairs on the ball and determine very accurately the direction of the ball from your location. Some weeks or months later you adjust the telescope at exactly the same setting and look at the ball. It is still there but is about 10 cm off center (Figure 1-7). How concerned would you be?

Astronomers wondered why Uranus' actual position was so far from its predicted position. Among the suggested explanations were (a) gravitational attraction did not vary exactly as the square of the distance, (b) the calculated value of gravity was incorrect, (c) Uranus was being affected by the attraction of an unknown planet.

A student at Cambridge University, John Adams, spent four years calculating where another planet would have to be to explain the behavior of Uranus. No one bothered to look at the point in the sky indicated by his results. In 1846 a French astronomer, Leverrier (*luh vehr YAY*), completed and published calculations that agreed almost exactly with those of Adams. An astronomer in Germany looked at the place where the new planet was predicted to be and sure enough there it was. The newly discovered planet was named Neptune.

The discovery of Neptune did much to prove that the law of universal gravitation was a fact. Within half a century after its discovery, Neptune was not precisely where it was predicted to be. Early in the twentieth century, astronomers calculated the location of yet another planet. It was not, however, until 1930 that Pluto could be identified against the background of many stars visible in that part of the sky.

Thought and Discussion

1. What were the principal contributions of Copernicus to the developing model of the solar system?
2. What recent tests of the law of universal gravitation serve to further confirm our understanding of it?

1-7 Proving that the Earth Rotates

If you watch the northern sky on a clear night you can see one star that seems to stand still. The other stars appear to move slowly in circular paths around the one that does not move. Watching the sky in a direction at right angles to the line from the observer to the pole star reveals another pattern of motion. The stars appear to move in straight lines from east to west across the sky. (See Figures 1-8 and 1-9.)

One of the problems with the Ptolemaic system relates to this apparent motion of the stars. Ptolemy had everything—sun, moon, and stars—revolving around the earth. The star in the northern sky remained fixed throughout the year, whereas the sun's path shifted north and south through an angle of 47°. Why should the sun be so different? Some observers over the years preferred to believe that the apparent motion of the stars could best be explained by having the earth rotate on an axis. The behavior of the sun could only be explained if the earth also revolved about the sun. As we have seen, it took Copernicus, Newton, and many others to establish the actual arrangement of objects in the solar system and their motions.

Until the middle of the nineteenth century there was no satisfactory proof that the earth rotates. Newton's first law of motion made it clear that a friction-free pendulum would swing in the same path indefinitely. A French scientist, Jean Foucault (*foo KOH*), reasoned that one way to prove rotation would be to

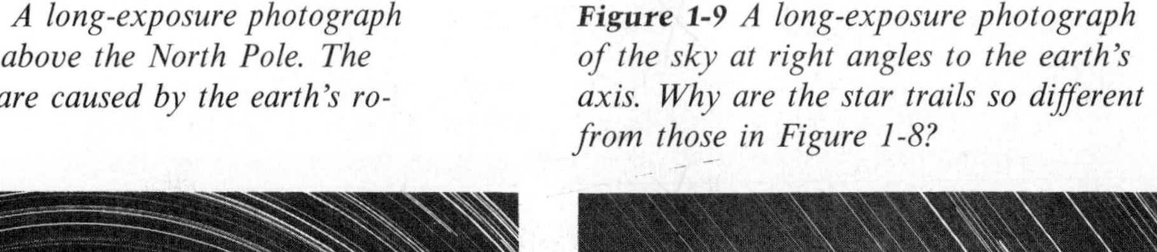

Figure 1-8 *A long-exposure photograph of the sky above the North Pole. The star trails are caused by the earth's rotation.*

Figure 1-9 *A long-exposure photograph of the sky at right angles to the earth's axis. Why are the star trails so different from those in Figure 1-8?*

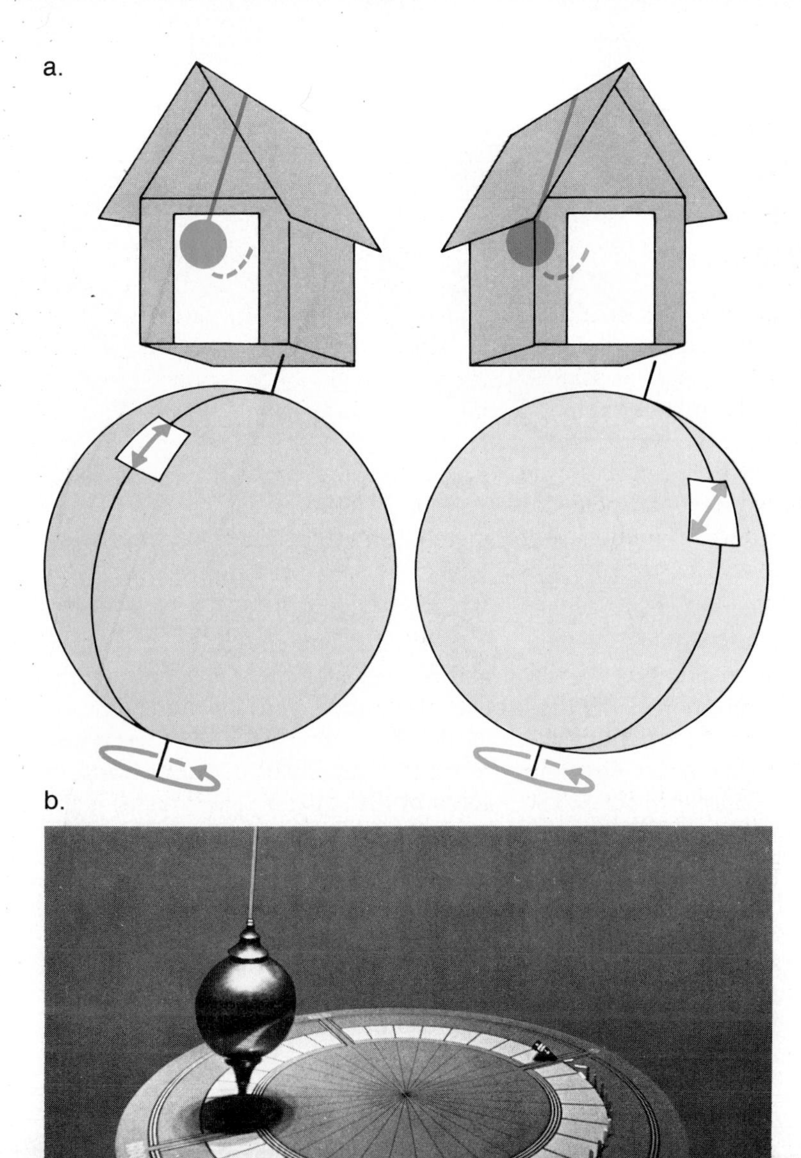

Figure 1-10 a. *Which changes direction, the house or the pendulum?* **b.** *The Foucault Pendulum in Boston's Museum of Science.*

suspend a pendulum and observe its behavior. This he did under the dome of a large public building in Paris.

Foucault's pendulum consisted of a cannon ball suspended on a 67 m wire fastened to a low-friction swivel in the dome. Fine sand was spread evenly over a low platform just below the suspended ball. A wire extending below the ball traced the path of the swinging ball in the sand. When the pendulum was swung in a straight line across the platform, the direction marked in the sand changed slowly as time went by. Many tests demonstrated that the change was always in the same direction and at the same rate. (See Figure 1-10.)

Foucault predicted that similar tests

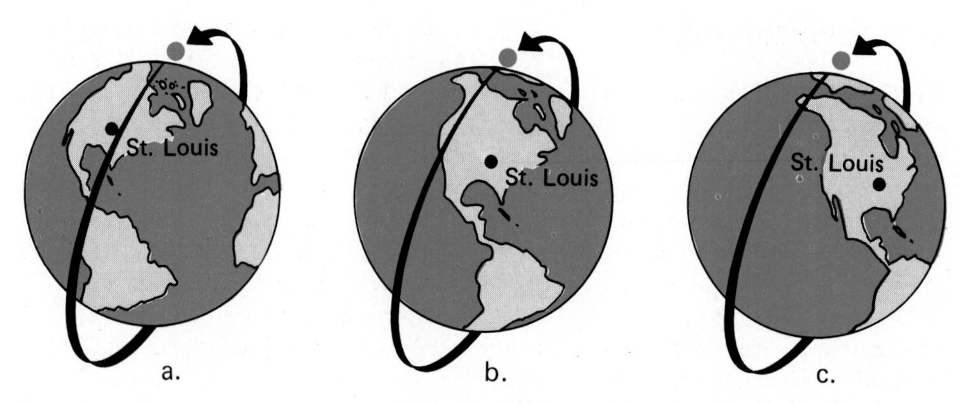

Figure 1-11 *A satellite in polar orbit does not change its path around the earth, but it appears to shift westward with each passage.*

farther north and south would show that the behavior of the pendulum was directly related to the position of the pendulum between pole and equator. The apparent rate of change in the direction of swing would be 360° ÷ 24 hrs = 15° per hour at the pole. At the equator there would be no change. In his day, Foucault's experiments added, for many people, convincing evidence that the earth did rotate on an axis. We now know that satellites launched in polar orbits appear to shift westward with each passage over the earth below. (See Figure 1-11.) The shift amounts to 15° times the period of revolution of the satellite in hours.

A rotating earth provides two natural bases for the definition of points and areas on the earth's surface. One is the axis of rotation between the poles. The North Pole and South Pole are points where the axis of rotation intersects the surface. The other base is the equator—the intersection of the earth's surface with a plane at right angles to the axis.

The equator and the poles of rotation and two kinds of lines make possible the use of latitude and longitude on the earth. (Refer to Figure 1-12.) **Latitude** refers to locations or distances north or south of the equator. The equator is the line of zero latitude and the poles have a latitude of 90° north and south, respectively. Lines marking degrees of latitude extend east and west around the earth and are called "parallels" because they are parallel to each other and to the equator. **Longitude** refers to points or distances east or west of the **Prime Meridian,** which is the zero meridian. Lines marking longitude extend from one pole to the other and are called **meridians.**

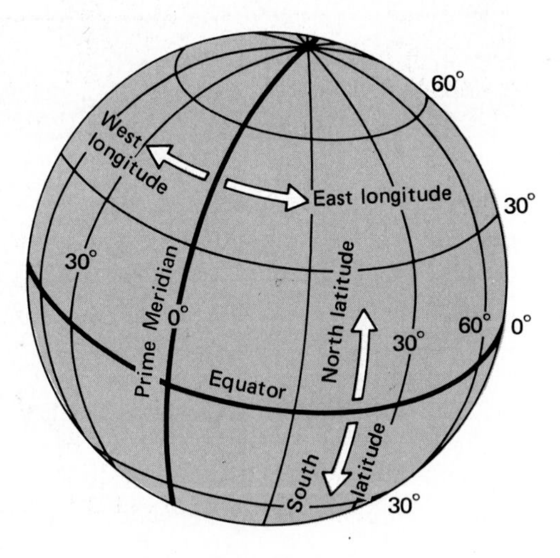

Figure 1-12 *Points on the earth are located by latitude north and south of the equator and by longitude east and west of the Prime Meridian.*

1-8 What Time Is It?

The Prime Meridian was established by international agreement in 1884. Before the agreement, anyone could select any north-south line as a zero meridian and use it to describe the location of cities and other points and to draw maps. Remember that longitude is also a measure of time. The twenty-four hours of each day represent one complete rotation of the earth. Each hour equals 15° of longitude. Increase in travel by ships and by train led to complete confusion in schedules so long as each community could set its watches as it pleased.

Determination of latitude and longitude depends on precise astronomical observations. This meant that the Prime Meridian should, if possible, pass through a good observatory. In 1884 there were four well known observatories well qualified to be chosen. These were located in Potsdam, Germany; Paris, France; Greenwich, England; and Arlington, Virginia.

A second requirement was that opposite the Prime Meridian there must be another meridian 180° east or west to serve as an **International Date Line.** This was necessary because, except for an instant each day when it is noon along the Prime Meridian, there are parts of two days somewhere on Earth. There must be an arbitrary line along which, except for one instant each 24 hours, it is one day of the week (or month) in areas east of this line and the next day of the week west of the line. It was desirable that this line cross as little populated land area as possible. To place the 180th meridian mainly in the Pacific Ocean meant that the Prime Meridian should pass through Greenwich, England.

There is also international agreement on time zones around the earth. (See Figure 1-13a.) In general, each time zone

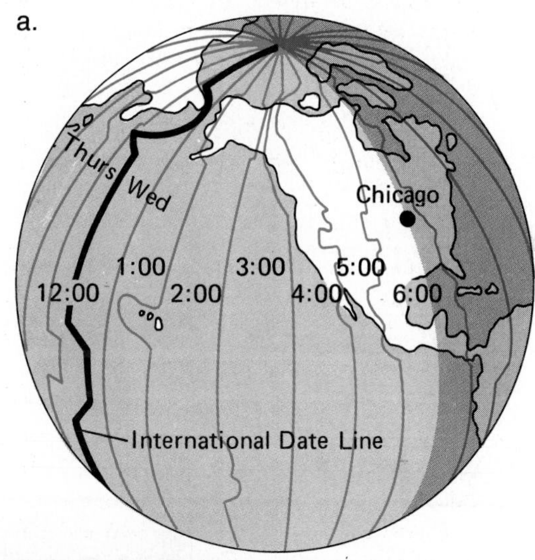

Figure 1-13 a. *Each time zone is more or less 15° wide.* **b.** *The cities you are calling in the telephone game.*

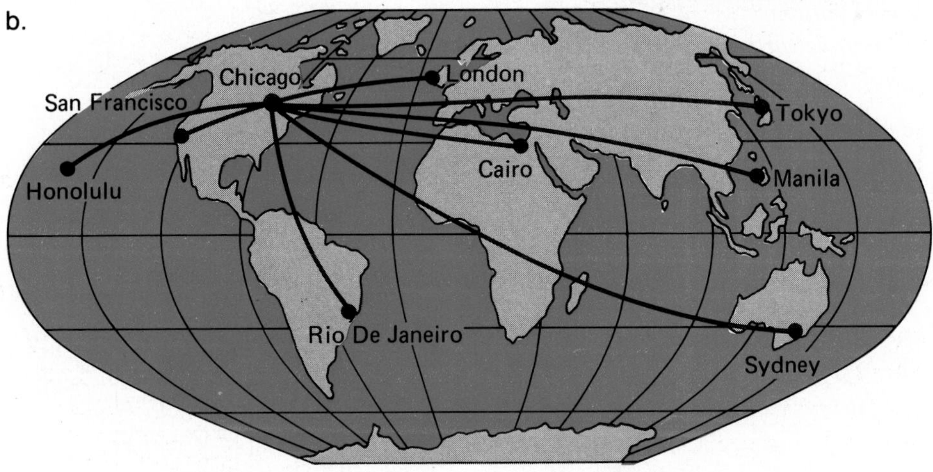

is 15° wide. For the contiguous states of the United States there are four time zones. Canada has a fifth, in the Maritime Provinces (except Newfoundland, which has its own time zone). Canada also has a bit of a zone that borders on Alaska. Alaska and Hawaii are in another zone.

Time Zone	*Meridian*	*Meridian Passes Near*
Atlantic	60°	Sydney, Nova Scotia
Eastern	75°	Philadelphia
Central	90°	New Orleans-Memphis-St. Louis
Mountain	105°	Denver-Pikes Peak
Pacific	120°	Central Washington and Oregon, along the Nevada-California border
Alaskan	150°	Anchorage

Boundaries between these time zones are usually irregular to avoid having neighboring cities separated by an hour in time. The actual area of each time zone is wider west of its meridian than east of it and this difference is greater in the northern part of the country. Why should this be so?

ACTION

Now imagine a telephone game in which you are calling from Chicago at 6 P.M. on Wednesday. (See map, Figure 1-13b.) The sun is just setting in the west. Turn the globe to this position. You rapidly call San Francisco, New York, Rio de Janeiro, London, Cairo, Manila, Tokyo, Sydney, Australia, and Honolulu. On each call you ask the *time* and *day.* What is the hour and day in each city? Where is it noon on Wednesday? Where is it 3 A.M. on Thursday?

1-9 Maps and Mapping

Latitude and longitude provide a basis for describing the location of places on the earth. They also provide the basis for the construction of maps of all kinds. Maps drawn by Eratosthenes and other ancient observers did show parallels and meridians. These were drawn through prominent places such as Alexandria as the author wished. The spacing was described in stadia.

A map is a model of the earth's surface or a part of it. The best model of the whole earth is a globe. For many reasons, however, we want maps that are flat pieces of paper. A road map could be made that kept the curved shape of the globe. But a curved surface does not hang on the wall or fold smoothly.

It was long ago determined that any

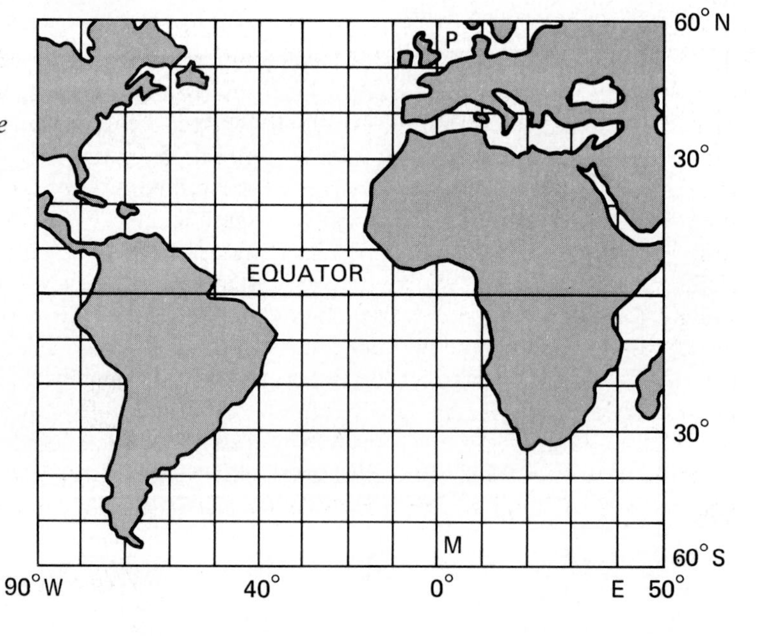

Figure 1-14 *A simple grid pattern of latitude and longitude, used as early as 100 A.D. Why is it not an accurate model of the earth?*

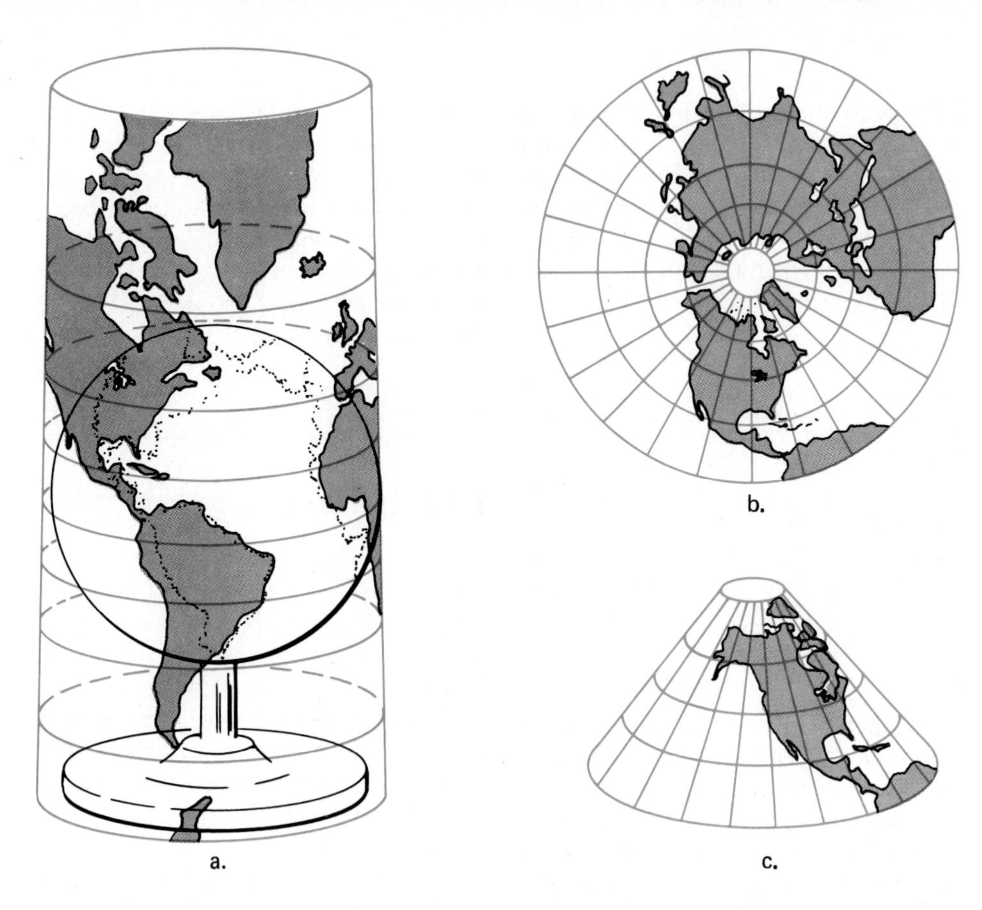

Figure 1-15 *Three ways of making a map from a globe.* **a.** *A cylinder touching the equator.* **b.** *A sheet touching the North Pole.* **c.** *A cone touching a parallel of latitude.*

large area of the earth's surface could not be shown with complete accuracy on a flat map. Any flat map will distort the size or shape of areas or the direction or distance between points. There are ways of projecting the global surface on flat surfaces so that features will have either the right shape or the correct size relative to other areas, but not both at the same time. We can choose one or the other or some compromise between them. We can also make flat maps on which any straight line will show true directions or the shortest distance between points, but not both. In general, we can produce flat maps to meet all our needs but they are never precisely accurate in all respects.

Figure 1-14 shows a simple, arbitrary grid pattern of longitude and latitude first used about 100 A.D. The parallels of latitude (shown only to 60°) are equally spaced in 10° units and the meridians are also spaced in 10° units along the equator. Away from the equator (north and south) the meridians are parallel instead of converging as they do on the globe. The poles, which are points on the globe, would be represented by a line the same length as the equator. Land areas along the equator are shown in proper shape and size. Away from the equator the land areas are increasingly distorted in shape *and* size. This map is useful only for areas along the equator.

A number of map projections represent the features of the globe projected onto a sheet of paper, as shown in Figure 1-15a, b, and c. These are actually pro-

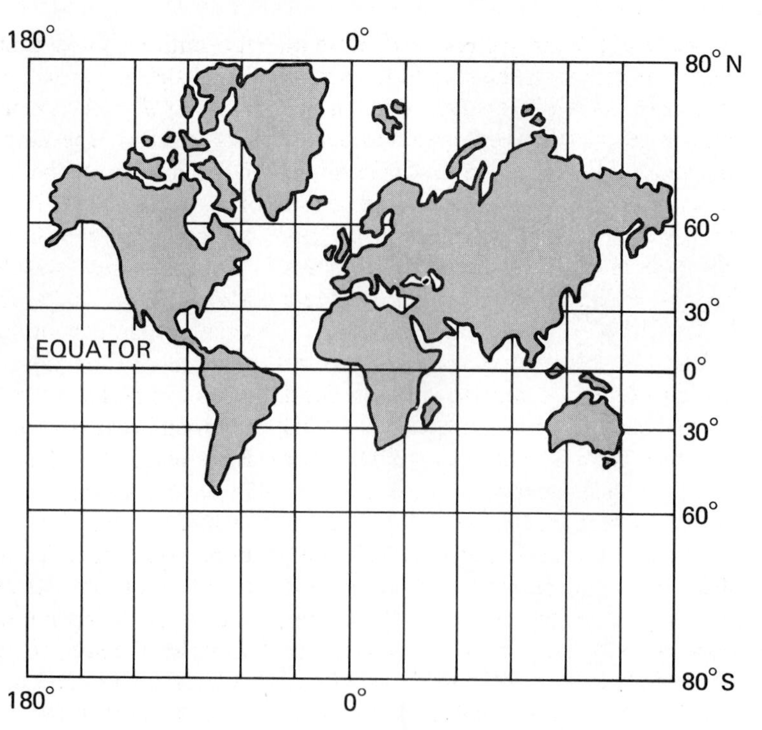

Figure 1-16 *The Mercator projection, a mathematically adjusted cylinder that shows true directions.*

duced by the use of mathematical equations and each yields a map base with special values. None is quite as accurate as a globe but for areas smaller than continents these maps have high standards of accuracy.

Study the examples of map projections shown in Figures 1-14 and 1-16. The distortion of familiar land areas in Figure 1-14 is obvious. You may conclude that Figure 1-16 is relatively free of distortion. Look at the data given in this table and study the map again.

Area	*Size (square kilometers)*
Africa	30,381,000
Australia	7,716,800
Greenland	2,184,000
North America	24,492,000
South America	17,862,000

The map in Figure 1-14 represents an equal area projection but shapes of areas in the higher latitudes are highly distorted. The map in Figure 1-16 is a projection in which shapes are acceptable in all areas but size is greatly distorted in the high latitudes. This is the Mercator (*mur KAY tur*) Projection much used in navigation and world-wide wind maps because straight lines anywhere on the map show true directions.

Maps of separate continents and smaller areas do not usually involve the distortions illustrated above. There are, however, certain conventions that are used in making almost any map. Sooner or later you will probably see a map from which you cannot interpret the information shown. That map does not include some of these conventions.

Directions (north, south, east, west) on the map must be indicated. "Up is north" is considered to be a convention in map making but do not count on it. Because most of the land areas (and most of the map makers) of the earth are north of the equator, it has become customary to have north at the top of the map. Even so, there should be a symbol somewhere on the map indicating north, north and south, or all four directions.

Scale of the map refers to the ratio of distances shown on the map to the actual distances on the earth. One unit on the map is equal to some larger number of the same units on the ground. Common units used to measure distances on a map are inches or centimeters. The corresponding units on the earth are miles (1 mile = 63,360 inches) and kilometers (1 kilometer = 100,000 centimeters).

Early in this century there was an international agreement to map all land areas on a scale of 1 to a million. This ratio can be written either 1:1,000,000 or $\frac{1}{1,000,000}$. This means 1 centimeter = 10 kilometers and 1 inch = 15.782828 miles on the ground, respectively. Inches, feet, yards, and miles are not convenient units to work with. Under the agreement each country was to do its own thing about getting the "millionth" map made. The United States decided to use map scales or ratios which had fractional relations to 1:1,000,000. The most common map scales have been 1:500,000 (states), 1:125,000, 1:62,500, and 1:31,250 (smaller areas). To make the millionth map we could just reduce the other maps ½, ¼, ⅛, $\frac{1}{16}$, and $\frac{1}{32}$, respectively. Our official mapping program was based on areas defined by latitude and longitude. The basic unit was an area one degree of latitude by one degree of longitude. These were normally mapped on a scale of 1:125,000 or one inch equals approximately two miles but exactly eight times the millionth. More detailed maps cover areas thirty minutes (30′) latitude × 30′ longitude, 15′ × 15′, or 7.5′ (that is, seven minutes thirty seconds or 7′30″ × 7′30″).

The map projection and care used in preparing these maps have been the best used anywhere. Each map is a flat piece of paper but if one glues a number of them together to get a map of a larger area, the assembled map will not hang flat against the wall. It will try, in a droopy way, to reflect the curvature of the earth. We have completed maps of the entire United States on a scale of 1:250,000 (approximately 4 miles per inch). If all of the sheets in this map were put together, the map would be about 19 m long and about 11 m high. If it were hung on a wall it would droop like the mainsail of a ship becalmed.

Maps are used widely in earth science. Many of the maps produced by earth scientists are part of the public information service of government. Weather maps are perhaps the best example and are found regularly in most newspapers. Knowledge of the landscape is important in urban development, flood control, land-use regulation, highway construction, exploration for mineral resources, and recreation. This information is shown on topographic maps produced and distributed by the United States Geological Survey and the several state geological surveys. In other countries similar agencies provide this service. These maps are now made from aerial photographs, which are also available for purchase.

In recent years the United States has put several surveying satellites in orbit around the earth. These provide a variety of specialized images of the earth, which are increasingly essential to almost every kind of investigation in earth science. Other kinds of maps that you will find in later chapters are geologic, geophysical (earthquakes, gravity, magnetism), ocean currents, soils, drainage patterns, climates, and star charts.

1-10 Investigating Maps as Models

When you have completed this investigation, you will have examined a kind of model that is very useful to earth scientists and other people. Remember that models can be mental or physi-

Figure 1-17 *A ground level view of an area near Morrison, Colorado.*

Figure 1-18 *An aerial view that includes the area shown in Figure 1-17.*

cal. As you do the investigation, see if you can recognize the type of model you are working with.

If you were standing in a field near the town of Morrison, Colorado, looking north toward Red Rocks Park, you might see the view shown in Figure 1-17. This represents the point of view you most commonly have of the world in which you live. That view is from near the earth's surface.

If you were asked to make a model of the earth's surface as you see it in this picture, how would you do it? What other information would you need to complete your model? How does what you see in Figure 1-18 provide you with more information? Where do you think the photographer was when he or she photographed this view?

Now imagine you are flying directly above the area and looking straight down as shown in Figure 1-19. What

Figure 1-19 *The hills near Morrison, from directly above.*

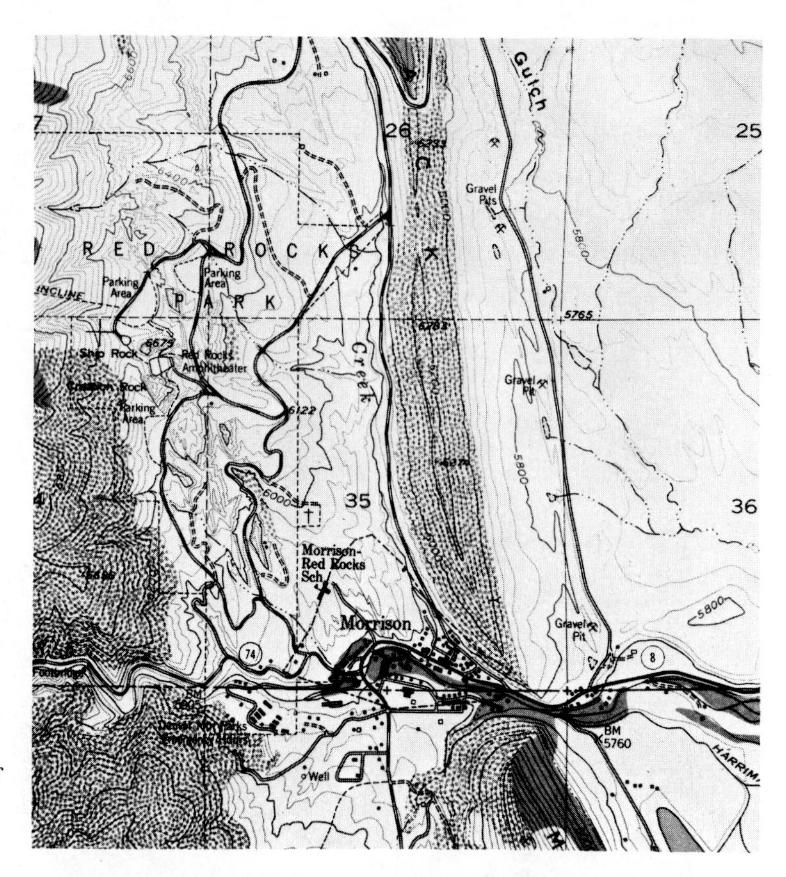

Figure 1-20 *A topographic map of the Morrison area.*

additional information does this view provide? If you have a stereoscope, examine Figure 1-19. Identify as many features as you can. Use a sheet of clear plastic and a marker and make a map of the area with the white line around it. Compare your map with the map of the same area in Figure 1-20. This is called a **topographic** (*tahp uh GRAF ihk*) **map.** How does a topographic map represent the hills and valleys?

MATERIALS
grease pencil, model mountain, plastic sheet, transparent box with cover

To better understand the way topographic maps show hills and valleys you will need a transparent box, a model of a mountain, and a grease pencil. Use the equipment as shown in Figure 1-21. Make a series of marks 1.5 cm apart up one side of the box. Place the model

Figure 1-21 *Draw lines around a model mountain as you add water. You can then use the lines to make a topographic map of the mountain.*

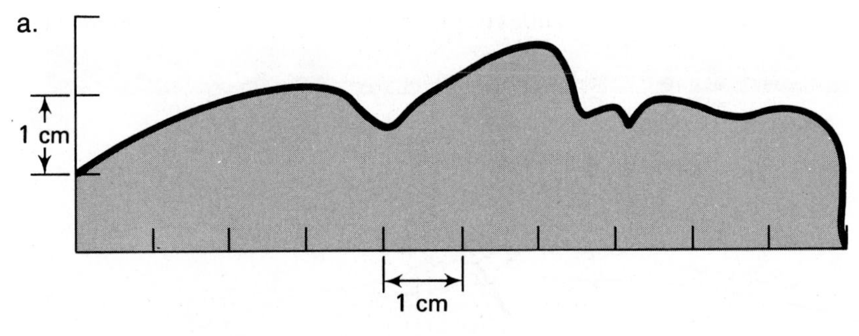

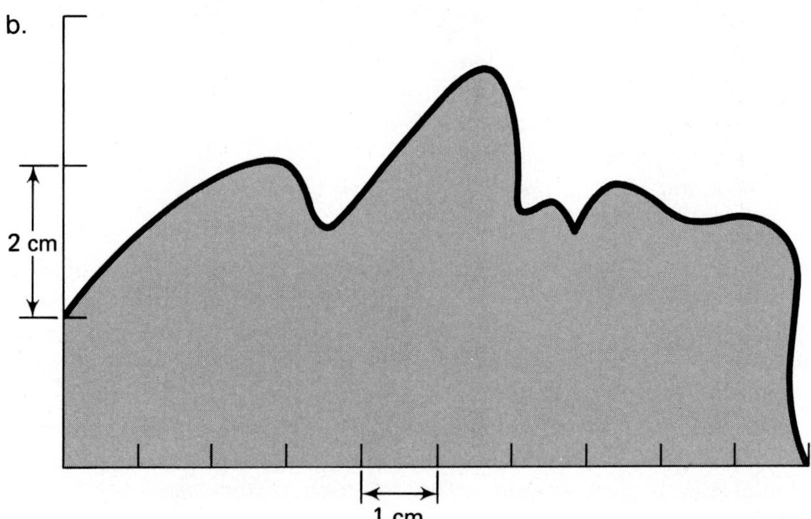

Figure 1-22 *A profile of a face drawn to two scales.* **a.** *Vertical-horizontal 1:1 (the same).* **b.** *Vertical-horizontal 2:1.*

mountain in the box and pour in water up to the first mark. Draw a line around the mountain at the water line. Add more water up to the second mark and repeat the procedure. Continue doing this until the mountain is covered with water.

When you have finished drawing the lines, put a clear plastic cover on the box. Trace the contour lines on a plastic sheet as you see them from above. If you close one eye, it may be easier.

How does your map of the model mountain compare with the hills on the topographic map? Do you think the statement "A map is a paper model of the real world" is a true one?

1-11 Profiles

All scientists use diagrams and pictures as well as maps to illustrate models of objects. Earth scientists use pictures, stereographic pictures, and two and three dimensional diagrams. All of these illustrations should be to some scale and the scale should be indicated. Many diagrams show a vertical section through some feature on the earth's surface. Such diagrams usually show the "lay of the land" (a profile of the land surface).

Figure 1-22a is a profile of an actual human face drawn to scale. Note that the scale in each direction is the same—1 cm : 2 cm or 1 : 2. (1 cm on the drawing equals 2 cm on the face.) Figure 1-22b is a profile of the same face but note that one scale is just twice the other. The facial features are exaggerated, when compared to the original. The owner of the profile shown in (a) would object to being represented by the profile in (b). Suppose you saw only (b) and knew the amount of exaggeration. Could you

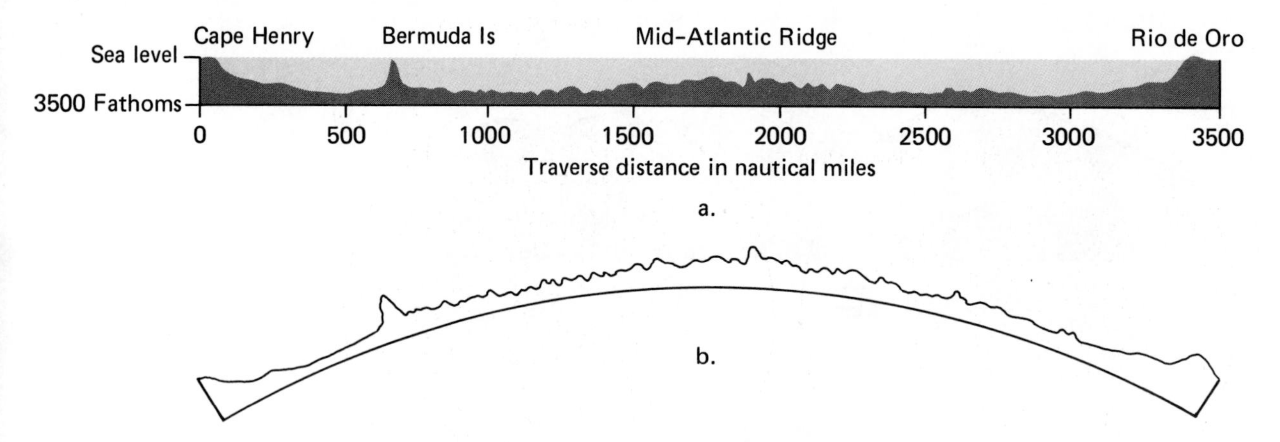

Figure 1-23 a. *A profile of the Atlantic Ocean floor from Cape Henry, Virginia to Rio de Oro, Spanish Sahara, Africa.*
b. *The same profile as in* **a.** *but showing the curvature of the earth.*

imagine how the actual profile would look? In other words, could you think to scale?

We will now borrow another profile. This one is of the floor of the Atlantic Ocean from Virginia to Spanish Sahara (Africa). (See Figure 1-23a.) You can see that the horizontal scale is given in nautical miles and the vertical scale in fathoms. No one could easily decide the ratio of one to the other. If we change both to the metric system, the horizontal scale becomes 1 mm = 47 km and the vertical scale is 1 mm to 1 km. This is an exaggeration of 47 times! You can imagine what the profile in Figure 1-22a would look like if it were exaggerated 47 times. If the ocean profile were drawn without exaggeration, 1 mm = 47 km vertically as well as horizontally, the whole diagram would be thinner than the bottom line. If the diagram is to show anything about the irregularities of the sea floor there must be a lot of exaggeration. The viewer must know the exaggeration in order to think to scale.

This Atlantic profile can illustrate another important aspect of profiles as models. In profile 1-23a the distance from end to end is 6600 km. This is just less than one-sixth ($\frac{1}{6}$) of the distance around the earth. If the earth is spherical, this profile represents an arc of 60°. Figure 1-23b shows this profile curved to represent the spherical earth. It is easier to draw profiles straight as in Figure 1-23a but the viewer must relate it to a real, spherical earth. In earth science there are two important suggestions. *Think to scale. Think spherical.*

1-12 The Layered Earth

Maps and profiles are made by people who are studying parts of the earth that they can see or measure. Some scientists are interested in the behavior of material that they cannot see or experiment with. This is especially true of scientists who study the stars, including our sun, or those who investigate the interior of the earth.

Early observations of the earth led to a rather simple model of its interior. People used to believe that from a once molten state the earth was slowly cooling and would end as a frozen wasteland. The outer part was solid rock—a crust. The interior was still molten but cooling, solidifying, and shrinking. The cold crust was always getting too big for the interior so it wrinkled here and there to make mountain ranges. Investigations of

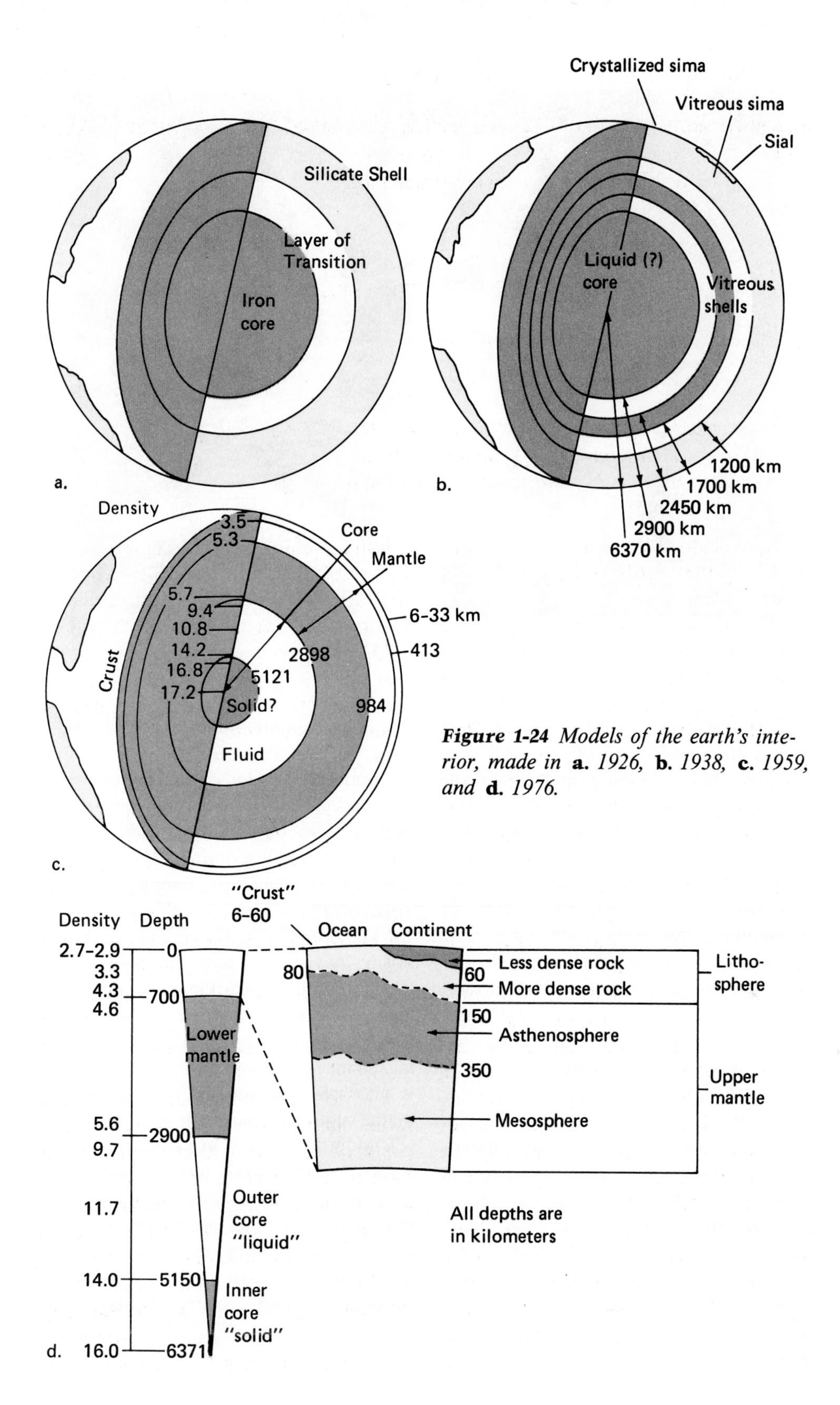

Figure 1-24 *Models of the earth's interior, made in* **a.** *1926,* **b.** *1938,* **c.** *1959, and* **d.** *1976.*

CARTOGRAPHER

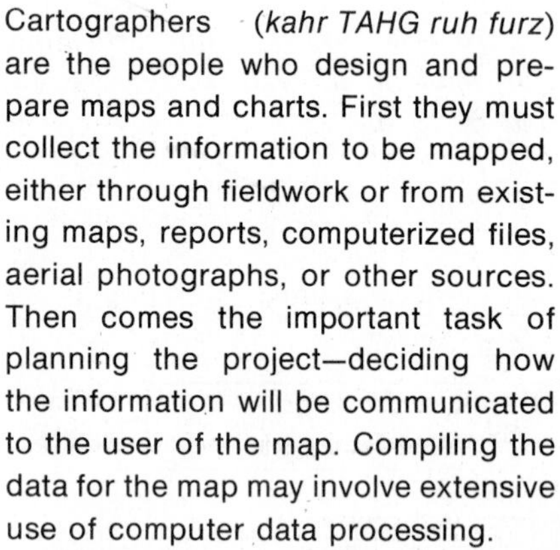

Cartographers (*kahr TAHG ruh furz*) are the people who design and prepare maps and charts. First they must collect the information to be mapped, either through fieldwork or from existing maps, reports, computerized files, aerial photographs, or other sources. Then comes the important task of planning the project—deciding how the information will be communicated to the user of the map. Compiling the data for the map may involve extensive use of computer data processing.

Most cartographers are employed by various government agencies at national and local levels, working on such projects as topographic and geologic maps, soil surveys, aeronautical and nautical charts, weather maps, and various types of statistical maps. In private industry, cartographers are employed by oil companies, utility and construction companies, and environmental and engineering firms. In addition, map and atlas publishers provide job opportunities for cartographers.

A cartographer usually has an undergraduate degree in geography, specializing in cartography. Courses in computer science are becoming increasingly important. A high mechanical aptitude is helpful, as well as the ability to visualize how the data will look on the completed map. A cartographer who wants to specialize should also consider courses in the area of specialization. For example, a person who is interested in working on geologic maps might take courses in geology.

Cartographers are assisted in their work by cartographic aids and cartographic technicians, who do much of the actual drafting. Preparation for these positions is a combination of on-the-job training and up to two years of formal training after high school.

the earth and its interior in the last half-century have produced frequent and major changes in the model (see Figure 1-24a, b, c, d). Most of what you will learn about the interior of the earth and the changes occurring in the seafloor was not known in 1960.

The shells of the solid earth shown in Figure 1-24d will be described in greater detail in later chapters. The **lithosphere** (rock sphere) and especially its upper part, the **crust,** is the portion of the solid earth about which we know most.

Of course the **atmosphere** (*atmos* = gas) and the **hydrosphere** (*hydro* = water) are as much a part of the earth as is the lithosphere. Both the atmosphere and hydrosphere are layered but the layers are too thin or irregular to show to scale as an upward continuation of Figure 1-24d. As you investigate these spheres in later chapters keep one thing in mind. A very narrow zone 10 km above and below the earth's surface is about the limit of life's environment.

Thought and Discussion

1. How would you describe the location of a point on a sphere without some system such as latitude and longitude?
2. What are the two most important conventions to be observed in preparing a map?
3. What convention should be observed in preparing a profile?

Unsolved Problems

This chapter is an account of discoveries about the earth over a period of three thousand years or so. You may have felt that it is an account of how and when a few people came up with answers to questions no one was asking. Actually, the people identified were only a sample of the many who contributed directly or indirectly to the answers. The questions were limited to the shape, size, motions, and models of the earth and of the solar system. During the same centuries other curious people were contributing to knowledge about many aspects of the changing earth. In the process, all of the sciences—astronomy, biology, chemistry, physics, and others—were invented and developed. Still there is no end of questions.

In spite of all these discoveries more people than any of us suspect still believe that the earth is flat and standing still. There are strongly held opinions about the limits of earth resources, food, fuels, metals and other minerals, even air and water, for our use. These lead to other opinions about an absolute limit, if any, to the human population that can be supported.

The more we know the more we know we don't know. This is illustrated at every point in this chapter. Now we may have come to a final need to know. How does all of life survive on this planet? The question of survival may be most important to humanity because humanity must make the decisions for all of life. Decisions require knowledge, a logical approach to conclusions, and a sense of responsibility to others.

Earth science can contribute to the solution of our survival problems. The necessary information will come from investigations of the many types of changes occurring at or near the earth's surface. Investigating those changes is the subject of this book.

How do we insure that everyone shall have the necessary knowledge? How do we insure that everyone shall be logical in reaching conclusions? How can we ever expect everyone to have a sense of responsibility to everyone else on earth?

This will be a long journey but every journey must start with a first step. Each of us can prepare ourselves to be knowledgeable, logical, and responsible. This is your most important unsolved problem—and mine!

Chapter Summary

Our knowledge of the solar system has grown over a period of many centuries. The ancient Greeks noted that the earth is spherical. One of them, Eratosthenes, calculated the earth's circumference with surprising accuracy. You can do the same with a shadow stick, if you know the angle of the cast shadow and the distance between it and a stick that would cast no shadow.

Ptolemy concluded that the sun, moon, and planets revolved around the earth. He thought the earth was smaller than Eratosthenes did.

As time went by and people made more and more observations, the modern understanding of our solar system developed. Copernicus (1543) thought the earth rotated on a north-south axis and revolved around the sun, along with other planets. After the invention of the telescope, Galileo (1610) announced the discovery of moons revolving around Jupiter.

Newton's laws of motion led to his discovery of the force of gravity. He concluded that gravitational forces hold the planets in their orbits. For an object to remain in orbit, it must be moving at just the right speed and be moving horizontally.

The understanding of gravitational forces led to the discovery of Neptune, which was affecting the orbital speed of Uranus. Pluto was discovered in the same way.

The rotation of the earth, that Copernicus recognized, helps us mark the passing of time. The Foucault pendulum illustrates the actual rotation.

All points on the earth can be located by the use of latitude parallels north and south of the equator and longitude meridians east and west of the Prime Meridian. The Prime Meridian passes through the poles and Greenwich, England. Longitude is also useful in marking the passage of time. There is a time zone for about every 15° of longitude, but the boundaries are irregular.

Maps and profiles are models of the earth's surface, or part of it. In making a map, there is no way to show the curved surface of the earth perfectly on a flat piece of paper. Various projections are used, as if the paper were tangent to the globe at a point or parallel. A good map must have directions and scale indicated. One of the most useful kinds of maps to an earth scientist is the topographic map, which uses contour lines to show differences in topography.

Profiles are useful because they show topography, but you must know what the exaggeration is (if any) and think to scale.

The solid earth is made up of layers or shells that are denser toward the center and less dense toward the outside. Most of the layers are solid, but there is a weak zone in the upper mantle and an outer core that seems to be liquid.

Questions and Problems

A

1. What is the difference between rotation and revolution?
2. What evidence can you give that the earth rotates?
3. What evidence can you give that the earth is revolving?
4. What are Newton's laws of motion?

B

1. If you observe one particular star, or planet, for an hour, in what direction will its position change in relation to you?
2. Can you explain how observation of the "moons" of Jupiter from the earth at different times of our year might provide a method of measuring the speed of light?
3. How do maps and profiles aid in learning about the earth?
4. What was the discovery that finally made it possible to predict planetary positions with accuracy?

C

1. Important processes in science are observation, analysis, and interpretation. From what you know of the work of Eratosthenes, Galileo, and Newton, by which process did each make his major contribution?
2. If it is 4:00 P.M. Wednesday at longitude 165° west, what time and day is it at longitude 165° east?
3. When the ships of Magellan returned to Spain after sailing westward around the world, the crew discovered that the ship's log was incorrect by one day. Was the log ahead or behind? How would you explain this error?
4. If the ancient Greeks were convinced that the earth is spherical and knew about directions, the equator, and the Tropic of Cancer, why did they not adopt latitude and longitude as we know them?

Suggested Readings

Beveridge, William I. *Seeds of Discovery.* New York: W.W. Norton Co., 1980.

Chapman, Clark R. *The Inner Planets.* New York: Charles Scribner's Sons, 1977.

King, Elbert A. *Space Geology.* New York: John Wiley & Sons, 1976.

Mutch, Thomas A., and others. *The Geology of Mars.* Princeton, New Jersey: Princeton University Press, 1976.

Sagan, Carl. "The Solar System." *Scientific American,* September 1975, pp. 22–31.

Short, Nicholas M. *Planetary Geology.* Englewood Cliffs, New Jersey: Prentice-Hall, 1975.

Tweney, Ryan D.; Michael E. Doherty; and Clifford R. Mynath. *On Scientific Thinking.* New York: Columbia University Press, 1981.

Wilford, John N., ed. *Scientists at Work.* New York: Dodd, Mead and Co., 1979.

unit II
The Water Cycle

Energy and the Water Cycle

FOUR HUNDRED YEARS AGO, *the great artist-scientist Leonardo da Vinci observed that "the air moves like a river, carrying the clouds with it." When you look at clouds streaming across the sky on a windy day, you can imagine them as part of a limitless river that originates in the ocean and empties on the land. This endless river, fed also by the water evaporating from soil, fields and forests, city streets, rivers, lakes, and streams, makes up the water cycle in the air.*

How does water get into the air? What makes the atmosphere give up or hold back its water supply? What moves the rivers of the air?

Imagine a square tank of water, 10 km on each side and 10 km deep. This quantity of water, with a mass of about one million million metric tons (a metric ton equals 1000 kg), rises every day from the earth's surface to a height of almost 3 km in the atmosphere. Your everyday experience with gravity tells you that lifting so much water so high must take a lot of energy.

After a summer shower, you may have seen thin clouds of moisture rising from warm city streets or from a freshly ploughed field. Soon the shallow puddles are gone. But after a winter rain, the puddles may remain for several days. The disappearance of a puddle of water from a paved street on a warm day may seem to be magic—a change without effort. However, the energy used to evaporate a puddle of water, as it disappears into the air, is about a hundred times greater than the energy you would need to lift the same puddle to the height at which it might become part of a cloud.

AFTER COMPLETING THIS CHAPTER, YOU SHOULD BE ABLE TO:

1. **describe the general make-up of the atmosphere at sea level.**
2. **describe the changes of water within the water cycle, and identify the processes that make the changes.**
3. **identify what causes clouds to form and tell differences between the two main types of clouds.**
4. **give examples of potential and kinetic energy.**
5. **identify three methods of energy transfer.**
6. **describe the causes of heat sources and sinks in the earth's atmosphere.**
7. **identify factors that affect the rate at which evaporation occurs.**
8. **describe the relationship between heat energy and temperature that exists when water changes phase.**
9. **determine and demonstrate the dew point and relative humidity.**
10. **determine the height at which clouds will form when the temperature and amount of water vapor are known.**
11. **describe the various types of precipitation, identifying the conditions under which each may form.**

Clouds and Energy

2-1 The Composition of Air

When water boils, bubbles appear and rise to the surface. These bubbles are made up of **water vapor,** which is an invisible gas. You see the holes where the water isn't in its liquid form. The steam that you can see rising from food cooking on the stove is made of water droplets, not vapor. Soon the steam evaporates and becomes part of the air, which is also invisible. Since people can't see air, early scientists had to plan special experiments to study its properties.

The two most abundant gases in the air were discovered 200 years ago by Antoine Lavoisier (*an twahn lah vwah ZEEAY*). In one experiment, Lavoisier heated an open bottle of mercury in a large jar that contained 800 cubic centimeters (cm^3) of air. He kept the mercury at a temperature just under its boiling point for 12 days. At first, red particles formed on the mercury. The particles stopped forming before the 12 days were up. Lavoisier calculated that the original 800 cm^3 of air in the jar had been reduced to between 670 and 685 cm^3. He reasoned

Figure 2-1 The chemical composition of the atmosphere.

Name	Chemical Composition	Percentage by Volume
nitrogen	N_2	78.1
oxygen	O_2	20.9
argon	Ar	0.9
carbon dioxide	CO_2	0.03
other materials		0.07
total		100.00

that 115 to 130 cm^3 of air had somehow been taken up by the mercury to form the red particles.

He then found that the gas remaining in the jar was less dense than ordinary air. It also put out the flame of a candle and quickly suffocated a mouse. Lavoisier called this gas "azote," from Greek words meaning "no life." You know it as nitrogen.

Next, Lavoisier collected the red particles and heated them to a high temperature. They gave off between 115 and 130 cm^3 of gas, the amount previously removed from the air. This gas made candles flame more brightly and did not suffocate a mouse. He called it "air eminently respirable, pure or vital." It was later named oxygen. The red particles were mercuric oxide, a chemical combination of mercury and oxygen.

Studies by later scientists have shown that there are other gases in the atmosphere besides the nitrogen and oxygen discovered by Lavoisier. Figure 2-1 shows the make-up of air at sea level.

As you can see in Figure 2-2, there are great changes in the make-up of the air,

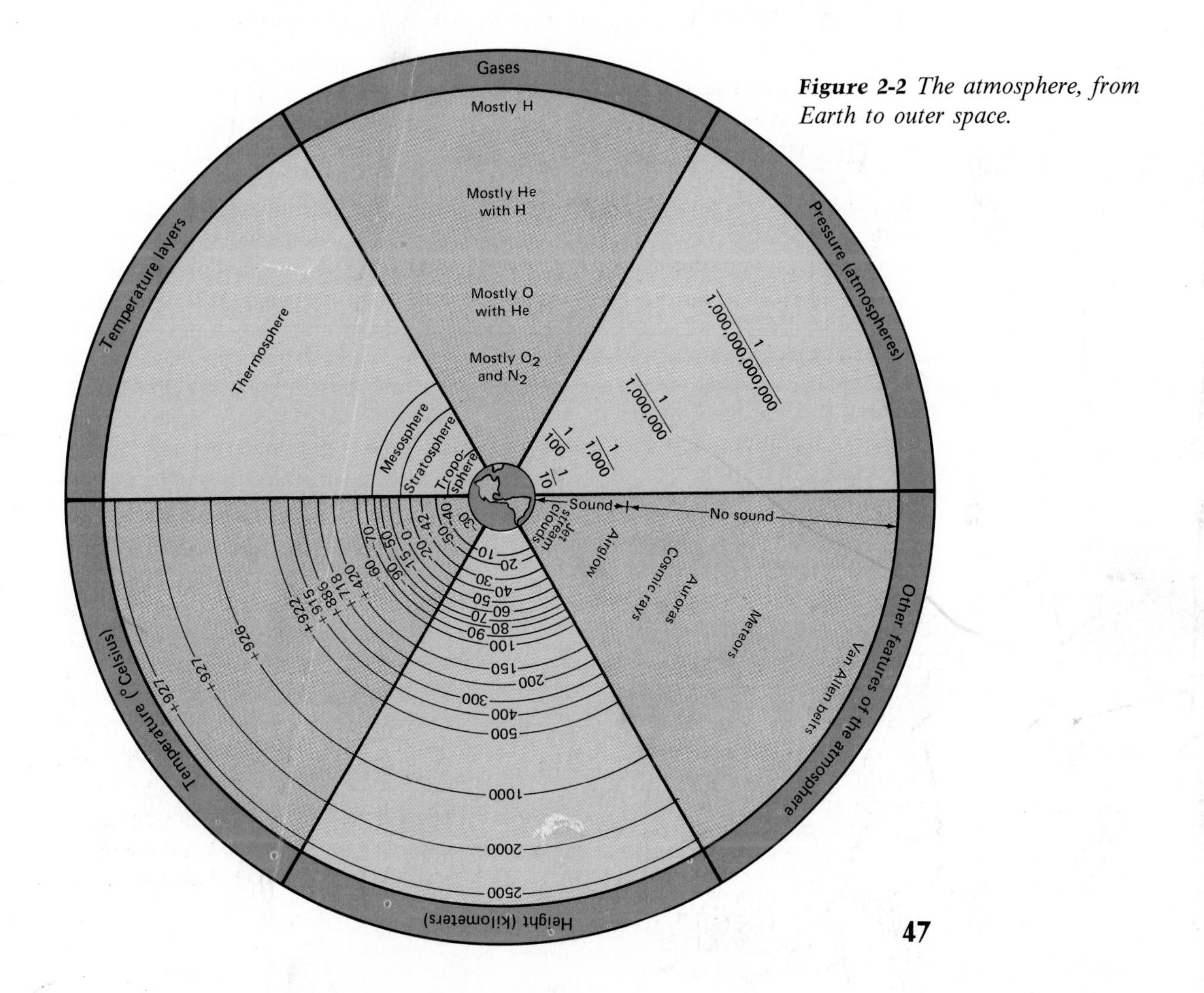

Figure 2-2 The atmosphere, from Earth to outer space.

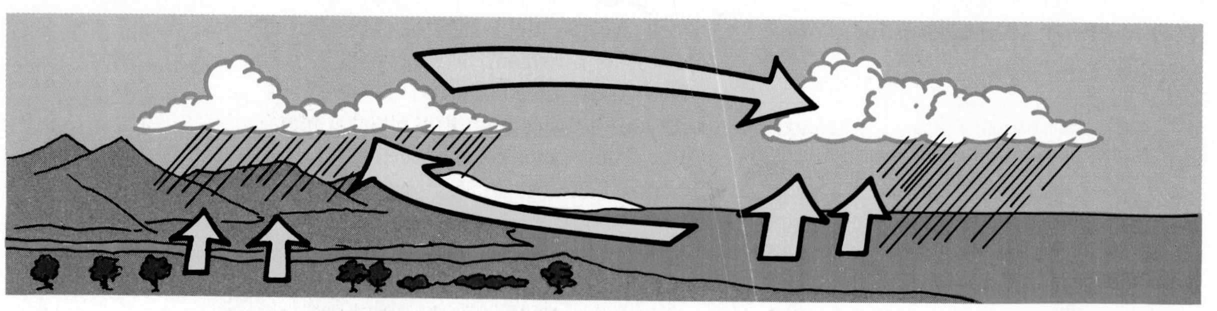

Figure 2-3 *In the water cycle, water follows many paths through the ocean, air, and land.*

from the earth's surface to the edge of outer space. Not only does the composition of the atmosphere vary, but the temperature and pressure also vary as you go outward from the earth. Some of these characteristics of the atmosphere will be discussed in later sections.

2-2 The Water Cycle

In the preceding section, no mention was made of the water vapor in the atmosphere. Under very humid conditions, water vapor can make up as much as three percent of the air at sea level. The **water cycle,** illustrated in Figure 2-3, is the exchange of water among the ocean, the air, and the land. Water molecules pass into the atmosphere as vapor from the ocean and land surfaces of the earth. When conditions are right, the vapor changes to clouds. This is the process called **condensation;** the vapor particles come together to form larger water droplets. When the cloud particles come together, water falls to the earth's surface as rain or snow. In this way, the water cycle is completed. Some of the water that falls on land flows in the earth's rivers and underground streams back to the sea. You will investigate this part of the water cycle, the interaction of land and water from the atmosphere, in Chapters 7 and 8.

Sometimes the water cycle is simplified in the following way. Water evaporates from the ocean into the atmosphere, where the winds carry it over to the land. Over land, the water vapor molecules condense into clouds and fall as **precipitation.** The rain and melted snow run off into the rivers or soak into the ground and flow back to the ocean, completing the cycle.

But this picture is too simplified. Water molecules lead a much more complicated life. For one thing, rain and snow fall on the ocean as well as on the land. And water evaporates from the land as well as from the ocean. Only about one-third of the water that falls on the earth's land surface flows directly back to the sea in rivers and underground streams. What do you think happens to the other two-thirds?

The water cycle occurs on many scales. A puddle of water that you watch evaporate today could fall on you tomorrow in a thundershower. On the other hand, the vapor molecules that condensed to make the shower may have started their trip from the surface of a distant tropical ocean the week before. Many large and small exchanges of moisture among land, sea, and air make up the water cycle.

2-3 Observing Clouds

ACTION

You can illustrate how clouds form by using the simple equipment shown in Figure 2-4. Put a little water into a two-liter wide-mouthed jar. Cover the jar and allow it to stand for a few hours. Then light a match, blow it out, and hold it for a few seconds in the jar. Re-cover the jar tightly with a rubber sheet that can be stretched easily. After a few minutes pull the sheet sharply. Describe what happens.

What happened to the air in the jar when you lifted the rubber sheet? Would the cloud have formed if you had not put smoke into the jar?

MATERIALS
two-liter wide-mouthed jar, cover, matches, rubber sheet, rubber band

Most clouds are caused by the cooling of air as it rises. Since the air pressure decreases rapidly with increasing elevation (Figure 2-2), rising air expands. When a gas expands, the molecules become more widely separated. The gas loses energy, since it uses some energy to expand into a region that is already occupied (by more gas), and since it takes some energy for each molecule to move farther away from those nearby. The energy that molecules (or other objects) have due to their moving about is called **kinetic** (*kih NEHT ihk*) **energy.** (The concept of kinetic energy, and that of potential energy also, is illustrated in Figure 2-5. Study the illustration carefully.) As rising air expands, then, its kinetic energy decreases.

Since **temperature** is a measure of the average kinetic energy of moving molecules, the temperature of expanding air drops. If you let the air out of an inflated inner tube or basketball, you can feel the drop in temperature of the outrushing air.

Water vapor condenses when it is cooled, but at ordinary temperatures it requires solid surfaces to condense on. For example, you may have noticed the vapor in your breath condensing on the cold windows of a closed automobile. In the atmosphere, the solid surfaces that water vapor condenses on to form clouds or fog are particles called **condensation nuclei** (*NOO klee eye*). Unfiltered air always contains some particles on which water vapor can condense. However, water vapor condenses more rapidly on some kinds of particles than on others. Smoke particles are especially good for condensation. In the preceding ACTION, if you had not put smoke into the jar you might not have been able to form a cloud. The cooling caused by expansion probably wouldn't have been sufficient for a cloud to form.

Figure 2-4 *You can cause a cloud to form inside a jar.*

Figure 2-5 a. *The energy system is at rest. The man has stored chemical energy in his muscles. It is potential energy.* **b.** *Chemical energy is being converted to work. As the pillar moves up the hill, it gains potential energy.* **c.** *The potential energy of the pillar now equals the work done to carry it up the hill against the force of gravity.* **d.** *As the pillar rolls down the hill, its potential energy is changed to kinetic energy.*

MATERIALS
bicycle pump

ACTION

What do you predict should happen to the temperature of air if you compress it? You can find out with a bicycle pump or a household insect sprayer. With the bottom hole open, pump ten times in about ten seconds and then feel the lower portion of the pump tube. Repeat this procedure, but this time hold your finger airtight over the hole at the bottom of the pump. How does the tube feel now?

Just as expanding air loses energy and cools, contracting air gains energy and warms. Two factors account for the fact that contracting air gains energy. For one thing, air does not normally contract unless energy is supplied to it in the form of work. For another, potential energy, which is greatest when molecules are far apart, is converted into kinetic energy as molecules move closer together. As the kinetic energy increases, the temperature also increases.

Thus, moist rising air forms clouds, but sinking air usually leads to clear skies. As air sinks, its temperature rises, and any clouds in the sinking air evaporate.

Because cloud droplets are so small and light, they almost float in the air. Slowly rising air will keep them from falling. Of course, the up and down motions of the air vary a lot. Different upward air movements produce different kinds of clouds.

There are many varieties of clouds, but they fall into two main types: stratus and cumulus. **Stratus clouds** are sheets or layers of cloud particles, covering a large portion of the sky. They can contain either water droplets or ice crystals. **Cumulus** (*KYOOM yuh luhs*) **clouds** usually appear as separate puffs or towering masses. They are made up mainly of water droplets. Many clouds have both stratus and cumulus features, illustrated in Figure 2-6.

The thin wisps of clouds high in the atmosphere are usually made of ice crystals. If you have ever seen a halo around the sun or moon, it was probably caused by such a cloud. **Contrails,** the white trails made by high-flying jet planes, are

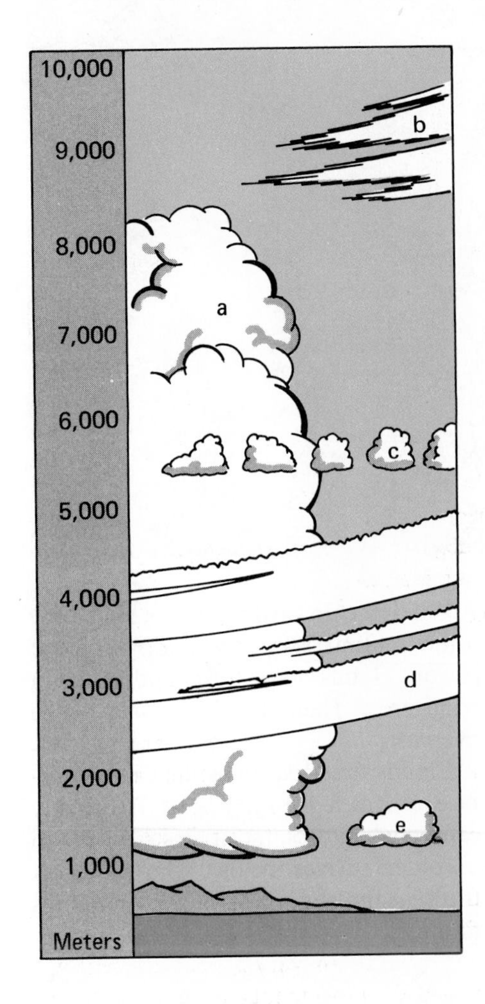

Figure 2-6 *The principal types of clouds and the heights at which they commonly occur.* **a.** *Cumulus (cumolonimbus or thunderstorm type),* **b.** *cirrus,* **c.** *high-level cumulus,* **d.** *stratus, and* **e.** *fair weather cumulus.*

Figure 2-7 *Long-lasting contrails (condensation trails) are caused by water vapor from jet plane exhausts.*

ice-crystal clouds (Figure 2-7). Water clouds, though white around the edges, usually have at least faint smudges of gray at their bases.

Cumulus clouds are formed by fast upward movements. The air may rise a meter or more per second. When air rises this fast, it tends to sink again nearby, as shown in Figure 2-8. This sinking air creates clear spaces between cumulus clouds.

When the upward motion of air is only a few centimeters per second, a layer of stratus clouds may form and cover a wide area. Then the sky is said to be overcast.

Clouds that rest upon the earth's surface are called **fog.** Fog is frequently caused when cool, moist air loses heat to a colder surface below. The air cools further, and the water vapor condenses into fog.

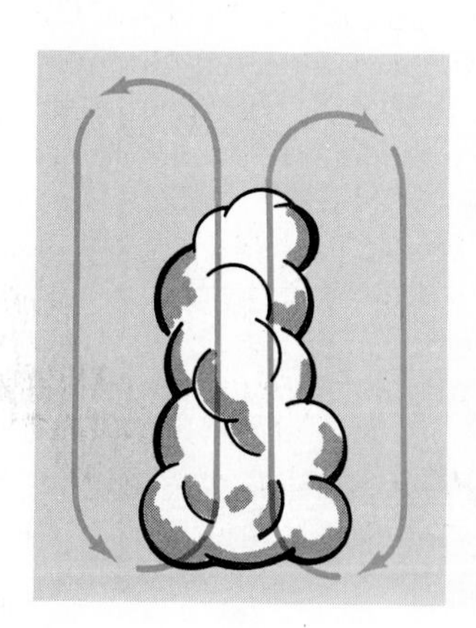

Figure 2-8 *A cumulus cloud above the Henry Mountains in Utah. Such clouds form in upward-flowing air currents. Where is air descending in this photograph?*

Figure 2-9 **a.** *Heat transfer by direct contact is conduction.* **b.** *The girl is being warmed by radiation.* **c.** *Warm air from the floor register will rise to the ceiling in a convection current.*

2-4 Energy

Cloud formation involves rising air. Why does the air rise? An answer might be, "Because of an input of energy." Air rises because heat energy is put into it.

On hot, muggy days, when there is a lot of moisture in the air, people may complain of not having the energy to do anything. But on fine days when the air is both cool and dry they often feel "full of energy" and ready to work. Your instinctive feelings about work and energy are probably not too different from the scientific concept shown in Figure 2-5. There are many kinds of energy, but all have one thing in common—*the capacity to do work.*

Heat is a form of energy with which everyone is familiar. Touch an object with your finger. If the object feels warm, heat is flowing into your finger from whatever it is you're touching. If the object feels cold, the heat transfer is in the opposite direction. When heat flows into the mercury or alcohol of a thermometer, the liquid expands and the top of the liquid column moves up. It is only through such changes, either in yourself or in the world about you, that you can find evidence of heat energy being used. Even if you can't see it, you can detect what it does.

Heat is transferred in three ways, illustrated in Figure 2-9. In all three ways, heat moves from material at a higher temperature to material at a lower temperature.

Two of the ways heat is transferred are no doubt obvious to you. One, heating by **conduction,** involves actual contact. A warm ocean heats cool air by conduction. The molecules in the warm ocean surface batter the molecules of the cool air into faster motion. Another way to say the same thing is that the higher-energy molecules transfer energy to the lower-energy molecules.

Secondly, heating by **radiation** is illustrated in Figure 2-9. As atoms and molecules vibrate, they send out waves of energy that travel through space. All objects, even icebergs, send out some radiation. (The iceberg doesn't warm you because you radiate more heat than it does.)

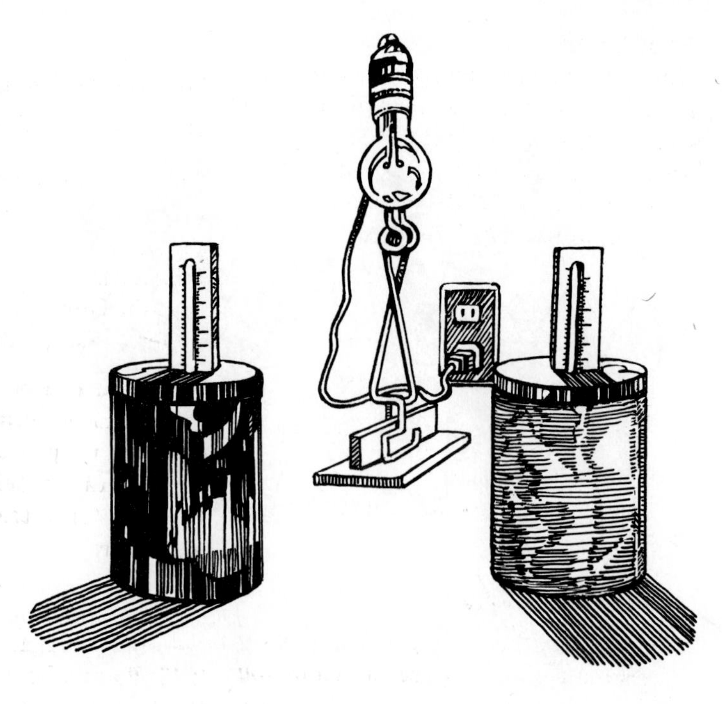

Figure 2-10 *The setup for investigating flow and change in energy.*

2-5 Investigating Flow and Change in Energy

You cannot see energy, but you can observe and analyze the flow of energy and its change from one form to another.

When you have completed this investigation, you should be able to describe how the color of a material is related to energy transfer.

MATERIALS
graph paper, two insulating lids with slits (for cans), 200-W lamp bulb (without reflector), metal can (black), metal can (shiny), two thermometers (−20°C to 50°C)

PROCEDURE
Use the equipment shown in Figure 2-10. Turn on the light and record the temperatures each minute for ten minutes. Turn the light off, remove it without disturbing the cans, and record the temperatures each minute for ten minutes. Make a graph of your data for each can. Label the horizontal axis in minutes and the vertical axis in degrees.

DISCUSSION
1. Which can heats faster?
2. Which can cools faster?
3. Which can absorbs energy better? What is your evidence?
4. Which can loses energy faster? How do you know?
5. How was the energy transferred in this investigation?
6. List and describe the forms of energy you observed.

2-6 Investigating Convection

There is a third way that heat is transferred from one place to another (Figure 2-9c). Heating by **convection** involves the actual movement of heated substances.

When you have completed this investigation, you should be able to explain and demonstrate the movement of air that is heated by radiation from a dark heated surface.

PROCEDURE

You will need a setup like the one illustrated in Figure 2-11a. Turn on the light and record the temperatures in the three areas of the aquarium that are marked A, B, and C. Record the temperatures every minute for five minutes.

DISCUSSION

1. Which area heated the most?
2. Which area heated the least?
3. How do you explain the heating at point B?
4. Why is the temperature at C different from that at B?

PROCEDURE

Do you think there is a circulation of air inside the aquarium? Use the setup you used for the previous activity, without the thermometers (Figure 2-11b). Introduce smoke, such as punk smoke, into the bottom of the aquarium at the end opposite the black paper. What happens?

DISCUSSION

1. Describe a **convection current** and how it starts.

MATERIALS

black construction paper, container with clear lid and opaque lid, insulating support for thermometer C, 200-W lamp bulb (with reflector), three thermometers (−20°C to 50°C)

MATERIALS

funnel, matches, punk (or other smoke source), rubber tubing, same setup as for first procedure (without the thermometers)

2-7 Investigating Land and Water Temperatures

The island of Bermuda is both a winter and a summer resort. People from the United States travel to Bermuda in the summer to escape the heat, and people from the United States and Canada go in winter to escape the cold.

When you have completed this investigation, you should be able to explain why people who live along seacoasts often enjoy cooling breezes from the ocean during hot weather.

PROCEDURE

Set up the materials as shown in Figure 2-12. Place a light so that the bottom of the bulb is no higher than 30 to 40 cm directly above the containers. One contains water and one contains soil. Turn the light on and record the temperatures of the containers each minute for ten minutes. After ten minutes have passed, turn the light off and again record the temperatures each minute for ten minutes.

MATERIALS

two containers (for soil and water), dry sand or soil, 200-W lamp bulb (with reflector), ring stand, four thermometers (−20°C to 50°C), water

Figure 2-11 *Setups for investigating convection currents in an aquarium.*

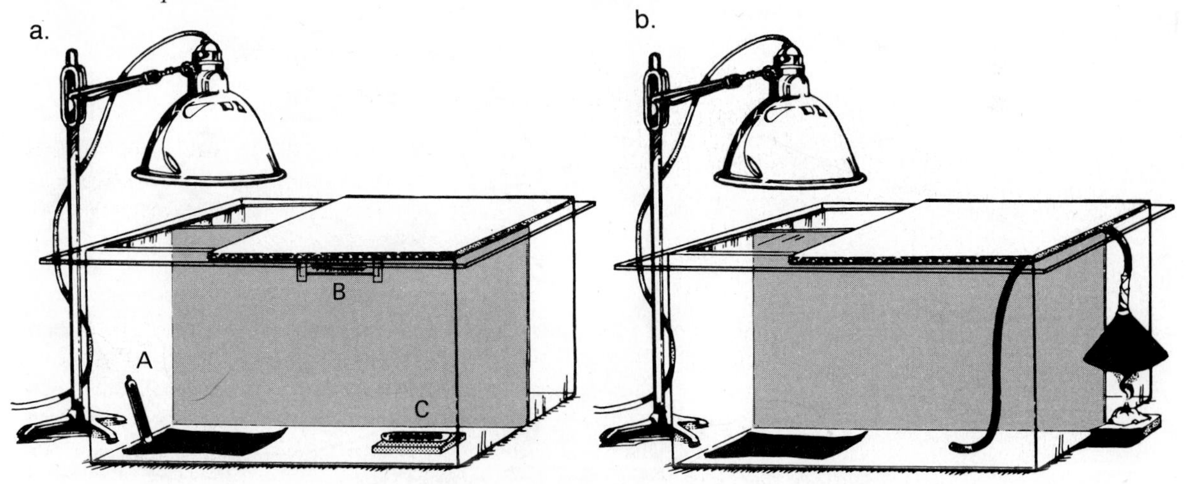

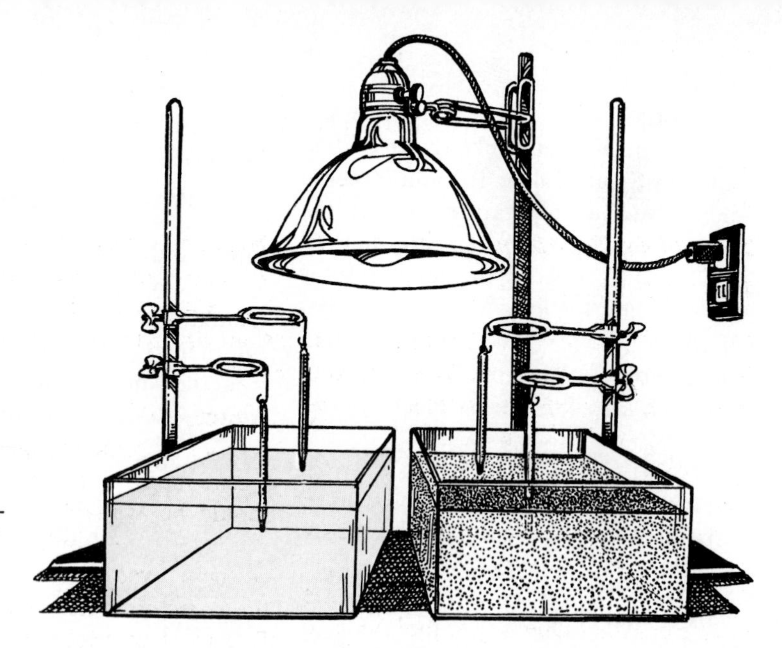

Figure 2-12 *Compare the heat absorption and radiation of soil and water.*

DISCUSSION

1. Did the air heat up faster over the soil or the water? Why?
2. Why did the rate of temperature change in the soil differ from that in the water?
3. Which received more heat from the lamp, the soil or the water? Why?
4. Which lost heat faster, the soil or the water?
5. The atmosphere gains heat from a **heat source** and loses heat to a **heat sink.** Which might be considered a heat source during the winter: soil or water?
6. Figure 2-13 illustrates the air motion in a sea breeze. Where is the heat source? the heat sink?

Thought and Discussion

1. Which form of energy, potential or kinetic, is best illustrated in the following stages of the water cycle: water vapor, floating clouds, falling rain, flowing streams?
2. Explain how a sea breeze might be a link in the water cycle.

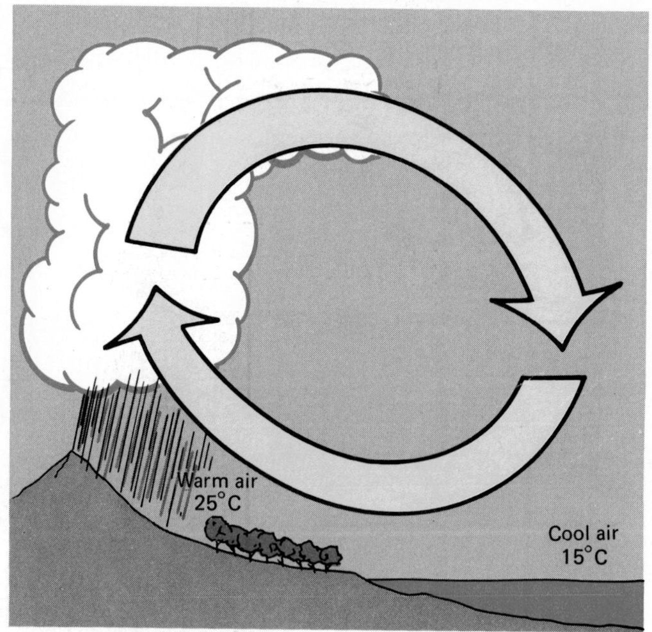

Figure 2-13 *What happens when the sun heats land and sea?*

2-8 Investigating Evaporation

Cloud formation involves rising air, but it also requires moisture. Moisture gets into the air by the process of **evaporation.** How would you define it?

Clothes on a line dry faster on a sunny day than on a cloudy day. Yet when the day is sunny, the air may also be windy or calm, moist or dry. How do these factors affect evaporation and influence the water cycle? One way to find out is to place a pan of water in the open air. You can then note the amount of water that evaporates during different types of weather. However, when several factors are acting at the same time, such as air moisture, wind, and sun, it is difficult to find out the effect of each factor by itself. In the classroom laboratory you can study the effect of one variable at a time on evaporation.

When you have completed this investigation, you will be able to identify the important factors that influence evaporation in nature.

Procedure

A balance, some sponges, a lamp, a fan, plastic bags, and hot and cold water should be available. (See Figure 2-14.) Use whatever supplies you need to investigate evaporation. Try to determine the effect of one variable at a time. Record your procedures and observations.

MATERIALS
equal-arm balance, fan, food coloring, graph paper, hot water, light (with reflector), plastic sandwich bags, sponge (or other absorbent material), transparent plastic sheets

Discussion

1. What factors influence the rate of evaporation?
2. Which of these factors has the greatest effect?
3. How do these factors operate in nature?

Figure 2-14 *You can investigate evaporation with equipment like this.*

Figure 2-15 The setup for investigating energy changes during evaporation.

2-9 Investigating Energy Changes During Melting and Evaporation

When a substance evaporates, it changes from a liquid into a gas. When a substance melts, it changes from a solid to a liquid. These changes are called **changes of phase.** You know that heat energy is involved in both cases. But the processes are not as simple as they seem.

When you have completed this investigation you should be able to describe the energy changes that occur during melting and evaporation.

MATERIALS
beaker, burner (or other heat source), crushed ice, graph paper, ring stand with ring (or tripod) and wire gauze, thermometer (−10°C to 110°C)

PROCEDURE
You will need a beaker of crushed ice, a burner or a hot plate, and a high-temperature thermometer. Set up the equipment as shown in Figure 2-15. While stirring the ice *gently* with the thermometer, read and record the temperature at one-minute intervals. Add heat until the water boils and then make three more readings. Make a graph of your results as you did for Investigation 2-5.

DISCUSSION
1. How does the energy going into the beaker affect the temperature?
2. When did the greatest temperature change occur? At that time what was the condition (phase) of the water?
3. What do you think caused the changes in the slope of the line on your graph?

2-10 Latent Heat and Change of Phase

During melting and boiling in Investigation 2-9, the temperature stopped rising, even though you kept adding heat. The heat was used to change the ice to water, and later, to change the water to vapor. The temperature did not increase because all of the energy added during melting and boiling was being used to force the water molecules farther apart.

The molecules of solids and liquids are bound together by strong forces of attraction. Separating the molecules requires an enormous amount of energy or work. The molecules of water in its solid form, as ice, are held tightly together. As a liquid, water molecules move about freely and slip and slide over each other. As a vapor in the atmosphere, the water molecules are so far apart that they exert hardly any force on one another, except when they collide.

The energy taken up by a substance during melting and evaporation is known as **latent heat.** *Latent* (*LAY tuhnt*) means "hidden" or "potential." When heat is absorbed by ice and the ice melts, that heat energy is stored in the water. And when the water evaporates, the heat energy that is used to bring about the change is stored in the water vapor. In moving farther apart, the molecules acquire greater potential energy. When water vapor changes back to liquid water, and liquid water to ice, the molecules give up energy to the surroundings. In this way the latent heat stored within water vapor is released when it condenses into the liquid particles of clouds or fog.

2-11 Air Pressure and Humidity

All of the gases in the air, including water vapor, make up the atmosphere. The atmosphere is held to the earth by the force of gravity. Each layer of air presses down upon the layer below with a force equal to the weight of all the air above. The pressure decreases upward very rapidly in the lower atmosphere.

The air pressure at sea level is called "one atmosphere." This is enough pressure to push mercury, which is a very heavy liquid, 76 cm up into a sealed glass tube. At the top of the earth's highest mountain, Mt. Everest, the pressure is less than one-third what it is at sea level.

There is little pressure in the upper layers of the atmosphere, for the air density at that height is very small. The atmosphere is "thin" with very few molecules in a given volume. Consequently, lighter gases like hydrogen and helium are able to escape from the earth's gravitational pull, since there are so few other molecules to bounce them back into the atmosphere.

The **pressure** of the air is the force with which its molecules strike a surface. The pressure depends partly on the number of molecules that strike a certain area. (Think of blowing up a balloon.) Pressure also depends on kinetic energy. The greater the kinetic energy of the molecules, the greater the pressure.

In a liquid, the molecules move in all directions, and with differing speeds. Near a liquid surface, the faster (more energetic) molecules are able to overcome the attractive forces of other slower ones and escape as a vapor. However, the escaping molecules collide with others and sooner or later may be knocked back to the liquid surface. Evaporation is the net transfer of molecules from the liquid to the gas phase.

ACTION

MATERIALS
grease pencil, small water glass, large jar, lamp or other heat source, ice cubes

Use a grease pencil to put a mark on the inner surface of a small water glass, near the edge. Make the mark a short line parallel to the edge of the glass. Put water into the glass, up to the mark.

Place a large jar over the glass. Arrange a lamp so it will heat the water. Leave it until evaporation ceases (when the water level in the glass stops falling). What happens? How do you explain your observations?

Now, turn the lamp off, and observe what happens when the jar and its contents cool off. You can speed up the cooling by putting ice cubes on the outside of the bottom of the jar. Explain any changes you see.

When you heat water in an enclosed space, evaporation continues only up to a point, depending on the temperature of the vapor and the liquid. When the temperature becomes constant, molecules leave the water surface and return to it at the same rate. Then, no further evaporation of water from the glass takes place. If you heat the water to a higher temperature by adding another lamp, evaporation will continue again until another balance is reached between the escaping and returning water molecules. The balance is called **saturation.** The air is saturated with water vapor.

The temperature at which saturation is reached is called the **dew point.** When saturation occurs in the atmosphere, the water vapor will start to condense. Explain why dew forms on grass after nightfall, when the temperature begins to drop. If the temperature drops to freezing before saturation occurs, water vapor changes directly to ice. In this way **frost** is formed.

What you observed in the ACTION will suggest to you that air at some temperatures has the capacity to hold more water than air at other temperatures. Which can hold more water vapor, warm air or an equal volume of cold air? Study Figure 2-16.

You know that air can vary in the amount of moisture it contains. At the same time, air in some places can be very moist (but not saturated) and the air in other places can be dry. Could the dew point be the same for both? Some deserts cool off greatly at night. Suppose you were in a desert where the temperature fell from 30°C to 10°C. Would you expect to find dew?

A common measure of the moisture in the air is relative humidity. **Relative humidity** is the ratio of the actual amount of moisture in the air at a certain temperature to the most moisture it could hold at the same temperature. It is expressed as a percentage:

$$\text{Relative humidity (\%)} = \frac{\text{actual moisture content} \times 100}{\text{saturation moisture content}}$$

In hot weather, relative humidity is a simple indicator of human comfort. When the air temperature is close to the body's temperature, people are very sensitive to differences in air moisture. High relative humidity slows up the evaporation of perspiration from your skin. Why does this make you uncomfortable?

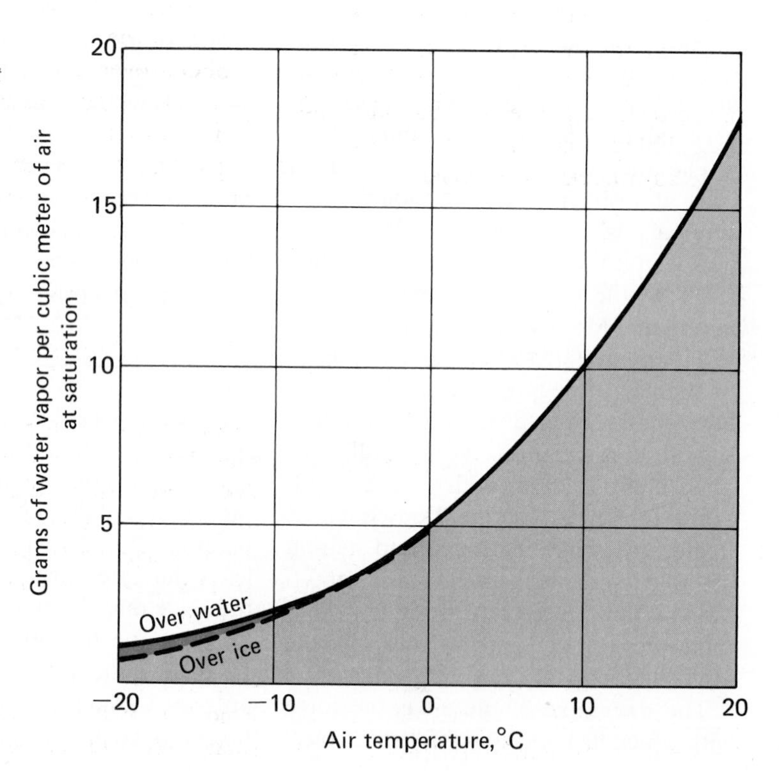

Figure 2-16 *The water vapor content of saturated air over ice and water.*

2-12 Condensation

Water vapor changes to liquid water in the air by condensation. You know that this phase change is just the reverse of evaporation. The addition of heat to water increases the energy of the molecules and allows the water to evaporate. When water vapor is cooled, molecules condense faster than they evaporate.

Condensation usually occurs in the atmosphere when the relative humidity is near 100 percent. In the ACTION in Section 2-11, the glass jar cooled more quickly than the air inside. Water vapor condensed on the glass when the air near the glass cooled to the dew point. If the air inside the jar cooled all at once, the vapor would condense on tiny salt or smoke particles in the air. Then you would see a cloud or fog fill the jar.

Water vapor collects on some kinds of particles more readily than on others. It collects most easily on salt crystals left drifting in the air from ocean spray that has evaporated. Water vapor condenses on these absorbent particles at low relative humidity. All the smog and many of the fogs you see are examples of condensation occurring at relative humidity less than 100 percent. Dust particles also serve as condensation nuclei.

Air near the earth's surface does not often become saturated with water vapor. Even though you have seen moisture condense on the side of a glass of ice water, you know now that the condensation does not require saturated air.

Condensation is one of the necessary steps in the water cycle. It results in the formation of cloud droplets. However, cloud droplets fall so slowly that condensation alone does not remove much water from the atmosphere. Practically all of the water in the air is removed by precipitation.

Thought and Discussion

1. Why do you suppose that the maximum amount of water vapor in the air is only about 3 percent of the total gases?
2. The temperature decreases as one goes up into the atmosphere. Above the equator, at a height of 17 km, it plunges to about −80°C. Above that altitude the temperature increases again. The cold part of the atmosphere over the equator has been called a "trap." How does this cold trap prevent much water vapor from getting into the warmer layer above?

Clouds and Rain

▣

2-13 Investigating Cumulus Cloud Formation

In many parts of the country, cumulus clouds appear in the sky on a warm afternoon or on a cold, windy day after a rain. When they first form, cumulus clouds look like large heaps of cotton. The flat base of the clouds marks the level where condensation begins.

If you know the temperature and the dew point at the earth's surface, you can calculate the height at which the two become equal. This is approximately the height of the flat bases of the cumulus clouds.

PROCEDURE
Find the dew point by slowly adding small pieces of ice to a can of water.

MATERIALS
crushed ice, dew-point temperature chart, metal can (shiny), sling psychrometer, thermometer (−10°C to 50°C)

Figure 2-17 *Finding the dew point in a classroom.*

Gently stir the mixture with a thermometer. (See Figure 2-17.) Record the temperature when drops of water begin to condense on the outside of the can. Repeat this several times to make sure of the temperature.

You can find the dew point indirectly by using the sling psychrometer as described in Appendix B.

DISCUSSION

1. Compare the results obtained with the psychrometer and can methods. How close should they be? What could cause differences in the results from the two methods?
2. When air rises it cools about 10°C for each kilometer it rises. The air's dew point also decreases, as it rises, at the rate of about 1.7°C per kilometer. At what height will cumulus clouds form on the day of your observations? (See Figure 2-18 to find out how to make your determination.)

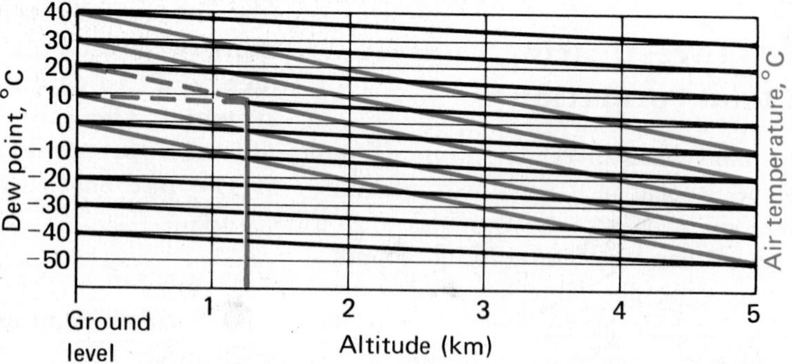

Figure 2-18 *Sloping lines show how the temperature and dew point of dry (unsaturated) air change when it rises in the atmosphere. You can estimate the height at which cumulus clouds will form by finding where the surface temperature and the dew point lines intersect. In the example, the ground temperature is 20° and the dew point is 10°.*

2-14 Observing Precipitation

Most of us are not really aware of the water cycle until it rains or snows, or unless it hasn't rained for so long that there is a drought. Rain, snow, and other precipitation complete the airborne part of the water cycle.

The size of precipitation depends on the kinds of clouds involved. The largest raindrops fall from cumulus clouds. They can reach the size of a medium-sized pea. Raindrops larger than this can form, but they quickly break apart as they fall, because of air resistance. Drizzle is composed of drops about the size of a period on this page. Drizzle falls slowly from low stratus clouds or from fog.

Sometimes snowflakes occur as beautiful six-sided or hexagonal ice crystals. (See Figure 2-19.) Usually, the single crystals clump together into larger snowflakes. The crystals grow when water vapor accumulates on solid particles called ice nuclei. The water vapor changes directly to ice. Particles of **sleet** (ice pellets) form when raindrops or partly melted snowflakes fall through a layer of cold air and freeze.

Hail is another kind of solid precipitation. It falls from cumulus clouds. Hailstones are balls or irregular lumps of ice. (See Figure 2-20.)

The tiny droplets that make up clouds condense on small, solid particles. It is natural to suppose that raindrops grow by further condensation on cloud droplets. However, most raindrops cannot form in this simple way. Observation shows that it takes about a million cloud droplets to make a raindrop. Condensation on a cloud droplet takes place much too slowly to increase its size a million times before the drop falls to earth.

One theory of how raindrops form involves ice crystals. Droplets and ice particles can exist in the same cloud, at temperatures well below freezing. The evaporation from an ice particle is less than the evaporation from a cloud droplet at the same temperature. Water vapor in the surrounding air tends to crystallize on the ice particles rather than condense on the water droplets.

The loss of vapor in the air causes more water to evaporate from the cloud droplets. The vapor continues to crystallize on the ice particles, and they grow rapidly and become snowflakes. The

Figure 2-19 *Ice crystals show a variety of shapes.*

Figure 2-20 *A hailstone that fell at Carrollton, Missouri, in 1975.*

flakes may melt in lower, warmer regions of the atmosphere and fall as rain. If they do not melt as they fall, the particles reach the ground as snow.

The temperature of many clouds is above freezing. There can't be ice crystals in these warm clouds, yet rain often falls from them. Raindrops can form in warm clouds if the cloud droplets collide and stick to each other. Raindrops grow faster when there are drops of many different sizes colliding. Collisions between droplets are more likely when the clouds are tall and convective movements are strong.

Thought and Discussion

1. Meteorologists can make some cumulus clouds grow very rapidly by dropping silver iodide particles into the cloud tops. The silver iodide particles serve as nuclei on which ice crystals can form. In effect, the experimenters are turning part of the water cloud into ice. Meteorologists point out that this change of phase makes the cloud warmer and causes it to develop. Explain how this can happen.
2. How is precipitation different from condensation?
3. If the earth's surface were all ice, what would the water cycle be like?

Unsolved Problems

A major problem in earth science is the explanation of the make-up of the atmosphere. How did the air reach its present composition? Is the atmosphere changing now?

The atmosphere consists largely of nitrogen and oxygen, with much smaller concentrations of other gases, called "trace" gases. Scientists believe that the atmosphere has changed drastically during its long history. Nitrogen, the most abundant gas today, probably made up only a tiny proportion of the earth's original atmosphere. Atmospheric oxygen (0_2) was probably not present at all. Instead, gases found today only in small concentrations—water vapor, carbon dioxide, and some of the trace gases—made up the great mass of the earth's early atmosphere.

Oxygen and carbon dioxide are examples of earth materials that, like water, undergo well-defined cycles. Oxygen can be formed by the action of the sun's ultraviolet light on water vapor (H_2O) when water gets into the upper atmosphere. Some scientists believe that the oxygen in our air was formed by this action. Others believe that the oxygen was brought to its present level and is maintained there by the life processes of plants. Carbon dioxide is a product of respiration, combustion, and organic decomposition.

The percentage change in atmospheric oxygen is so slight that the oxygen cycle is of mainly theoretical interest. But the level of atmospheric carbon dioxide (CO_2) is increasing about 1% a year due to the increasing use of petroleum, coal, and natural gas. The carbon dioxide cycle is therefore of greater concern. Some scientists predict levels of atmospheric CO_2 that might cause a drastic change in the earth's climate. Can you name other waste gases that cause more obvious harmful effects? Would you consider CO_2 an air pollutant?

Many substances that get into the atmosphere, whether naturally or from the burning of fuels, unite easily with water molecules. Most are at least partially water soluble. So precipitation has dissolved in it some substances that can be harmful to living organisms, as well as some that are beneficial. The increasing acidity of rain and snow is part of the general air pollution problem that is now getting more attention.

Chapter Summary

We look to the reservoir in the air for our supply of fresh water. This reservoir is constantly refilled by evaporation of water from the earth's land and water surfaces. The ocean is the main reservoir of the water cycle. The earth's water cycle is the *total interchange* of water among the land, the sea, and the air.

The atmosphere acts as a pump and condenser of water vapor in the water cycle. Evaporation is controlled primarily by heat. Adding heat to water gives the molecules more energy and thus increases the rate at which they can escape into the atmosphere. However, air motions are necessary to remove the water vapor so that evaporation can continue. Upward motions lift the water vapor into the atmosphere, where it is transported by the winds. The upward transport of water vapor is caused primarily by convection. Convection represents a transfer of potential energy into kinetic energy. As the vapor rises into regions of lower pressure in the atmosphere, it cools and condenses. The energy stored as latent heat in the vapor is released, raising the temperature of the atmosphere. This released heat intensifies convection, adding more "fuel" to the water cycle.

Water vapor condenses to form clouds. There are two main types of clouds. Cumulus clouds are usually produced by the small-scale, rapid updrafts caused by convection. Stratus or layered clouds are usually produced by the gradual upward movement of air.

After cloud droplets are formed, the water still has to collect in large drops or crystals to form precipitation. Raindrops can be formed in two ways. Vapor in the clouds can crystallize on ice particles that melt as they fall to Earth. In warm clouds, only the collisions of droplets can account for the growth of raindrops. This process may also operate in colder clouds.

The energy that powers the atmosphere's circulation and the water cycle comes from the sun.

HYDROGRAPHER

Information about the flow of water in our rivers and streams comes from measurements made by hydrographers (*hy DRAHG ruh furz*). Hydrographers measure the rainfall and river flow at set metering locations. They measure the depth of the water and the rate of flow. They also take samples of the water for analysis. Later the hydrographer gathers all the data together and prepares graphs and charts that show waterflow patterns.

Hydrographers often work at dams or on canals, where accurate information about the waterflow is essential. There they may take samples of silt from the water to be analyzed and measured. Hydrographers make frequent measurements of the amount of silt and rate of waterflow in the desilting (*dee SIHLT ihng*) basin. The desilting basin is a lake or basin that has been made to prevent a canal from filling up with silt.

Training for work as a hydrographer usually consists of two years of education after high school. On-the-job training may take the place of some formal education. A hydrographer should like working outdoors, have good vision, and be able to do moderately strenuous work. Skill in mathematics is useful in preparing the required charts, graphs, and reports.

Questions and Problems

A

1. Distinguish between potential and kinetic energy. Give examples of each.
2. Identify the three methods of energy transfer. Give an example of each that might be observed within the next hour.
3. List the processes that occur within the part of the water cycle that you studied in this chapter. In other words, what, in general, can happen to a water molecule in the atmosphere?
4. List the common elements of the atmosphere at sea level, in the order of their abundance.

B

1. Suppose you are using a psychrometer, and the dry bulb reads 25°C and the wet bulb stabilizes at 15°C. What are the dew point and relative humidity under those conditions? Use Appendix B to find your answers.
2. The air temperature is 20°C and the dew point is 0°C. Approximately how high must the air be lifted to reach saturation?
3. A rapidly moving mass of cold air is pushing into an area of warm, moist air. What kind of clouds would you expect to find in the air just ahead of the advancing cold air?

C

1. Explain why the earth's atmosphere at sea level is not always the same temperature at different locations at the same latitude.
2. What factors affect the rate at which evaporation takes place? Give examples of these factors as they occur at your school or home.
3. Describe the effects of energy input and temperature changes as water changes from ice to liquid to vapor, at sea level.
4. List the different kinds of precipitation and explain the conditions under which each kind might occur.

Suggested Readings

Hanson, Kirby. "Carbon Dioxide." *Weatherwise,* December 1980, pp. 253–258.

Schaefer, Vincent J., and John A. Day. *A Field Guide to the Atmosphere.* Boston: Houghton Mifflin Co., 1981.

3 Weather and Climate

THERE IS SOME *water in the air everywhere. Even over the Sahara desert, the water in the air could supply a heavy rainfall. If the atmosphere were forced to give up all its water, there would be, on the average, 2.5 cm of rainfall over the entire earth.*

Of course, the atmosphere doesn't deliver its water equally everywhere. In some areas, the atmosphere supplies plenty of water to the land and makes it productive. In other places the water supply comes in huge downpours. In March, 1952, on the island of Réunion off the eastern coast of Africa, over 1.8 meters of rain fell in 24 hours. In many places where there is usually enough rain for crops and the other needs of people, there are occasional droughts lasting a season, a year, or several years. In the early 1970's, a long drought brought great suffering to the people who farmed the lands bordering the Sahara desert. What causes a drought? What causes a desert?

Climatologists looking for answers to these questions start with the water cycle and with air motion—with wind. From early times people have looked for orderly patterns of motion beneath the apparent aimlessness of the winds. Aristotle believed that the winds are produced by a kind of "breathing" of the earth. A 17th century English pirate and explorer named William Dampier (DAM pee ur) *wrote about the trade winds and* tuffoons (*typhoons or hurricanes*) *of the Atlantic Ocean. The trade winds, which blow in zones on either side of the equator, cover about half the earth's surface. In some places they blow moist and create typhoons; in other places they blow dry and form deserts.*

One person who read what Dampier wrote about winds was an English scientist named Edmund Halley. Eventually, Halley published the first map of the trade winds and explained them as part of a great world-wide circulation of the atmosphere. The breeze you may have felt on the way to school today is a part of that same global circulation.

AFTER COMPLETING THIS CHAPTER, YOU SHOULD BE ABLE TO:

1. **construct an isothermal map to identify temperature patterns of your environment.**
2. **use weather map symbols to identify various weather events.**
3. **describe the effects of the earth's rotation on the direction of air currents.**
4. **distinguish between the effects of high and low air pressure air masses and describe the ways they help to form weather fronts.**
5. **identify the causes of vertical air motions and explain their role in the formation of various kinds of storms.**
6. **use a series of weather maps and a sequence of weather data for a certain location to make a weather forecast for a particular day.**
7. **describe and locate the major wind belts of the earth and the general pattern of air circulation in the atmosphere.**
8. **explain what is meant by the term "greenhouse effect."**
9. **identify the different layers of the earth's atmosphere and describe their relationships to each other.**
10. **distinguish between zones of high and low moisture content on the earth's surface and identify factors that create these differences.**
11. **list various factors that have an effect on climate and explain how these factors can influence the climate of a particular area.**

Weather

3-1 Investigating Temperature Patterns

What does the word "pattern" mean to you? Perhaps the first thing you think of when you hear the word is a design on cloth or the orderly arrangement of some objects. But there are many kinds of patterns. Scientists often use maps of patterns when attempting to describe differences in conditions within a given portion of space.

Think of a three-dimensional portion of space, such as the room you are in or some part of the atmosphere surrounding your home. How would you describe to someone else the pattern of certain things that occur there, but can't be seen, like different temperatures, air pressures, or even the loudness of sounds? You would find it very difficult unless you could use a device called an isoline map to show how these quantities change from place to place within that portion of space. An **isoline** is a line on a map that connects points where a particular condition, such as temperature, pressure, or sound, is the same.

When you have completed this investigation, you should be able to identify temperature patterns in your classroom.

PROCEDURE

Measure and record the temperatures of several places in your room at: (a) floor level, (b) desk-top level, and (c) two meters above the floor (at or just above head level). Your teacher will give you the necessary instructions for making and recording these measurements.

Make a map of the room temperatures for each level (see Figure 3-1) that shows the lines that pass through points where the temperature is the same. If the isolines are used to connect places where temperatures are equal, they are called **isotherms** (*iso-* from the Greek word meaning "equal" and *-therm* from the Greek word for "hot"). Connect all the points of equal temperature with smooth isotherms. Do this for each temperature value on your map for each level. Color the spaces between the lines so that areas of the same temperature range are all the same color. Make different temperature ranges different colors. Use the same color key for all three levels. When your maps are completed, stack the sheets for each level so that your "model" resembles the situation in the room at the time the temperatures were taken. Note that isotherms might also run vertically, from level to level (if we had some way of drawing them) as well as just along each level.

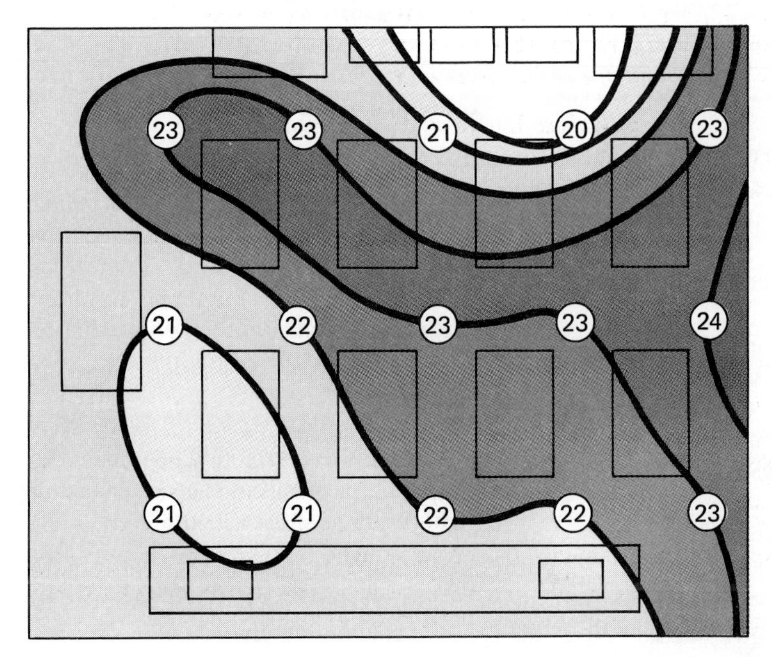

Figure 3-1 *A map of isotherms in a classroom.*

MATERIALS

marking crayons (four colors), room-plan maps, thermometers

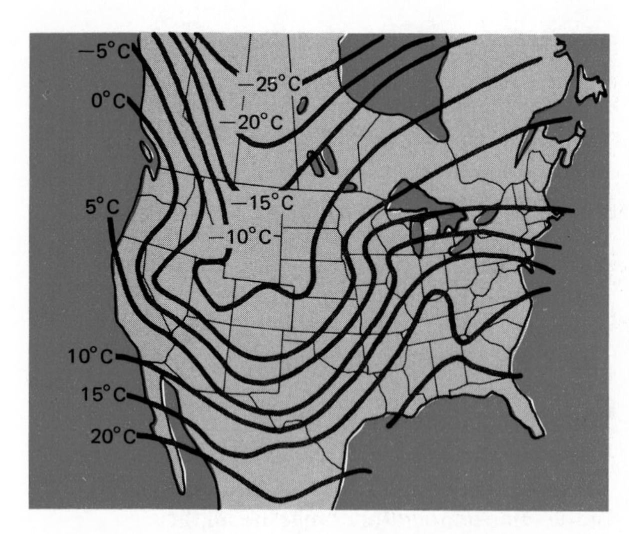

Figure 3-2 *Winter isotherms in North America. Is this a relatively cold or a relatively warm day for winter?*

DISCUSSION

1. What are some of the factors that caused the different temperatures within the room? What energy sources (or energy sinks) in the room can be detected by the isotherm patterns on your maps?
2. Would the temperature patterns in the room be the same tomorrow at the same time?
3. What are some other variables in your classroom that could have been measured at the same time you did the temperature pattern? How could they be detected and described?

Figure 3-2 shows an example of isolines on a large scale—isotherms across part of North America on a winter day. What do you think the temperature was in Regina, Saskatchewan? in Mexico City, Mexico?

Within a portion of space you can measure many variables that tell you about conditions in that space. You can measure the density of the air. You might measure the density of a liquid, like water, or a solid, like rock under the earth's surface. You would find a definite value at every point for the quantity you measured, as you have just done in your investigation of the temperature pattern in your classroom. When you were finished, you could show the values and their pattern on a map.

3-2 Investigating Surface Weather Maps

You may be familiar with the patterns made by **isobars,** the lines connecting points of equal pressure on a map. As you have seen in Section 3-1, other isolines can be drawn for other quantities.

Other kinds of lines and symbols are also used to show the patterns of weather on maps. **Streamlines** are lines along which the wind is blowing at a given moment. **Cold fronts** show where cold air is replacing warm air. **Warm fronts** indicate where warm air is replacing cold air.

When you have completed this investigation, you will be able to identify weather patterns and the ways in which these patterns are related to each other.

MATERIALS
four base maps of North America, with data plotted at selected stations

PROCEDURE
Use the plotted weather maps supplied by your teacher to draw isobars, isotherms, streamlines, and isolines of cloudiness and precipitation. To draw streamlines, make your lines parallel to the wind arrows. (See Figure 3-3.)

DISCUSSION

1. Where is the air coming from that is spreading out from the center of the high pressure area in the West?
2. Where are the winds converging (coming together)? What happens to the converging air?

3. What is your explanation for the rainy and cloudy areas?
4. In which areas are streamlines crossing isotherms? Where are the winds bringing in colder air? warmer air?
5. Where would you draw a cold front? a warm front?

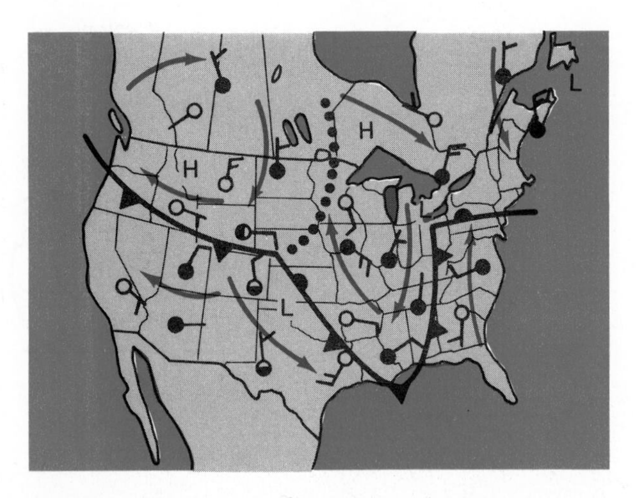

Figure 3-3 *Streamlines and fronts in North America. What is the relationship between them?*

3-3 Investigating the Coriolis Effect

If the earth did not rotate, the wind would blow directly across the isobars from high to low pressure. But the earth does rotate, causing the air to be deflected or bent away from a direct path toward low pressure. The deflection of objects moving freely over the earth was first explained by Gaspard C. Coriolis (*kawr ee OH lihs*) in 1835, so it is called the **Coriolis effect.**

When you have completed this investigation, you will have modeled what happens to air moving over a rotating earth.

PROCEDURE

Use the equipment shown in Figure 3-4. Imagine that the circular tray represents the Northern Hemisphere seen from above the North Pole. The center of the tray would be the North Pole; the outside edge, the equator. By turning the tray slowly counterclockwise, you can duplicate the earth's rotation.

Prepare an unmarked surface. Let the ball roll down from the top of the ramp. It should make a track on the surface. Keep the ramp in the same place. Now slowly turn the tray and roll the ball down again. Experiment with the equipment for a while.

DISCUSSION

1. What force set the ball moving?

MATERIALS
circular "magic slate" surface (or sand or carbon paper), circular tray (or large pie tin), ramp, steel ball, turntable

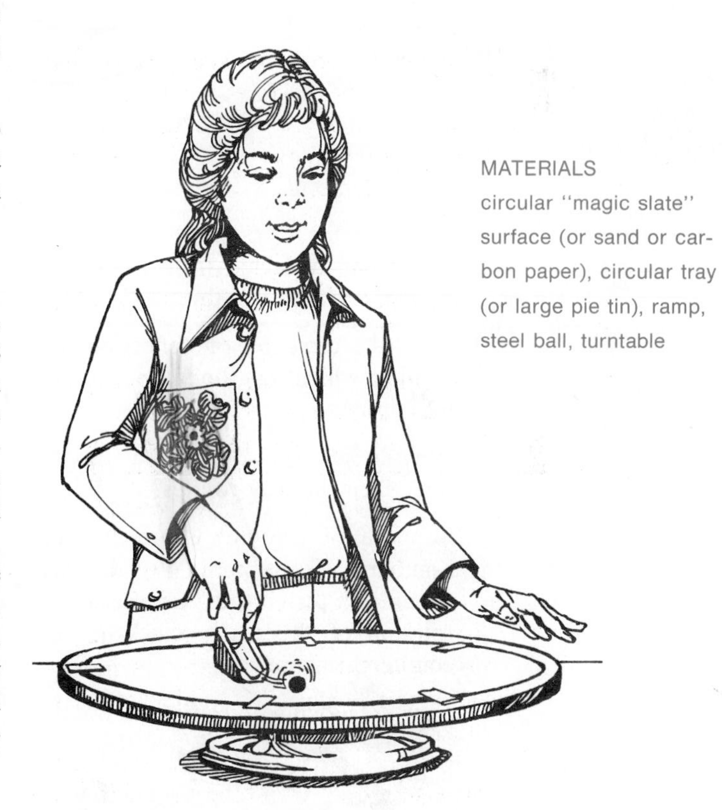

Figure 3-4 *A setup for investigating the Coriolis effect.*

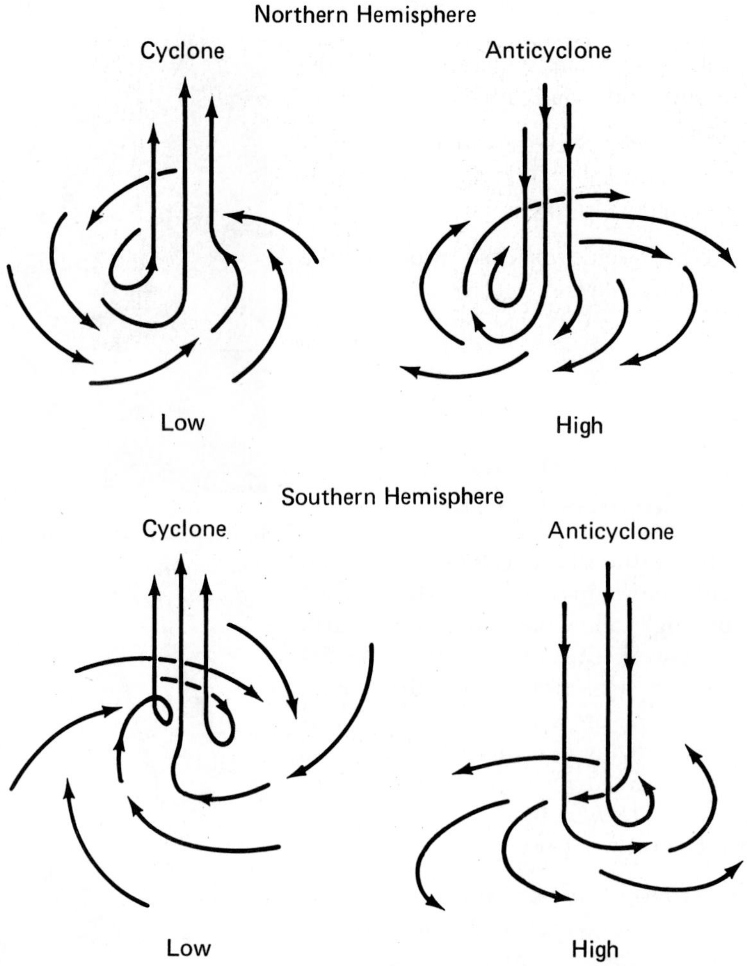

Northern Hemisphere

Cyclone

Anticyclone

Low

High

Southern Hemisphere

Cyclone

Anticyclone

Low

High

Figure 3-5 *In the Northern Hemisphere, air in a low rises and moves inward counterclockwise. Air in a high sinks and spreads out clockwise. Compare this with the Southern Hemisphere circulation.*

2. Is the ball deflected from a straight path everywhere on the tray?
3. Does the direction of deflection depend on which way the ramp is pointing?
4. Why doesn't the ball roll in a straight line when the tray rotates?
5. Suppose the tray now represents the Southern Hemisphere. Which way should you turn it to model the actual rotation of the earth, clockwise or counterclockwise?
6. How are moving objects in the Southern Hemisphere affected by the earth's rotation?
7. Refer to a weather map of the United States or Canada. Air in a high pressure system spirals out away from the center. Air in a low spirals inward toward the center. How does the Coriolis effect explain wind directions around highs and lows?

3-4 Cyclones, Anticyclones, and Fronts

In a Northern Hemisphere low pressure area, the inward-flowing air moves upward as it converges near the center. This pattern of air motion forms a **cyclone** that can be hundreds of kilometers in diameter. In an **anticyclone,** the wind spirals outward from a high pressure center, and around it, in a clockwise motion. The outward-flowing air is replaced

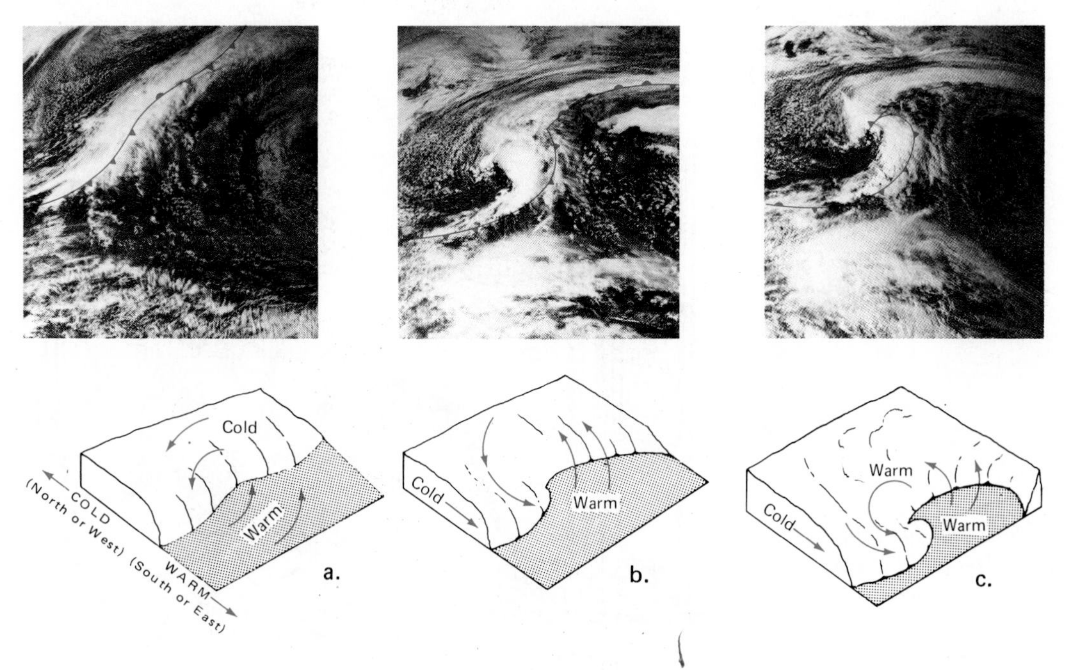

Figure 3-6 *The development of a frontal cyclone in the Northern Hemisphere.* ***a.*** *Warm and cold air push against each other.* ***b.*** *The warm air glides up and over the cold air, and the cold air moves under the warm air.* ***c.*** *The cold front overtakes the warm front. The photos show how cloud patterns change during cyclone development along a nearly stationary cold front over the North Pacific Ocean in winter.*

by air sinking down from above. Figure 3-5 illustrates the direction of rotation around cyclones and anticyclones in the Southern Hemisphere. Does it agree with your findings in Investigation 3-3?

ACTION

With a rectangular container such as a plastic tank or a glass baking dish you can make a model of a **front,** or boundary, between fluid masses. Fill the container with water and make a plastic or cardboard wall to divide the tank into two parts. Pour salt into one half of the tank. Use about 1 g of salt for each 10 ml of water. Then add food coloring to the same side. Allow the water in the tank to become calm. Carefully draw up the separating wall. Try not to disturb the water. Describe what happens in the tank.

In the ACTION, a front forms between two water masses of different densities. Because the dense mass slides under the lighter mass, the front becomes inclined, or angled. In time, the front in the tank will become horizontal. In the atmosphere, the cold, dense air remains inclined like a wedge under the warm air until a cyclone forms.

As shown in Figure 3-6, when a Northern Hemisphere cyclone develops, warm air blows against a wedge of cold air. The warm air glides up and over the cold air, forming a warm front. The cold air current to the northwest curves and pushes the warm air ahead of it. That part of the cyclone is called the cold front. At the same time the whole cyclone moves more or less eastward in the same direction as the strong winds high in the troposphere (review Figure 2-2 for the position of the troposphere).

MATERIALS

plastic tank or glass baking dish, cardboard, salt, food coloring

MATERIALS
hollow plastic cylinder, hot water, ice, smoke source

***Figure 3-7** The setup for investigating the effect of temperature differences in a column of air.*

Since the cold air moves faster than the warm air, the area of warm air narrows. (See Figure 3-6c.) The cold dense air sinks and spreads out, forcing the less dense air upward. As the cold air near the center of the cyclone overtakes the warm front, all of the warm air between the two fronts is squeezed upward. In time, when the area of warm air has disappeared, the cyclone begins to break up, for the temperature contrasts that created its energy have been destroyed.

A cyclone moves most rapidly when it is first developing. An older cyclone moves less rapidly.

3-5 Investigating Vertical Air Motions

In Chapter 2 you learned that clouds and precipitation occur in rising air. When the atmosphere resists either rising or sinking motion, it is said to be **stable.** When air continues to move and gains speed if it is given a slight push upward or downward, it is **unstable.**

When you have completed this investigation, you will be able to explain what causes the atmosphere to be stable or unstable.

PROCEDURE

Set up the equipment pictured in Figure 3-7. Use two thermometers to show how temperature varies with height. At first, the temperature should be about the same at each height.

Pour water containing small pieces of crushed ice into the container at the bottom. Read the thermometers every three minutes until the temperature stops changing at both levels. Make a graph of the temperature change at each height. Label the horizontal axis in minutes and the vertical axis in degrees. When the temperature has stopped changing at both levels, let smoke drift through the rubber tubing into the bottom of the cylinder. Let the smoke fill about one-third of the cylinder.

DISCUSSION

1. Where in the cylinder is heat transferred? In what direction is the heat transferred?

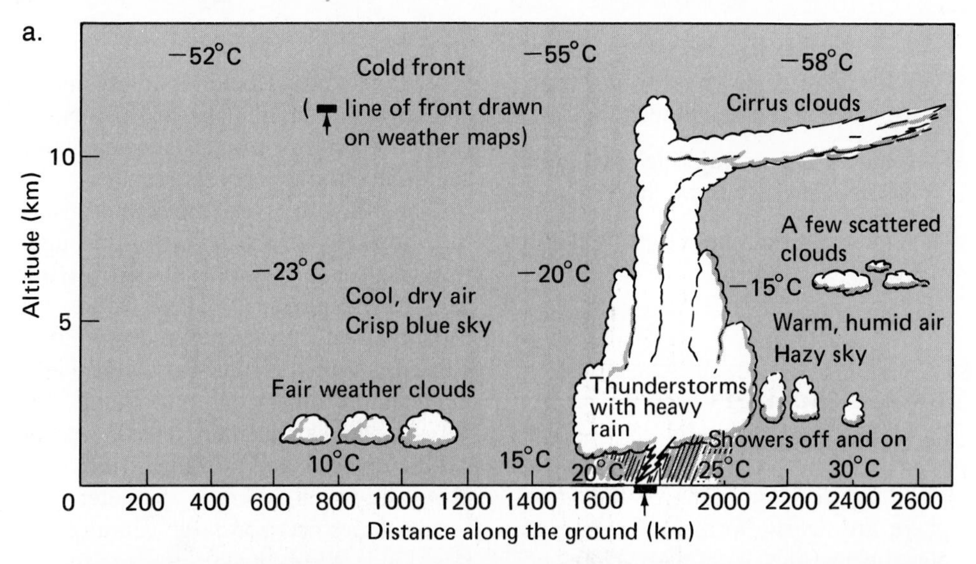

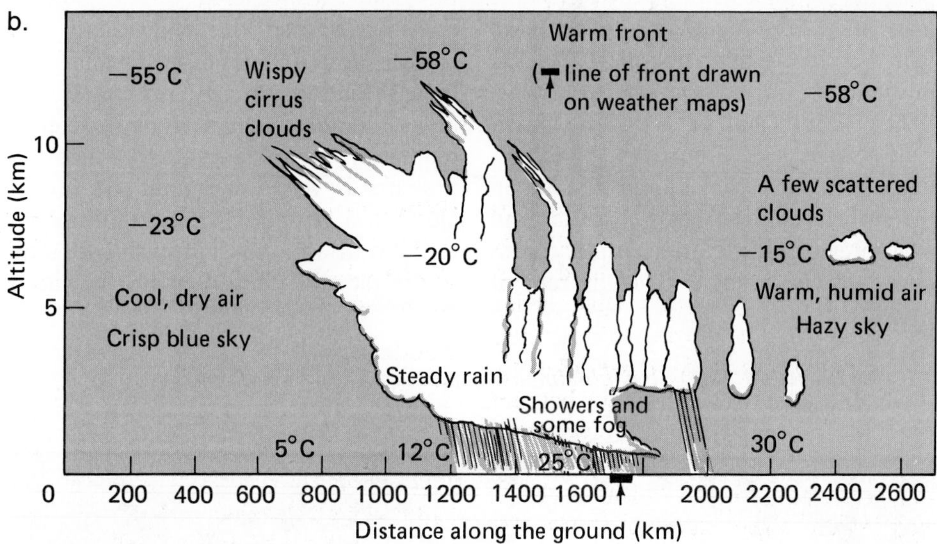

Figure 3-8 a. *Cross section of a cold front.* **b.** *Cross section of a warm front. There is a great difference between the vertical scale and the horizontal scale so that you can see the details more easily.*

2. Does the smoke rise in the cylinder or rest on the bottom? What is the direction of air motion?

PROCEDURE

Exchange the can of iced water for a can of hot water. Follow exactly the same procedure as before. Read and record the temperatures at the beginning and at three-minute intervals. Watch the smoke in the cylinder as the temperatures change.

DISCUSSION

1. In what direction does the air in the cylinder transfer heat?
2. How does the movement of the smoke suggest how cumulus clouds might be produced?

3. What kind of air mass would probably have cumulus clouds?
4. What air mass would increase air pollution over cities and industrial areas?
5. What conditions help carry pollutants away?

3-6 Air Motions and Weather

The large-scale motions of the atmosphere are nearly horizontal. For example, the motions within a cyclone take place in a disc of moving air no more than 8 or 10 km thick but often 2500 km in diameter.

The warm front of a cyclone has a gentle slope. (See Figure 3-8.) The motion of the warm, moist air as it blows up the surface of the front is gradual and widespread. This motion usually produces stratus clouds and steady rain or snow. The cold air behind an advancing cold front often lifts the warm air abruptly. Cumulus clouds, showers, and sometimes thunderstorms result.

When the air is very moist and unstable, cumulus clouds develop into huge, towering clouds called **cumulonimbus.** (See Figure 3-9.) These clouds are accompanied by thunder and lightning. Cumulonimbus clouds are wider and taller than ordinary cumulus clouds. Severe storms can contain several sets of strong upward and downward air currents and reach 50 km in diameter.

Tornadoes occur in large cumulonimbus clouds. Most North American tornadoes occur over the midwestern and eastern United States in spring and summer. Tornadoes are small rotating storms several hundred meters to a kilometer in diameter. (See Figure 3-10.) The air pressure is very low within the funnel cloud of a tornado. What could happen to the walls of a house if it were directly in the path of the center of a tornado?

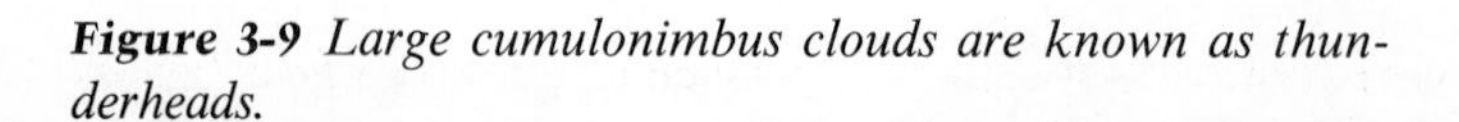

Figure 3-9 *Large cumulonimbus clouds are known as thunderheads.*

Figure 3-10 *Tornadoes over water and land. A tornado over water is called a waterspout; the one here was in the Florida Keys. The other tornado occurred in Oklahoma.*

A tornado's winds can reach speeds of 400 km per hour or more. They may be caused by intense convection at some place within a cumulonimbus. Some large cumulonimbus clouds rotate slowly. When the rotating air rushes inward to replace the air drawn out by convection, this rotating air spins faster and faster as it converges. You can see this effect for yourself by twirling a weight tied to a string around your finger. The weight moves faster as the string winding around your finger becomes shorter.

ACTION

Create a vortex, or whirlpool, in your wash basin or bathtub. Partly fill it with water. First, direct the water toward one side as you let it pour in. Then release the drain. As the water flows out, note the direction in which the vortex rotates. The next time, direct the water toward the other side. The water has spinning motion to begin with. It spins more rapidly in the same direction as it converges toward the drain, because it makes smaller and smaller circles. Can you explain why tornadoes rotate in the same direction as cyclones?

Tropical cyclones form over the warm waters of the tropical ocean when cumulonimbus clouds become organized around a low pressure center. Tropical cyclones are called **hurricanes** in the Western Hemisphere and **typhoons** in the Western Pacific Ocean. Figure 3-11 is a spectacular view of a hurricane taken from space. The bands of cumulus clouds spiral inward toward the eye of the hurricane. Around the center, the clouds develop into the tall cumulonimbus clouds that ring the eye of the storm.

Figure 3-11 *A photograph of a hurricane off the Florida coast, taken by astronauts. The eye of the storm is hidden by a pancake of cirrostratus clouds. Why can't you get a picture of an anticyclone?*

The hurricane's rotating winds, which may exceed 250 km per hour, result when air flowing inward toward a low pressure center is deflected by the earth's rotation. The strong winds and low pressure of the storm cause mountainous waves and unusually high tides. Although the hurricane winds are very destructive, the greatest damage and loss of life are caused by the waves and tides that accompany the storm.

MATERIALS
daily weather maps and weather data from Weather Watch (or the equivalent)

3-7 Investigating the Movement of Weather

At middle and high latitudes there is an endless succession of cyclones and anticyclones. (See Figure 3-12.) At heights a few kilometers above the earth's surface, these closed patterns of air circulation give way to high-level westerly winds. These winds, blowing from west to east *on the average,* control the weather at the earth's middle latitudes. The waves in the upper westerlies are related to the cyclones and anticyclones below, as you can see in Figure 3-12. High in the troposphere air flows through the waves fast enough to slow down or speed up a jet plane as much as 300 km per hour. The waves usually drift toward the east, also, but much more slowly than the wind blowing through them at these high levels. This strong current within the westerly winds is called a **jet stream.**

Weather predicting is based on the careful plotting of the movement of weather systems: the cyclones, anticyclones, and associated fronts.

When you have completed this investigation, you will be able to predict future weather and observe the accuracy of your predictions.

Procedure

As part of your Weather Watch, you have been posting daily weather maps (see Figure 3-13) while at the same time recording your own weather data. Observe the patterns made by the isobars on a series of weather maps. Note the areas of high and low pressure ("highs" and "lows"), cold fronts, and warm fronts. Track their movements from day to day. Use your own recorded weather data to see what relationships you can observe between your data and the weather movements over your area as indicated by the sequence of weather maps.

Discussion

1. Describe any consistent relationship you can find between the pressure patterns and the wind directions plotted at various places on the maps. What is the wind direction around a

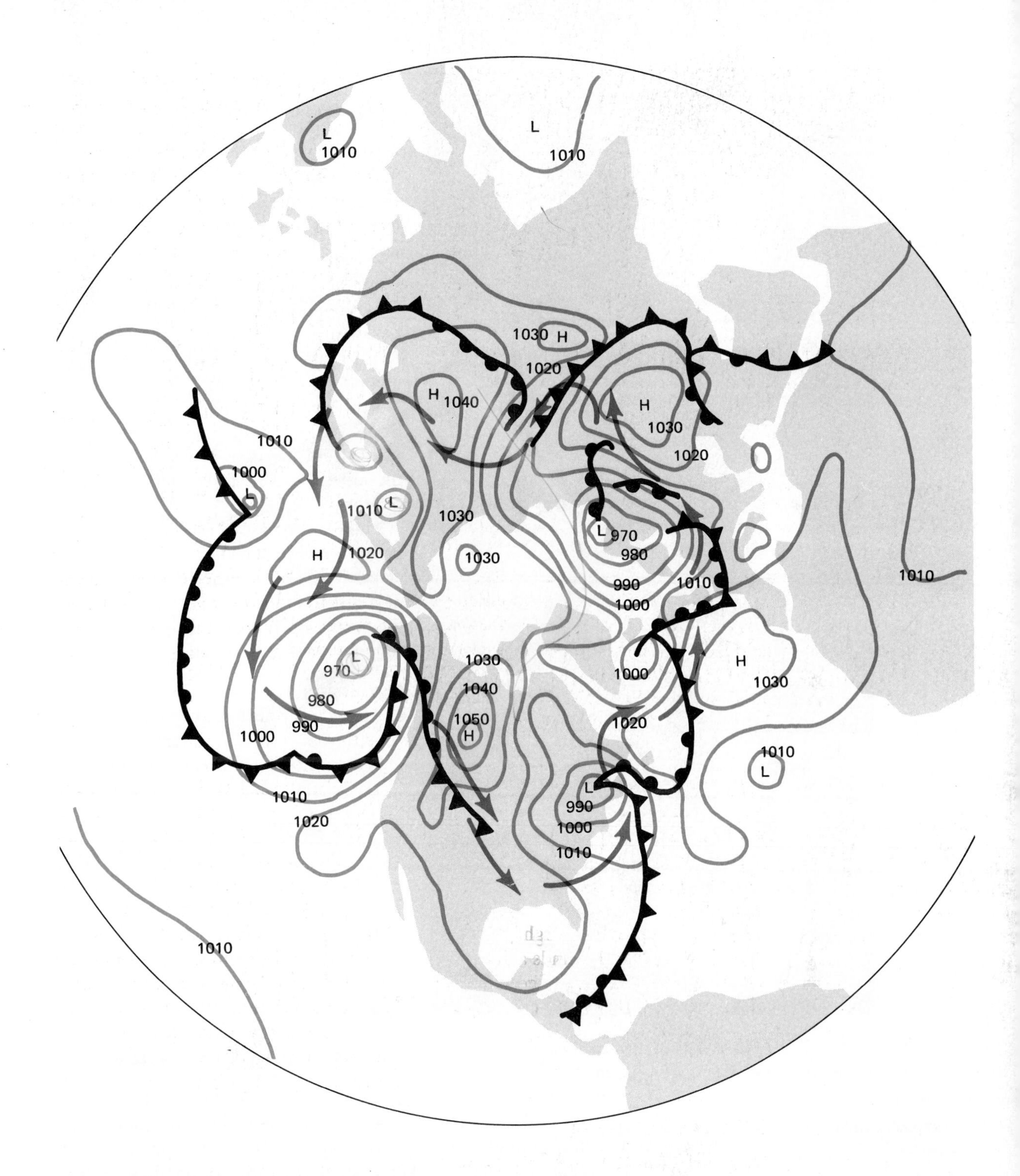

Figure 3-12 *A Northern Hemisphere weather map, showing streamlines of the upper air currents above the surface weather pattern. Note the southward bulges of the higher air-stream.*

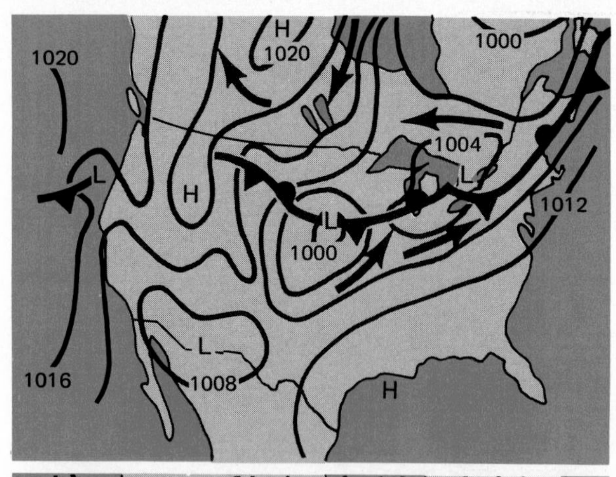

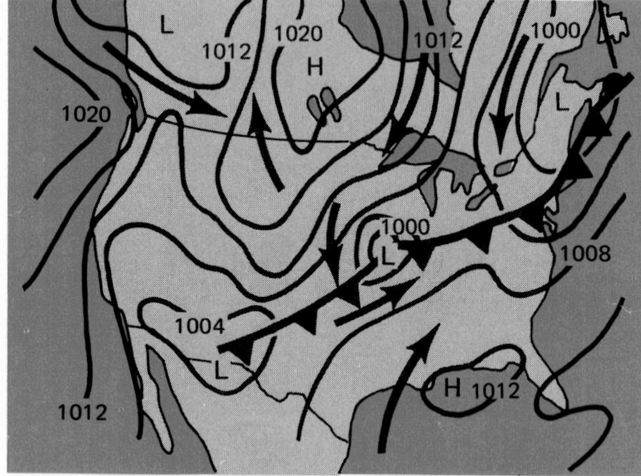

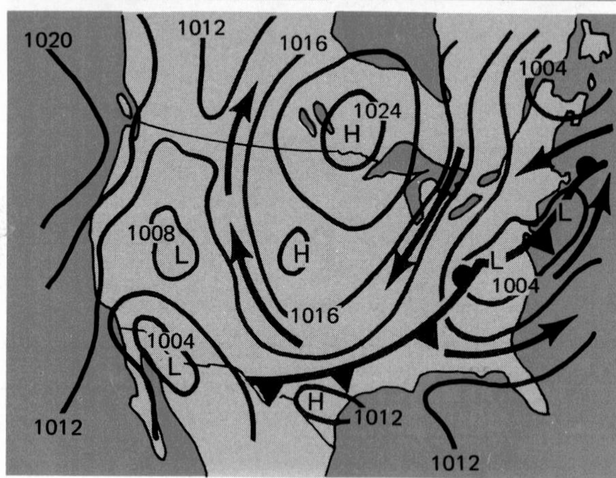

Figure 3-13 *A typical movement of weather across North America.*

MATERIALS
base map of North America, several weekly sets of National Weather Service maps

center of low pressure? around a center of high pressure?

2. What was the direction of movement of the highs and lows as they passed your longitude?

3. What changes of weather occurred, according to the Weather Watch data, as these highs and lows moved past?

4. If a cold front or a warm front, or both, passed over your locality, describe the weather changes that occurred.

PROCEDURE

Examine a sequential series of U.S. or Canadian Daily Weather Maps. Select any weather station. Note the positions of lows, highs, and fronts for those days at that city. Then, on a blank map of North America, sketch the isobars and fronts as you think they might be 24 hours from the time of the last map of the series. Weather forecasters call this a prognostic map.

Make a forecast of the pressure, temperature, wind direction, sky condition, cloud type, and weather for that station 24 hours from the time of the last map of the series. Be prepared to give the reasons for your prediction.

DISCUSSION

1. How does your forecast map compare with the observed pattern of isobars and positions of fronts on the next day's map?

2. Compare your weather forecast with the actual weather data recorded on the next day's map for that station. How did you do as a forecaster?

3. If your predictions were not entirely correct, what factors contributed to any errors in your predictions?

Thought and Discussion

1. Weather systems like cyclones and thunderstorms deliver great quantities of heat into the atmosphere. What part does condensation play in this upward transfer of heat?

2. Frontal, or middle and high latitude, cyclones usually move from west to

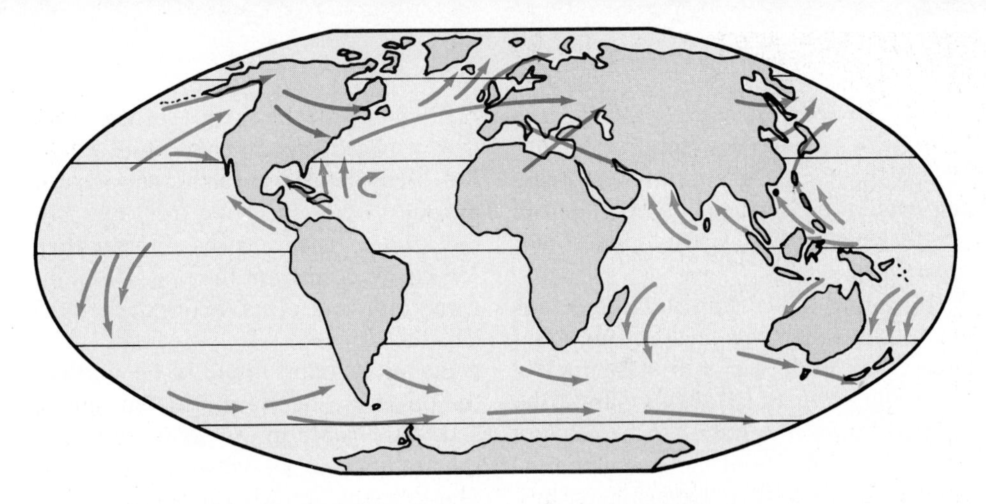

Figure 3-14 *Paths of middle latitude cyclones are shown in blue and paths of tropical cyclones are shown in red.*

east, though in a very general sense. On the outline map of an imaginary continent that your teacher will give you, draw the tracks that you think cyclones would most often follow.

Examine Figure 3-14. Which way do the tropical cyclones drift? the higher-latitude cyclones? Explain why the tropical and frontal cyclones move in opposite directions.

The Circulation of the Atmosphere

3-8 Worldwide Patterns of Air Motion

From the time of Columbus onward, sailors traveling westward to the New World found the most favorable winds at low latitudes, south of the Canary Islands. You can find the Canary Islands in Figure 8-9. These winds blew from the east or northeast day after day with little change in speed. They promised a reasonably safe trading voyage, so sailors called them the **trade winds,** or trades. Another belt of easterly, or southeasterly, winds lies in the Southern Hemisphere. The trades of the two hemispheres come together in a region of light, changeable winds and hot, humid air. Here, a sailing ship might drift for days without enough wind to push it. Another danger within the tropics was the dreaded hurricane. Sometimes one of these tropical cyclones, drifting slowly westward in the steady trade wind circulation, brought what had seemed a promising voyage to a disastrous end. (See Figure 3-14.)

To Edmund Halley, who published the first map of the winds of the world in 1686, the uniform trade wind belts of the two hemispheres suggested an orderly, worldwide circulation of the atmosphere. Halley was the first scientist to propose that the average or typical circulation of the atmosphere is caused by convection.

The sun's heat is greatest in the zone of the equator, where the rays always strike the earth's surface more or less directly. Halley pointed out that the heated air near the equator should rise, move toward the poles, and then sink as it cools in the polar regions. The rising air would be replaced by the denser trade winds, blowing from the colder parts of the earth.

If heating and cooling at different latitudes were the only factor, you would expect the trade winds to blow directly from the poles. But Halley's map of the winds showed the trades to be easterly winds—blowing from the east or northeast in the Northern Hemisphere, and from the southeast in the Southern Hemisphere.

Fifty years later, George Hadley showed why the trades are easterly winds. Hadley said that, in moving toward the equator, the air would begin to blow toward the west because of the earth's rotation. You know this phenomenon as the Coriolis effect. (Hadley worked out this part of it 100 years before Coriolis.)

ACTION

To see how air drifting toward the equator becomes an east wind, make this simple calculation.

At the latitude of New Orleans, the eastward speed due to the earth's rotation is about 35,000 km a day. But a point on the equator moves eastward at a speed of 40,000 km a day. Imagine that air moves from New Orleans to the equator. Assume, too, that the air keeps its original eastward velocity, 35,000 km a day, as it moves south. What will its motion along the equator be, relative to the earth, when it arrives? (You can use the Coriolis effect setup to help you visualize what's happening.)

What would be the motion of air, originally at rest at the equator, if it were moved to New Orleans? Assume the air is not slowed down by friction or pressure forces.

Friction with the earth's surface slows the westward speed of the trade winds. But high above the earth's surface, air moving toward the poles from low latitudes often reaches speeds greater than 150 km an hour from the *west*. Does this agree with your observations of the Coriolis effect?

Early navigators found, to the north of the trades in the Northern Hemisphere and to the south in the Southern Hemisphere, belts of "brave westerly winds." These winds, blowing from the west, helped speed European sailing ships on their return voyages across the Atlantic. The westerly winds, however, were not nearly so steady and reliable as the easterly trades. They were often strong, but especially in the North Atlantic they sometimes shifted all around the compass. Organized weather observations were not made until the 19th century, so no one knew why the westerlies were so changeable. Your investigation of weather maps should show you why. What is the reason?

Hadley explained the west winds of middle latitudes, like the easterlies in the tropics, as an effect of the earth's rotation. When air rising near the equator moves toward the poles, it eventually becomes a wind from the west because of the earth's rotation. When this air sinks to the earth's surface, Hadley said, it keeps its direction from the west, but its speed is slowed down by friction with the earth's surface.

3-9 The General Circulation of Air

The general circulation of air by which Hadley tried to explain the trade winds and the westerlies is today called the **Hadley cell.** This cell is a circulation in which the air rises at one place and sinks at another. Hadley pictured one such cell in each hemisphere. According to Hadley's picture, the air in the Hadley cell should make a closed circuit.

But the atmosphere's general circulation—its average or typical pattern of motion—is much more complex than Hadley thought it was. For one thing, there appear to be three vertical cells, instead of one. (See Figure 3-15.) So there are two zones within each hemisphere where upward motions are frequent and two where downward motion is typical.

Look at the satellite view of the Northern Hemisphere in Figure 3-16 to see if you can find evidence of such zones. Where the air is moist, upward motions produce clouds. In zones of sinking air, the sky is normally clear. The "general circulation" of the atmosphere refers to its typical motions over a long period of time. Individual daily maps or photographs such as the one in Figure 3-16 are the best way to picture it. The traveling cyclones and anticyclones are an important part of the general circulation, and these are lost in averaging a great many daily maps.

In Figure 3-16, the zone where the trade winds of the two hemispheres meet, now called the **Intertropical Convergence Zone,** is marked by a broad, nearly continuous band of clouds near the equator. Sometimes two such bands are observed. Quite often the cloud bands are made up of many huge cumulonimbus clouds. Strong upward motions in these towering clouds contribute rising air to the Hadley circulation. As the air rises, it moves toward the poles, forming the upper level westerlies. Some of this air sinks within 15 or 25 degrees of the equator, and becomes the tropical easterlies, or trade winds.

Another part of the sinking air continues toward the poles to become the surface winds of middle latitudes. When frontal cyclones develop, the warmer air moves upward and toward the poles. At higher latitudes, or in winter over the interiors of the continents, the air cools

Figure 3-15 *A simplified diagram of wind patterns and pressure belts.*

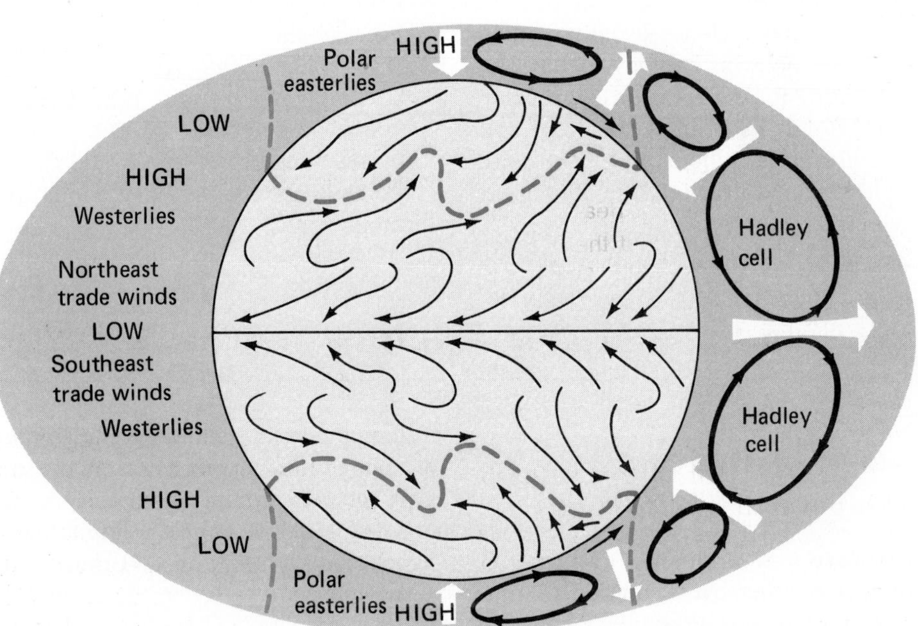

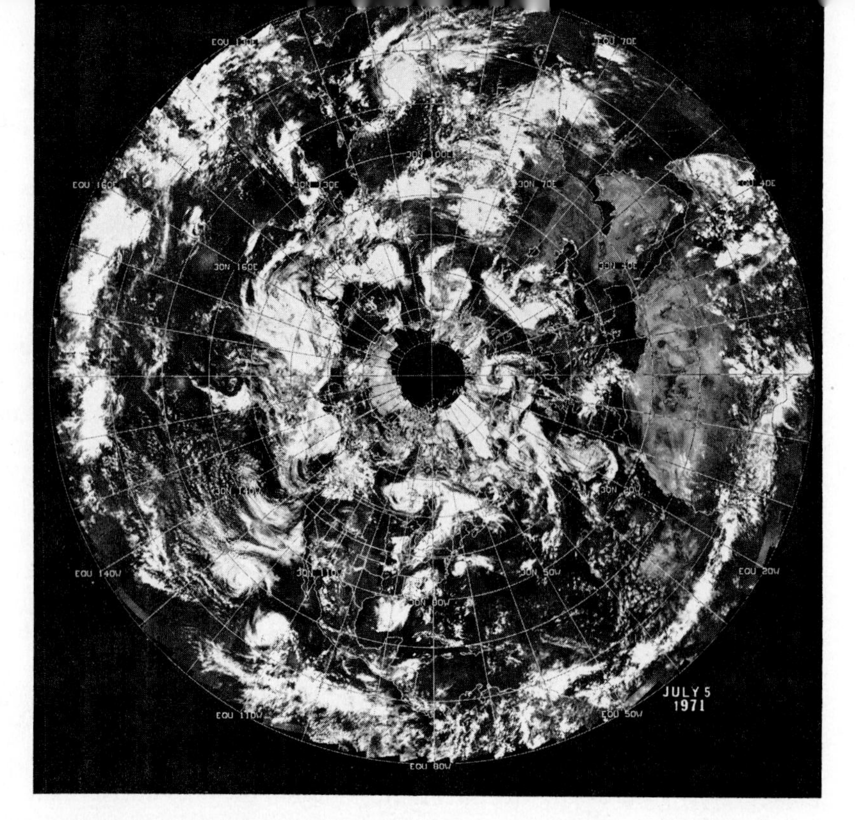

Figure 3-16 *Cloud patterns in the Northern Hemisphere as seen from a weather satellite.*

and sinks. When new cyclones form, this cold air streams toward the equator, sinking and spreading out as part of the surface westerlies.

The frontal cyclones usually drift toward the poles as well as to the east, accounting for the areas of low pressure near latitude 60 degrees. The cold anticyclones usually join the subtropical highs centered near latitude 30 degrees, helping to keep pressure high.

Thought and Discussion

1. On the map of an imaginary continent, draw streamlines representing average air motions based on Figure 3-15. Shade the zone where the streamlines converge. Where would you expect to find upward motions, on the average? Where would the air be sinking? Identify dry winds and moist winds.

Climate and the Water Cycle

3-10 Insolation and Temperature

Climate may be thought of as the history of weather over a period of time. The most important single factor controlling the earth's climate is the amount of energy the earth receives from the sun. This is called the incoming solar radiation. For convenience, climatologists have coined a word for it—**insolation.**

Another vital factor in the earth's climate is the atmosphere itself. The atmosphere acts like a greenhouse, keeping the earth's surface temperature relatively constant and much warmer than it would otherwise be.

Very hot objects like the sun emit short-wave radiation. Cooler objects emit long-wave radiation. About half of the solar energy is short-wave radiation that we see as light. This energy passes through transparent substances. In contrast, the earth emits long-wave radiation. Long-wave radiation is felt as heat. You can feel heat radiating from pavements, the soil, and other surfaces in hot weather.

A window pane transmits sunlight. It is nearly transparent, and much of the short-wave energy passes through. Only a little energy is absorbed to heat up the glass. However, the walls and furniture inside a room absorb a large part of the solar radiation coming through the window. The energy radiated from the furniture, unlike the original solar energy, is all long-wave radiation. Much of it is unable to pass out through the window pane. This is why the car seats get so hot on a hot, sunny day when all the windows are closed. Try putting a piece of glass in front of a hot object to see how the heat waves are cut off. A greenhouse traps energy in this way (when the sun shines) and so does the atmosphere.

The atmosphere, like glass, transmits short-wave solar radiation freely, but traps the earth's long-wave radiation. It is mainly the water vapor concentrated in the lower part of the atmosphere that acts like window glass. The water vapor is nearly transparent to sunlight coming in. However, the vapor absorbs much of the long-wave radiation going out and coming in. The vapor sends some of this radiation back to the earth's surface. Carbon dioxide in the atmosphere contributes to the greenhouse effect also, but less than water vapor does.

High in the atmosphere is another absorbing gas called ozone (O_3). It is vitally important to us because it absorbs practically all of the **ultraviolet radiation** from the sun. This is very short-wave radiation of high energy. Even the small amount of ultraviolet radiation that passes through the atmosphere can give you a severe sunburn. If too much ultraviolet radiation reached the earth's surface, life as we know it could not exist.

A large part of the atmosphere is sandwiched between these two good absorbers, the earth's surface and the ozone layer. Notice in Figure 3-17 how the temperature first decreases above the

Figure 3-17 *Temperature varies greatly with height in the atmosphere.*

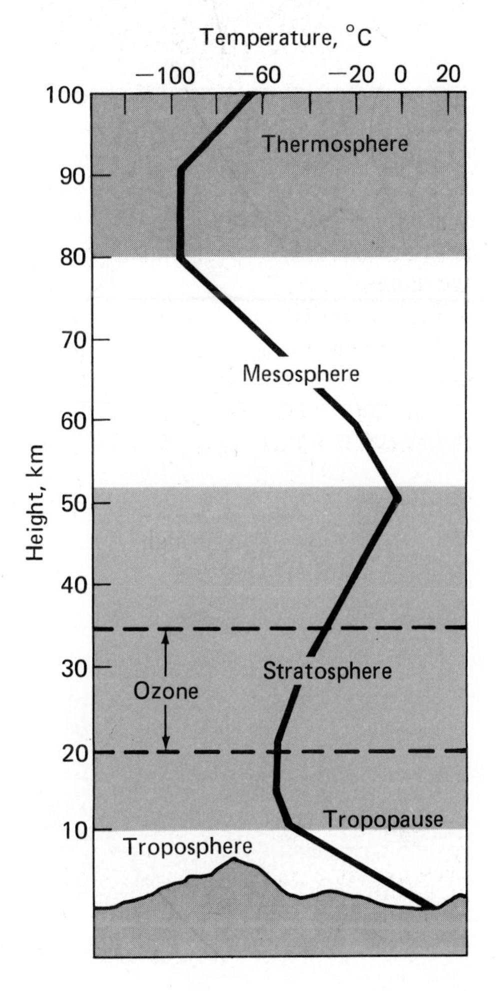

earth's surface and then starts increasing to the top of the ozone layer. The top of this layer absorbs most of the ultraviolet radiation first, so the temperature is high there. It is also high at the earth's surface. In between, where the atmosphere is largely transparent to sunlight, the temperature is low.

The temperature changes at different heights mark different layers of the atmosphere (review Figure 2-2). The troposphere, the layer next to the earth's surface, is warmed at the bottom. When air is heated from below, it rises. Ordinarily, before it reaches the top of the troposphere, it cools enough to stop rising and begins to fall. The air in the troposphere usually becomes thoroughly mixed by these rising and falling motions.

The stratosphere, above the troposphere, is relatively warm at the top and colder at the bottom. It tends to be quite stable with little vertical air motion. The stratosphere acts as a lid on the unstable troposphere below. Relatively little water vapor gets into the higher layer, so the stratosphere plays only a small part in the water cycle.

The word "climate" comes from the Greek *klima,* meaning "sloping surface of the earth." The Greeks knew that the climate zones of the earth were related to the angle at which the sun's rays strike the earth. Figure 3-18 shows why. The earth is a sphere, so most of its surface is not at right angles to the sun's rays. The higher the latitude, the greater the surface over which a given amount of energy is spread.

Since the amount of insolation received depends on the latitude, the isotherms on world temperature maps are roughly parallel to the latitude circles. (See Figures 3-19 and 3-20.) As the earth moves around the sun on its tilted axis, the insolation received at each latitude varies. (See Figure 3-21.) The length of the period of daylight also changes with the season. In which direction does the earth's temperature pattern move between January and July? (See Figures 3-19 and 3-20.)

The ancient Egyptians, Greeks, and Druids all concluded that the seasons were caused by changes in the path of the sun as it revolved around the earth. Recall that Eratosthenes knew that once a year the sun would shine on the bottom of a well near the present Aswan, Egypt. During the next six months the sun, at noon, was lower in the sky each day until it was 47° below the vertical at Syene and only 43° above the southern horizon. Then it moved back toward the vertical at Aswan to complete its annual migration. An Earth revolving around the sun provided a more satisfactory and simple explanation of the seasons.

Although the average isotherms run more or less parallel to the latitude circles, there are places where they bend north and south. Try to explain why.

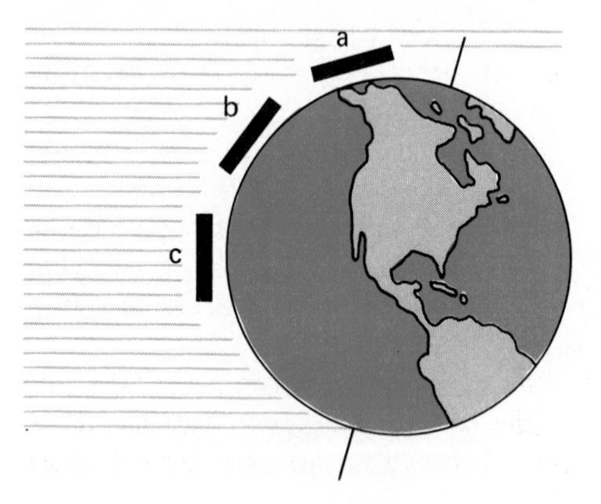

Figure 3-18 *Areas a, b, and c are all the same size. How many more "rays" of sunlight does equatorial area c receive than does polar area a?*

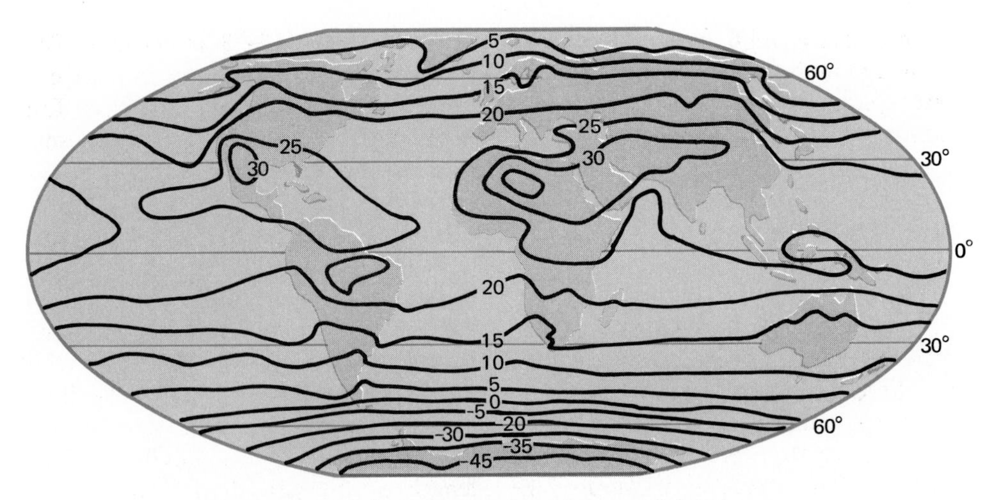

Figure 3-19 *Average July temperatures (degrees Celsius).*

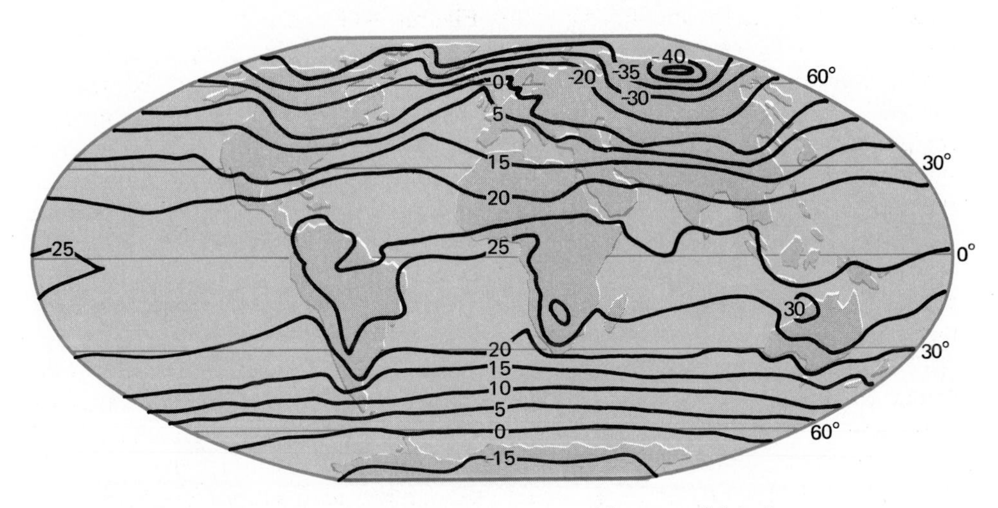

Figure 3-20 *Average January temperatures (degrees Celsius).*

3-11 Moisture Zones of the Earth

The air contains more moisture in some areas than it does in others. One reason is that air temperatures vary greatly from place to place around the world. The amount of water vapor the atmosphere can hold depends on the temperature of the air. Look at the evaporation graph in Figure 3-22. At what latitudes is the evaporation lowest? Where does evaporation exceed precipitation? What is the principal factor affecting evaporation at the earth's surface?

Figure 3-23 is a map of world precipitation. Where are the zones of low rainfall located? How does the map compare with the evaporation graph? At what latitudes would you expect to find deserts?

The trades come from the sinking air of the subtropical highs. They are therefore very dry and can absorb vast

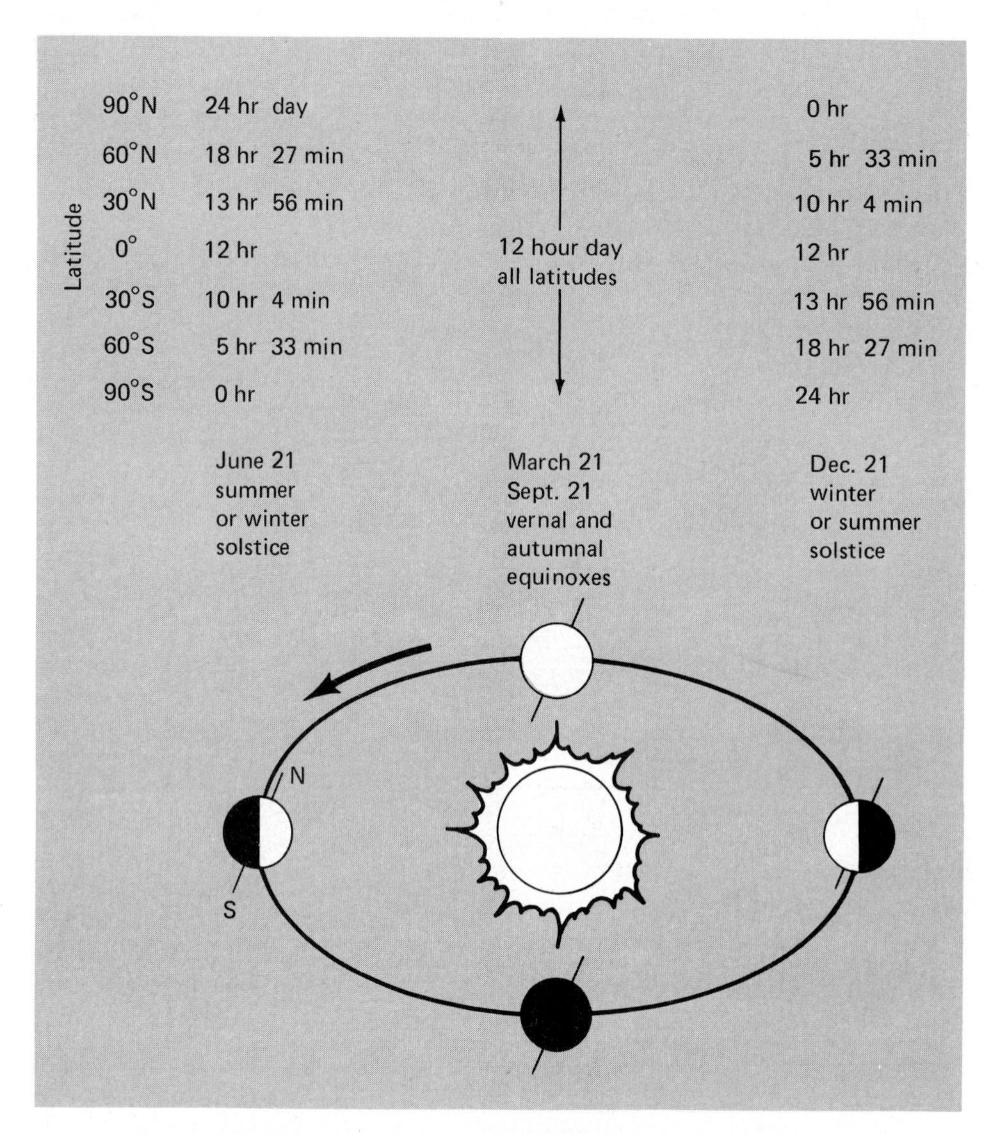

Figure 3-21 *As the earth moves around the sun, the length of daylight varies with latitude and the seasons.*

amounts of moisture. As you might expect, the earth's dry climates occur in the regions where evaporation exceeds precipitation.

Deserts are generally found in the centers of dry regions. If you compare Figures 3-23 and 3-24, you will notice that areas of light precipitation extend from the deserts of North America and Asia into the polar region. The northern parts of these continents receive the same amounts of rainfall as the desert regions, but they are not deserts. The temperature, and therefore evaporation, is so low that even the small amount of precipitation that does occur is greater than the evaporation.

Notice the belt of heavy precipitation near the equator. Here precipitation is greater than evaporation. The trades

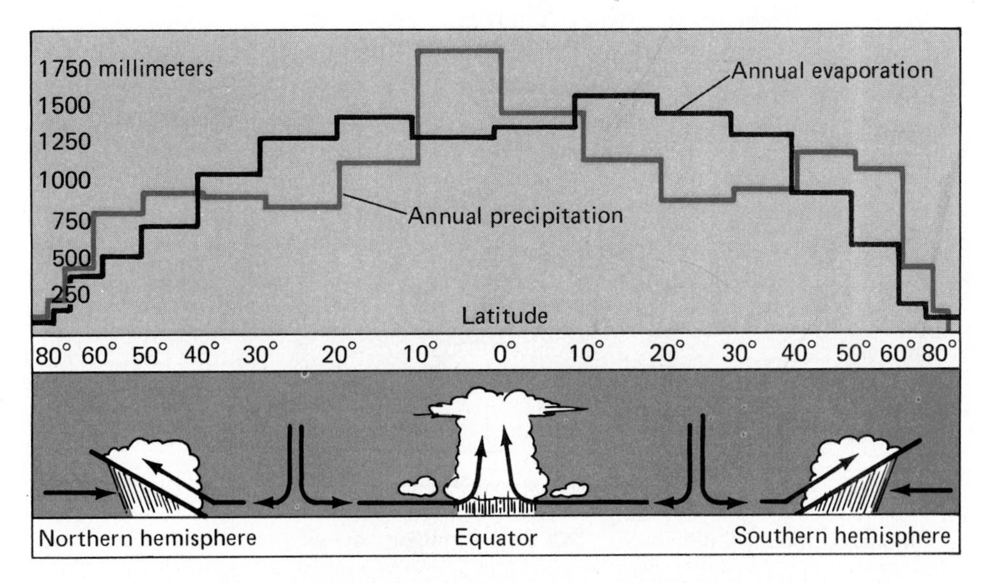

Figure 3-22 *The annual precipitation and evaporation vary with latitude. How do the differences relate to the air movements in the lower part of the diagram?*

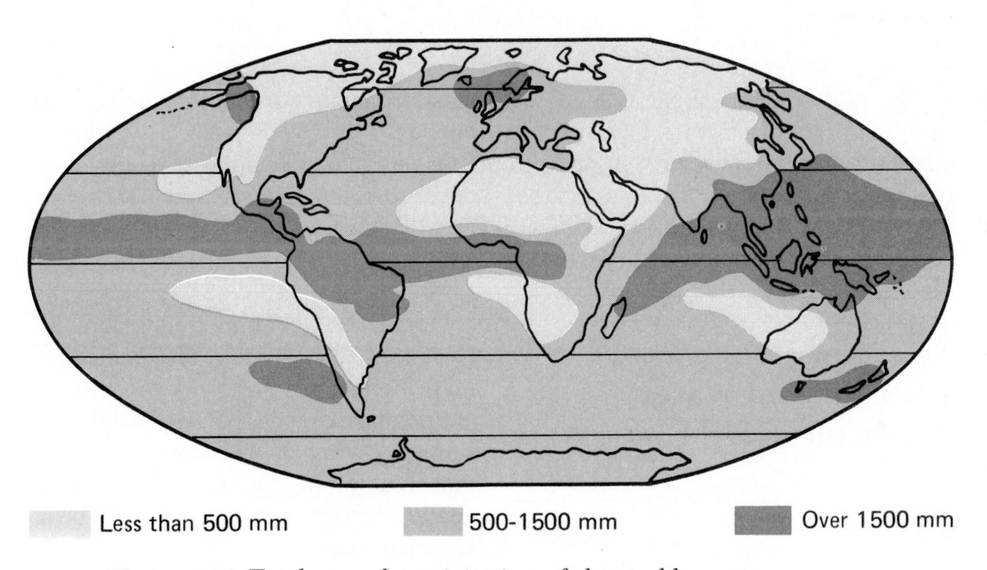

Figure 3-23 *Total annual precipitation of the world.*

carry moisture into the Intertropical Convergence Zone where the air rises and the moisture condenses. On the average, the convergence zone between the trades is located north of the equator. What kind of climate would you expect in this region?

Now notice the latitudes of the areas of heavy precipitation in the Northern and Southern hemispheres. What do

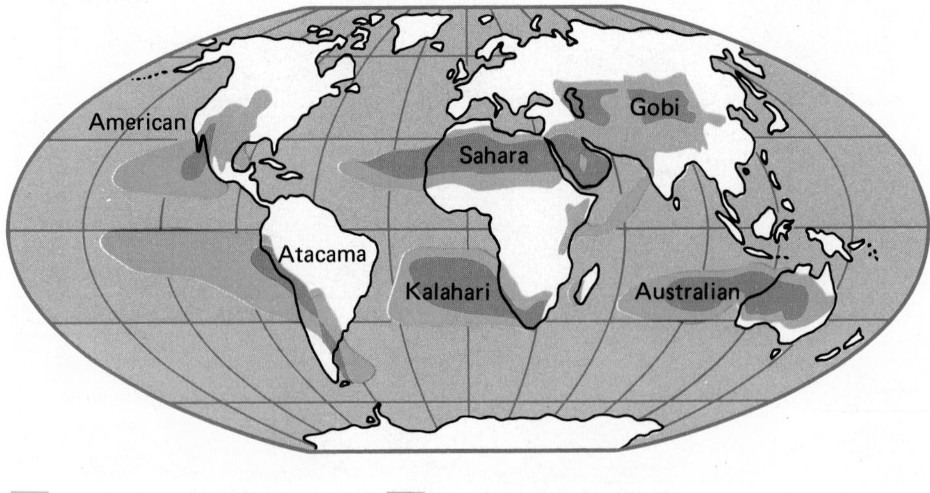

Figure 3-24 *The areas of the world that have moisture deficits, or dry climates. Each area contains a major desert.*

you know about air circulation in the middle latitudes that suggests a reason why there should be extra precipitation at those latitudes? This zone is not continuous around the earth because land, oceans, and mountains influence the air circulation and the amount of rain that is carried.

3-12 Land and Water Affect Climate.

ACTION

On Figure 3-19 follow the 60 degrees north latitude circle. Write down the July temperatures over the Gulf of Alaska, central Canada, the Atlantic Ocean north of Great Britain, and eastern Asia. Write down the values for the same points in January. (See Figure 3-20.) Now compute the differences in temperature from July to January for each of the four points. Which location has the greatest annual range of temperature? Which has the least? Compare the seasonal temperature changes along 60 degrees south latitude. Compare the temperature range at 60 degrees south latitude and 60 degrees north latitude.

Oceans and continents interrupt the basic latitudinal pattern of climate. The map in Figure 3-25 shows the average temperature changes between winter and summer. What do you conclude about the climates of the interiors of continents? They are said to have **continental climates.**

The size of a continent affects both the temperature range and the amount of moisture in the interior. The larger the continent, the greater the effect. **Marine climates** have smaller temperature changes from winter to summer.

At latitude 40 to 50 degrees, cyclones usually are carried from west to east by the strong high-altitude westerly winds that sweep around each hemisphere.

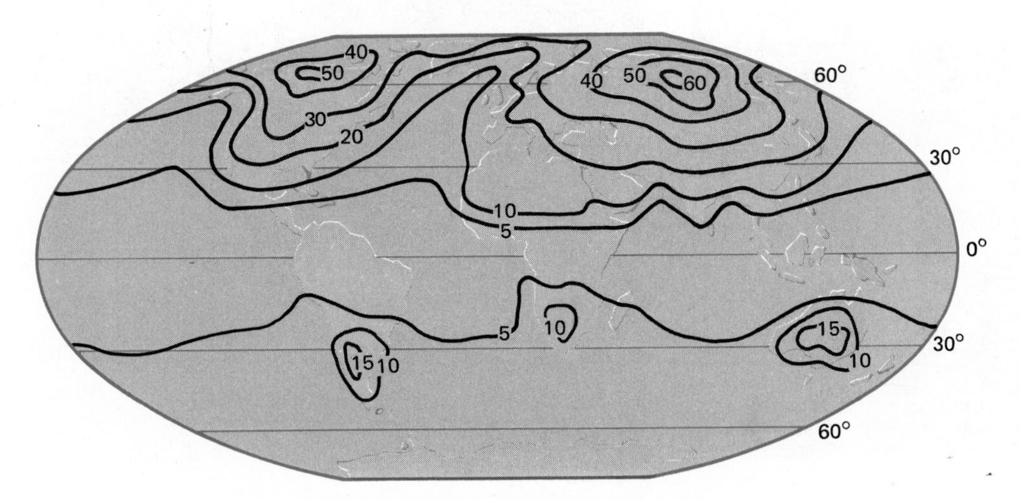

Figure 3-25 *A map of annual temperature ranges. The numbers are the differences in average temperatures from the warmest to the coldest months (°C). Notice the effects of land masses—compare the values for Iceland and Siberia.*

Therefore, rainfall is heaviest on the western sides of the continents. (See Figures 3-23 and 3-26.) Air from the Gulf of Alaska carries moisture to western Canada and the United States. In summer the air is relatively cool, and in winter relatively warm, but it is always moist. Therefore, coastal areas of the northwestern United States have high humidity and abundant rainfall. Conditions are similar in northwestern Europe.

Cyclones moving eastward across a continent lose most of their moisture before reaching the interior. The climate becomes progressively drier farther inland along the same latitude. In the deep interior of the continent, far from ocean sources of moisture, there is little precipitation and therefore little evaporation. The air remains dry and the skies clear. One example of a dry continental climate is the Gobi Desert in Central Asia. (See Figure 3-26.)

We can now summarize the causes of dry climates. The pattern repeats from continent to continent. Look at the Northern Hemisphere in Figure 3-24. The desert regions on the western sides of the continents at 25 to 30 degrees north and south latitude are caused by sinking, dry air from subtropical high pressure areas. The dry air is cooled and stabilized by cold ocean currents. The coastal deserts extend into the continental interiors. These interior deserts are too far from ocean sources of moisture to receive much precipitation. The warm ocean currents and the onshore flow of air keep the eastern sides of North America and Asia moist. The effects of ocean currents will be covered in greater detail in Chapter 5.

3-13 Mountains Further Modify the Climate.

The shift from a dry region into a wet one is generally gradual. However, mountains can cause sharp boundaries between different climatic zones. Mountains channel air movements and affect

a.

b.

Figure 3-26 a. *Middle latitude cyclones bring moisture from the Gulf of Alaska to the Yukon Territory of Canada.* **b.** *The Gobi Desert in Eastern Asia receives less than 10 cm a year.*

fronts and cyclones. For example, during most of the winter, cold air moving down from Canada is held east of the Rocky Mountains as it flows southward. But polar air can sweep across the central plains of North America into and beyond the Gulf States, sometimes freezing the citrus crops of Florida. The Rockies usually prevent such polar air masses from damaging citrus crops in California.

Mountains lying near the shores of oceans prevent marine air at the earth's surface from penetrating far inland. Such a barrier exists along the Pacific coast of North and South America. In Asia the Himalayas help keep the Indian winters mild, while northern China and Korea are not protected from cold winds blowing out of the continental interior.

Highlands also have a marked influence on the amount and distribution of precipitation. In mountains most precipitation falls on the slope the wind is blowing against when the air rises to flow over a mountain range. Little rain falls on the other side of the range where the air descends and dries out.

Thought and Discussion

1. Your teacher will give you an outline map of an imaginary continent. Locate on it the areas that you think would have humid climates and those that you would expect to be dry. Where would you expect to find a desert?

2. On the map of the imaginary continent, which areas might be affected by tropical cyclones? Which areas would be affected by higher latitude cyclones?

Unsolved Problems

In recent years, meteorologists have made much progress in understanding weather and climate. Big computers have made it possible to solve the mathematical equations that describe the atmosphere. However, highly accurate predictions are still not possible.

Very little is known about the connections between the upper and lower atmosphere. The thin upper regions of the atmosphere are strongly affected by energy changes observed on the sun's surface. But it is not clear how or to what extent these changes affect our weather. Computers make it possible to determine in much greater detail than formerly how waves in the atmosphere carry energy from one region to another. Studies of atmospheric waves are expected to explain interrelations not now understood.

The atmosphere is strongly coupled with another fluid at its base—the ocean. What takes place at this boundary is especially important for long-range predictions. The ocean stores not only most of the earth's water—all but about 3%—but also carbon dioxide, containing about 99% of the terrestrial CO_2. Thus the exchanges of water and carbon dioxide at the ocean surface are of critical importance in the cycles of both.

As with water molecules, the balance between the CO_2 in the air and the sea surface depends on the temperature. In general, carbon dioxide is taken from the air into the ocean in cold waters, and released from ocean to air in warm surface waters. But the details of the exchange are quite complex. Apparently, about half the carbon dioxide added to the atmosphere by the burning of fossil fuels remains there.

Scientists are trying to improve their models of the earth's climate. A primary objective of the experiments is to find out the "sensitivity" of climate to different factors, or variables. When the CO_2 concentration in a typical experimental model atmosphere is doubled over its present atmospheric concentration, the average temperature of the model earth rises several degrees Celsius. The polar regions become ten degrees warmer. The experiments make use of highly simplified models of the real earth, so scientists don't take these results completely at face value. But neither can they dismiss them. During the 1980's, more complex models made possible by larger computers may make scientists more confident of predictions of the CO_2 effect on climate.

Chapter Summary

Most of the energy that our planet receives from the sun passes through the atmosphere and heats the earth's surface. The surface emits long-wave radiation upward to heat the atmosphere. The heat energy is absorbed by carbon dioxide and water vapor in the air. These gases radiate part of the absorbed energy back to the surface again. As a result, the earth's surface is normally warmer than the air above, and very much warmer than it would be without an atmosphere.

The earth's spherical shape causes more energy to be received at the equator than at the poles. The resulting temperature difference causes a circulation in each hemisphere between equatorial and polar latitudes. The earth's rotation turns this circulation so that the winds tend to blow parallel to the latitude circles. The earth's rotation and shape also cause the circulation in each hemisphere to break up into three cells. So, on the average, there are two zones in each hemisphere where upward motions are frequent and two zones where downward motions are the rule. Because of these motions, and also because the incoming solar energy depends on the latitude, the earth's climate is primarily zonal. But the latitudinal belts of climate are strongly modified by continents and oceans. These, together with mountain ranges, produce marked changes in the underlying zonal patterns of temperature and precipitation.

METEOROLOGIST

Most people picture a television weather forecaster when they think of meteorology. But meteorologists do much more than predict weather. Some meteorologists do research on specific chemical and physical properties of the atmosphere. Others specialize in the study of climates. Determination of the general weather pattern of an area can be useful to many different types of people, like farmers and building designers. There are also meteorologists who specialize in developing and using instruments that measure and record data on atmospheric processes, such as the size of cloud droplets or the direction and speed of winds. Meteorologists even work in the space program, studying the atmospheres of other planets!

Most meteorologists work either directly or indirectly for national governments. Government agencies provide basic, public weather observations, forecasts, warnings, and climatological information according to standards set by the United Nations' World Meteorological Organization. In the United States, about two thirds of all meteorologists are either employed by the National Oceanic and Atmospheric Administration or are on duty with the military as weather officers. Most of the remainder work for research and consulting firms, as radio and television weather broadcasters, or as university teachers and researchers. In Canada, the agency responsible for basic weather service is the Atmospheric Environment Service.

Formal training in meteorology is acquired in several ways. Some universities give bachelor's degrees in meteorology. Others award only advanced degrees in meteorology and suggest another physical science as an undergraduate major. The armed services give and support meteorological training. Meteorologists are assisted by weather technicians and weather observers, who outnumber professional meteorologists by about two to one. The best preparation for these positions is a combination of formal study, such as two years of technical school, and on-the-job training.

Questions and Problems

A

1. Explain how water vapor and carbon dioxide play a role in the "greenhouse effect" that helps to keep the earth's surface warm.
2. Where on the earth would you expect to find the greatest annual range of temperatures? Explain your answer.
3. Why do cyclones rotate in opposite directions in the Northern and Southern hemispheres?
4. Where are the wet and dry belts in the basic climate pattern of the earth? How are they produced?
5. List the various layers of the atmosphere and indicate how they differ from one another.

B

1. A rapidly intensifying cyclone has just passed over your area. Would you expect the forward movement of the storm to speed up or slow down during the next 24 hours? Explain.
2. Classify the following into two groups. Select those that can be mapped using isolines and those that cannot: air pressure, sound, light intensity, the force of gravity, magnetism, water temperature, altitude above sea level, density of a fluid, odor, humidity. Be able to give reasons for your decisions.
3. Describe the stages in the development of a frontal cyclone.
4. Write down a station model like the ones on U.S. Department of Commerce daily weather maps that will correctly illustrate the following weather data: air temperature 24°C, barometric pressure 1024.4 mb (up 0.8 mb over the past 3 hours), dew point 14°C, sky condition partly cloudy, winds NW at 5 knots per hour.
5. The jet stream has been north of its usual position for the past month. Along the usual position of the jet stream, was the rainfall during this period greater or less than normal? Give reasons for your answer.

C

1. Small whirlwinds called "dust devils" rotate either clockwise or counterclockwise, although tornadoes rotate only cyclonically. Suggest an explanation.
2. Why do tropical cyclones usually lose force when they move over a large land area?
3. When the air is very dry, the temperature just before sunrise tends to be lower than it is when the air contains a lot of water vapor. Explain why this happens.
4. Very little precipitation occurs in the polar regions, yet these regions are not deserts. Why?
5. How do mountain ranges affect the climate where the wind blows moist air against them?

Suggested Readings

Likens, Gene E.; Richard F. Wright; James N. Galloway; and Thomas J. Butler. "Acid Rain." *Scientific American,* October 1979, pp. 43–51.

Olson, Steven. "Computing Climate." *Science 82,* May 1982, pp. 54–60.

Schaefer, Vincent J., and John A. Day. *A Field Guide to the Atmosphere.* Boston: Houghton Mifflin Co., 1981.

Schwoegler, Bruce, and Michael McClintock. *Weather and Energy.* New York: McGraw-Hill Book Co., 1982.

4 Waters of the Land

THE EARTH *has been recycling its water for about three billion years. The distribution changes, but the total supply remains constant.*

When it rains, people get involved in the water cycle. Rivers and lakes are our main source of fresh water for drinking, cleaning, and irrigation. Before the invention of the steam engine, waterways were the major routes for the exchange of people, goods, and ideas. For many of us, fresh water also provides recreation and natural beauty.

In some river valleys, such as the Nile in the past, people have depended on flooding to enrich the soil each year. But in many regions floods are natural disasters. Spring floods are caused by melting snow and seasonal rains over great river basins.

Throughout much of history, the dependence of people on water has led them into great efforts to control the waters of the land. Some of these attempts have not turned out well because all of the consequences could not be predicted.

When rain falls, some of it flows over the surface of the earth into streams, some evaporates, and some seeps into the ground. The picture on the opposite page shows a visible part of the land-based water cycle. You do not see the invisible water vapor moving from the land back into the atmosphere. And ordinarily people are unaware of the reservoir of water beneath the earth's surface. But the storage and flow of water beneath the surface is one of the primary concerns of earth scientists.

AFTER COMPLETING THIS CHAPTER, YOU SHOULD BE ABLE TO:

1. **identify the sources of fresh water that are found on and under the earth's surface.**
2. **list examples of how fresh water is stored near the earth's surface.**
3. **describe how water infiltrates the ground and becomes ground and capillary water.**
4. **predict what effect a change in the size of soil particles will have on the amount of water the soil can hold and the rate at which water will move through it.**
5. **explain how runoff depends on the form and intensity of precipitation and on the storage conditions beneath the surface.**
6. **explain why evaporation and transpiration depend on both water and the energy available.**
7. **use water budget graphs and stream flow charts to identify periods of water shortages and excesses at certain locations.**
8. **describe, in general terms, how flood forecasts are made.**
9. **explain the effect of organic wastes on the existence of fresh water bodies, such as lakes and rivers, on the earth's surface.**

Moisture Income and Storage

4-1 Where Does Fresh Water Come From?

During the time of the Roman Empire, some people in North America and the Middle East drank dew. They used large piles of rocks located over catch basins to collect their drinking water. Dew condensed on the rocks and dripped into the basins below. In the ancient city of Theodosia on the shores of the Black Sea, these basins, also called "surface wells," collected about 60,000 liters of water a day.

Fog is another source of moisture. In some places, dense fogs form as warm, moist air is cooled by ocean currents. The fogs blow into the coastal hills and mountains. Moisture collects on trees and shrubs and drips to the ground below. This moisture keeps plants growing in areas where rain seldom falls, as in

Figure 4-1 *Trees and shrubs benefit from the moisture brought by fog rolling into a forest on the Oregon coast.*

the coastal desert of Peru. Along the coast of California, fogs add to the rainfall in the dense redwood forests. (See Figure 4-1.)

Rain and snow are the major sources of fresh water for living things. Precipitation varies from place to place, from one season to another, and from year to year. The amount that falls in a period of time also varies. Figure 4-2 shows the world's record amounts of rainfall observed for different time periods. Some rainfalls occurred with thunderstorms, some accompanied tropical cyclones, and others were produced by monsoon rains.

Figure 4-2 *Extreme rainfalls of the world. In short periods of time, they usually occur as thunderstorms. The Cherrapunji extremes were caused by monsoon rains.*

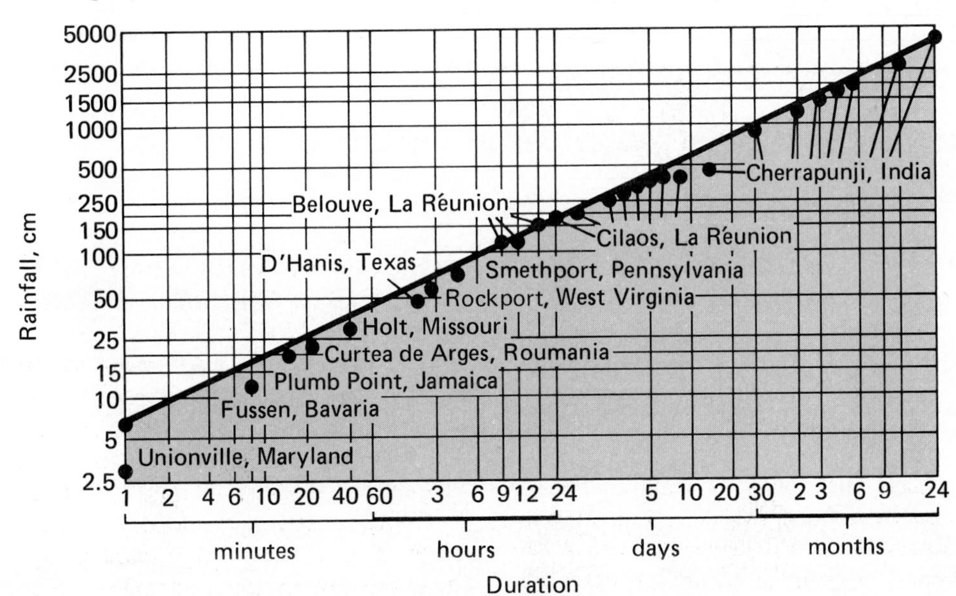

In some deserts, rain is so uncommon that the natives do not have a word for it. Yet in tropical areas such as the Amazon basin it rains nearly every day. Do you know how often it rains where you live? What are some of the ways you could find out?

Figure 4-3 shows what happens after precipitation reaches the earth. Consider a single raindrop. It might soak into the ground, run off, or evaporate. You might think that rivers and streams are the main carriers of water from the land. However, only about one-third of the precipitation falling on the continents runs off in streams and rivers to the ocean. About two-thirds returns to the atmosphere by evaporation from soil and plants.

4-2 How Fresh Water is Stored

Most of the fresh water stored on the earth is frozen in glaciers. **Glaciers** are huge, slowly flowing masses of snow and ice. (See Figure 4-4.) There are about 30,000,000 km^3 of ice now on the earth's surface.

In addition to glaciers, the snowfields of the world store a considerable amount of water. If all the ice and snow on the earth were melted, the level of the ocean would rise an estimated 30 to 60 m. Every 10 to 15 cm^3 of snow yields one cm^3 of water, depending on the density of the snow. Snow accumulates during the cold season in each hemisphere. The greatest amounts of snow fall at high

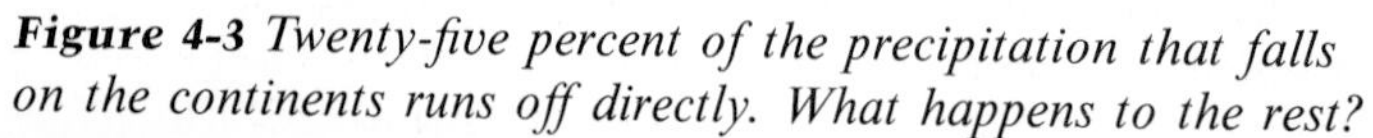

Figure 4-3 *Twenty-five percent of the precipitation that falls on the continents runs off directly. What happens to the rest?*

Figure 4-4 *Large amounts of fresh water are temporarily stored in glaciers like this one in the Yukon Territory.*

elevations and on the borders of the cold polar regions. Most of this snow melts in the spring. Only small amounts fall in the coldest areas of the earth, because the cold air contains little water vapor. However, most of the water within the polar regions is held as snow and ice because evaporation and melting are slow.

Much of the water melted from snow runs into streams and lakes. When the soil under the snow is frozen, almost all of the meltwater runs off. If the soil beneath the snow is not frozen, meltwater soaks into the soil and adds to the water supply in the ground. Of course, rainfall also soaks into the ground. If the rainfall is light, most of it may evaporate from the soil surface. If the rainfall is heavy and continuous, a considerable amount may filter down into the soil.

Rainwater moves downward into pores (holes) in the loose soil until it reaches rock. The "solid" rock below the soil is called **bedrock.** Bedrock can have pore spaces (tiny holes), fractures, cracks, and other openings that store additional water. Some regions, such as New England, have only a thin layer of soil. Most of their subsurface water is found in the pore spaces or fractures in the bedrock.

Rocks that have many pore spaces are referred to as **porous.** Sandstone and other porous rocks can store large amounts of water. The lake behind the Aswan Dam in Egypt did not fill up as quickly as planners had hoped it would. One reason is that the 500-km-long western bank is largely sandstone and absorbs seemingly endless amounts of water.

Finally, lakes are an important form of water storage in many parts of the world. Lakes form in low areas that extend below the level of the water in the surrounding soil and rock. They control streamflow and provide water for cities and farms. Many natural lakes have been changed for greater storage capacity by controlling the amount of water that flows out of them. New lakes are also created by damming rivers.

ACTION

Find out where your drinking water comes from. What steps are taken to treat the water before you use it?

4-3 Investigating the Movement of Water in Soil and Rock

MATERIALS
two 600-ml beakers, 100-ml graduated cylinder, meterstick, paper towels, plastic cylinders filled with different sizes of beads, ring stand and clamp, rubber band, sand (coarse and fine) for investigation of capillarity

You have probably noticed water seeping or flowing from a hillside (Figure 4-5).

When you have completed this investigation, you will be able to answer these questions: Can water move upward through soil? What affects the flow of water underground? Do you have any idea how fast ground water moves through pore spaces in earth materials?

PROCEDURE
Set up a column as shown in Figure 4-6. Your teacher will provide you with several samples, each sample consisting of particles of the same size. Place 100 ml (milliliters) of uniform-sized particles in the column. Make sure the wire screen is in position to prevent the particles from running out. Find the volume of water necessary to just cover the upper surface of the particles. What percent is it of the total volume (100 ml) occupied by the particles? This is the percentage of space between the particles and is called **porosity.**

Open the clamp and allow the water to run out. Record the amount of water retained by the particles. Add 300 ml of water to the cylinder holding the grains. Record the time required for all the water to drain through the particles. This is a measure of **permeability,** the rate at which water can pass through a porous material. Repeat the procedure using the other sizes of particles.

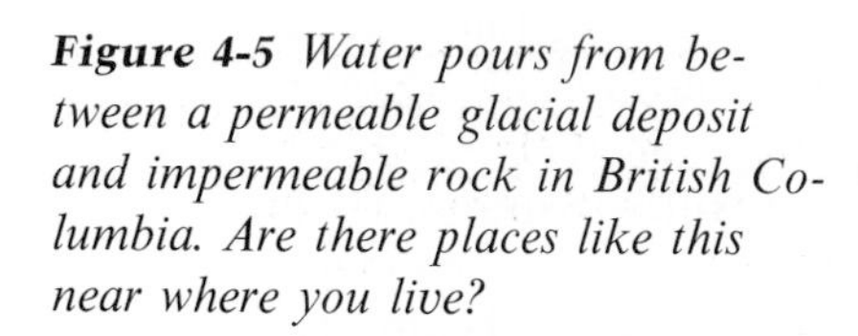

Figure 4-5 *Water pours from between a permeable glacial deposit and impermeable rock in British Columbia. Are there places like this near where you live?*

Figure 4-6 *The setup for investigating porosity and permeability.*

Figure 4-7 *How much water will rise through the sand in the tube?*

DISCUSSION

1. How does the diameter of the particles affect the porosity of the material?
2. How does the grain size affect the amount of water retained in the column after draining?
3. How does the grain size affect the rate at which water flows through a material (300 ml ÷ time)?

PROCEDURE

To see if water can rise through the soil, set up the apparatus as shown in Figure 4-7. Use 200 ml of fine dry sand in the tube. When your partner is ready to time, lower the tube into the water so that the base is just beneath the water surface. Keep the tube perpendicular to the table. Time and record any change in water level in the tube at 30-second intervals. Repeat the process with 200 ml of coarse sand.

DISCUSSION

1. What do you think accounts for movement of water upward in the tube?
2. Suggest a reason why the water moves through the two kinds of sand at different rates of speed.

4-4 Water Moves into the Ground.

Water moves downward rapidly in sand because both the sand grains and the spaces between the grains are relatively large. Clay, however, has very little pore space and very small grains. Most soils contain both clay and sand. Many of these clay and sand particles are held together in clusters. Raindrops hitting bare ground tend to break up the clusters. The soil particles then form a more closely-packed layer that does not soak up water quickly.

Where plants cover the soil, their leaves absorb the impact of raindrops. Therefore, soil clusters remain unbroken, and the large pores in the soil are preserved. Rainwater quickly penetrates. At a research station in Ohio it was observed that only 0.7 cm of water seeped through one kind of bare soil in an hour. By contrast, when the same kind of soil was protected by a layer of straw, 5.6 cm of water passed through in an hour—a flow eight times faster.

Farmers know the importance of having plant cover when the most intense rainstorms occur. Plants on the land, whether on a farm or in the mountains, greatly lessen erosion (see Chapter 7) and flooding. Decaying plants are a source of organic matter for the soil that can make the soil sponge-like, so that water sinks into it more easily. More precipitation enters the soil instead of running off.

Many soils that are mixtures of sand and clay have pore spaces of varying sizes, just as a bag of marbles of different sizes would have spaces of varying sizes. The larger openings permit water to move rapidly to lower levels under the pull of gravity. At each point of contact between particles, however, tiny droplets of water are held back by surface tension and by the molecular attraction between the water and the solid particles. The water stored in this way is called **capillary water.** Capillary water cannot be drawn down by gravity. It can be removed only by evaporation into the air through soil openings or by absorption into plant roots. The supply of capillary water in the upper layer of soil allows plants to survive long periods between rainfalls. The pores in the soil also serve as channels for the circulation of air, which is vital to plant growth.

The amount of capillary water in saturated soil depends on the size of the soil particles. (See Figure 4-8.) Clay soil has very small particles and small pore spaces. Sandy soil has larger grains, with more space between them. A layer of saturated clay soil can hold two to four times as much capillary water as a similar layer of sandy soil. It takes longer to saturate a clay than a sand, but the clay will hold water longer than the sand.

A mixture of sand and clay is a **loam.** Usually loam also contains organic material. Where would grass burn up most quickly during a dry spell: in sand, loam, or clay?

4-5 Water Is Stored at Lower Levels.

In most places, some of the rainwater filters down to lower and lower levels in the soil. Water accumulates below the

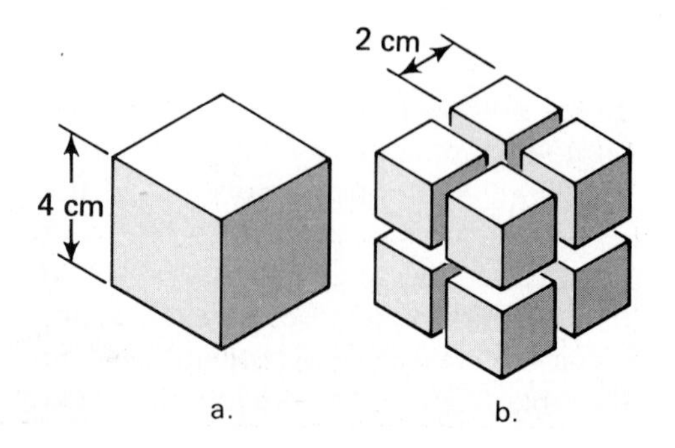

Figure 4-8 *Dividing an object creates new surface areas. Suppose the cube in* **a.** *is 4 cm on each side. What is the area of the surface? If the cube is divided into eight smaller ones, as in* **b.,** *what is the surface area of all the cubes? If each is divided again into 1 cm cubes, what will be the total surface area?*

Figure 4-9 *Water seeps down into the ground water zone. What factors might affect the height of the water table?*

top layer of soil and fills the available pore spaces. The top of this saturated zone is the **water table.** The surface of a lake or river is a part of the water table that you can see. Water below the water table is known as **ground water.**

When moisture income is greater than moisture loss (by evaporation and by plants), the water table usually rises. When the water table is higher than nearby streams and valleys, the ground water flows into them. Figure 4-9 illustrates the water movement.

Streams may be supplied continually by ground water moving into stream channels. During a dry period when there is no incoming moisture, ground water may continue to flow into rivers. In this way the water table is lowered. In desert areas the reverse can happen: the water table is often lower than nearby stream beds. When this happens, water moves from the river into ground water storage, often leaving the stream channel completely dry. (See Figure 4-10.)

Pore spaces greatly influence the rate at which ground water moves. Ordinarily, ground water flows much more slowly than surface water. Speeds commonly range from about 3 to 30 m per day. In rare instances, buried ancient river channels exist. These channels contain loose mixtures of sand and rocks (gravel). The ground water flows much more rapidly through these channels than through most earth materials.

Some earth materials transmit and store large amounts of water and are called **aquifers** (*AK wuh furz*). Sandstone, for example, usually is an excellent aquifer. It usually has abundant pore spaces that may contain air or

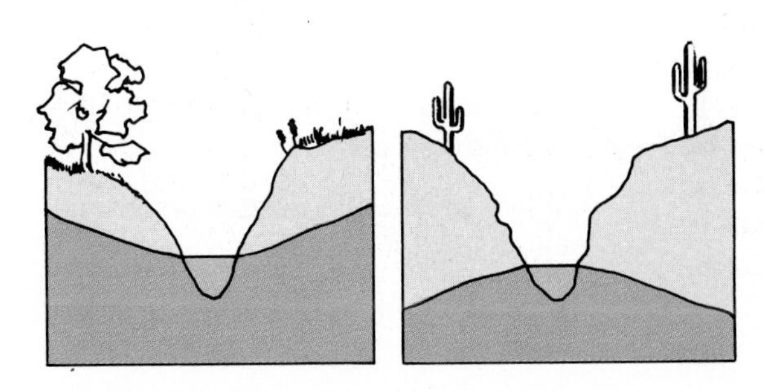

Figure 4-10 *Ground water may move into streams. Or streams may add water to the ground water zone. It depends on the season and the location.*

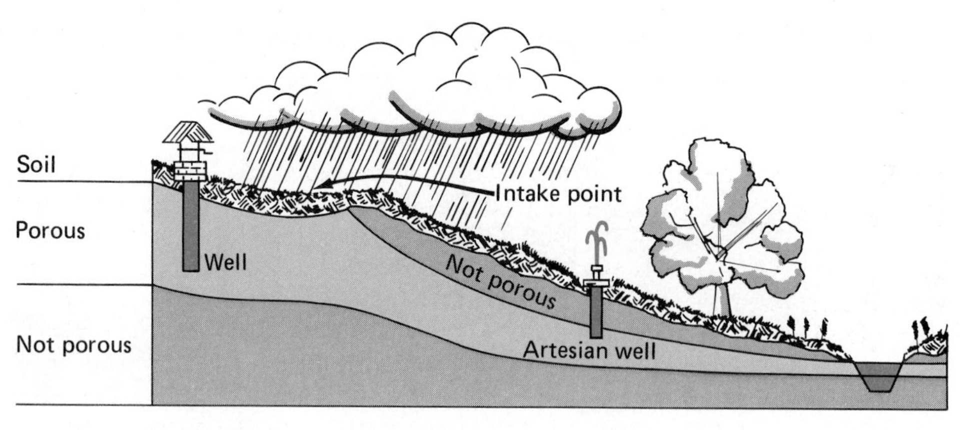

Figure 4-11 *What is the difference between the two wells in this diagram?*

water. Water seeps into sandstone easily but moves much more slowly through rocks composed of finer particles, such as shale, which is a solidified clay or mud. Igneous and metamorphic rocks (defined in Chapter 6) usually are not very porous. The number of cracks and fractures determines the amount of water these rocks can hold.

In many parts of the world, porous rock layers carry water from higher to lower elevations. Where the porous layer is between two nonporous layers, an **artesian system** may be created. Study Figure 4-11. The nonporous layers act like the walls of a pipe. They prevent the release of the water, and pressure builds up. Wells tapping this aquifer may flow continuously at the surface, provided there is adequate rainfall and great enough pressure. The water pressure depends on the difference in height between the well and the intake point—the point where water enters the artesian system.

Part of the underground moisture remains in storage, part of it moves into rivers on the way back to the ocean, and part of it is used by people. Water may actually be used several times before it returns to the atmosphere or the ocean. For example, in some areas industries are required to pump the water they circulate through machinery into the ground, to recharge the ground water supply.

On islands and in coastal areas, the ground water may consist of both fresh and salt water. The fresh water, being lighter, floats on top of the salt water. The surface between the fresh and salt water layers rises and falls with the tide, and some mixing occurs. When fresh water is pumped too rapidly, the fresh water table is lowered and salt water can flow into wells.

Thought and Discussion

1. How does plant cover influence runoff?
2. Distinguish between capillary water and ground water.
3. How does a river continue to flow during dry spells?
4. What happens to ground water storage during a drought?
5. Must the outlet of an artesian well be higher than the intake area? Why?

4-6 Evaporation and Transpiration

You have seen moisture evaporate from wet surfaces when the sun comes out after a rain. Evaporation removes water from swamps, lakes, and wet soils.

Plants withdraw water from the root zone of the soil. This moisture is carried through the plants from the roots to the leaves, where it changes to vapor and escapes to the atmosphere through leaf openings. This process is known as **transpiration.** Most of the capillary water withdrawn by plant roots is transpired. A very small part is used by plants to build new tissue.

ACTION

To see evidence of transpiration, obtain a green potted plant, such as a geranium. Place a plastic bag around the plant so that it is airtight. The following day observe any changes in the bag.

To record the amount of moisture lost, place two potted plants on a balance. Cover one of them with a plastic bag, first. Balance the plants, by adding gram masses to one side if necessary. The following day record any differences in the adjustment of the balance.

Repeat the ACTION using two cactuses, and explain any differences in the results.

MATERIALS
two potted plants, plastic bag, balance with gram masses, two cactuses

Sometimes a single word, **evapotranspiration,** is used to refer to evaporation *and* transpiration. Evapotranspiration is the only means by which capillary water can be removed from the soil.

Evapotranspiration from the earth's surface is difficult to measure accurately. The water loss from tanks and pans can be measured, but these do not duplicate natural conditions exactly. Some tanks used for estimating water loss contain soil and plants from the area under study. The amount of water that is added to the tanks can be measured. This gives an estimate of the water that could be lost through evaporation and transpiration. The water loss from such tanks depends mainly on air temperature. What other weather factors would be important in determining water loss?

4-7 The Water Balance

An estimate of the evapotranspiration that could occur with an unlimited supply of water is useful in obtaining the water balance of an area. The water balance or **water budget** accounts for the income, storage, and loss of water over an area.

Water budget graphs like those in Figure 4-12 show how the water balance changes during the year. Most places have a surplus of water for part of the year (usually the cool season) and not enough water at another season (usually the warm one). Deserts have a year-round deficit (shortage).

Soil moisture usually decreases during the hot months because of high evapotranspiration. Therefore, stored water is normally at a minimum at the end of the warm season. (In some regions, like the southern United States, tropical storms in late summer and early autumn cause temporary surpluses of water.) As the days get shorter and the sun's rays become less direct, moisture income normally begins to exceed water loss to the atmosphere. During the winter, soil and air temperatures are low. Some plants stop growing, and moisture loss from the

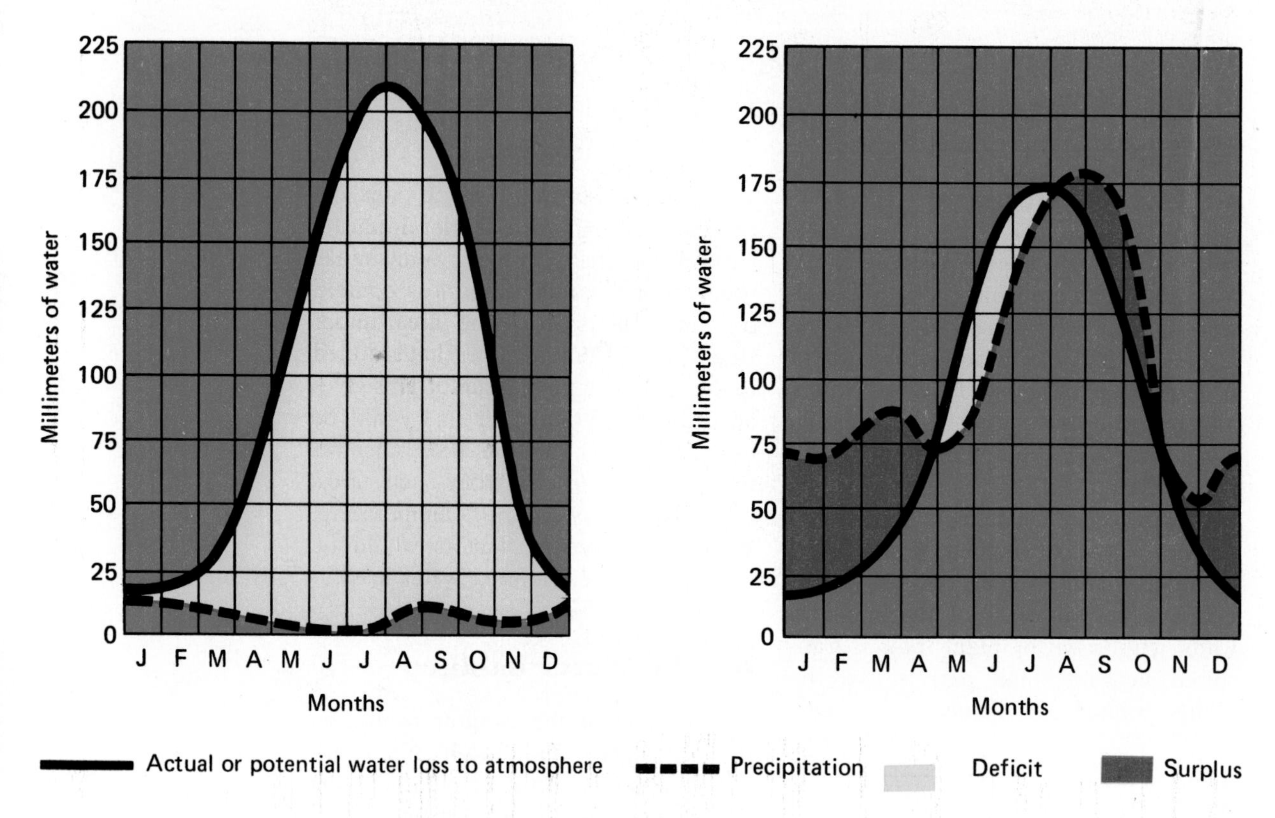

Figure 4-12 *Water budget graphs made for Yuma, Arizona and Savannah, Georgia. Which graph is from Arizona?*

ground is slight. In mountainous areas, snow may build up.

In late winter and early spring in middle latitudes, the melting snow cover releases large quantities of water. The surface layers of soil are soon saturated, and some of the water flows into the underground aquifers where it is stored. Much of the water from the melting snow joins the surface runoff.

In most humid areas, such as Savannah, Georgia (Figure 4-12), the annual cycle of moisture surplus and deficit repeats itself in much the same way year after year. Within the larger annual cycle of income, storage, and outgo are a number of smaller cycles that are repeated each time it rains. With the passage of each storm there is a period of soil moisture recharge followed by withdrawal.

Drought or floods are part of the annual pattern of the water balance in many regions. In much of eastern North America, floods occur each spring. Melting snow and spring rains water the land in greater amounts than the soil can absorb. Widespread runoff swells rivers beyond their capacities. **Runoff** refers to water that flows on the surface or through the ground into streams and lakes.

In most dry or arid regions, long periods of summer drought are the rule. Evaporation increases with summer heat, and rainfall remains low for many months. The occasional thundershower that does come is often so heavy that the water rushes across the land, washing away soil and cutting gullies. Some of the most obvious examples of land being washed away are in dry regions.

Figure 4-13 *The Arkansas River flood of June 20, 1965. You can see normal river flow at the lower right. Three minutes after this photograph was taken, water covered all but the treetops.*

4-8 Runoff

Floods are caused primarily by excess surface runoff. The flooding of the Arkansas River (Figure 4-13) followed a series of steady rains. Long after all the soil openings were filled to capacity, rain continued to fall. Water began to collect in streams and flow over the banks, covering many square kilometers of land.

Many intense storms die out before maximum runoff occurs. Look at Figure 4-14. About how many hours after the heaviest rainfall did the greatest runoff occur? There are three reasons for a delay between rainfall and its appearance as streamflow. First, the earliest rain is captured by the pore spaces in the soil. A second reason for the delay is that it takes time for water to run over the ground and collect downslope in stream channels. Finally, as the water table gradually rises, ground water flows slowly into nearby streams.

Figure 4-14 *A graph of rainfall and runoff. Why are the three curves so different?*

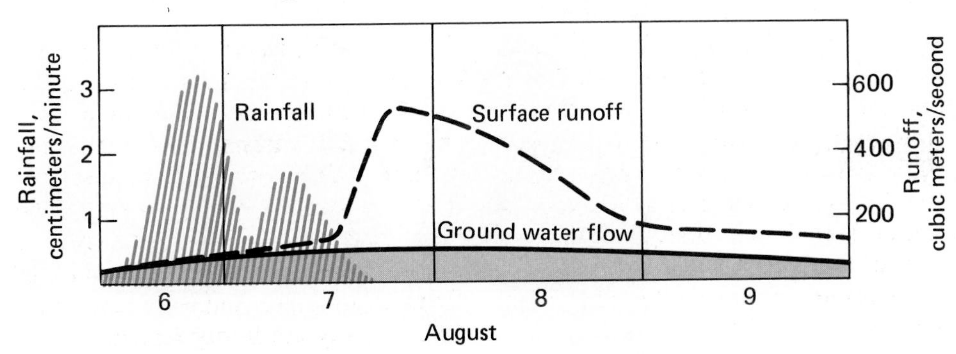

MATERIALS
graph paper

Ground water makes up more than half the yearly flow of many rivers. In most rivers, ground water flow regulates streamflow. Storage in swamps, marshes, and lakes also regulates streamflow. Large storage areas such as these can keep the depth of a stream nearly constant. When swamp lands are drained or filled, the runoff and streamflow patterns are changed. Would surface runoff increase or decrease?

Figure 4-15 shows the volume of flow in two major rivers over several years. How did the actual flow in November, 1975 compare with the average flow for that month? What factors could have caused this difference?

4-9 Investigating a Flood

When you have completed this investigation, you will be able to predict a flood, using real data.

Imagine that in late summer a tropical cyclone crosses the coast, bringing hurricane winds, storm tides, and heavy rains. As the storm moves inland, away from its source of moisture, the wind and rain decrease. The storm becomes a weak low pressure area, changes its course, and moves across a mountain range toward the sea. Suddenly, weather radarscopes begin to show rain echoes from towering

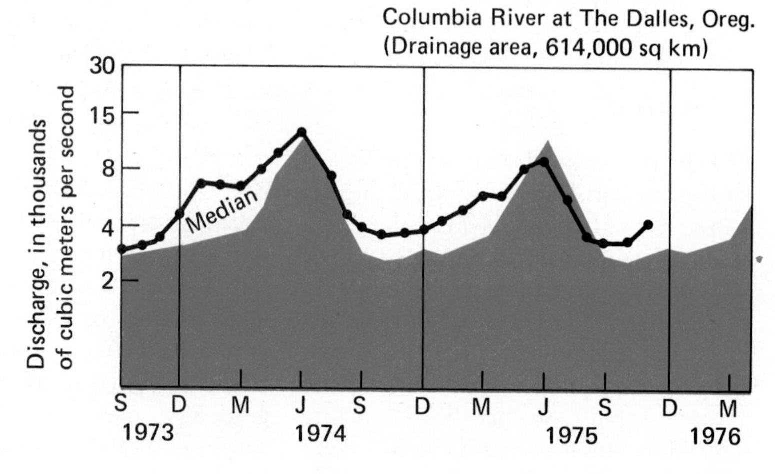

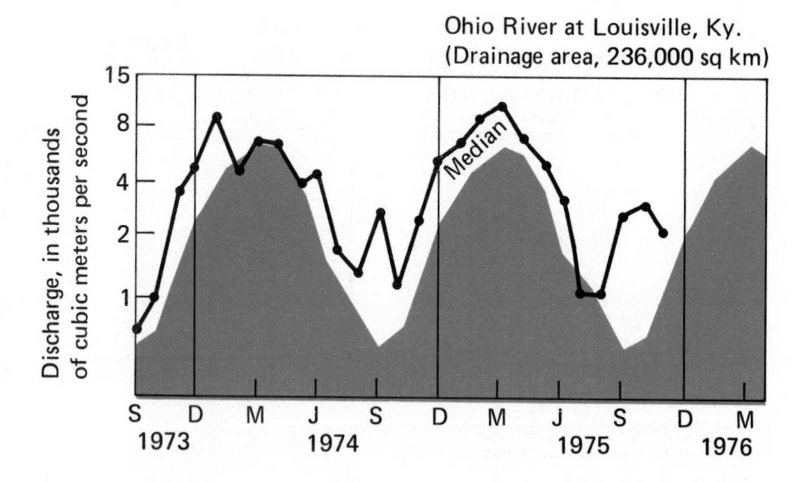

Figure 4-15 *Hydrographs of two large rivers.*

convective clouds. The observing stations in the path of the storm report very heavy rains. One rainfall station reports 70 cm of rain in five hours!

The National Meteorological Center predicts that the storm will pass directly over a river basin that has not had a disastrous flood within the memory of people living there. A quick inspection of weather records shows that the temperature and rainfall during the past month have been about normal.

PROCEDURE

Figure 4-16 contains data that will make it possible for you to estimate the peak of the flood and the time the flood crest will arrive at a town downstream. The data are based on river gauge records from previous storms in the same river basin.

Each person in your group may plot a different graph. In each case draw smooth curves through the plotted points. The four graphs can then be assembled so that your group can discuss the forecast and answer the questions. Be sure to add a title to each graph, to tell what it shows.

I. Plot the time from the beginning of rain to peak flow (on the horizontal axis) versus the duration of the rainstorm (on the vertical axis). This graph will help you predict when the river will crest and how much time there is to warn the town.

II. How high will the water rise? To determine the height of the flood from the new storm, you will need to know what discharge the storm will produce. The **discharge** is the amount of water that flows past the stream gauging station in the town each second. Plot the height of the river (vertical axis) versus the discharge (horizontal axis). The flood stage is 10 meters. Draw a line on graph II to represent the flood stage.

III. The discharge depends upon the length of the storm. Plot the discharge for one centimeter of runoff (vertical axis) versus the duration of the storm (horizontal axis). Remember that this graph only tells you how much discharge

Figure 4-16 *Data from ten previous storms, for Investigation 4-9.*

I		*II*		*III*	*IV*	
Time, from Beginning of Rain to Peak Flow, Hours	*Duration of Storm, Hours*	*River Height, m*	*Discharge m^3/sec*	*Discharge for 1 cm of Runoff, m^3/sec*	*Storm Runoff, cm*	*Storm Rainfall, cm*
					(Summer, normal soil moisture)	
11	5	6.5	3,500	680	0.1	3.3
15	14	6.5	3,700	480	0.5	4.5
21	18	7.2	3,900	410	1.0	3.2
26	18	8.2	4,600	500	1.3	4.0
28	24	9.0	5,000	400	1.3	5.4
29	28	9.7	6,200	300	0.8	6.0
35	31	10.7	7,000	310	3.0	8.0
42	37	11.1	8,200	280	7.0	13.0
46	43	11.8	10,000	230	9.5	16.5
48	49	12.2*	11,600	220	12.7	20.0

*High water mark, record flood to date.

can be expected from one centimeter of runoff.

IV. The total runoff from the storm depends on the amount of rainfall. Plot the runoff (vertical axis) versus the rainfall (horizontal axis). The predicted rainfall over the area is 27.5 cm. This is more than has ever been recorded in the river basin. You will have to extend the curve on the graph beyond the last two points to estimate the amount of runoff from the predicted rainfall.

DISCUSSION

Now assemble the four graphs and answer these questions:

1. If the discharge doubles, do you expect the river stage (height) to double?
2. What is the proportion of runoff to rainfall for storms of 5 cm? 20 cm? Why are the proportions different?
3. If the weather last month had been hot and dry, would you expect more or less runoff from the storm? What is the reason for your prediction?
4. What is the runoff from the predicted storm rainfall of 27.5 cm?
5. Assuming the storm lasts for five hours, how much discharge will the 27.5 cm of rainfall produce?
6. How high will the river rise? Use the extended part of the curve in graph IV to make your prediction.
7. If the rain begins at 5:00 P.M., when will the flood crest probably occur at the gauging station?
8. A sudden, violent flood after a storm is called a **flash flood.** Should you warn the people of the town now about a possible flash flood? Or should you wait until the rainfall reports come in six hours from now?

4-10 People Change the Water Cycle.

Natural changes are always going on in lakes, rivers, and estuaries (where rivers meet the ocean). For example, heavy rains may erode the land and wash sand and mud into rivers and lakes. In some cases, the activities of people speed up these natural changes. The ways people use the land can increase natural erosion. Farm and pasture lands usually erode more than land having a natural cover of vegetation. In areas bulldozed for road construction, the rate of erosion may be two thousand times greater than over forested areas. After the road is built, what can be done to prevent erosion?

Without sewage and industrial waste, the natural lifetime of a lake may be thousands of years. In the first stage, the lake may be deep and have little aquatic life. In time, soil materials are washed into the lake, it begins to fill in, and the water becomes shallower. Nutrients also flow in. The lake begins to support more plants and animals. The final stage begins as nutrients become abundant. Algae grow into huge blooms. When they die and decompose, the bacteria that decompose them use up much of the dissolved oxygen in the water. The lack of oxygen kills many aquatic animals. In time, the lake becomes a swamp and finally a land area. This aging process is known as **eutrophication** (*yoo trohf ih KAY shuhn*).

Wastes from industrial, municipal, and agricultural sources may speed up the eutrophication of lakes. Nutrients from sewage hasten the eutrophication of lakes and rivers by fertilizing the plant life. Heat added to the water from power plants may also cause changes in the aquatic life and speed up the growth of algae.

As the plants increase and more organic matter decomposes, the amount of

TECHNICAL WRITER

Technical writers put scientific and technical information into language that can be easily understood. They research, write, and edit technical materials and also may produce publications, sales, or audiovisual materials. In addition to clarifying technical information, some technical writers use their writing skills in marketing, advertising, and public relations work.

Before they begin a writing assignment, technical writers must learn all they can about the subject by studying reports, blueprints, specifications, and product samples. They may read technical journals and consult with the scientists and engineers working on a project. The draft of the document may go through several revisions before being accepted in final form.

Technical writers work in a variety of industries, including electronics, computer, and communications firms. Many work for publishers of business and trade publications, professional journals, or scientific and technical books. The rapidly growing information industry provides employment for technical writers, as do a number of government agencies.

There are no rigid requirements for a career as a technical writer. A combination of good writing skills and scientific or technical knowledge is required, and this combination may be acquired in a variety of ways.

Technical writers should be logical and intellectually curious. They should be able to work with others, since they are often part of a team of scientists and technicians. Freelance writers must be self-starters.

oxygen in lakes and ponds may be reduced. When all the oxygen is used, the bacteria that do not require oxygen begin to thrive. These produce the gas hydrogen sulfide, which turns the lake or pond dark and causes foul odors.

Mixing in the hydrosphere is not nearly as effective as in the atmosphere. Water pollution lasts much longer than air pollution. Sometimes the processes set in motion, like rapid aging, cannot be reversed without immense spending of energy and money.

Thought and Discussion

1. Runoff takes place over the surface of the soil and also as ground water flow. What would you consider the most accurate way to measure the runoff from a river basin?
2. The seeding of clouds that are already overhead is a possible method of controlling the runoff cycle. If it were possible to produce precipitation reliably in this way, would it speed up the runoff cycle?

Unsolved Problems

The distribution of the world's fresh water often does not match human needs. We have not yet found ways to overcome water shortages (or population excesses) at reasonable cost. The problem has been approached in many ways: seeding clouds to produce rainfall, extracting fresh water from seawater, decreasing evaporation from reservoirs, building reservoirs to hold seasonal surpluses, transferring water from one basin to another, using water more efficiently, keeping existing rivers pure, and planting types of vegetation that will hold back runoff and produce a more even flow of water. It has been suggested that coastal cities might get their water by towing icebergs from polar regions and melting them. Some of these proposals work well in certain local areas, but none offers a complete solution.

Chapter Summary

Although more precipitation falls on the ocean than on the land, water on the land is an important part of the water cycle. Water on the land can be considered in terms of income, storage, and loss. Precipitation is the main source of water on the land. The amount of rain and snowfall varies widely and depends on the season and geographic location.

Water reaching the land may be stored in snowfields, glaciers, lakes, and streams. It may be stored in porous rock and in the soil as capillary water in the root zone or as ground water at lower levels. Water is removed from the land largely by evaporation and transpiration. Two-thirds of the precipitation falling on the continents goes back into the atmosphere by this means. The other third leaves the land as runoff in surface streams and ground water.

In adjusting to the environment and altering it for the sake of convenience, people sometimes change the patterns of water on the land. The effects of such changes on local weather and climate have not been thoroughly investigated. However, poor use of water and land causes many undesirable changes in lakes, rivers, and estuaries. These changes endanger the supply of fresh water.

Questions and Problems

A

1. What are the main "reservoirs" that supply fresh water to the earth's surface?
2. What is a "pore space" in soil and how is water stored in these spaces?
3. How is bedrock able to store ground water?
4. What factors govern the penetration rate of water into the ground? What are some others that you can think of, in addition to those mentioned in the text?
5. What effect on the potential for flooding of local rivers and lakes occurs when an "undeveloped" area becomes increasingly built-up with new homes or shopping centers?
6. Where does the energy come from when water is evaporated or transpired? On a typical continental land mass, which will probably account for the greatest amount of water loss to the atmosphere: transpiration from plants, evaporation from soil, or runoff into streams and lakes?

B

1. Explain (using Figure 4-14) why the greatest increase in surface runoff does not usually occur at the same time as the greatest increase in precipitation.
2. Using the water budget graphs in Figure 4-12, identify the months at each location when (a) a water shortage might exist, and (b) the possibility of flooding would be greatest.
3. In Investigation 4-3 (Soil Water Movement) you used samples of uniform grain size. Predict the effect on the results of the investigation if a sample of "mixed" grain sizes had been used.
4. Look at the hydrographs of the Columbia and Ohio Rivers (Figure 4-15). Which river drains the area with the greatest amount of runoff for each unit of surface area? What factors would contribute to these results?

C

1. In which season of the year are evaporation and transpiration highest? Why?
2. Would water evaporate faster from a pan of hot water or a pan of cold water? How would the evaporation rate from large deep lakes differ from the rate of evaporation from small shallow lakes?
3. During humid periods in the spring, water condenses on snow surfaces. What effect does this have on the rate of melting? How would this affect the potential for spring flooding?
4. In a particular artesian system, the grains of the sandstone aquifer generally decrease rather uniformly in size as the distance from the source of the water supply increases. If all other factors are the same, how would this affect the amount of water in the aquifer and the rate at which the water moves through it? Support your answer with explanations.

Suggested Readings

Ambroggi, Robert F. "Water." *Scientific American,* September 1980, pp. 100-116.

Davis, Kenneth S.; Luna B. Leopold; and the editors of *Life. Water.* New York: Time, Inc. (Life Science Library), 1982.

Webster, Peter. "Monsoons." *Scientific American,* August 1981, pp. 108–118.

5
Water in the Sea

IN THE PHOTOGRAPH *on the opposite page, a hill of water rises behind a surfer, and his board slides down it. He races with the curl of the wave over his head. If he goes too slowly, the wave will catch up and flip him. If he goes too quickly and slides down off the wave, he'll stop moving. But the wave won't stop and he'll be flipped under tons of water. So he does his balancing act, walking up and down the moving board, adjusting his speed.*

The thrill of surfing is to feel the energy of the waves and learn to use it. Sea waves have enough energy to heat and light the largest cities, if we could only find a way to harness it. Making use of this constantly available energy remains a problem to be solved.

The water this surfer rides off the north coast of Oahu, Hawaii, has a warm temperature that is comfortable for surfing. Yet some of this water has flowed under the ice in the Arctic. The ways in which water moves from polar seas to the tropics and back provide clues to the energy system of the ocean.

If the surfer accidentally swallows a mouthful of seawater, he will find that it has an unpleasant, salty taste. Seawater is unfit for drinking, and people drifting at sea have died of thirst. If the surfer falls from his board, he will float in the ocean more easily than he would in fresh water. These qualities of seawater come from minerals that are dissolved in it.

The ocean covers most of the earth's surface and is the great reservoir of the earth's water. It dominates the earth's weather and climate. In fact, the ocean influences the total environment of the earth's surface.

AFTER COMPLETING THIS CHAPTER, YOU SHOULD BE ABLE TO:

1. **compare the compositions of fresh water and seawater and identify the major source of the materials dissolved in the oceans.**
2. **state the general range of salinities that occur in the oceans of the world and list the approximate ratios of the major salts found in seawater.**
3. **identify some of the trace elements that occur in seawater and explain why variations in trace element composition are common.**
4. **describe how energy and materials are exchanged between the ocean surface and the atmosphere.**
5. **trace the path of energy from the sun to the ocean waves and currents.**
6. **define the parts of an ocean wave and explain the movement of water particles within both deep and shallow water ocean waves.**
7. **determine the effect of temperature and salinity changes on water density and the velocities of deep ocean currents.**
8. **explain how the circulation of ocean currents has a basic effect on world climate patterns.**
9. **predict specific local effects that prevailing ocean currents may have on the climatic patterns of an adjacent land mass.**
10. **select the most representative climate-graphs for certain geographic locations from a number of examples, when you are given data concerning factors that affect the climate for those areas.**

The Ocean in the Water Cycle

5-1 The Water Cycle Makes the Sea Salty.

In many ways, water is a special substance. Water exists as a solid, liquid, and gas at temperatures ordinarily found on Earth. So it is continually moving about in the atmosphere, on land, and in the ocean. It absorbs energy at some places, releases energy at others, and moves materials over the surface of the earth.

If you were to follow a water molecule around, you would learn that on the average it is in the ocean 98 out of every 100 years. The water molecule would spend twenty months as ice, about two weeks in lakes and rivers, and less than a week in the atmosphere. To a water molecule, going through ice, lakes, and air is like taking a rare vacation from the ocean.

The water in the ocean contains a variety of other substances, some of which make it salty. Energy from the sun evaporates seawater and starts the water cycle (Figure 5-1). Water from rain and snow becomes part of the surface and ground water supply. From there it flows into streams and rivers, and rivers flow back to the sea. So some water that evaporated from the sea returns to the sea. In flowing across the land, the water picks up tiny particles of soil and rock and becomes muddy. The water also dissolves materials from the soil and rock. Almost every earth material can dissolve to some extent in water. The dissolved material usually occurs as ions in the water. An **ion** (*EYE ahn*) is an atom or group of atoms that has an electrical charge because it has gained or lost at least one electron. Read on, for the explanation.

To understand ions, you have to recall what you know about the make-up of matter. Think of a full salt shaker. It contains small crystals of table salt. Each crystal consists of billions of ions of the elements sodium and chlorine. An **element** is made up of very small particles called atoms, that are all the same. That is, all sodium atoms are alike, and all chlorine atoms are alike.

You remember that atoms consist of even smaller particles, the protons, electrons, and neutrons (and others). The protons and neutrons are all packed into a small nucleus at the center of the atom. The electrons move throughout the rest of the space occupied by the atom. The electrons the farthest away (on the average) from the nucleus take part in chemical reactions.

How does salt form from sodium and chlorine atoms? A sodium atom has one electron that it can easily give away. If it

Figure 5-1 *Water and energy enter the atmosphere from the sea.*

Figure 5-2 *Salt is removed from water at this solar evaporation field in Saltair, Utah.*

bumps into an atom that has the ability to attract electrons, the sodium atom will lose the electron. It will then be a sodium ion. (The symbol for sodium is Na, from the Latin word *natrium.*)

One Na atom (11 protons, 11 electrons) minus one electron forms one Na^+ ion (11 protons, 10 electrons). Note that when the neutral atom loses an electron, it becomes positively charged (+). This tells you that an electron is a negative charge (−).

The chlorine atom is one of those that attracts loose electrons. If it is bumped by a sodium atom, it will attract an electron from the sodium, and become a chloride ion. One Cl atom (17 protons, 17 electrons) plus one electron forms one Cl^- ion (17 protons, 18 electrons). The sodium ions and chloride ions are oppositely charged, so they attract each other. They form crystals consisting of billions of ions, held together by these attractive forces.

Some ions involve more than one element and more than one electron attracted or lost. The materials that make up rocks and soil are usually in the form of ions. Since ionized substances will dissolve in water, rivers can dissolve materials from their banks and beds.

Some areas have a high evaporation rate. There, water that collects in basins evaporates and leaves dissolved materials behind, as in Figure 5-2.

Because of the force of gravity, the heavier particles of rock and soil carried by the river eventually settle and are left behind on the river bottom. The ions remain dissolved in the river water, however, and most of them are carried to the ocean. The earth materials dissolved in seawater make it different from fresh water. We cannot drink it or use it to directly water our crops. Seawater also corrodes most metals in a short time unless they are protected from it.

People have known for thousands of years that if seawater is evaporated or boiled away, some white crystals will be left behind. This was one of the early ways that people living near the sea got salt for cooking. If one kilogram of water from almost any part of the sea is evaporated, about 35 grams of solid materials will be left. (See Figure 5-3.) The number of grams of dissolved material in 1000 grams of seawater is the salinity (*suh LIHN uh tee*) of the seawater. The salinity of average seawater is about 35 grams per kilogram, or about 3.5 percent of seawater by mass.

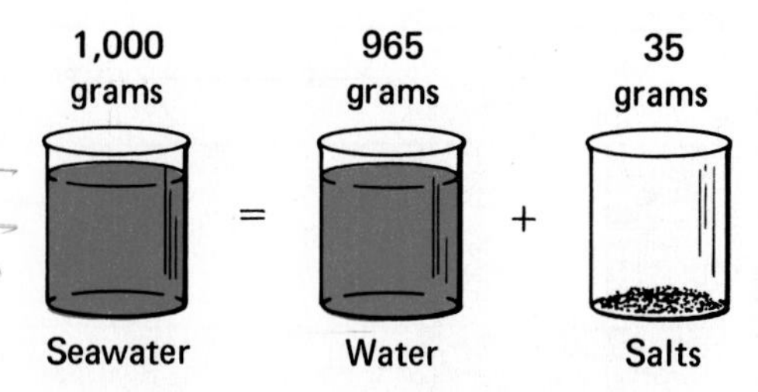

Figure 5-3 *Each kilogram of seawater contains about 35 grams of dissolved salts.*

5-2 Materials Dissolved in Seawater

A simple test of seawater shows that it contains the compound sodium chloride, which we use as table salt. Sodium and chloride ions make up about 85 percent of all ions dissolved in seawater. Chemical tests can measure the amounts of about 60 other elements in seawater. However, ions of just six elements make up more than 99 percent of the total mass (Figure 5-4).

Some of the elements in rocks and soils are much more easily dissolved than others. Once they get into the ocean, they are removed from the water by living organisms. Silicate compounds and calcium carbonate are washed into the ocean in very large amounts. There, clams and other organisms take up the calcium and silica to make their shells and skeletons. Since organisms take so much out of the water, some elements that are common in ocean plants and animals have low percentage in seawater.

Sodium and chloride ions are so soluble that they accumulate in seawater more than other ions do. The ocean could hold much more of them than it does now. Large amounts are carried from the land but they are little used by sea organisms. Most ocean-living organisms have biological systems to keep extra salt out of their bodies.

Figure 5-4 *The relative amounts of the common ions in salt water. Each number represents a percent of all the ions in the water.*

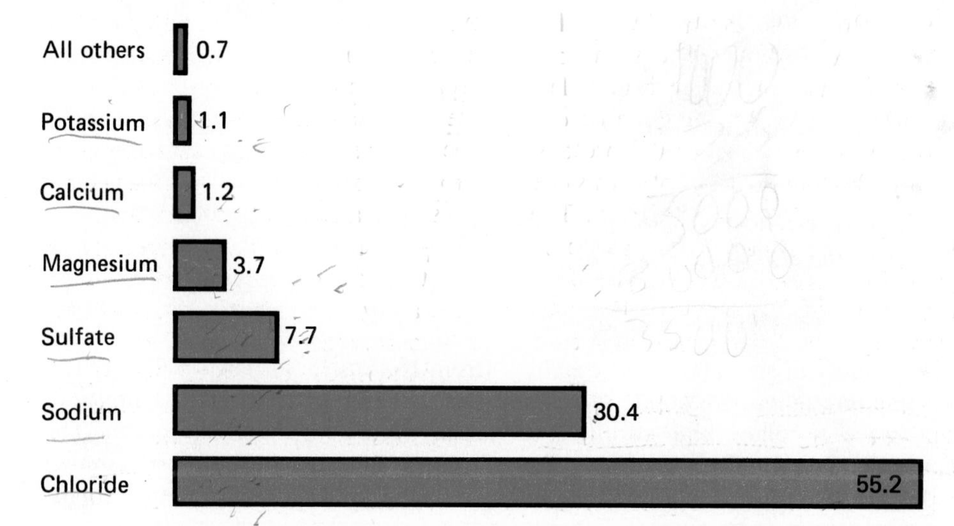

REPORT ON THE COMPOSITION OF OCEAN-WATER. 25

Challenger Number.	Date.	Station.	Latitude.	Longitude.	D.	δ.	Per 100 grms. of total Salts. Sea Water.	Chlorine.	SO_3.	CaO.	MgO.	K_2O.	Na_2O.	Alkalinity per kilo. units of V place.	Sulphuric Acid per 1 grm. of Chlorine.	Laboratory Number.
	1874.															
962	July 12	252	37°52′ N	160°17′ W	2740	850	2911·3	55·431	6·372	1·725	6·227	1·316	41·429	260	·11496	50
963	.. 12	252	37°52′ N	160°17′ W	2740	B—100	2940·0	55·450	6·371	1·811	6·209	1·391	41·261	218	·11490	51
1151	,, 16	...	...	...	...	200	2873·8	55·519	6·388	1·664	6·194	1·316	41·446	149	·11506	62
...	,, 17	254	35°13′ N	154°43′ W	3025	...	...	...	...	...	...	...	...	...	...	...
...	,, 27	260	21°11′ N	157°25′ W	310	...	...	...	...	...	...	...	...	...	...	...
907	,, 28	...	...	...	...	B	2895·5	55·281	6·369	1·689	6·207	1·343	41·603	399	·11521	61 & 61A
1100	Sept. 2	269	5°54′ N	147° 2′ W	2550	25	2862·1	55·412	6·437	1·706	6·251	1·331	41·367	221	·11617	343
1106	,, 2	269	5°54′ N	147° 2′ W	2550	B	2900·6	55·549	6·434	1·717	6·216	1·355	41·261	79	·11582	344
1155	,, 16	276	13°28′ S	149°30′ W	2350	B	2861·7	55·437	6·428	1·726	6·242	1·319	41·358	207	·11595	345
1221	Oct. 14	285	32°36′ S	137°43′ W	2375	B	2858·3	55·440	6·471	1·721	6·200	1·278	41·401	157	·11672	346
1259	,, 25	290	39°16′ S	124° 7′ W	2300	B	2897·1	55·478	6·429	1·701	6·209	1·336	41·366	151	·11588	347
1300	No	295	38° 7′ S	94° 4′ W	1500	B	2873·5	55·424	6·434	1·713	6·187	1·333	41·409	189	·11609	348
Mean,								55·414	6·415	1·692	6·214	1·333	41·433	225	·11576	
Mean, excluding Number 871 (Chall. No.).								55·420	...	...	...	...	...	220	...	

DISCUSSION OF THE PRECEDING TABLE.

In going over the 77 reports embodied in this table, we see that although the concentration of the waters is very different, the percentage composition of the dissolved material is almost the same in all cases; the mean values being as follows:—

(*In* 100 *parts of Total Salts.*)

Chlorine,*	55·420
Deduct basic oxygen equivalent to this chlorine,	− 12·503
Muriatic acid, $Cl_2 - O$	42·917
Sulphuric acid, SO_3	6·415
Lime,	1·692
Magnesia,	6·214
Potash,	1·333
Soda,	41·433
	100·004

* Excluding the abnormally low value in Challenger number 871.

(PHYS. CHEM. CHALL. EXP.—PART I.—1884.) A 4

Figure 5-5 *What important finding is illustrated on this page from the HMS* Challenger *Reports?*

The salinity of seawater is not exactly the same throughout the ocean. It varies from place to place and at different depths. The first studies of seawater were made after the cruise of HMS *Challenger* (from 1872 to 1876). The *Challenger* Reports gave an analysis of 77 samples of water from all parts of the ocean. The findings showed that the most common ions always occur in the same proportions. (See Figure 5-5.) These results show how completely mixed the ocean is in all of its basins. If the amount of just one of the common ions is measured (chloride for example), the amounts of the other common ions in the sample can be determined.

The results were confirmed one hundred years later at the centennial celebration of the *Challenger* expedition. The most modern methods showed the common ion proportions to be the same in all oceans and at all depths. Electronic equipment aboard the new space shuttle *Challenger* can measure the ocean salinity rapidly to show small changes.

Less common ions in seawater, such as phosphates and nitrates, are not always in the same proportions relative to each other. Near the surface of the sea, tiny drifting marine plants that use them grow and are eaten by equally tiny animals. In this way, the less common ions in the surface waters are used up. The ions are carried to deeper water when the remains of the dead tiny animals and plants sink to the bottom. The deep water is enriched in the uncommon ions from the decaying organisms.

In contrast to the nearly constant salinity, trace elements do vary. **Trace elements** are those that are found in very small quantities in seawater. They make up the other one percent (see the first paragraph of this section).

Variations are caused by the differences in the minerals brought to the sea by streams or added by human activities. Such variations are always small when compared with the total volume of the ocean and are noticed only near their sources. Precise chemical tests from thousands of samples are needed to detect these small additions to the ocean. These tests must be from samples taken over many years from all parts of the vast ocean.

Pollution caused by the activities of people has not caused any great change in the ocean as a whole, so far. But in local areas near sources of pollution, such as bays, there have been significant changes in the water. These changes are under continuous study.

Because seawater contains almost all of the known elements, there have been many dreams of "mining" gold and other precious metals from the sea. People have obtained ordinary salt from seawater for many centuries. Economical methods have now been found to remove magnesium and bromine. All attempts to obtain other minerals from seawater have been fruitless.

5-3 The Sea and Atmosphere Exchange Matter and Energy.

Most dissolved solids remain in the sea when water at the surface evaporates. However, some salt does move from the sea into the atmosphere. Breaking waves toss water droplets into the air. The smallest droplets may completely evaporate before they can fall back into the sea. The salts that were in these droplets remain in the air and are carried by the wind as tiny crystals. Eventually, moisture in the atmosphere may condense on the salt crystals and form raindrops or snowflakes.

Gases as well as solids move back and forth between the sea and the air. The most important gases exchanged are oxygen and carbon dioxide. All plants and animals, in the ocean as on land, must have oxygen to live. Normally, the surface waters of the sea are saturated with oxygen. When fish or other animals use oxygen from the water, oxygen from the air replaces it. Oxygen can also pass from the sea to the atmosphere. Marine plants release the oxygen that they do not use for their life processes, just as land plants do. Ocean currents can carry oxygen to great depths in the ocean.

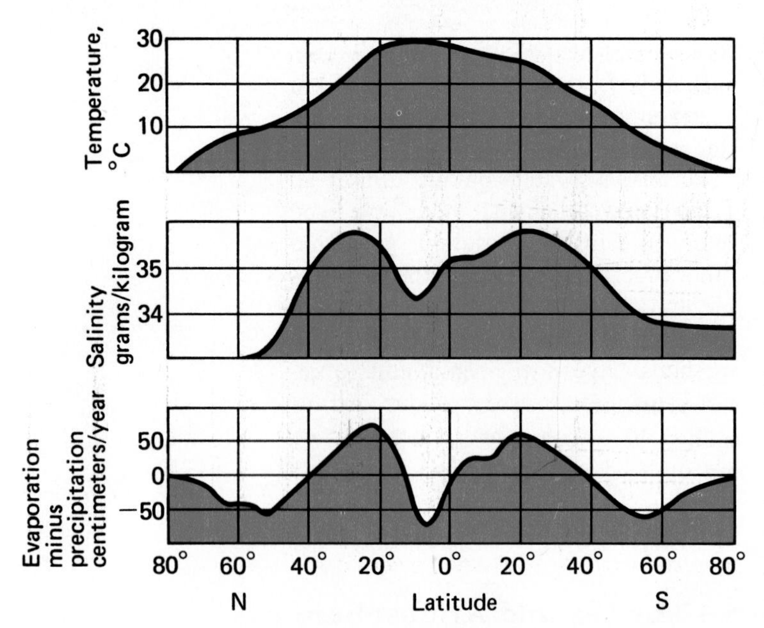

MATERIALS

tall glass or bottle, heat source

Figure 5-6 *The surface water temperature, salinity, and evaporation minus precipitation vary with latitude. How are they related?*

ACTION

You can see for yourself that water contains dissolved gases. Fill a tall glass or bottle with cold water and put it in a warm place, or heat it gently (but do not boil it). Explain what you see happening.

The most abundant material that moves between the sea and the air is water. The atmosphere gets about 80 percent of its water vapor from the evaporation of seawater. When the sea is warmer than the atmosphere, water vapor passes rapidly from the sea to the air. Much of this water vapor eventually condenses and falls as precipitation on some other part of the ocean or on the land.

George Wüst (*Voost*), a German oceanographer, explained that the exchange of water between the sea and the air must affect the ocean's salinity. The water that evaporates from the sea contains no salts. Evaporation would increase the salinity, the percentage of salt, in the seawater that remains. Rain would dilute the salt water and reduce its salinity. If Wüst's idea is correct, the surface of the sea should have higher-than-average salinity at latitudes where evaporation is greater than precipitation. Where would the salinity of the surface waters be lower than average? Look at 30° north and south latitudes in Figure 5-6. Is Wüst's idea supported?

Figure 5-7 *Even in tropical regions, the water in the ocean depths is very cold.*

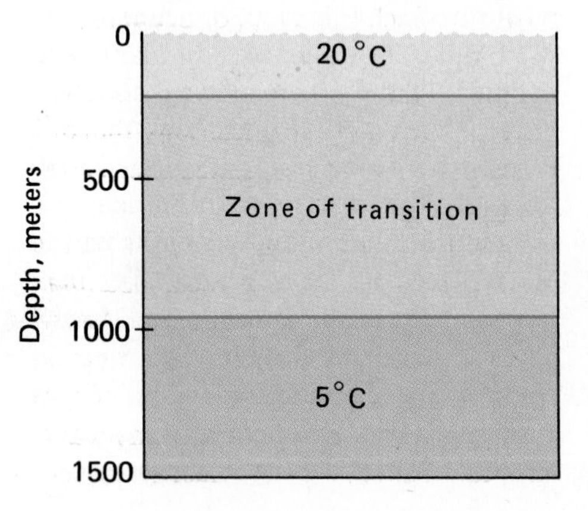

MATERIALS

alcohol

ACTION

Wet the back of one hand with water and wet the back of your other hand with alcohol. Can you explain the difference you feel between the two? Repeat the action and blow on the back of your hands. Is there a difference? Where did the energy come from to evaporate the water and alcohol?

About half of the energy coming from the sun is reflected back into space or absorbed by the atmosphere before it reaches the earth's surface. Some of the energy that does reach the earth's surface is reflected and further heats the air. Most of the energy that finally reaches the ocean is absorbed in the top few millimeters. Figure 5-6 shows how the surface water temperature varies with latitude.

Notice in Figure 5-7 that there is a surface layer many meters thick with nearly uniform temperatures. This layer is made when the heat absorbed at the surface is carried downward as the waters are mixed by waves. Only in this upper layer is the ocean really warm. Most travelers sailing on warm tropical seas do not realize that less than a kilometer away from them (straight down) the water is nearly as cold as ice.

Thought and Discussion

1. How does energy from the sun change seawater?
2. Why is the composition of sea salt so different from the composition of the earth's crust?
3. What happens to most of the energy coming from the sun?
4. Why is water the most abundant material exchanged from the sea to the air?

The Sea in Motion

5-4 Waves Carry Energy.

Have you ever seen water completely still at the seashore, or in a large lake, or even a pond? If you have, it was not for long. Large bodies of water are constantly in motion. Mostly you notice the waves.

The waves get their energy from the wind and carry the energy across the ocean. The spectacular effects of this energy are seen on seacoasts. There, waves erode the cliffs and pound the rocks into fine sand (Figure 5-8). If you have ever been knocked over by a breaking ocean wave, you know its great energy.

Water waves that are caused by the wind form in the same way as the ripples you can make by blowing on water in a pan. When you stop blowing, the ripples stop. The wind blows on the sea for a much longer time than you can blow on the water. So, at sea the ripples grow into small waves. The longer and harder the wind blows, the larger the waves become. Waves more than 30 m high have been measured from ships. Waves will keep moving even after the wind has stopped. The wind has given them enough energy to travel hundreds of kilometers across the ocean. You may see large breakers at the shore even on a windless day.

Waves on a sea or lake are usually a mixture of different heights and lengths. No two waves look exactly alike. Some things about all waves are the same, however. Each has a top, or **crest,** and a bottom, or **trough** (Figure 5-9). The height of the crest above the trough is called the **wave height,** and the distance between crests is called the **wavelength.**

If you watch a group of waves pass a marker or wash over a rock, you may notice that the time between crests is always about the same. This time is called the **period** of the waves. If you know the length and period of any wave, you can figure out how fast it travels.

Figure 5-8 *Waves constantly pound and help destroy the rocks of a rocky coast.*

Wave speed equals wavelength divided by period. A wave with a length of 156 m and a period of 10 sec travels at a speed of 156/10 or 15.6 meters per second.

Over deep water, waves with long wavelengths travel faster than those with short wavelengths. The longer waves race ahead, leaving the shorter waves behind. They may even run ahead of the storm which causes them. Long waves crashing on a beach are usually a warning of an approaching storm.

The water riding up and down in the crests and troughs does not move across the ocean with the waves. The water particles move, but in circular paths. Try the following ACTION to help you understand the principles of wave motion.

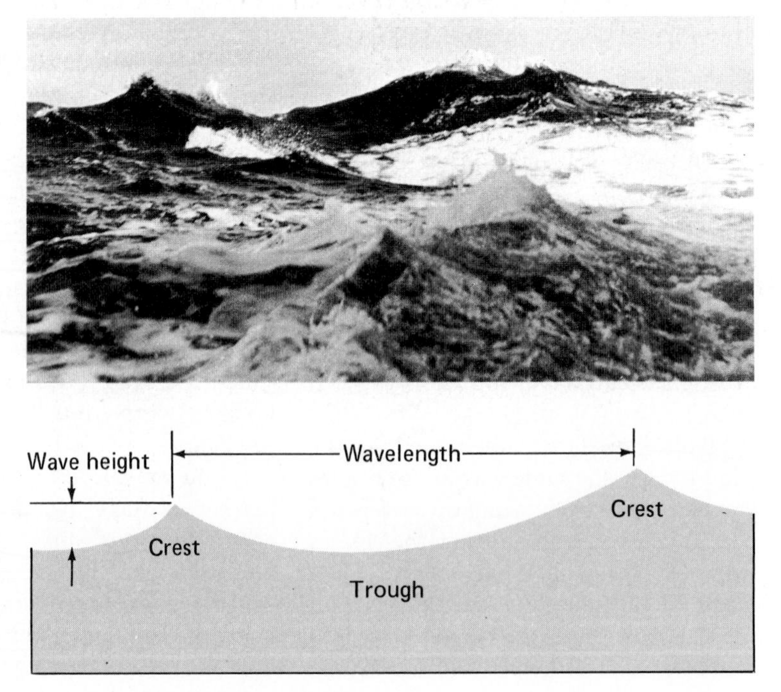

Figure 5-9 *Terms used to describe waves.*

ACTION

Tie one end of a rope that has a knot in the middle to a doorknob. Extend the rope to its fullest length and jiggle it up and down. Waves will pass along the rope toward the doorknob. How does the rope itself move? What is the motion of the knot? Is this a good model of the motion of water particles in a wave?

MATERIALS
rope

The water particles in waves move differently in deep water from the way they move in shallow water. Where the water is deep—deeper than one-half the length of the wave—the water particles move around in circles. The diameters of the circles at the surface of the sea are the same as the wave height. (See Figure 5-10a.) Below the surface, the water particles move in circles of smaller and smaller diameter. There is no longer any detectable motion at a depth of about one-half the length of the wave. Submarines dive below the surface and ride in the smooth ocean depths. Divers also reach calm waters within a few meters of the surface.

In water shallower than one-half wavelength, the circular motion of water particles is changed by the sea floor. The result is the elliptical paths shown in Figure 5-10b. The bottom of the wave drags on the sea floor and moves more slowly than the top of the wave. This causes the wave to "break" on the beach. Breakers, or surf, throw water up on the beach, carrying sand and other sediment that has been stirred up from the bottom.

The drag on the sea floor also causes the waves to slow down. Waves coming from deep water start to catch up with the waves in shallow water. They do not overtake the waves ahead, but the length becomes shorter and shorter.

Figure 5-10 *In deep water waves, water particles move in circular orbits.* **a.** *The diameter of the circle decreases with depth.* **b.** *In shallow water, the bottom of a wave drags on the sea floor. The orbits of the water particles become flattened.*

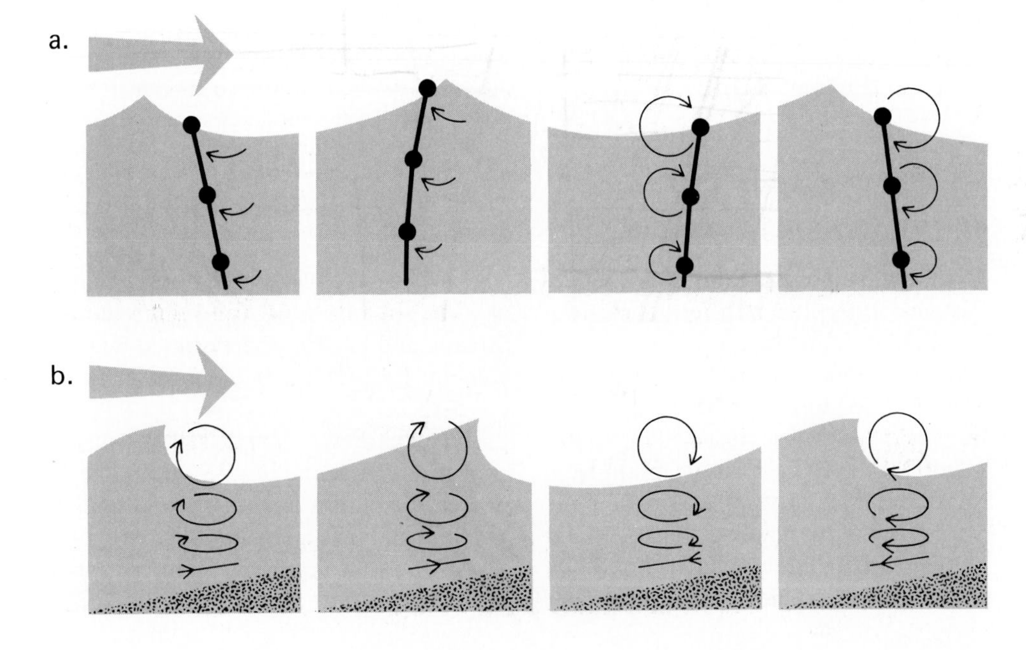

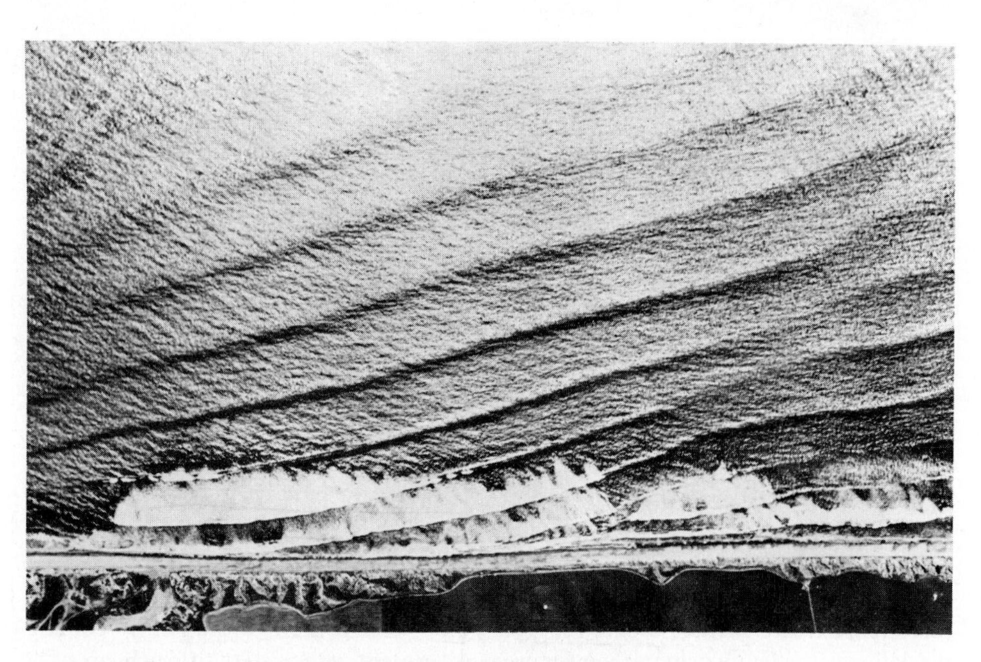

Figure 5-11 *Waves bending at a beach. The part of a wave that is in deep water moves faster than the part in shallow water. This makes the wave swing parallel to the beach.*

Most waves approach a beach at an angle. This means that part of a wave is in deep water and part in shallow water. The part in deep water travels faster than the part in the shallow water, and the result is that the waves bend toward the beach. When waves finally reach the beach, they are almost parallel to the shoreline (Figure 5-11).

5-5 Winds Cause Currents at the Ocean's Surface.

When Benjamin Franklin was Deputy Postmaster for the northern American colonies, it came to his attention that mail ships took two weeks longer than whaling ships to make the voyage from England to America. He asked his cousin Timothy Folger, a whaling captain from Nantucket, Massachusetts, to explain the extra speed of the whaling ships. Captain Folger replied that the whalers knew of a place in the ocean where the water flowed like a river. He went on to say that whaling captains:

> . . . are well acquainted with the stream because in our pursuit of whales, we run along the side and frequently across it to change our side, and in crossing it have sometimes met and spoke with those packets [ships] who were in the middle of it and stemming [going against] it. We have informed them that they are stemming a current that was against them to the value of three miles an hour and advised them to cross it, but they were too wise to be counseled by a simple American fisherman.

Franklin asked Folger to draw a chart of this stream, now called the Gulf Stream, and had the chart printed by the General Post Office. It is shown in Figure 5-12.

Since earliest times, sailors have known of currents in the ocean, and steered their ships to use or avoid them. Yet information on currents was not collected in an organized manner until 1855. A United States Navy Officer,

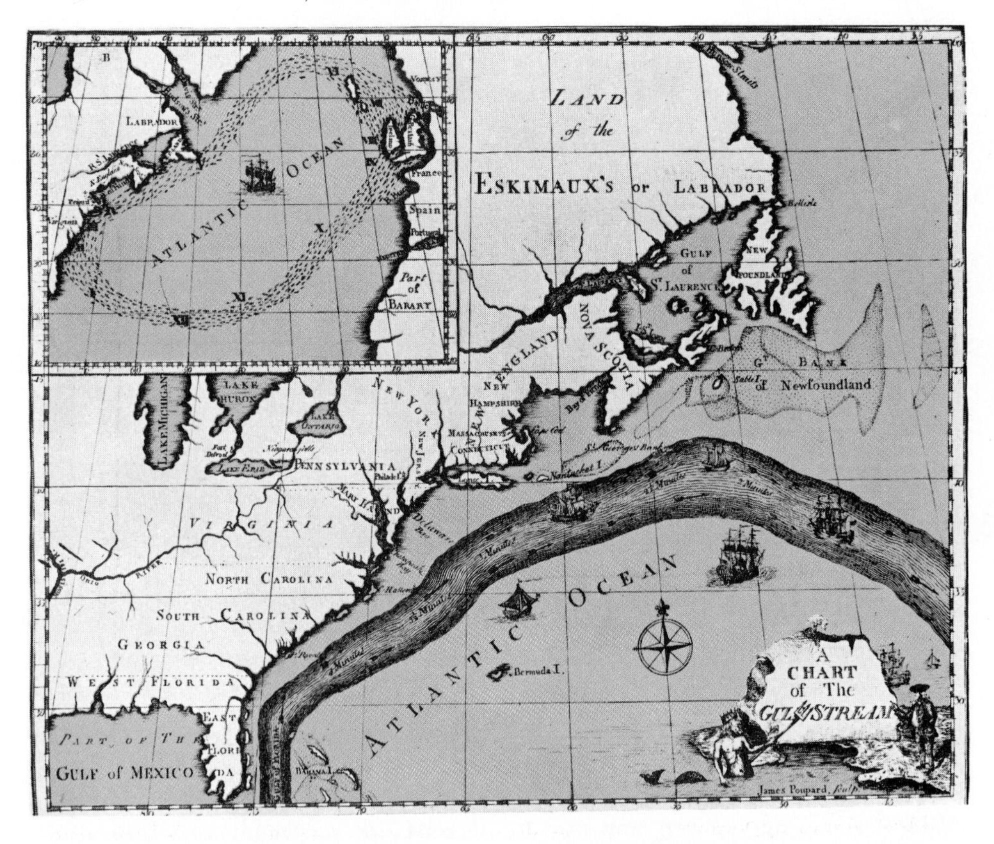

Figure 5-12 *Captain Folger's chart of the Gulf Stream.*

Matthew Fontaine Maury, gathered and published data on winds and currents in that year.

Certain patterns in ocean currents became obvious from Maury's studies. You can see from Figure 5-13 that the currents in each of the ocean basins are similar. The water moves in large, almost circular paths north and south of the equator. These currents are strong in some places and weak in others. Currents flowing away from the equator carry tropical waters to higher latitudes. (In much the same way, winds carry heat from equatorial regions toward the poles.) Ocean currents flowing along the eastern shores of an ocean basin carry cold water from polar latitudes to the tropics. You can swim in Florida waters that have temperatures warmer than 30°C. At the same latitude off Baja California, the sea temperature is likely to be 15°C. The difference is caused by the sources of the waters along the shores of the two peninsulas.

The sun is the basic source of energy for ocean currents. It is the wind that turns the sun's radiant energy into the kinetic energy of the currents. As winds blow over the sea, they exert a force on the water, just as they exert forces on trees, houses, or blades of grass on land. The water is pushed ahead of the wind in the same direction the wind blows.

There are two important winds that supply energy to the large currents of the ocean. These are the trade winds, which blow from the east on either side of the equator, and the westerlies in the mid-latitudes. Of the two, the trade winds are more constant. They pile up water on the west sides of the oceans, that is, against

Figure 5-13 *The general pattern of surface currents in the world's oceans. How does the Northern Hemisphere circulation compare with that of the Southern Hemisphere?*

the east coast of South America in the Atlantic Ocean and against the Philippine and Solomon Islands in the Pacific. Water does not pile up well, however. It flows downhill and escapes in large currents moving away from the equator. The Gulf Stream in the Atlantic Ocean and the Kuroshio Current in the Pacific Ocean are caused by this piling up of water, aided by the Coriolis effect.

Even though these currents move huge volumes of water, they flow slowly compared to the wind or to large rivers such as the Mississippi and the Amazon. In mid-ocean the speed of a current is usually less than 2 km per hour. Oceanographers and sailors sometimes find currents faster than this. When they do, it is usually in a narrow strait such as the one between Florida and the Bahama Islands.

5-6 Investigating Currents

Wind causes the currents at the surface of the sea, but differences in water density cause circulation in the deep ocean.

MATERIALS

artificial seawater solution, two 100-ml beakers, food coloring, heat source, hot water, ice, two large test tubes, thermometer (−10°C to 110°C)

When you have completed this investigation, you will have seen some of the factors that affect the density of seawater, and how density differences cause water to move in the deep ocean.

PROCEDURE

Your teacher will give you a sample of artificial seawater. Without adding anything to your sample, change its density. Keep a record of the different methods you use and of the evidence that you actually have made a more or less dense solution.

DISCUSSION

1. Identify some ways of determining that the densities of the sample seawater have actually changed.
2. What are some of the ways that the density of the samples are changed without the addition of any other material?
3. What processes in nature could cause differences in the density of seawater?

PROCEDURE

Set up the equipment as shown in Figure 5-14. Your teacher will provide you with three solutions of different salinities. This will be your artificial seawater. Add a drop or two of food coloring to each solution. Pour a test tubeful of one of the samples of "seawater" into the large sloping tube filled with fresh water. Measure the rate at which the salt solution travels down the tube.

Use the other two solutions, along with your methods of changing "seawater" density (from the first part of this investigation) to investigate what effect these factors have on subsurface ocean currents. Refill the sloping tube with fresh water each time you repeat the testing procedure.

DISCUSSION

1. What factors had to remain the same in order for your results of each test to be of any significance?
2. If you observed evidence of currents, what was the evidence and how did the currents behave?
3. What caused the currents?
4. Which of the factors tested appeared to have the greatest influence on creating the currents? Would your answer apply equally well to the earth's oceans? Explain.
5. How could these currents, which occur under the surface of the water, have any effect on the currents that occur at or near the surface of the earth's oceans?

5-7 Density Differences and Deep Currents

Surface currents can be measured directly in much the same way that winds are measured. Buoys with current-measuring instruments are anchored in the ocean and left for months at a time. The speed and direction of the currents are recorded on magnetic tape for later study by oceanographers. Deep currents, on the other hand, are not easily measured. Buoys and current meters have been anchored below the sea surface, but recovering them is tricky. As a result, there are only a few hundred measurements of deep currents. This contrasts with the many thousands of measurements of currents of the upper ocean.

Through the middle of the twentieth century it was thought that the depths of the ocean were still and lifeless. Oceanographers have now taken pictures of, and even captured, animals in the deepest parts of the ocean. This means that there must be some way for water rich in oxygen to reach these great depths.

Less is known about deep ocean currents than those at the surface of the ocean. They cannot be seen and they are difficult to study. In 1951, a subsurface current at the equator was discovered in the Pacific Ocean and named the Cromwell Current. A deep-flowing stream 5000 m deep in the Antarctic Ocean was

Figure 5-14 *The setup for investigating different water densities.*

MATERIALS

artificial seawater, two beakers (100- or 250-ml), meterstick, plastic column, ring stand and clamp, sponge, two or three large test tubes, one or two thermometers (−10°C to 50°C)

measured for the first time in 1971. In 1975, an oceanographer, returning to the Cromwell Current, learned that subsurface waves with lengths of 1000 km moved through the ocean, causing great changes in the speed of the current. Then, in 1979–1980, several special subsurface buoys were placed in the tropical Pacific Ocean to measure deep water movements. A wide variation in current speeds and directions was encountered. This variation showed that deep currents have "bursts" of speeds up to 8 km per hour for two or three days at a time.

Deep currents flow because of differences in density from one place in the ocean to another. The density of seawater depends on temperature and salinity. The transfer of matter and energy between the air and the sea changes both temperature and salinity. Oceanographers have learned the origin of the bottom and deep waters of the sea by carefully measuring the temperature and salinity at many thousand places.

The densest water in the ocean is formed around Antarctica. Freezing leaves the salt in the unfrozen water. When seawater freezes at the surface, the salinity of the water below greatly increases. The high salinity and low temperatures produce extremely dense water that sinks to the bottom. This Antarctic water flows beneath the slightly less dense North Atlantic water, as shown in Figure 5-15. Oceanographers think that the water at the bottom of all the great ocean basins of the earth is formed in the two polar regions near Greenland and Antarctica.

The Gulf Stream carries relatively warm, salty water into the northern Atlantic Ocean off the eastern coast of Greenland. There the water cools and mixes with cold waters flowing south from the Arctic Ocean. The result is an increase in the density of the Gulf Stream water. This causes it to sink toward the bottom of the ocean. The newly formed current moves southward into the deep ocean basins at a speed of about 20 kilometers per year. Figure 5-15 shows the general flow of the subsurface waters of the Atlantic Ocean.

As the deep water moves gradually toward the equator, it mixes with the less salty water above it and becomes less dense. As the water flows south past the equator, what was once deep water moves upward toward the sea surface. Finally, near Antarctica some of the mixture rises to become part of the Ant-

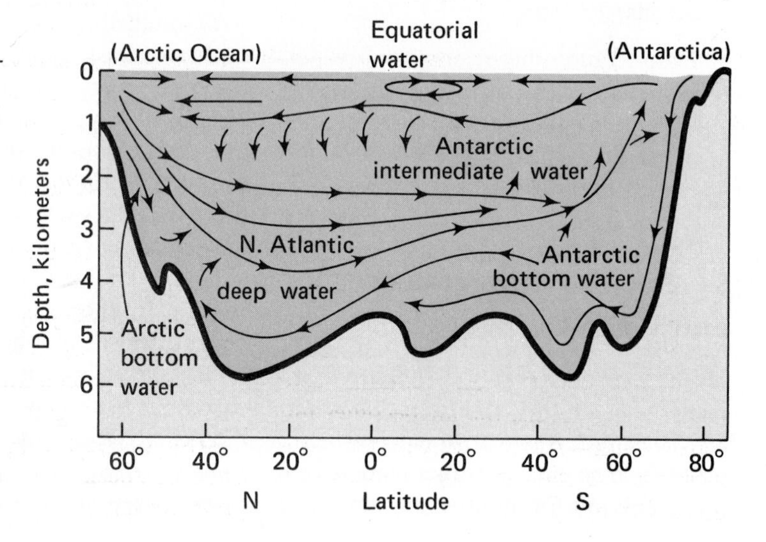

Figure 5-15 *A north-south cross section of the Atlantic Ocean showing the movement of currents.*

arctic Intermediate Current. It then flows northward. In this way, water that was once at the bottom of the North Atlantic returns to the North Atlantic Ocean with surface currents.

You have seen that the upper part of the North Atlantic deep water mixes with the water above it. The lower portion mixes with the bottom water that was formed in the Antarctic Ocean. The resulting current flows around Antarctica much like the surface currents shown in Figure 5-13. Some of it eventually moves into the Pacific and Indian oceans. In 1971, oceanographers found that the mid-level water in the Pacific Ocean is a mixture of water from two areas. Part of it comes from the North Atlantic, near Norway, and part comes from the Antarctic Ocean. In 1974, the same scientists learned that this water flows through the deep channel east of Samoa to reach the North Pacific. This is striking evidence of the continuous mixing of water throughout the ocean.

ACTION

The North Atlantic Deep Current flows from the coast of Greenland downward through the Atlantic Ocean at about 20 km per year, and the water begins to rise at about 30° south latitude. How long does it take the water to travel this distance? If a typical ocean has been on the earth's surface for 4 billion years, how many such trips could an individual water particle make? Would this imply that the ocean is well or poorly mixed?

Thought and Discussion

1. Where do waves obtain their energy?
2. How does wave motion change when waves enter shallow water?
3. How does the energy from the sun generate ocean currents?
4. What is the source of deep water currents in the ocean?

Return to Climate

5-8 Oceans and Climates

Remember the imaginary continent? Now that you have studied oceans, you are in a position to say more about its climate. Think about the effects that the ocean can have on the climate of an island or a continent.

Oceans store heat during the summer and release it slowly in the winter. Warm currents flow toward the poles along the eastern sides of continents. Cold air moving over the warm Gulf Stream, for example, will be warmed at its lower level. Because the air remains cold in the upper levels, the air becomes unstable as it is warmed from below. Cloudiness and rainfall develop in the moist, unstable air.

Cold ocean currents flow toward the equator along the western shores of continents. Air moving toward the coast of California is cooled by the cold water offshore. Since it is cooled from below, the air is stable and tends to resist vertical motions. How does this explain the famous fogs of San Francisco? Review Figure 3-24. Notice that the great deserts of the world tend to be located on the eastern sides of the oceans in the subtropics.

What would you expect to happen when a relatively warm current passes by an island at a high latitude? (Why is the "Emerald Isle" green?)

Why do you suppose the southern coast of Alaska has a great amount of precipitation?

You know that the Sierra Nevada in California is a mountain range with forests, lakes, and rivers. California is at a latitude where you would expect frequent westerly winds. What, then, is the source of much of the Sierra Nevada precipitation?

Why are there "rain forests" in the state of Washington?

If you live in the eastern half of the United States, why is the back side (western part) of a high pressure air circulation usually moist? On the other hand, why is the northerly flow around a high (from Canada) usually drier?

Why is a middle latitude seacoast city probably 10° warmer in winter and 10° cooler in summer than its suburbs 25 km inland?

Remember, the ocean influences the total environment of the earth's surface.

MATERIALS

colored pencils or crayons, imaginary continent maps

Figure 5-16 A map of the imaginary continent, to use in Investigation 5-9.

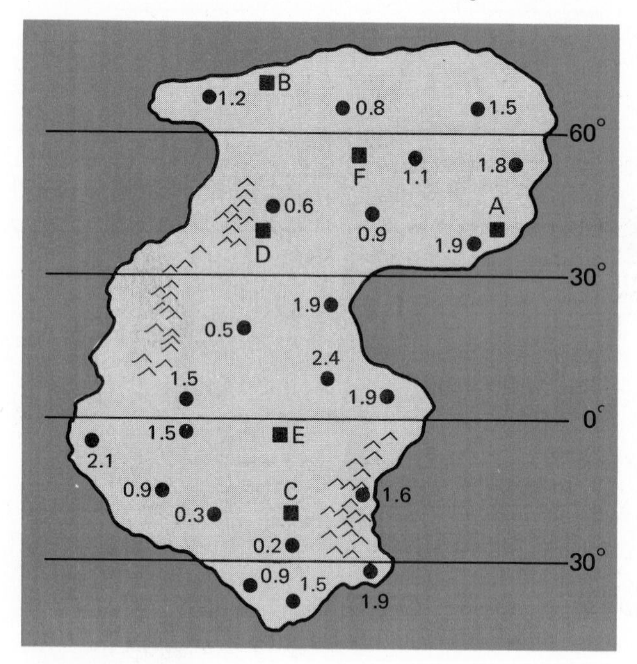

5-9 Investigating Climates of an Imaginary Continent

You have investigated a number of factors in Chapters 2, 3, 4, and in this one that may have an effect on the climate of an area.

When you have completed this investigation, you will have seen how two or more of these factors might act together to produce a particular climate at a given geographic location.

You are aware that many areas of the world have climates that are quite different from the one in which you live. Examples of such climates are deserts, rain forests, and ice caps. Which influencing factors are most important in causing such differing climates? Can you predict what the climate would be like if you are given the important information?

PROCEDURE

To start with, you need an enlarged copy of Figure 5-16, which is a map of an imaginary continent. There are a number of ways in which this continent is very similar to actual continents. The numbers found by each point in Figure 5-16 represent ratios found by dividing the annual precipitation by evapotranspiration for each location. We can call this ratio the P/ET. (Refer to sections 4-6 and 4-7 for review.) You will use the ratios and the lettered locations later in the investigation.

You can work alone, or with a group. Consider the major climate influencing factors (latitude, altitude, land forms, nearness to large bodies of water, surface winds, and ocean currents). It might help to draw ocean currents and average wind directions on your map. Outline on it as complete a pattern of differing climates as you can. Reach an agreement of opinion within your group as to what the various climates might be, *before* going on to the next part of this investigation.

DISCUSSION

1. Defend your group's (or your) prediction as to what the climate patterns of the imaginary continent might be, in a class discussion. Which of the factors are the most significant in determining what the climate of an area will be?

PROCEDURE

Now, consider the P/ET values. If the P/ET value is above 1.0, how does the evapotranspiration compare with the precipitation? In that case, is the climate relatively moist or relatively dry? Compare the P/ET values with the climate areas of your map.

DISCUSSION

1. How close were your predictions of dry and moist (or humid) areas? If any of your predictions were wrong, can you tell why?

MATERIALS

colored pencils or crayons, imaginary continent maps

PROCEDURE

Figure 5-17 consists of a climate graph for each of six different locations. A climate graph combines a line graph of average monthly temperatures with a bar graph of average monthly precipitation totals for any location. On the map of the imaginary continent, each of the six points lettered A through F has a climate that corresponds to one of the six climate graphs.

Choose the climate graph in Figure 5-17 that describes the climate at each of the lettered points on the map. Reach an agreement of opinion within your group.

MATERIALS

model climate graph form (for making climate graph from local data)

Figure 5-17 *Climate graphs of six areas to use with the imaginary continent map.*

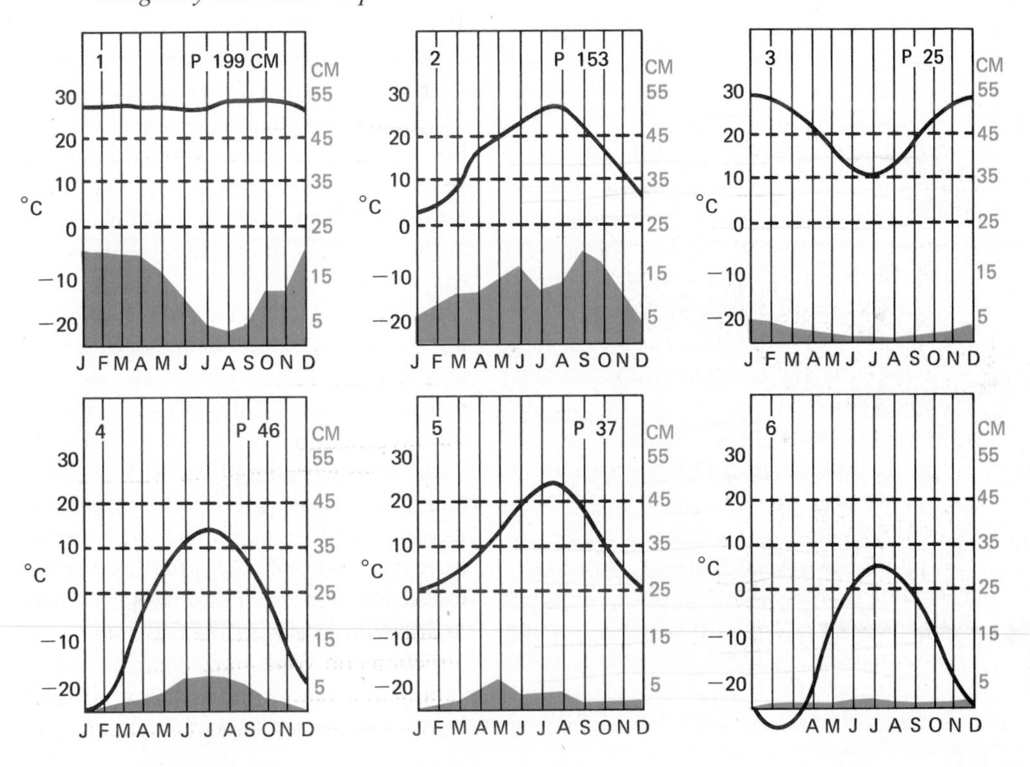

DISCUSSION

1. Each of the climate graphs that you used in this investigation represents an actual location on the earth. On a map of the world, identify one or more places for each climate graph that you think could be the actual location. Explain the influence at each place of the significant factors that have an effect on climate.

Thought and Discussion

1. Which continent has a climate pattern most like that of the model continent you have just worked with? Which continent has a completely different pattern? Why is it different?
2. What climate changes would probably occur in northwest Europe if the Gulf Stream stopped flowing?

Unsolved Problems

Motion in the deep ocean was thought to be slow and gentle until a number of measurements were taken over periods of two or three weeks in 1970 and 1971. It appears that deep water may move in "fronts" and "storms," much like air masses in the atmosphere. Many more fronts were measured in 1975, in both the Atlantic and Pacific oceans. Then, in 1980–81, extensive experiments were conducted throughout the Pacific, by scientists from eight oceanographic institutions. The "fronts" and "storms" are still puzzling, because the source of the energy for these ocean motions has yet to be learned.

Water becomes denser because of evaporation and freezing. It sinks to the bottom of the ocean to form the deep and bottom currents. If water is sinking in these great volumes, somewhere there must be equal volumes rising toward the surface. So far, no one has discovered these rising masses of water. Or does the water rise to the surface in many small masses?

The most studied of all ocean currents, the Gulf Stream, still holds many mysteries beneath its tropical water. Most ocean current charts show it as a broad stream, as you saw from the chart prepared by Captain Folger. Recent research has indicated, however, that it is actually a narrow, winding stream with many loops, as shown in Figure 5-18. For reasons that are not yet understood, the position and speed of the Gulf Stream change from month to month.

A perplexing problem became the subject of much research in the early 1970's and is still being studied in the 1980's. New methods showed the presence of heavy metals and materials manufactured by human beings in many marine animals and in ocean waters. Before the mid-1960's, marine scientists knew that heavy metals must be distributed in the ocean, but they had no ways of measuring the small quantities. The new techniques showed that metals (such as lead and mercury) occurred in certain marine animals in amounts previously unsuspected (0.5–2.0 parts per million, or 0.002 percent). Further, chemists noted that compounds made by people, such as chlorinated hydrocarbons (DDT) and polychlorinated biphenyls (PCB), were in the tissues of animals in the ocean as well as on land.

In 1971, a large effort was started by American and West European scientists to collect and analyze the water and animals in the Atlantic and Pacific Oceans and their adjacent seas. Petroleum products were included in the study, along with the heavy metals and the compounds made by people. Results through 1980 indicated that crude oil dumped into the sea poses no permanent threat to the ocean. Although damage to beaches, sea birds, and marine organisms on shore may be great for a while, the ocean bottom recovers within two years. Animals and plants come back,

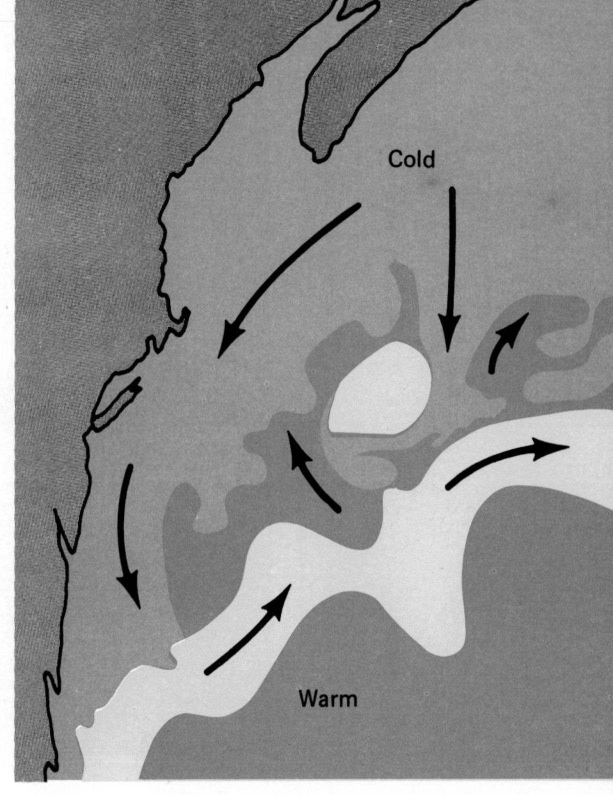

Figure 5-18 *How does this photograph of meanders in the Gulf Stream compare with Folger's map in Figure 5-12?*

and swimming organisms show no effects at all. Also, the amount of mercury in fish away from the coasts is minor and apparently normal.

Chapter Summary

The quality that makes the earth unique in the solar system is the large amounts of water on its surface. This water exists in all three phases: solid, liquid, and gas. Individual water molecules spend most of their time in the great world-wide ocean basin. As it goes through the water cycle, water is continually changing from one phase to another, using and releasing energy.

Water is in constant motion, carrying materials from the land to the sea. The salinity of the ocean varies from place to place, but the most common salts always occur in the same ratios.

Waves get their energy from the winds blowing across the surface of the water. The same transfer of energy from the winds causes ocean surface currents. These currents, like the circulation in the atmosphere, carry heat from equatorial to polar regions and thereby help to determine climates. The density of surface waters increases with cooling in high latitudes. This denser water sinks and produces deep ocean currents.

OCEANOGRAPHER

Oceanographers (*oh shuhn AHG ruh furz*) study the movements, properties, and life of oceans. Some oceanographers work on research ships, making tests and observations about ocean tides and currents. Others study the underwater mountains or the sediment on the sea floor. They use special instruments to measure and record their findings. Sometimes they use underwater cameras to photograph the ocean floor or the plants and animals that live there. There are many areas of specialization in oceanography, and new ones are opening up. New areas of interest include discovering methods of mining resources from the sea floor, studying the oceans for new sources of food, and trying to develop ways of providing fresh water from the sea.

Aquatic research stations and marine laboratories employ many oceanographers. Others work for federal and state or provincial agencies, for private industrial firms, or in the science departments of colleges and universities. An advanced degree in oceanography is required for high-level jobs. There is a lot of competition for jobs in oceanography among people with less education. People with bachelor's degrees may be research assistants.

There are also job opportunities in oceanography for technicians. Oceanographic technicians assist in many phases of the work. Instrumentation technology is a fast-growing field that is important to oceanography, since oceanographic research depends on a variety of complicated equipment. Preparation for these jobs usually consists of two years' technical education plus on-the-job training.

Questions and Problems

A

1. What is salinity? How is it usually expressed?
2. In what ways does seawater differ from fresh water?
3. Discuss two ways in which energy leaves the ocean.
4. What is meant by the wavelength and period of water waves?
5. Where is most solar energy absorbed in the sea?
6. What causes ocean surface currents?
7. What causes currents in the depths of the ocean?

B

1. How do salts get into the sea?
2. About how much seawater must be evaporated to obtain 500 g of sea salts?
3. An oceanographer testing a sample of mid-ocean surface water finds almost twice as many phosphate as nitrate ions. Of what significance is this discovery?
4. Name several things that are transferred from the air to the sea or from the sea to the air.
5. What is the speed of travel of waves with period 6 sec and wavelength 56 m?
6. If one-half the radiation striking the sea surface is absorbed in the first meter, and one-half the radiation that passes through the first meter is absorbed in the second meter, and so on, what fraction striking the sea surface reaches a depth of 5 m? 10 m?
7. What is the relation between salinity and density of seawater?

C

1. How much seawater is needed to yield 500 g of magnesium?
2. Trace the flow of energy from the sun to a wave breaking on a beach.
3. What are some differences between deep and shallow water waves?
4. Are Northern Hemisphere south-flowing currents warm or cold currents? Describe the climatic effect they may have as they move past adjacent land masses.
5. How would the speed of currents in the ocean compare with that of atmospheric winds? Explain.
6. What determines seawater density?
7. How can the hypothesis that the densest water in the ocean all comes from around Antarctica be tested?
8. Construct simple climate graphs for the following pairs of locations:
   (a) St. Louis, Missouri, and San Francisco, California (both 38°N)
   (b) Nome, Alaska, and Reykjavik, Iceland (both 64°N)
   (c) the North Pole (90°N) and the South Pole (90°S).

   Identify and discuss reasons for the differences and similarities in the climate graphs for the locations within each pair.

Suggested Readings

Barton, Robert. *The Oceans.* New York: Facts on File, 1980.

Wylie, Francis E. *Tides and the Pull of the Moon.* Brattleboro, Vermont: Stephen Greene Press, 1979.

unit III
The Rock Cycle

6 Earth Materials

A VOLCANO *like the one on the opposite page is the only place where you can see solid rock form before your eyes. Lava flows that emerge from it will solidify in a few hours once they stop moving. And lumps of lava that are thrown out of the crater cool and solidify even before they hit the ground.*

Lava is a mixture of rock material and gases at a temperature generally above 1000° C. The most abundant gas is water vapor, which creates great clouds (commonly called smoke) over volcanoes. The molten material in volcanoes is believed to come directly to the earth's surface from depths of 10 to 100 km. So we see only the final stage of this rock-making process.

In this chapter you will investigate the nature of the materials of which all rocks are made. You will discover that the outer layer of the earth is really simple in chemical composition. Only eight elements are present in amounts greater than one percent. These elements combine to form the dozen or so compounds (called **minerals***) that make up more than 95 percent of the mass of the visible solid earth. The basic materials of the solid earth are called* **rocks***. The processes by which these rocks form will be investigated in later chapters.*

Finally you will read about the great variety of natural resources that come directly or indirectly from the earth. Mineral resources, especially ores of metals, are produced as part of the processes that form rocks.

AFTER COMPLETING THIS CHAPTER, YOU SHOULD BE ABLE TO:

1. **identify the important interfaces between earth materials.**
2. **distinguish between: a.) atom and molecule, b.) element and compound, c.) compound and structural units such as SiO_4, CO_3, SO_4, d.) crystal and crystalline structure, and e.) mineral and rock.**
3. **explain the relationships between the eight most abundant elements and the predominant minerals in the earth's crust.**
4. **account for the close relationship between specific gravity, hardness, cleavage, and streak and the composition, atomic structure, and bonding in minerals.**
5. **explain why deposits of most metals are rare and usually small compared to deposits of iron and aluminum ore.**

Some Basic Ideas

6-1 Types of Earth Materials

If you were to start toward the earth from space, you might think of yourself as "zooming in" as a T.V. camera often does. Imagine you are taking a space-to-earth trip and heading for the Superstition Mountains, east of Phoenix, Arizona. As you start, the earth looks as it does in Figure 6-1. In this photo taken from space you can see the colors of three different kinds of earth materials. Patches of white clouds swirl above the surface of the earth. Openings between the clouds reveal blue patches of ocean, and the land looks brown or red.

We can classify or group the earth materials into three basic types. The word used to describe the rocks in the solid outer crust is lithosphere. (The Greek word *lithos* means stone.) The water in the oceans, lakes, rivers, and ice-fields is the hydrosphere. (*Hydro* is Greek for water.) The air moving in currents around the earth forms the atmosphere. (*Atmos* is the Latin word for vapor.)

Any place where two or three of the spheres meet is called an **interface.** Try to identify the interfaces in Figure 6-2. Interfaces are not always distinct. The atmosphere may contain both water and solid particles. Rocks and soils contain water and air, and streams and oceans can contain air and solid materials.

6-2 The Shells of the Earth

Having landed near the Superstition Mountains (Figure 6-3a) you can walk over to the mountains to see what they are made of. Even from a distance it

Figure 6-1 *A view of the earth from space.*

Figure 6-2 *What interfaces can you see in this picture?*

a.

b.

c.

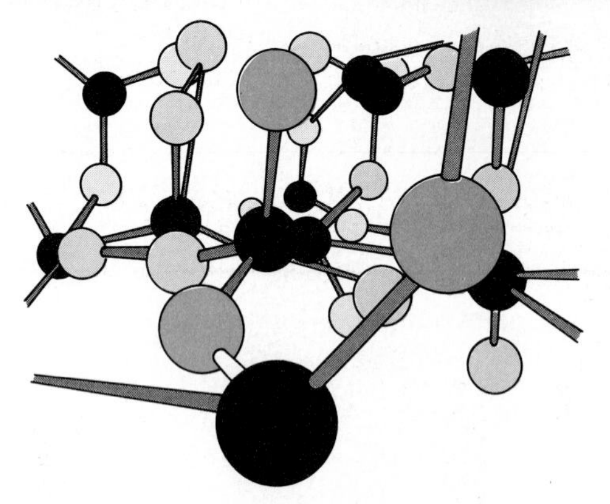

d.

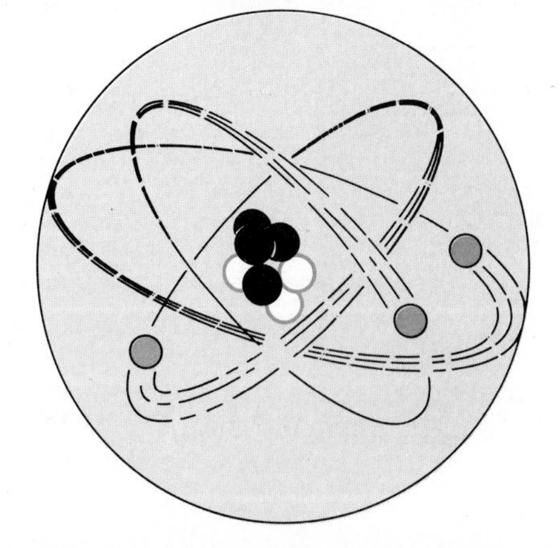

e.

Figure 6-3 *Focusing on earth materials at different scales.* **a.** *A ground view of the Superstition Mountains.* **b.** *A rock specimen from the Superstition Mountains.* **c.** *A photo taken through a microscope of a thin section of rock, showing mineral grains.* **d.** *A model of the internal structure of a mineral in the rock—atoms bonded together into a compound.* **e.** *A model of an atom.*

should be obvious that they are made of rocks. When you get there you can pick up a loose piece that looks like the one shown in Figure 6-3b. What kind of rock is it? You may have seen many different kinds of rocks. Have you ever examined rocks carefully? Do they all seem to be the same kind?

When you begin to study rocks, you may ask yourself a number of questions. For example, why are there so many kinds of rocks and how did they form? To answer these questions you must know a little about what the earth is like between the surface of the ground and the earth's center.

The solid earth consists of a number of layers or shells. Review Figure 1-24. The outermost shell, the one you have already learned about, is called the lithosphere. The upper part of the lithosphere, including the rocks we see at and near the earth's surface, has long been called the crust. The crust is about 5 to 10 km thick under most oceans and 24 to 60 km thick under the continents. The lithosphere below the crust is solid and consists mainly of materials more dense than surface rocks. The total thickness of the lithosphere varies from 100 to 200 km. Below the lithosphere is a zone of material which (because of the high temperatures at that depth) seems to be so weak that it flows like thick tar. This zone is called the **asthenosphere** (weak sphere). Below the asthenosphere is a very thick shell extending to a depth of 2900 km. It behaves like something hard or solid. This is called the **mantle.** Below the mantle is the **core,** which extends to the center of the earth 6371 km below the surface. The outer core, from 2900 to 5150 km, behaves like a liquid, and the inner core behaves like a solid.

The rock you have picked up near the Superstition Mountains comes from which layer of the solid earth?

6-3 Investigating Rocks and Minerals

After completing this investigation, you should be able to list the characteristics that are important in distinguishing one rock from another. You will determine how many materials occur in amounts greater than ten percent in the rocks you examine.

MATERIALS
magnifier, rock samples (including crushed rock sample), teasing needle

PROCEDURE

To find out how rocks differ from each other, take the ones your teacher has set out and describe each one as carefully as you can. Make a separate list of words that describe the characteristics of each rock.

In the next part of the investigation you will examine a sample of crushed rock. Use a magnifier and a teasing needle to separate the crushed material into piles of similar particles. Now make a list of descriptive terms for each pile you made.

DISCUSSION

1. Are some terms better than others for identifying rock properties?
2. Compare the lists for crushed materials with the ones you made for the rocks. Which was easier to describe? Why?

6-4 Minerals

Your study of the rocks in Section 6-3 should have shown you that at least some rocks are made of several different materials. One of the ways that geologists identify rocks is to study the different materials of which they are made. Look at Figure 6-3c.

The solid distinct materials that make up the rocks in the earth's crust are the minerals. Some minerals occur in the form of crystals. Crystals are solids with regular geometric shapes and smooth flat surfaces called **faces.** The faces have a definite pattern (see Figure 6-4). Sometimes the crystal form does not show because the mineral is made up of many small irregular grains. When many grains grow very close together the crystal faces cannot form.

The mineral cinnabar (*SIHN uh bahr*) can be separated into two other substances: mercury and sulfur (see Figure 6-5). The mercury and sulfur cannot ordinarily be broken down. They are elements. Cinnabar is called a compound because it is made up of two elements. Most minerals are compounds (see Figure 6-3d), but some, such as gold, silver, and diamond, contain only one element.

Figure 6-4 *Two crystals of a substance can look different, but the angles between the faces are always the same.*

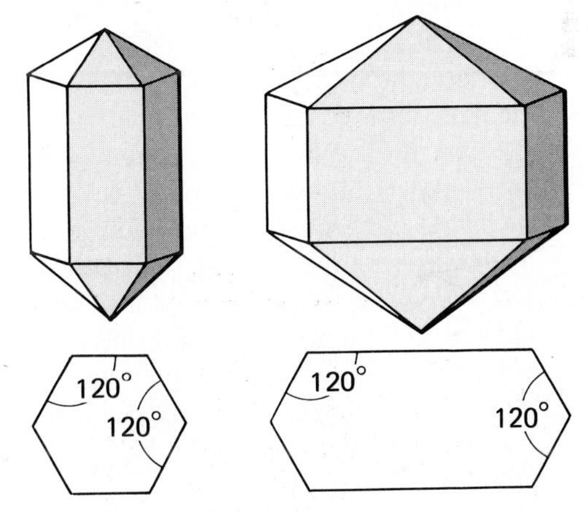

a.

b.

c.

Figure 6-5 **a.** *The compound cinnabar.* **b.** *The element sulfur.* **c.** *The element mercury.*

Whether they are compounds or elements, they are called minerals if they were formed in or on the earth by nature. The same chemical compounds or elements are not called minerals if they are made by people.

6-5 Elements

The term "element" has a long history. Over 2000 years ago the Greek philosopher Empedocles (*ehm PEHD uh kleez*) taught that matter consists of the four "elements" earth, air, fire, and water. He believed that all materials were made by combining these four elements.

It had been known for a long time that heating a certain soft red rock with charcoal would change it to a much harder gray material called iron. This change was thought to be a rearrangement of the elements of fire, water, earth, and air in the red rock. Why, then, asked some early experimenters, couldn't the gray metal be changed into shiny yellow gold? Thus was born the "science" of alchemy (*AL kuh mee*), the search for a way to change common earth materials into valuable gold. Of course, no one was able to find a way. Suppose the discovery had been made. How valuable would gold be now?

Alchemy was practiced for many centuries, partly as a swindle. The alchemists did make many valuable scientific discoveries and were the fore-runners of our modern-day chemists. They realized, for example, that certain substances like iron, mercury, gold, and sulfur could not be further broken down or changed into other elements by any methods they knew. Alchemists also decided that since they were able to get iron from certain red rocks, the iron must have been in the rocks in the first place.

The next step in our "zoom-in" from space (Figure 6-3e) will let us take a very short look at the insides of atoms. The alchemists thought atoms were the smallest particles into which matter could be divided. However, it wasn't until 1808 that the English scientist John Dalton laid the foundation for the modern atomic theory. He presented four ideas to explain the differences between elements and to account for the behavior of gases.

1. All substances are composed of small, solid indestructible particles called atoms.
2. The atoms of a given substance have the same size and shape.
3. The atom is the smallest particle of an element that enters into chemical changes.

4. Compounds are formed by combinations of the atoms of two or more elements.

John Dalton believed that the atom was the smallest particle of matter and that it could not be further broken down. However, you know John Dalton was wrong. The atom *can* be further broken down, under certain conditions. And when it is, some of the particles that formed the atom are found to be electrically charged. Electrons have negative charges and protons have positive charges.

You have already learned that most minerals are compounds composed of two or more elements. For example, the mineral galena (*guh LEE nuh*) is composed of lead and sulfur. The mineral quartz is composed of silicon and oxygen. A ruby is made of aluminum and oxygen. Electrical forces hold these atoms together. The atoms are so close to each other that the paths of some of their electrons overlap. These charged particles are "shared" by the atoms and bond them together. The nature of atomic bonds is not easy to understand or to explain briefly. The idea of overlapping paths and shared electrons should give you a rough idea of bonding.

See Appendix D for a list of known elements and the symbols that scientists use for them. All atoms of a particular element may not be exactly the same. They will all have the same number of electrons and protons, but the number of neutrons can vary. For example, if a neutron is added to a hydrogen atom, deuterium is formed. It behaves chemically like ordinary hydrogen, but it is almost twice as heavy. These atoms of any element that differ only in mass are called **isotopes** (*EYE suh tohps*). Scientists have discovered that there are isotopes of many elements in nature. Some of the isotopes of uranium, strontium, and thorium are unstable and are known as radioactive isotopes. They "decay" or break down to form other isotopes. Radioactive isotopes have been used to determine the ages of rocks, to destroy cancer cells, and to make nuclear bombs. The alchemists' dream, gold made from common material, has not yet come true, but scientists have succeeded in making some elements from others.

Thought and Discussion

1. The near surface shells of the earth are called spheres. Can you suggest names for the mantle and core that will also end in "sphere"?
2. How would you define "rock"?

Rocks and Minerals

6-6 Rock-Forming Minerals

More than a hundred chemical elements are known. Eight of these make up more than 98 percent of the crust by weight. We may mention a few others, but many with which you are familiar are found in such small amounts that we can ignore them. Study the table in Figure 6-6.

What can you learn from this table without remembering all the numbers? The eight elements together make up more than 96 percent of the crust in terms of volume, mass, or numbers of atoms. Oxygen is the most abundant element by mass. It composes 60 percent of all atoms in the crust and 94 percent of the total volume. It may be said that

Element	Symbol	% of Mass	% of Atoms	% of Volume
oxygen	O	46.71	60.50	94.24
silicon	Si	27.69	20.50	0.51
aluminum	Al	8.07	6.20	0.44
iron	Fe	5.05	1.90	0.37
calcium	Ca	3.65	1.90	1.04
sodium	Na	2.75	2.50	1.21
potassium	K	2.58	1.40	1.85
magnesium	Mg	2.08	1.80	0.27
		98.58	96.70	99.93

Figure 6-6 *Common elements in the earth's crust.*

the crust is a pile of oxygen atoms with atoms of all other elements in the holes between them. Silicon is the second most abundant element by mass and number of atoms, but is very low in volume. This means that the most common atoms between the oxygen atoms must be those of silicon. It is also clear that an oxygen atom must be much larger than a silicon atom. You should also expect that nearly all chemical compounds in the crust will be composed of these eight elements. The most abundant atoms in these compounds will be those of oxygen and silicon.

In solid compounds all of the atoms in the substance are arranged in a fixed pattern. We should expect then to find that one pattern of oxygen and silicon atoms is most common. This is exactly what we do find. The neatest and most common pattern is one silicon atom surrounded by four oxygen atoms. Each of the oxygen atoms is touching the other three, and all four of them are touching the silicon atom.

MATERIALS
four balls of the same size, one smaller ball or clay

ACTION

To see what this means, do an experiment. Take four balls of the same size—ping-pong, baseballs, or basketballs will do. Place three of them in a triangular position so that each touches the other two. Now place the fourth ball on top of the three over the hole between them. You should now have the only possible arrangement in which each of the four will touch the other three. Now find a smaller ball that will just fit in the hollow space between the four. If this ball is too big it will prevent all four of the big ones from touching each other. If it is too small it will

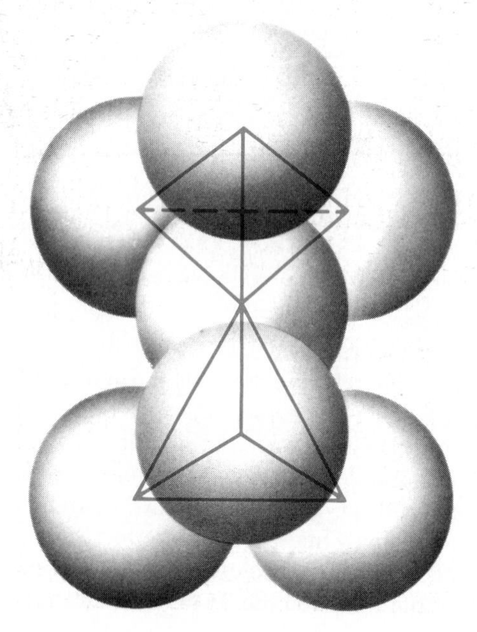

Figure 6-7 *Silicon-oxygen groups can join by sharing oxygen atoms.*

fall through a hole between three of the large ones. You can make one the right size by molding some paper or clay into a sphere.

You have just made a model of the most common pattern of atoms in the earth's crust! Minerals that have this pattern of oxygen and silicon atoms as part of their atomic arrangement are called **silicates.** Silicates compose more than 96 percent of the crust. There are many silicate minerals, but you need to remember only a half-dozen important types. These types differ from each other in two ways—the abundance and arrangement of silicon-oxygen groups, and the other elements present. See Figures 6-7 and 6-8.

In Figure 6-9 the silicate types are indicated by mineral names. These names represent mineral groups that are closely related in chemical composition and physical properties.

For convenience, the rock forming silicates are divided into two groups. **Sialic** (*sy AL ihk*) **silicates** are those rich in silicon and aluminum. These include the feldspars, quartz, and white mica. Together they compose about 85 percent of the crust (Figure 6-9). **Simatic silicates** are those richer in magnesium and silicon. They include pyroxene (*pih RAHK seen*), amphibole (*AM fih bohl*), olivine, and black mica. These are dark colored (black, dark gray, or greenish) and together make up less than 15 percent of the crust. Remember that the silicates compose more than 96 percent of the crust and are the most abundant minerals in most rocks. (Review Figure 1-24b and d. Where in the crust are sialic silicates common?)

Figure 6-8 *In these common silicate minerals, the silicon-oxygen groups occur in different patterns.*

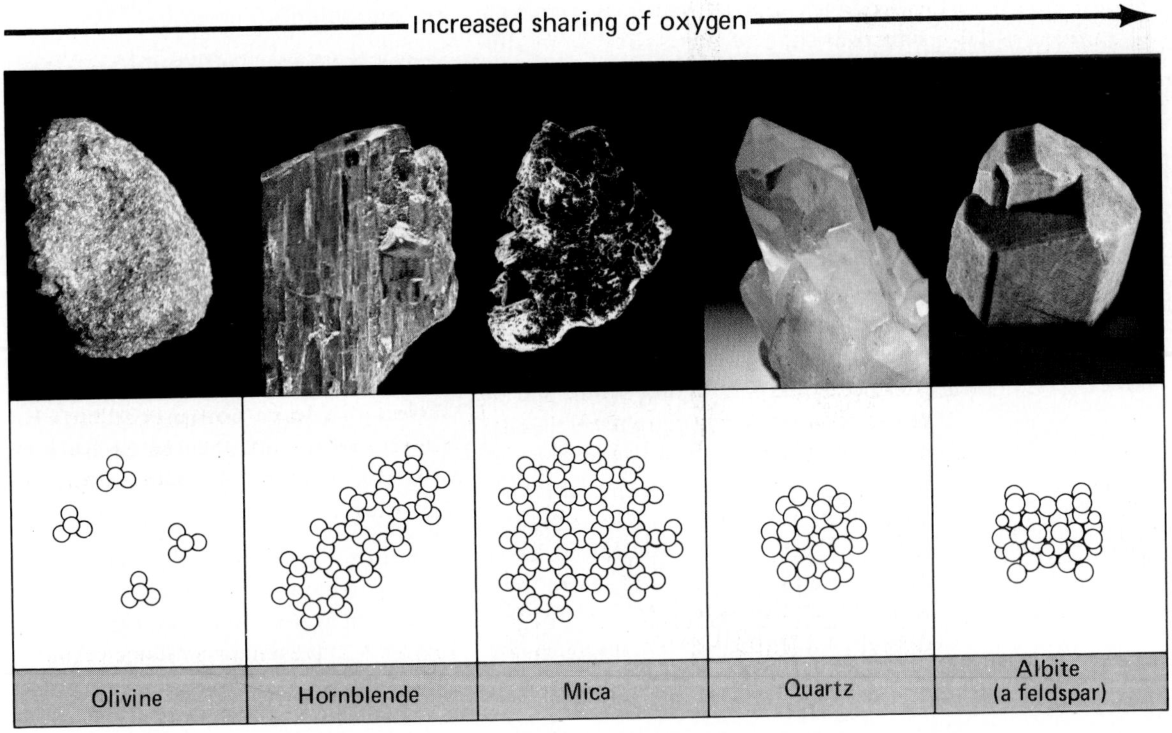

Increased sharing of oxygen

Olivine	Hornblende	Mica	Quartz	Albite (a feldspar)

Mineral	Ratio Si to O	% of Si and O in Mineral (Mass)	Other Elements present	Volume % of crust
olivine	SiO_4	65	Mg, Fe	1.5
pyroxene (augite)	SiO_3	65	Mg, Fe, Ca	4.0
amphibole (hornblende)	Si_4O_{11}	67	Mg, Fe, Ca, Al	5.0
micas: biotite (black), muscovite (white)	Si_2O_5	71	K, Al, Mg, Fe	4.0
feldspars: orthoclase, plagioclase	Si_3O_8	75	Al, Ca, Na, K	64.0
quartz	SiO_2	100	none	18.0
				96.5

Figure 6-9 *Rock-forming silicates.*

There are several important minerals that are not silicates. The more common ones include the following.

Carbonates are compounds containing an atomic unit consisting of one carbon atom surrounded by three oxygen atoms—the carbonate group. The most common carbonate is calcium carbonate, the mineral calcite ($CaCO_3$). It is the principal mineral in a rock called limestone.

Sulfates are compounds containing a unit consisting of one sulfur atom surrounded by four oxygen atoms—the sulfate group. Gypsum ($CaSO_4 \cdot 2\,H_2O$) is the most abundant rock-forming sulfate. It is the principal mineral in the rock of the same name.

Halides are compounds containing chlorine or fluorine together with sodium, potassium, or calcium. Halite ($NaCl$) is the mineral in rock salt and in ordinary table salt.

Oxides are compounds composed of oxygen and one or more other elements. The most common oxides are those of iron and aluminum. Hematite (Fe_2O_3), magnetite (Fe_3O_4), and limonite ($Fe_2O_3 \cdot H_2O$) are the principal oxides of iron. These minerals are called ore minerals. An **ore mineral** is one that has enough of a particular element in it to be worth mining. Hematite is the principal ore of iron, magnetite is the magnetic iron ore, and limonite forms a low-grade ore. The most abundant ore of aluminum is a rock called bauxite, which consists of a number of aluminum oxide minerals. For present purposes you can consider bauxite to be a mineral.

Sulfides are minerals containing one or more metals combined with sulfur. These are common ore minerals. Examples include pyrite, iron sulfide (FeS_2); galena, lead sulfide (PbS); and sphalerite, zinc sulfide (ZnS).

6-7 Investigating Mineral Properties

MATERIALS
balance, beaker (or other container for weighing an object in water), copper penny, glass, iron nail, mineral samples, scissors, unglazed porcelain

Each mineral has distinctive physical properties because of the kinds of elements present, the arrangement of the atoms, and the strength of bonding between atoms or ions.

When you have completed this investigation you should be able to identify many of the common minerals.

Specific gravity is the mass of the mineral compared to that of an equal volume of water. As water can be considered to have a density of one (1) this is really another way of saying "density." You may determine density as you did in the earlier investigation or determine specific gravity as follows: 1. Weigh the specimen in air and record the weight. 2. Weigh the specimen submerged in water and record the weight. 3. Calculate according to the equation

$$\frac{\text{weight in air}}{\text{weight in air} - \text{weight in water}} = \text{SpG}$$

The numerical result should be the same if you determine density.

Hardness is the resistance of the mineral to scratching. The hardness of minerals is expressed in a scale from 1 to 10. The following common objects can be used for testing hardness.

Object	*Hardness*
Fingernail	2.5
Copper penny	3.5
Iron nail	4.5
Scissors	5.5
Glass	6.0

Cleavage is the tendency of some minerals to break along smooth, flat, parallel surfaces. Study Figure 6-10.

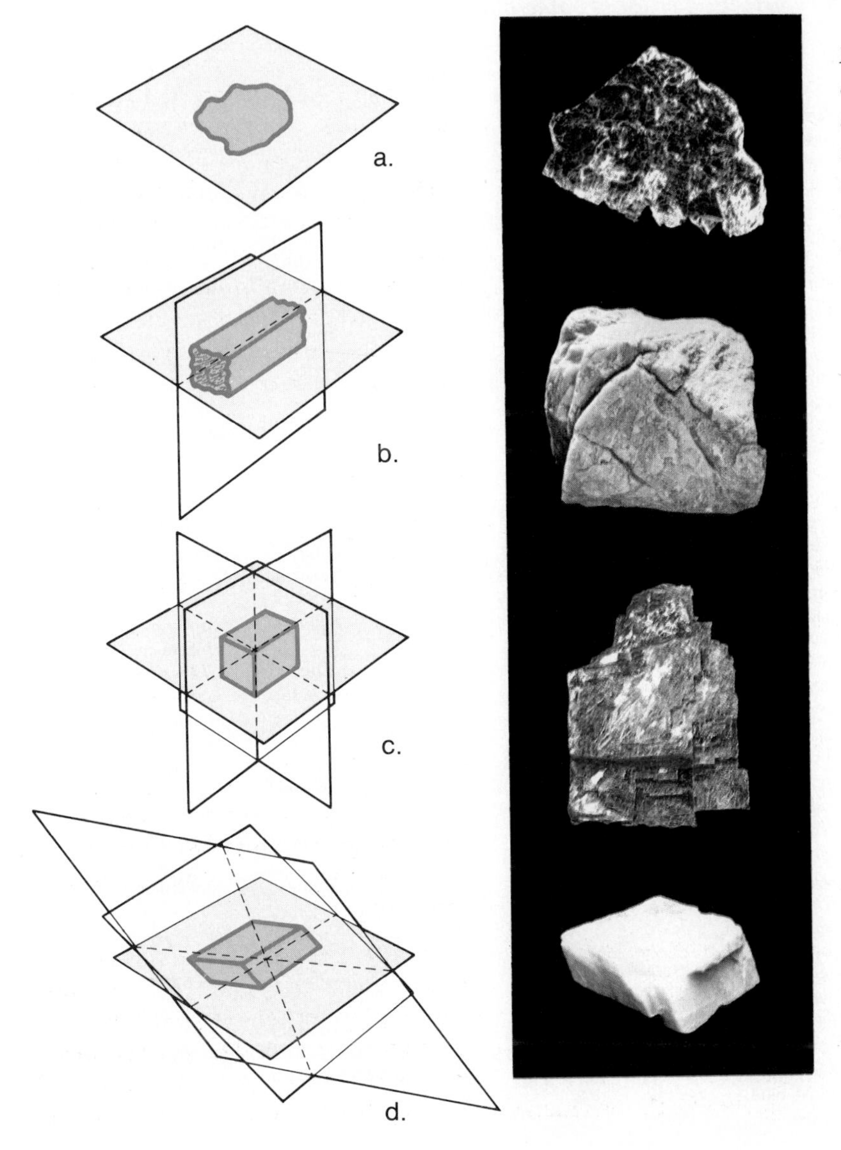

Figure 6-10 *Cleavage planes are characteristic of some minerals.* **a.** *Cleavage in one plane; black mica.* **b.** *Cleavage in two planes at right angles to each other; feldspar.* **c.** *Cleavage in three planes at right angles, defining a cube; galena.* **d.** *Cleavage in three planes not at right angles; calcite.*

Some minerals show cleavage in one direction; some cleave in two directions; others cleave in three, four, or six directions. The cleavage surfaces may look like crystal faces, but broken mineral specimens rarely show crystal faces. The quality of cleavage is also characteristic of individual minerals and may be described as perfect, good, distinct, or poor. Breaking other than cleavage is called "fracture" but is not very distinctive.

Streak is the color of the powdered mineral. Simple streak tests can be made by rubbing a sharp point of the specimen on a piece of unglazed porcelain. A 1 cm streak is as good as a kilometer—don't overdo it.

Luster is the appearance of the mineral in reflected light. Most rock-forming minerals have lusters that can be described as glassy, greasy, earthy, dull, or pearly. Many minerals show metallic

Figure 6-11 *Properties of some minerals.*

Mineral Name	*Color*	*Streak*	*Luster*	*Hardness*	*SpG*	*Cleavage*	*Remarks*
apatite	green or brown	white	glassy	5	3.2	1-poor	used in making fertilizer
biotite	black	colorless	glassy, shining	$2\frac{1}{2}$–3	3.1	1-perfect	black mica, fractures in very thin plates
calcite	colorless, white	colorless	glassy, pearly	3	2.7	3-rhomb. perfect	rock forming mineral
cinnabar	red	bright red	glassy, earthy	2–$2\frac{1}{2}$	8.1	2-perfect	mercury ore
corundum	brown, pink, blue	none	sparkling, dull	9	4.0	none	gem stone, used as an abrasive
diamond	grayish-black	none	sparkling, dull	10	3.5	4-perfect	gem stone, industrial saws
feldspar	white, gray, flesh-red	white	glassy	6	2.5 to 2.7	2-good 90°	common rock-forming minerals
fluorite	light purple, yellow, green	colorless	glassy	4	3.2	4-perfect	used in steel and glassmaking
galena	lead gray	gray-black	metallic	$2\frac{1}{2}$	7.6	cubic	lead ore
graphite	steel gray to iron black	black to gray-black	metallic, earthy	1–2	2.2	1-perfect	feels greasy, used as a lubricant
gypsum	colorless, white, gray	colorless	silky	2	2.3	1-perfect	used in making plaster of Paris
halite	white, red, blue	colorless	translucent glassy,	$2\frac{1}{2}$	2.2	cubic perfect	common salt, tastes salty
hornblende	dark green to black	colorless to gray	glassy	5–6	3.2	2-perfect 120°–60°	an amphibole, a common rock mineral
magnetite	iron black	black	metallic	$5\frac{1}{2}$–6	5.2	none	magnetic
muscovite	tan, green, yellow, white	colorless	glassy, silky	2–$2\frac{1}{2}$	2.8	1-perfect	white mica, flakes in thin sheets
olivine	olive to gray, green, brown	colorless	glassy	$6\frac{1}{2}$–7	3.4	1-poor	green rock-forming mineral
pyrite	pale brass, yellow	green or brown-black	metallic	6–$6\frac{1}{2}$	5.0	none	"fool's gold"
pyroxene	black, dark green	black or dark green	glassy, dull	5–6	3.3	2-poor to good	accessory in igneous rocks
quartz	colorless, white	none	glassy	7	2.6	none	gem stone, common rock-forming mineral
talc	white, green, gray	colorless	glassy, pearly	1	2.8	1-perfect	greasy feel, used in talcum powder
topaz	clear yellow, pink, blue, green	none	glassy	8	3.5	1-perfect	gem stone

luster. This property is less distinctive than the ones described above.

Color is self-explanatory, and you describe colors as you see them. Some minerals always show the same color but most minerals show a variety of colors. This property is less distinctive than luster.

Other physical properties include taste, feel, magnetism, transparency, and fluorescence, but these are not generally useful.

PROCEDURE

Determine the specific gravity (density), hardness, cleavage, and streak of the mineral specimens provided. Record the data as appropriate.

DISCUSSION

1. How many of the specimens appear to be rock-forming minerals?
2. Which minerals do you have? Use Figure 6-11 to assist you in identifying the minerals.

6-8 Some Things That Minerals Tell Us—Conditions of Formation

The rock from the Superstition Mountains (Figure 6-3b) is one of the main materials of the earth's crust. Knowing the minerals the rock is made of will help us to know how the rock was formed and what name to give it.

Some minerals are formed under very special conditions and are therefore quite scarce. The high value of precious stones, for instance, is due to their scarcity as well as their durability and their beautiful colors. Most gem stones were formed under conditions of temperature and pressure that are rare in nature.

The diamond (Figure 6-12) is perhaps the best known and most valued of the gem stones. It is the hardest mineral known and it has a brilliant luster. Most diamonds are found in kimberlite, an igneous rock pushed through the crust from deep within the earth. Concentrations of kimberlite and diamonds are found in India, Brazil, South Africa, and Arkansas.

In 1955, after many years of trying, scientists succeeded in producing artificial diamonds. Success in forming diamonds was reached by using extremely high temperatures and pressures. Knowing this, what can you conclude about past conditions in the part of the earth's crust where a diamond crystal is found?

Graphite is a mineral that is dark gray to black in color. It is so soft and flaky that it is used to make the "lead" in pencils and to make the slipperiness in some lubricants. Most of the early attempts to make diamonds resulted in graphite. Both diamond and graphite are made of the same single element—carbon. By using X-rays we can examine the arrangement of the atoms in diamond and graphite crystals. These arrangements are shown in Figure 6-13. The X-rays are reflected from layers of atoms in the crystals. We infer that the different crystal structures cause the varying physical properties of these two forms of carbon.

Minerals that form only at certain temperatures and pressures are important for the geologist who is working out the history of the earth. For instance, a geologist may learn from laboratory studies that a mineral forms only at temperatures above 300°C no matter what the pressure. When this mineral is found in a body of rock, we can be pretty sure that the temperature there was at least 300°C when the rock was formed.

6-9 Rock Classification by Origin

Centuries ago, some people observed that certain rocks occur in layers that are similar to the layers of sand and mud

Figure 6-12 a. *A diamond crystal from South Africa embedded in kimberlite.* **b.** *Hemimorphite crystals from Mapimí, Mexico.* **c.** *Wulfenite crystals from Arizona.* **d.** *Proustite crystals from Chanarcillo, Chile.* **e.** *Rutile crystals from Boiling Springs, North Carolina.* **f.** *Pyrite crystals from Colorado.* **g.** *Calcite crystals from Ohio.*

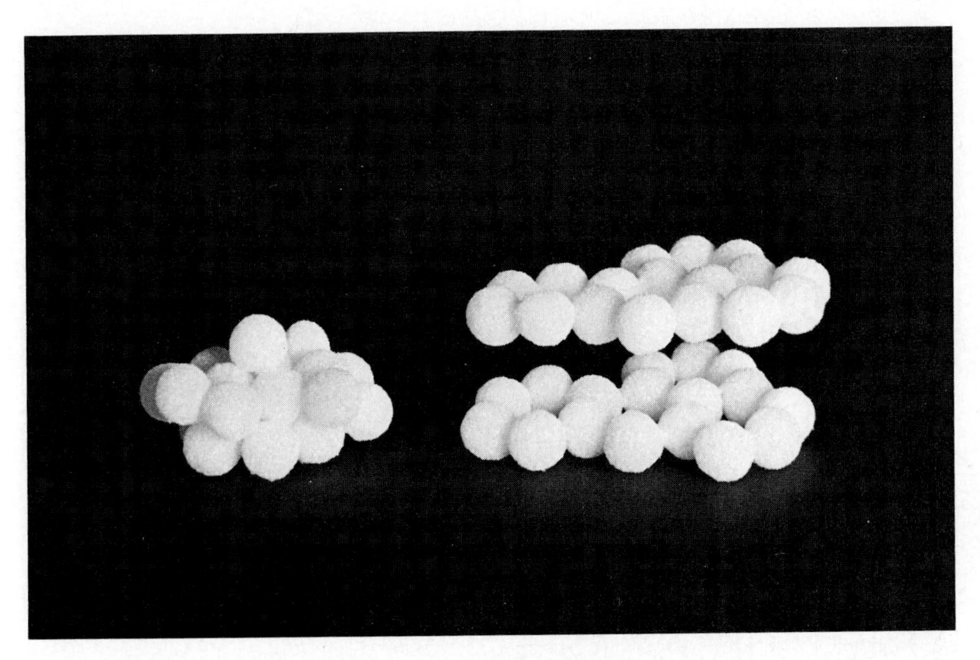

Figure 6-13 *Compare the arrangement of atoms in the diamond and graphite crystals.*

found today in lake bottoms and along the sea shore. Other rocks look like hardened lava from volcanoes. Rocks containing the same minerals may differ greatly in **texture,** that is, the size and arrangement of mineral grains. Eventually, these observations led to a classification system for rocks based on the way they were probably formed.

Three major types of rocks, on the basis of their probable origin, are now recognized by almost everyone. These rock types are called sedimentary rocks, igneous rocks, and metamorphic rocks. You will study the processes that form the various types in later chapters.

Sedimentary rocks are made of sediment (rock and mineral fragments) carried over the earth's surface mostly by streams and deposited mainly in layers on the ocean floor. **Igneous** (fire) rocks are those that solidify from a molten state. You can divide the igneous rocks into two types—**plutonic** (intrusive) ones, which solidify below the earth's surface; and **volcanic** (extrusive) ones, which solidify on the earth's surface. **Metamorphic** rocks (changed form) are sedimentary or igneous rocks in which the minerals or texture or both have been changed by high temperatures and pressures without melting. See Figures 6-14, 6-15, and 6-16.

Thought and Discussion

1. What is a silicate? a carbonate? an oxide?
2. Why should specific gravity, hardness, and cleavage be the most distinctive physical properties of minerals? Why should a streak color be more distinctive than the color of a mineral sample?
3. Why should a classification of rocks by origin be better than some other basis?

Figure 6-14 *The volcano Surtsey (above) rose above the surface of the Atlantic Ocean near Iceland in 1963. Solidified lava (left) near Kilauea Volcano in Hawaii.*

Figure 6-15 *What is the most noticeable feature of this sedimentary rock near Washington, Pennsylvania?*

Figure 6-16 *Compare this metamorphic rock near Wallingford, Vermont, with the rock in Figure 6-15.*

6-10 Natural Resources

Rocks and rock-forming minerals account for more than 98 percent of all the elements in the earth's crust and for about 96 percent of the total mass of the crust. If you will look again at Figure 6-6 you will see that only two of the familiar metals are included. Metals, important as they are, are only part of the great variety of natural resources we use.

All of our natural resources come from the earth, with the help of the sun. The four basic sources of materials are land, water, sunlight, and air. Figure 6-17 shows the relations between these sources and the major varieties of raw materials. The raw materials can be processed into the great variety of resources for industry and the home.

The metals and other elements that we use are in the earth's crust. The average amount of these elements present in the crust is very low, as you can see in Figure 6-18.

In Figure 6-18, the average crustal content of some common metals is shown in "parts per million" (ppm). Ten thousand parts per million equals one percent. Only iron and aluminum occur in amounts greater than one percent, and they are included in Figure 6-6. The

Figure 6-17 *What is the contribution of each of the basic resources to your daily life?*

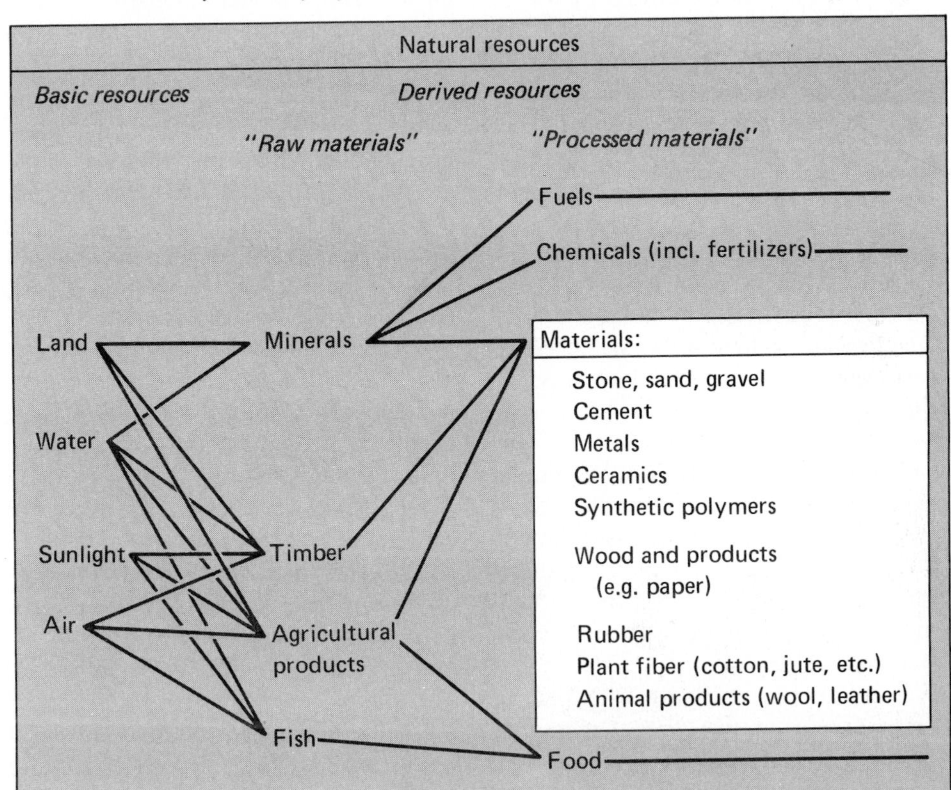

Element	Average Crustal Content (ppm)	Cutoff (ppm)	Approximate Enrichment Times Average Content
mercury	0.089	1,000	11,200
tungsten	1.1	4,500	4,000
lead	12	40,000	3,300
chromium	110	230,000	2,100
tin	1.7	3,500	2,000
silver	0.075	100	1,330
gold	0.0035	3.5	1,000
molybdenum	1.3	1,000	770
zinc	94	35,000	370
uranium	1.7	700	350
carbon	320	100,000	310
lithium	21	5,000	240
manganese	1,300	250,000	190
nickel	89	9,000	100
cobalt	25	2,000	80
phosphorus	1,200	88,000	70
copper	63	3,500	56
titanium	6,400	100,000	16
iron	68,000	200,000	3.4
aluminum	83,000	185,000	2.2

Figure 6-18 *Metal elements in the earth's crust.*

third column of Figure 6-18 shows the minimum content of each metal that must be present if we are to produce it by mining. Therefore, if you want to mine a metal, you have to find a place where there is an unusual accumulation of it. The last column shows the amount of enrichment over the average content needed to make a mineable ore. For instance, a nickel mine requires an enrichment of at least 100 times the average crustal content ($89 \text{ ppm} \times 100 = 8900 \text{ ppm}$). The lowest increases required are those for iron and aluminum. There is so much of these metals in the crust that an ore of iron or aluminum should be common. For the other metals the level of enrichment necessary to create an ore is so high that an ore of any of these should be rare, and they are. It takes special processes and conditions to produce ore deposits.

Every metal in Figure 6-18, except uranium, plus twenty other elements are found in an ordinary telephone handset. Imagine the total number of elements important to a modern industrial society.

No country on Earth can claim self-sufficiency in mineral resources. Two metals, iron and aluminum, must be available in large quantities in an industrial society. Iron ore requires large amounts of coke (which is made from coal) to produce pig-iron. Manganese, 7 g of it per kg of pig-iron, is necessary to make common steel. Production of aluminum from its ore requires large amounts of energy. Only a few countries have these relatively abundant resources in adequate amounts. Fewer yet have even the smaller amounts of the many other elements needed.

As technology changes, the needs for certain mineral resources change also. In recent years there has been growing concern over the shortage of **strategic minerals,** minerals essential to the national defense that must be imported. The four strategic minerals receiving the most attention are cobalt, niobium, chromium, and tantalum. These metals are widely used in the aerospace and elec-

Commodity	Billions of U.S. Dollars			
	Developing Countries	Non-Communist Industrial Countries	Communist Countries	Total World Exports
Agricultural raw materials	9.2	17.4	3.8	30.4
Ores and minerals*	5.1	7.1	1.7	13.9
Food	22.8	51.2	6.9	80.9
Fuels	43.0	13.7	5.8	62.5
Manufactured and processed products	27.9	296.0	36.6	360.5
Residue	1.5	6.9	3.1	11.5
Total†	109.4	392.3	57.9	559.7

*Nonferrous metals—as distinct from ores—are included in manufactured and processed products.
†Figures may not add to totals due to rounding.

Figure 6-19 *Exports of resources and derived goods.*

tronic industries, and over 90% of the supplies must be imported.

A coordinated three-point effort is being made to reduce dependence on strategic minerals. First, more widely available minerals can be substituted in some cases, for example, nickel for cobalt. Second, components are being redesigned to use less of the materials. And third, new classes of materials are being researched to replace some strategic minerals. For example, new ceramics are being developed in the hope that they can eventually replace some uses of strategic minerals.

Interdependence of people everywhere in regard to natural resources means that there must be a lot of exchange of goods. Figure 6-19 shows the world-wide exports of natural resources and derived goods. The categories of goods are generalized but clear enough for our purposes. Source countries are grouped as developing countries, non-communist industrial countries, and communist industrial countries. Do you think any one group of countries can get along without at least one of the others?

The imbalance of supplies of these materials among nations cannot be solved entirely or forever by finding additional deposits. The facts of occurrence and the technology of discovery, production, and processing are well known. Political, economic, and social problems must be resolved if there is to be any prospect of equal sharing by all.

As you investigate the formation of sedimentary, igneous, and metamorphic rocks in future chapters you will learn about the processes which can form valuable mineral deposits.

Thought and Discussion

1. Which of the "processed materials" in Figure 6-17 come from ordinary rocks?
2. In which type of rock would you expect to find the fossil fuels (coal and oil)? Why?
3. Which mineral resources do you think we are likely to run out of first? Can they be replaced with anything else? What steps could be taken during the next 50 years to properly use and conserve our mineral resources and plan for the future?

Unsolved Problems

Advanced industrial societies such as those in North America and Western Europe share an interdependence with

Mineral	% Imported	Major Foreign Sources
Strontium	100	Mexico, UK, Spain
Columbium	100	Brazil, Malaysia, Zaire
Mica (sheet)	99	India, Brazil, Malagasy
Cobalt	98	Zaire, Belgium-Luxembourg, Finland, Norway, Canada
Manganese	98	Brazil, Gabon, South Africa, Zaire
Titanium (rutile)	97	Australia, India
Chromium	91	USSR, South Africa, Turkey, Philippines
Tantalum	88	Australia, Canada, Zaire, Brazil
Aluminum (ores & metal)	88	Jamaica, Australia, Surinam, Canada
Asbestos	87	Canada, South Africa
Platinum group metals	86	UK, USSR, South Africa
Tin	86	Malaysia, Thailand, Bolivia
Fluorine	86	Mexico, Spain, Italy
Mercury	82	Canada, Algeria, Mexico, Spain
Bismuth	81	Peru, Mexico, Japan, UK
Nickel	73	Canada, Norway
Gold	69	Canada, Switzerland, USSR
Silver	68	Canada, Mexico, Peru, Honduras
Selenium	63	Canada, Japan, Mexico
Zinc	61	Canada, Mexico, Peru, Australia, Japan
Tungsten	60	Canada, Bolivia, Peru, Thailand
Potassium	58	Canada
Cadmium	53	Mexico, Canada, Australia, Japan
Antimony	46	South Africa, Mexico, P. R. China, Bolivia
Tellurium	41	Peru, Canada
Barium	40	Ireland, Peru, Mexico
Vanadium	40	South Africa, Chile, USSR
Gypsum	37	Canada, Mexico, Jamaica
Petroleum	35	Canada, Venezuela, Nigeria, Netherlands Antilles, Iran

Figure 6-20 *Imports supplied a significant percentage of the total U.S. demand in a recent year.*

the developing nations. The developing nations produce raw ores of metals and other essential natural resources. The importance of these countries to the United States is shown in Figure 6-20. Most developing countries, however, lack the balanced supply of the major industrial mineral resources needed to establish an industrial economy.

Managing industrial raw materials for the benefit of everyone is one of the major problems facing society today. Various nations have been working together to develop agreements in many areas. Among these are international law of the sea, trade in minerals and products, mineral leasing and royalty procedures, and exchange of information on research in resource development. Along with the development of resources, waste must be reduced. Recycling of mineral-derived materials must be increased.

Leadership in policy development will come through science and technology. The goal should be to form a rational and objective approach to decisions. Only careful management of resources will ensure supplies for a long time.

MINING ENGINEER

As development of mineral deposits increases all over the world, there is a need for more mining engineers. Mining engineers perform many types of work related to finding and extracting minerals for industries to use. Some plan the construction of mines and may be responsible for their safe operation. Other mining engineers develop new equipment. Recently, more and more mining engineers have become involved with the reclamation of mined land and the control of pollution caused by mining operations. New opportunities for mining engineers also include mining metals from the sea and developing oil shale deposits. The field is an expanding one with many new challenges.

Most mining engineers work for private employers in the mining industry. Others work for firms that manufacture mining equipment, for government agencies, or as private consultants. A bachelor's degree is required for engineering jobs, and an advanced degree is recommended for many. Technicians who assist mining engineers may advance to engineering positions by passing a professional licensing examination.

Mining engineers must be willing to work under all types of uncomfortable, sometimes hazardous, conditions. They should also have initiative and be able to work as part of a team. Aptitude in physical science and mathematics is necessary for success as an engineer.

Chapter Summary

The outermost solid layer of the earth is the lithosphere. It is composed of an outer crust with a layer of solid denser material below it. At the base of the lithosphere is the asthenosphere, which is weak and can flow like thick tar. Below the asthenosphere is a thick, solid spherical layer called the mantle. Inside the mantle is a liquid outer core and a solid inner core.

More than 95 percent of the mass of the visible solid earth is composed of a dozen or so solid chemical compounds called minerals. The minerals are combined in various ways to form rocks. Most minerals are composed of two or more elements, with the atoms bonded together.

The most abundant atoms in the earth's crust are those of oxygen and silicon. These occur in units of one silicon atom surrounded by four oxygen atoms. Minerals containing such units are silicates. The two major groups of silicates are those containing aluminum and those containing magnesium.

Other important minerals are calcite, which forms limestone; halite, or rock salt; and the oxides of iron, which are mined. Each mineral has its own set of physical properties, such as specific gravity, hardness, cleavage or fracture, and luster. Minerals such as diamond can be used to deduce conditions of temperature and pressure that existed in the past.

Rocks are classified according to the way they were probably formed. Sedimentary rocks consist of rock and mineral fragments deposited by moving water. Igneous rocks have solidified from a molten state. Metamorphic rocks are rocks that have been changed by high heat and pressure.

Our mineral resources are minerals that form a very small percent of the earth's crust. Ores of most metals are rare. All countries of the world must learn to share the planet's mineral resources.

Questions and Problems

A

1. How do the three main types of rocks form?
2. What is the definition of a mineral?
3. What is the difference between an element and a compound?
4. How can the shape of its crystal faces help to identify a mineral?
5. Why should the atmosphere and hydrosphere be simple and uniform in chemical composition compared to the lithosphere?
6. What subatomic particles form the nucleus of an atom?
7. Is ice more dense than water or less dense? Why?
8. How does the abundance of metals in the earth's crust compare with the abundance of oxygen?

B

1. What is the difference between an atom and an ion?
2. What keeps atoms together in a compound?
3. How was the practice of alchemy important to the development of modern science?
4. Why do minerals appear in the form of crystals?
5. Water has twice as many hydrogen as oxygen atoms, yet oxygen makes up

most of the mass of the hydrosphere. What is your explanation?

6. If the crust contains such small amounts of metals such as copper and lead, how can we obtain these materials?

C

1. John Dalton proposed that all substances are composed of small, solid, indestructible particles called atoms. Are atoms still considered to be small, solid, and indestructible? Explain.
2. What difference in structure accounts for the fact that mica cleaves in one direction whereas quartz breaks irregularly?
3. Look at Figure 6-6. Which atom has the larger volume, the Fe atom or the Na atom?
4. Why may it not be reasonable to expect large accumulations of metals in the oceans?

Suggested Readings

Brownlow, Arthur H. *Geochemistry.* Englewood Cliffs, New Jersey: Prentice-Hall, 1979.

Chesterman, Charles W., and Kurt E. Lowe. *Audubon Field Guide to North American Rocks and Minerals.* New York: Chanticleer Press, 1978.

Dietrich, Richard V., and Brian J. Skinner. *Rocks and Rock Minerals.* New York: John Wiley & Sons, 1979.

Hurlbut, Cornelius S., Jr. *Manual of Mineralogy.* 19th ed. Edited by James D. Dana. Somerset, New Jersey: John Wiley & Sons, 1977.

Hurlbut, Cornelius S., Jr., and George S. Switzer. *Gemology.* Somerset, New Jersey: John Wiley & Sons, 1979.

Pough, Frederick H. *A Field Guide to Rocks and Minerals.* 4th ed. Boston: Houghton Mifflin Co., 1976. (Paperback)

Prinz, Martin; George Hawlow; and Joseph Peters, ed. *Guide to Rocks and Minerals.* New York: Simon and Schuster, 1978.

7 Weathering and Erosion

IN CALIFORNIA *a construction worker jams down a plunger, blasting thousands of cubic meters of rock and soil from a hillside.*

A mud-red stream slices its way across Arizona, removing tons of silt and sand and leaving a great canyon behind.

In a cemetery in Massachusetts a tourist attempts to read the inscription on a tombstone, but its centuries-old message has been all but erased by the action of weathering agents.

The captain of an ore ship steaming across Lake Superior retraces the paths of ice age glaciers that gouged the lake basin out of solid rock.

In Peru an earthquake triggers an avalanche that roars down a mountainside, burying 20,000 villagers beneath tons of rock and mud.

The processes of weathering, running water, wind, glacial ice, and landslides are but a few of the geologic agents that have been changing the face of the earth for billions of years. Now, people are on the list. But regardless of which "tool" nature uses or how fast it works, the end result is always the same: parts of the land are gradually worn away and moved from one place to another.

AFTER COMPLETING THIS CHAPTER, YOU SHOULD BE ABLE TO:

1. **describe the two basic types of weathering processes.**
2. **explain why water is called the "universal solvent."**
3. **recognize weathering products and explain how they differ from the parent rock.**
4. **describe how soils develop and explain how mature soils reflect the climate conditions under which they were formed.**
5. **describe the role of gravity in erosion.**
6. **explain the relative importance of water, wind, ice, and organisms as geologic agents.**

Earth's Surface Wears Away

7-1 Weathering Changes Rock.

Most of the rocks of the earth's crust formed deep below the surface. The environment there is quite different from that at the earth's surface. Temperature and pressure increase with depth in the earth's crust, and the amounts of air and water decrease.

Surface rocks break down when they are exposed to air and water. This process of breaking down rocks is called **weathering.** Weathering does not occur only at the earth's surface; rocks may also be weathered at any depth reached by air and water.

Sand, mud, and other rock fragments are produced by weathering. Yet, the process is not entirely destructive. Although the rocks are destroyed, the matter they contain is not. In fact, new minerals can be formed in places where rocks are weathered. More important, when rocks are deeply weathered, they are reduced to soil, which is the support for much of life.

There are two basic types of weathering processes: physical and chemical. Physical processes make small rocks out of big ones without changing the minerals in the rock. (See Figure 7-1.) This type of weathering is called **disintegration.** One agent of disintegration is shown in Figure 7-2. How can plants be effective weathering agents?

Because water is a common liquid that expands when frozen, it is the main agent of disintegration. Water that freezes in the cracks of rocks will expand, forcing the rocks to break apart. In high mountains, where water repeatedly freezes and thaws, the expanding ice pries the rocks apart, and disintegration may be relatively rapid. (See Figure 7-3.)

Chemical weathering, or **decomposition,** is a slow but continuous process.

Figure 7-1 *Disintegrating columns of solidified lava in California.*

Figure 7-2 *Can you describe this cause of disintegration?*

The minerals in chemically weathered rocks are changed into new and often more valuable products such as soil. **Soil** is the loose material at the surface of the earth, in which plants can grow. Because of their physical and chemical properties, certain minerals dissolve more readily than others. When these minerals dissolve, pits and cracks are left behind. As time passes, the remaining material may become a skeleton of the original rock. Such rocks are usually soft and easily broken apart.

ACTION

Put several different rocks in an aluminum pie pan and heat them in the oven at "high" setting. Carefully remove the pan after 15 minutes and pour cold water on the rocks. Do some of the rocks break apart? Do others break up later, as the rocks cool? Is this an example of physical or of chemical weathering?

Since decomposition occurs on mineral surfaces, it is aided by disintegration. When a rock is weathered into smaller and smaller particles, the total surface area can increase thousands of times. (Refer back to Figure 4-8. How much will the surface area increase if a cube, one centimeter on each edge, is cut into eight equal parts?) The two weathering processes work together. Rock disintegration exposes new surfaces and speeds up chemical weathering. In turn, decomposed rocks break up more easily.

MATERIALS
several different rocks, aluminum pie pan, heat source

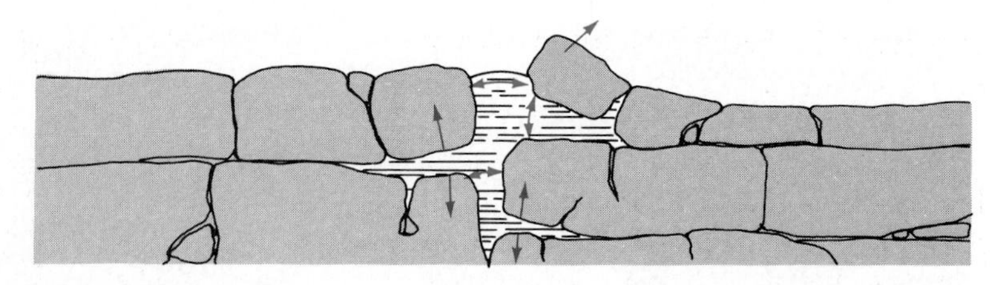

Figure 7-3 Alternate freezing and thawing causes rocks to break apart.

MATERIALS
two cups, sugar cube, granulated sugar

ACTION

Fill two cups with equal amounts of warm water. Place a sugar cube in one cup. In the other place a tablespoon of granulated sugar. Which type of sugar dissolved first? Why?

MATERIALS
steel wool, closed container, copper penny, nickel

7-2 Water—the Universal Solvent

Water, like many common objects, is taken for granted. Yet, water is one of nature's most effective agents of erosion, whether it is in its liquid, gaseous, or solid state. One of water's most important properties is its ability to dissolve salts, such as sodium chloride (NaCl).

Water molecules have the ability to separate the ions that make up the sodium chloride. As this happens, the salt begins to dissolve. This is the way in which water attacks rocks and soil. A compound like sodium chloride (which is called halite as a mineral) dissolves readily in water. Because most earth materials do not dissolve as readily as halite, chemical weathering works rather slowly.

Water alone is a powerful solvent. The addition of other chemicals, such as oxygen and carbon dioxide, makes it even stronger. For example, oxygen from the air dissolves easily in the thin film of water that surrounds certain rock particles in the soil. The oxygen can then combine with iron atoms exposed by the weathering of iron containing minerals such as olivine or magnetite. This is how rust (iron oxide) is formed. The iron oxide can exist as a coating or stain on mineral grains. Iron staining is a common cause of color in rock exposures. Some iron stains are yellowish- or reddish-brown; others are blue-gray. The final color depends on the ratio of iron to oxygen in the rust.

ACTION

Moisten some steel wool and wrap it in plastic or place it in a closed container so it will not dry out. Observe the steel wool for several days. What changes do you see? What has been formed? Is the "weathered" steel wool easier to break than the original steel wool? Is this an example of chemical or of physical weathering? Will a copper penny or a nickel change in the same way?

When carbon dioxide is dissolved in water, some of it combines with the water to form carbonic acid (H_2CO_3). This acidic water is more effective than pure water in dissolving certain ions such as calcium and potassium. Carbon dioxide is given off when the living cells of plants and animals use oxygen. (You know that carbon dioxide dissolves in water, for this is the gas that is found in every carbonated drink.) The calcium and potassium ions dissolved in the acidic water may then be absorbed through roots and used for plant growth. (See Figure 7-4.)

The lichen (*LY kuhn*) growing on exposed rock in Figure 7-5 releases carbon dioxide on the rock surface. How will this promote weathering of the rock? Lichens can grow where other plants cannot survive because of the absence of soil.

The remains of dead plants and animals are usually broken down by simple life forms such as bacteria, molds, and funguses. The result is called **organic matter.** Weathered rock and organic matter may accumulate and support plant life.

Decayed plant and animal matter is eventually reduced to fine particles that may stick to mineral grains and darken them. Only a small percentage of this organic matter is needed to color soil black or dark brown.

The large cavern shown in Figure 7-6 formed because the primary mineral in limestone dissolves readily in carbonic acid. This soluble mineral is calcite. The removal of soluble material by ground water moving through soil or rocks is called **leaching.** Leaching occurs as underground water moves through buried rocks and dissolves certain minerals from them. The dissolved minerals are carried downward in the crust where they may form new minerals or new deposits of the old ones at another location. For example, calcite leached out of buried limestone may later be deposited on the roof of a cavern to form stalactites (Figure 7-6).

Figure 7-4 *Roots give off carbon dioxide. It dissolves in water, and carbonic acid is formed. Carbonic acid speeds up chemical weathering.*

7-3 Investigating Products of Weathering

Although granite is considered to be a long-lasting rock, much soil has been produced by the weathering of granitic

Figure 7-5 *Lichens are the first plants to grow on exposed rocks when weathering begins. Lichens can exist for years at a time with very little moisture.*

MATERIALS
granite samples (coarse-grained and crushed), magnifier, plastic tubes with caps (or small pill bottles), soil samples (subsoil and topsoil), teasing needle, water

Figure 7-6 *Carbonic acid dissolved in water carved out Carlsbad Caverns in New Mexico. The deposits hanging from the roof of the cavern are called stalactites (they "hold tight" to the roof). The deposits that grow from the floor are stalagmites (they "might" reach the roof).*

rocks. The minerals of which granite is composed weather at different rates.

When you have completed this investigation you will be able to identify some of the products of weathering that remain after ions and other small particles are washed or blown away.

PROCEDURE

Examine the granite and the two soil layers provided for you.

DISCUSSION

1. In what ways are the two layers similar?
2. Can you identify two minerals that exist in both the granite and the two soil layers?

PROCEDURE

Put half a teaspoonful of each soil sample in a test tube with water. Shake the test tubes and let the materials settle. Discuss the results. The soil layers you examined were formed from granite.

DISCUSSION

1. Which of the two samples was taken from the top layer? How can you tell?

Minerals that weather slowly remain as rock fragments. Those that weather rapidly form ions and particles small enough to remain suspended in water for a long time. Ions and very small particles are more easily washed away than more solid rock fragments.

Aluminum and silicon, released from minerals during weathering, can combine with oxygen to form a group of silicates called clay minerals. These minerals frequently occur as very small particles. The formation of clay minerals in soil is one example of the creative or constructive aspect of weathering. We might compare the weathering process to wrecking a building and using the pieces to build a new structure.

Thought and Discussion

1. Why do rocks weather?
2. How does physical weathering aid in chemical weathering?
3. How is carbonic acid produced and how does it affect weathering?
4. Why are many earth materials reddish-brown? What element often produces color in earth materials?

7-4 How Soils Develop

Rock weathering not only destroys; it also produces the life-supporting soil layer on the earth's surface. Soil is the link between solid rock and the world of living things. Although a few plants such as lichens can grow on bare rock, most land plants need soil to survive. Plants, in turn, provide food for animals.

Physical and chemical weathering continue as soil is developed. The effect of weathering can be seen in the layers, called **horizons,** that develop in the rock debris. The boundary between horizons is not always distinct. A poorly-developed soil with only a few horizons is called **immature soil.** (See Figure 7-7.) Mature soils develop over a longer period of time and have many vertical layers. (See Figure 7-8.) All the layers in a particular soil make up the **soil profile.**

After rock debris begins to accumulate and soil is formed, the first important change is the addition of growing plants. Plants assist the weathering process, but also slow down the erosion of soil by wind and water. The soil that remains in place becomes more mature and thicker.

Figure 7-7 *How many horizons can you see in the immature soil?*

Figure 7-8 *How many horizons can you see in the mature soil?*

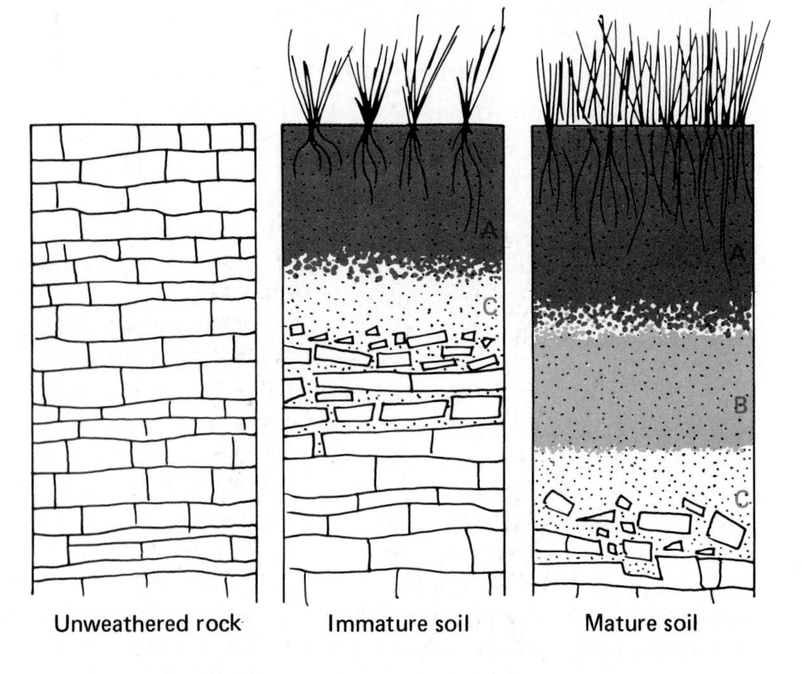

***Figure** 7-9 Stages in the development of a mature soil. The A horizon is topsoil, the B horizon is subsoil, and the C horizon is rock fragments. What could affect the time it takes to develop a mature soil?*

Organic matter from more or less decayed plants and animals forms **humus** (*HYOO muhs*). Humus is the primary source of the nitrogen needed for new plant growth. Humus collects in the uppermost layer of soil, called the **topsoil** or A horizon.

In an immature soil, the topsoil lies directly on top of the solid rock, or bedrock. A mature soil has another horizon separating the topsoil (A horizon) from the rock fragments (C horizon). (See Figure 7-9.) The A and B horizons are made up of several layers.

The B horizon, or middle layer, is called the **subsoil.** As water seeps through the topsoil, some minerals are dissolved. Other materials, particularly fine particles of clay and iron oxide, are carried downward by ground water. The smaller particles are deposited in the subsoil.

Because of its higher clay content, the subsoil is harder to plow than the topsoil. Clay particles may also clog the pores in soil, making it more difficult for air, water, and roots to penetrate. However, a certain amount of clay is necessary to hold the soil together. Equally important, the negatively charged clay particles attract and store positive ions of calcium, magnesium, and potassium. Many plants thrive in soils, called alkaline soils, that are rich in these elements.

7-5 Factors That Influence Soil Formation

Several factors influence soil formation, but climate is the most important. It is known that a given climate will produce similar kinds of mature soils from a variety of bedrocks.

In warm, moist climates, rocks decompose in a relatively short time. There is usually a rapid chemical breakdown, hastened by abundant plant and animal life. In deserts, there is little rainfall, the air is dry and hot, and bedrock weathers more slowly. In temperate regions, where there are strong seasonal variations in climate, rocks weather by a combination of processes. Frost action is dominant in winter when expanding ice may pry rock fragments farther apart.

Heat, rainfall, and leaching by ground water are more effective in warmer climates, especially in summer, spring, and fall. Chemical processes that hasten rock decay operate more effectively during these seasons. There is ample water and temperatures are high enough to speed chemical reactions.

Because of the importance of climate, the broad soil regions of the world follow the distribution of climates. Even so, soils and climate zones are not always identical. Climate is only one of five factors that control the development of soil. Time, the type of bedrock, the local land surface, and living organisms are other factors that affect soil formation.

The minerals in rocks do not weather at the same rate. For example, quartz is a very resistant mineral. When it finally does disintegrate, it forms sand and silt. It supplies no clay or plant food to the soil. Most rocks are mixtures of minerals, but few of the minerals are as resistent as quartz.

The original bedrock from which a soil forms is called **parent rock.** Soils, like children, do not always resemble their parents. This is particularly true of older soils. Soils normally develop very slowly. In some areas, a mature soil may take thousands of years to develop. Mature soils usually form on gently sloping lands with good drainage.

Immature soils usually develop on steep slopes where the upper soil is likely to be very thin because part of it is frequently washed away. They also occur in regions that are flooded regularly. New material is deposited with each flood and the soil horizons never get a chance to mature. Poor foresting or farming practices, which allow topsoil to be worn away, will also prevent the formation of mature soils.

ACTION

With the help of your teacher, select a site where you can examine an entire soil profile. Avoid locations near construction sites or where there is frequent flooding. If possible, compare two locations; for example, one at the top of a hill and one at the bottom.

You should wear old clothes and take along a shovel, a ruler, small paper bags for samples, and a pencil and notebook. It will probably be necessary to dig down one or two meters (or even more) in order to reach unweathered bedrock.

After the profile is exposed, make a sketch of the horizons. Are the boundaries between horizons distinct? Record the depth of each horizon and take a soil sample from each layer. Back home or at school you can compare the color, texture, and porosity (amount of pore space) in the layers. Is your profile mature or immature? Would you call it a forest, grassland, or desert soil? Compare your observations with the soil descriptions in Section 7-6.

MATERIALS
shovel, ruler, several small paper or plastic bags, pencil, notebook

---

7-6 What Kinds of Soils Are There?

None of the five factors of soil formation are constant over the entire earth. Because of this wide variation, there are hundreds of thousands of possible combinations of factors and possible local soil types. To sensibly study soils, we must group similar local types into broader categories.

Figure 7-10 shows the distribution of the major soil types in North America. What type of soil is typical of your part of the country?

Mountain soils (Figure 7-11) are usually rocky and thin. The slopes in mountainous regions are too steep to allow the formation of a mature profile. However, local pockets of other soil types are often found in regions of predominantly mountain soil.

Forest soils generally form in regions of high rainfall. The rainfall causes much leaching of the forest floor, which produces a soil low in certain substances, especially calcium and magnesium. Such

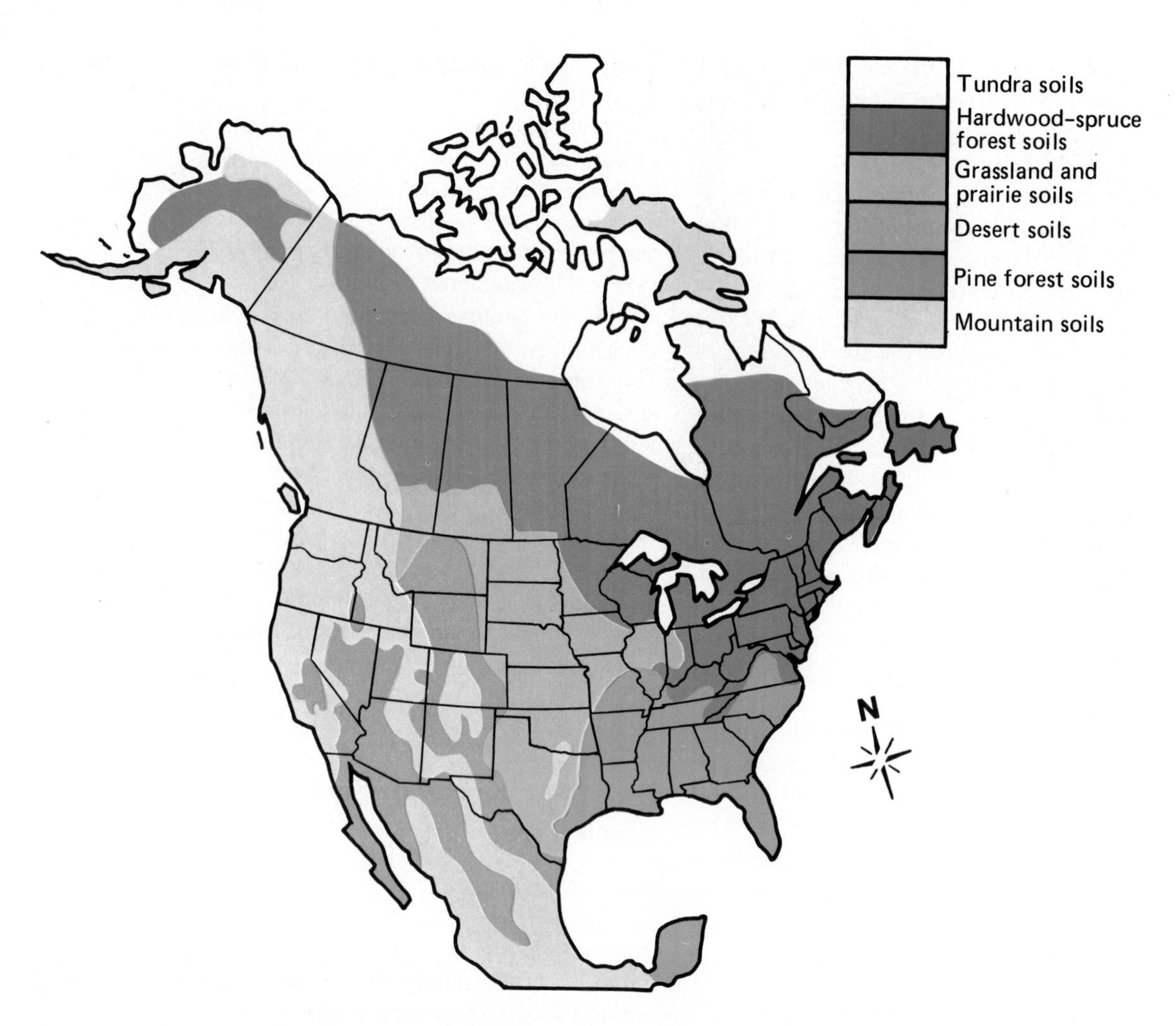

Figure 7-10 *This map shows the distribution of major soil types in North America. These soil types can be subdivided into numerous local varieties.*

a soil is **acidic.** Pine, hemlock, and spruce require little calcium and magnesium and therefore thrive in acidic soil. Falling leaves and needles provide most of the organic matter in the forest. A thin topsoil forms when the vegetation decays.

Trees can grow in leached soil because they mature slowly and their roots reach deep into the ground. A plant that matures in a single season requires a more fertile soil. Strange as it may seem, the lush rain forest in Figure 7-12 is growing in a poor soil. Little light reaches the forest floor. Therefore, few green plants grow there, and little humus is formed. It has been suggested that jungles should be farmed to increase the world's food supply. Unfortunately, only a few years of farming would completely strip a forest soil of the little plant nourishment that it can provide.

When properly cared for, **grassland soils** are more fertile than forest soils. The topsoil is thick and rich in humus from the decay of grass stems. In the United States, wheat is grown and cattle are grazed on the grasslands.

Figure 7-11 *How would this mountain soil in the Wallowa Mountains, Oregon, differ from soil in a warm and humid area?*

Figure 7-12 *This rain forest soil in Washington's Olympic National Park does not have enough nourishment to support intensive farming.*

Figure 7-13 *What features of cactuses help them survive in this Arizona desert?*

Figure 7-14 *Irrigation has turned this desert in Utah into productive farmland.*

Figure 7-15 *Tundra soil supports a variety of grasses, mosses, and small flowering plants.*

Prairie soils are a form in-between grassland and forest soils. Prairie soils resemble grassland soils because they have a deep topsoil rich in humus. But they also receive the high rainfall typical of forests. This unique combination makes them naturally fertile and very productive. Most of the "Corn Belt" in the United States consists of prairie soil.

Desert soils are very rich in minerals. Because of the shortage of moisture, they are only slightly weathered or leached. The lack of rainfall also limits plant growth. Thus, the soil is low in nitrogen and humus. Nevertheless, desert plants, such as cactuses (Figure 7-13), can grow in this poor soil.

With proper irrigation, a desert soil can be fertile (Figure 7-14). But irrigation is tricky. If the water is allowed to stand and evaporate, it deposits the salts that it carried in solution. The soil may then become too salty to grow anything. In recent years, parts of California's rich Imperial Valley suffered this fate. This is but one way that human activities have altered natural patterns of weathering and erosion.

MATERIALS
soil, glass container

ACTION

Place some soil in a glass container and slowly pour water over it. Does the soil absorb the water? Do air bubbles rise from the soil? How might this ACTION be related to the process of leaching?

In some parts of Alaska the combination of low temperatures, slight rainfall, and slow evaporation produces **tundra** (*TUHN druh*) **soil** (Figure 7-15). These regions are so cold that the deeper layers of soil remain permanently frozen. (They are known as **permafrost.**) The upper soil layer thaws, but has poor drainage during the summer. This water-saturated soil is black from slowly decaying humus.

Tundra is an easily damaged soil. Worse yet, climate conditions in tundra regions do not favor rapid "healing" when the fragile soil is disturbed. Because of this, special precautions must be taken when constructing roads, pipelines, and other works.

Thought and Discussion

1. How does subsoil differ from topsoil?
2. What is the difference between mature and immature soils?
3. In what type of climate would you expect rock weathering to be most complete?

7-7 Investigating Stream Erosion

How are the streams in Figure 7-16 different? Which one contains more water?

When you have completed this investigation, you will be able to recognize the effects of stream slope and stream volume on erosion.

MATERIALS
catch buckets, gravelly sand, protractor, siphon tube (with screw clamp), support to raise one end of trough, timer, trough, water

PROCEDURE
Use the equipment shown in Figure 7-17. Put 50 milliliters of a gravel and sand mixture in the trough. Erode the sand from the trough with running water. Repeat the procedure several times, varying the slope and stream volume. (The stream volume can be varied by loosening or tightening the screw clamp on the tube.)

DISCUSSION
1. How does stream slope affect the rate of erosion?
2. How does stream volume affect it?
3. How did the different sizes and shapes of the particles affect their movement?
4. How could stream volume and stream slope change in nature?

7-8 Gravity Drives Erosion.

Imagine that the products of weathering had collected where they developed for the billions of years since the earth formed. How would the earth's surface differ from what you now see? What evidence is there in Figure 7-18 that loose weathered material was removed?

Figure 7-16 *Which of these streams would carry more sediment?*

Figure 7-17 *A setup for investigating stream erosion.*

Figure 7-18 *How was the Grand Canyon of the Colorado River formed?*

Erosion (*ih ROH zhuhn*) is the movement of rock and soil particles from one place to another. Because of the force of gravity, every object on the earth tries to move toward the center of the earth. So gravity is the force that pulls objects downhill. Water, ice, and winds work along with gravity. The actions of organisms, including people, loosen earth materials. This makes earth materials more likely to move from higher to lower places.

A boulder like the ones in Figure 7-18 can roll downhill. If the boulder does not roll all the way to the sea, it may later move again. While it is moving, the boulder erodes itself and other objects in its path. The greater the distance that the boulder falls, the greater its speed and the more damage it causes.

Rock debris moves rapidly downhill during landslides. (See Figure 7-19.) **Creep** occurs when earth materials shift downward very slowly. (See Figure 7-20.) Water between the soil particles helps them slide downhill. Alternate freezing and thawing may also help this movement. Not all materials move as noticeably as a landslide. But a tremendous number of individual sand and soil particles gradually move to lower elevations each day.

Water vapor carried to high elevations by the air eventually ends up as water flowing down over the slopes into stream channels. In a stream valley, the erosion takes place mainly on the valley slopes. The stream is like a conveyor belt that carries away the material from the valley walls.

Figure 7-19 *The 1959 Yellowstone Earthquake caused this landslide.*

Figure 7-20 *What evidence of creep can you see here? What do creep and landslides have in common? How do they differ?*

a.

b.

Figure 7-21 a. *You can see the effect of a severe rainstorm on unprotected soil. Note the pinnacles of soil that were protected by stones.* **b.** *The impact of a raindrop.*

Gravity also controls the geologic work of ice and wind. Can you explain how?

7-9 Water, Ice, Wind, and People Erode the Land.

The erosive action of water, ice, and wind wears away the earth's surface and carries loose earth material from the land to the sea. Human activities such as road construction and farming may also speed up erosion. The seaward journey of most rock and mineral fragments is long and winding. A particle originally carried by a glacier might rest in one place for thousands of years before it is moved along again by a stream. This process of transportation and deposition may be repeated many times before material loosened from a mountainside eventually reaches the sea.

Soil may be thought of as a temporary deposit of weathered material. The amount of soil now on the land is the difference between the amount produced by weathering and the amount removed by erosion. (Why are mountain soils commonly thin and rocky?)

After a heavy rainstorm, a field without plants to slow down erosion might appear as shown in Figure 7-21. Erosion is a problem on many farmlands because the land does not have sufficient plant cover to protect the soil. The eroded material will eventually wash away or cave into a stream.

All streams carry an invisible load of ions in solution. These ions are carried along with suspended particles and eventually reach the sea. The size of this chemical load depends on the kind of rock and soil in the area that feeds the stream.

Particles of minerals and rocks that are suspended in a stream give it a muddy appearance. In a large river, this visible load of sediment may be tremendous. For example, the Mississippi River carries about two million metric tons of sediment to the Gulf of Mexico each day.

Not all of the sediments carried by a stream are suspended in the water. Larger fragments may be rolled or bounced along in the stream, as shown in Figure 7-22. When these rock fragments strike other particles or solid stream beds, they may break into smaller pieces that can be carried even farther. These rock particles are like tools, breaking and grinding each other and the surfaces of other rocks.

A stream's ability to erode or to deposit material is not always the same. A stream can carry more material and larger particles during a flood than during its normal flow. However, as flood waters drop and velocity decreases, the larger particles begin to settle out. Next, sand settles to the stream bed where it may be dragged along the bottom. Finally, only fine silt and clay remain suspended in the water. If the stream volume continues to decrease, some of these fine particles will also be deposited. Velocity can also vary along the length of a stream. Consequently, the stream may remove material in some places and deposit it in others.

Glaciers, moving sheets or "streams" of ice, are another agent of erosion. **Valley glaciers** (Figure 7-23) usually move downslope by following old stream-cut valleys. Rocks may be frozen into the ice

Figure 7-22 *How would the bed load of this stream in Washington be related to its velocity?*

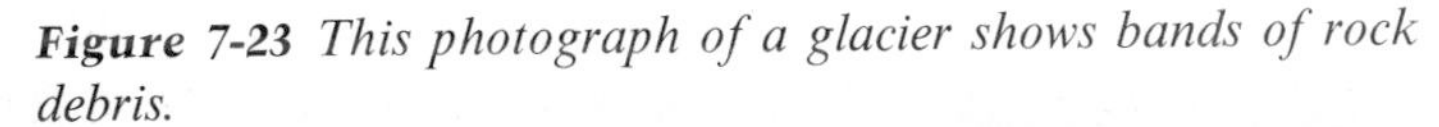

Figure 7-23 *This photograph of a glacier shows bands of rock debris.*

Figure 7-24 *These gouges in the bedrock on Hyers Island, Cross Lake, Manitoba, were formed by stones frozen into the bottom of a moving glacier.*

and torn from the valley floor or wall. Rocks falling on the glacier's surface add to the load. Each rock picked up by the glacier becomes a sharp tool that scratches and gouges the rocks over which it passes. (See Figure 7-24.) As glaciation continues, the originally V-shaped stream-cut valley is scooped out to form a U-shaped valley (Figure 7-25).

Ice sheets, or continental glaciers, are so large that they may cover entire continents (Figure 7-26). The largest of

Figure 7-25 *Which of these valleys was changed by a glacier?*

Figure 7-26 *The ice shelf in Admiral Byrd Bay, Antarctica.*

Figure 7-27 *Till is deposited when the ice of a glacier melts.*

these, the Antarctic ice sheet, blankets most of Antarctica, covering an area almost twice the size of the United States. In places, this great ice sheet is almost 4000 m thick. These huge glaciers spread slowly over the land, wearing away the earth's surface and transporting vast amounts of rock debris.

But like other geologic agents, glaciers must eventually deposit their loads. Most glacial deposition occurs when the ice melts, leaving jumbled masses of ice-transported rock fragments of all sizes. (See Figure 7-27.)

Wind is another geologic tool of erosion. Wind must have a much greater velocity than ice or water to move particles of the same size. Fine materials like silt and clay-sized particles are easily lifted by winds. Because wind usually cannot carry large particles, it sorts material by size. The pebbles and boulders that are left behind produce a rough, rocky surface. If there is a large supply of sand, dunes are formed in places where the wind loses energy and drops its load. The sand dunes in Figure 7-28 appear to be stationary. They are actually in constant motion because the grains roll over each other on the surface of the dunes.

During periods of extreme drought, wind erosion may strip the land of fertile topsoil (Figure 7-29). Dust-bowl conditions are often worsened by poor farming procedures and overgrazing of ranch and pasture land.

Destructive human activities have played an important role in erosion during recent years. However, conservation specialists and environmental geologists have done much to keep human-caused erosion to a minimum.

7-10 Running Water, the Master Leveler

The major agent of erosion is the water that falls on and runs off the land. Rainfall is most important in shaping all landscapes—even deserts. Cloudbursts in desert regions cause many gullies to form. Desert regions of soft clay beds that have been badly eroded in this way have a special name—**badlands** (see Figure 7-29).

The average total precipitation on all land areas each year is at least 125,000 cubic kilometers. Some of this water falls

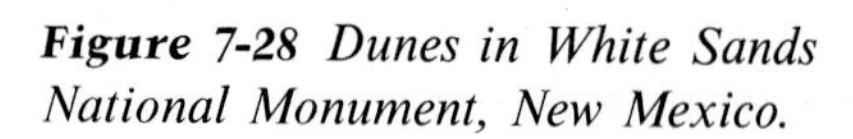

Figure 7-28 *Dunes in White Sands National Monument, New Mexico.*

Figure 7-29 *During the 1930's, a "black blizzard" passed over Springfield, Colorado, in what was called the "dust bowl." Total darkness lasted for half an hour. How could this have been prevented?*

gently as snow, but most of it lands as raindrops. Large drops can splash sand grains 30 cm or more into the air (review Figure 7-21). Where land is not protected by vegetation, raindrops have sufficient force to move particles to lower elevations.

About 75 percent of the total precipitation is either evaporated or remains in rocks and soil. The remaining 31,250 cubic kilometers runs off the land into the sea. The ability of running water to erode land is far greater than all other agents of erosion combined.

Figure 7-30 *Running water has carved this badlands landscape in Badlands National Monument, South Dakota.*

Figure 7-31 *Ocean waves cause erosion along coastlines such as this one in New Brunswick.*

In the United States, stream-gauging stations regularly are used to measure the load of suspended sediment. The earth materials carried in solution and dragged along the stream bed are also measured. If you measure the total load of a stream, it is possible to calculate the total amount of land removed. This amount ranges from less than 40 to over 2000 tons per square kilometer each year. At this rate, it is estimated that all the land now above sea level could be leveled in about 12 million years. However, this is unlikely because other earth processes work to uplift portions of the crust.

Ocean waves also erode the land, but only in a very narrow band along the shores of the continents. Assuming that the strip along the shoreline where waves are active is 200 m wide, waves acting on the entire coast would affect less than 0.2 percent of the land. (See Figure 7-31.)

Although glaciers can transport tremendous amounts of material, they are not a major leveling agent today. Large valley glaciers carved the spectacular landscape in Yosemite National Park. Still larger sheets of ice gouged out the basins occupied by the Great Lakes. Today, however, glaciers cover only about 10 percent of the land area of the world, mainly in Antarctica and Greenland. Even during the last Ice Age, glaciers covered only about 30 percent of the land.

Compared to running water, wind is a very minor erosional agent in most places. On land, wind continually moves fine particles without necessarily carrying them to lower elevations. In arid regions it may shift loose sand around, forming dunes, and making the land less level than before.

Wind blowing in from the sea can actually move material from the beaches inland, building dunes and thereby returning sediment to the land. On some unprotected shores, such as parts of Cape Cod, sand is blown directly out to sea. This is the only example of net erosion by wind alone. Although the wind plays a small part in eroding the land, its delicate sculpturing can create beautiful landscapes. (See Figure 7-32.)

Thought and Discussion

1. What are the causes of erosion?
2. How are particles moved by each agent of erosion?
3. What is a "dust bowl"? How does it develop?
4. How does erosion by glaciers differ from erosion by streams?

Unsolved Problems

People play an important part in the never-ending struggle between ice, wind, water, plants, and gravity and the materials of the earth. Some efforts such as wise methods of farming and flood control reduce erosion. More often, however, overfarming, overgrazing, and strip mining have exposed the soil to the erosive effects of wind and water. The soil must be protected if it is to continue to support life. Solving erosion problems and attempting to conserve our precious soil is a problem that should concern all nations.

Scientists are extending their explorations onto the sea floor. There, they find a whole new set of unsolved problems. Do processes that shape the land like weathering and erosion also operate under the sea? How do they differ from those that occur on land?

Figure 7-32 *The Three Sisters in the Goblin Valley, Utah. Were they produced by the same erosional forces as the features in Figure 7-31?*

Chapter Summary

Weathering is the geologic process that physically and chemically breaks down rocks. The amount and variety of minerals in the rock and the type of environment determine the rate of weathering.

If the loose rock fragments remain in place, soils may eventually develop. In time, these products of weathering are changed into mature soils whose structure depends on the climate and vegetation of the area. If the rock fragments are transported, resistant minerals become separated from those that weather easily.

Gravity exerts a constant force to move material to lower elevations. Gravity moves earth material slowly by creep and rapidly during landslides. Water, ice, and wind move materials long distances. Of these three, running water is by far the most effective agent of erosion.

Although most eroded rock eventually winds up in the sea, the route is long and full of detours. Rock and soil particles are moved, deposited, picked up again, and redeposited countless times before they reach the ocean. Weathering, soil formation, and erosion operate together to wear away the land. Human activities play an important part in weathering and erosion. People promote both erosion and the recovery of eroded lands.

Questions and Problems

A

1. What happens to rocks that are exposed to air and water?
2. Why is water such an important factor in weathering?

SOIL CONSERVATIONIST

Soil conservationists help farmers and other land owners to develop plans for conserving soil and water. Their assistance is of great value in planning land use. Soil conservationists may do detailed mapping of an area to record its soil, water, and vegetation. They recommend methods of preventing further soil erosion and stabilizing runoff. They frequently have to evaluate different plans in terms of cost and effectiveness. Once a conservation program has been decided upon, the soil conservationist gives the land manager continuing assistance in carrying out the program, and helps in solving any problems that may arise.

Most soil conservationists are employed by government agencies at the federal, state or provincial, or local level. There is increasing employment of soil conservationists by private industry as land-use planning grows in importance. A bachelor's degree in soil science or a related field is the minimum education required for work as a soil conservationist.

Soil conservationists should enjoy working outdoors, since a large part of their work is done in the field. The ability to write good reports is useful. In addition, a soil conservationist should be friendly and tactful and like helping other people.

3. How does weathering of rocks and minerals contribute to the well-being of people?
4. What are the products of the weathering of granite?
5. How do mature desert soils differ from soils of forest regions?
6. What is the role of gravity in erosion?
7. How are materials moved by streams?

B

1. Why are almost all the sand particles in the dunes around the Great Lakes and on the beaches of New England composed of quartz?
2. List some of the factors that affect the rate at which a rock will weather away. Which of these factors are most important?
3. A limestone contains 10 percent impurities, including some insoluble clay minerals. If this limestone weathers at the rate of 30 cm in a thousand years, how many years would be required to form 1.5 m of soil?
4. Compare the velocity of a stream to the amount of dissolved chemicals the stream carries.
5. How would you decide whether the soils in your area were formed from bedrock or sediment?
6. What evidence would establish that the loose material at a given location was a glacial deposit?
7. How could you tell whether a stream valley in the mountains was formed chiefly by a glacier or by running water?

C

1. A cube with an edge of one centimeter is cut into 1000 cubes of equal size. What is the length of one of the small cubes? How much surface area is exposed by one of the small cubes? What is the total surface area exposed by the 1000 smaller cubes? If this were a cube of earth material, what effect would the increased surface area have on the rate of weathering?
2. Would the weathering of 1.5 m of limestone and the weathering of 1.5 m of sandstone produce soils with the same thickness? Explain.
3. Material on the top of a hill was found to be 30 percent limestone fragments. Limestone bedrock exists 30 m below the surface. The closest limestone deposit is 160 km away. How could you account for the high content of limestone fragments in the soil materials? Assume that the material at the top of the hill had not weathered from the limestone bedrock.

Suggested Readings

Boyer, Robert E., and P. B. Snyder. *Geology Fact Book.* Northbrook, Illinois: Hubbard Press, 1973

Leet, L. Don; Sheldon Judson; and M. E. Kaufman. *Physical Geology.* 6th ed. Englewood Cliffs, New Jersey: Prentice-Hall, 1982. Chapters 4, 11, 12, 14,15, and 16.

Matthews, William H., III. *Geology Made Simple.* Rev. ed. New York: Doubleday and Co., 1982. Chapters 8–13. (Paperback)

Matsch, Charles L. *North America and the Great Ice Age.* New York: McGraw-Hill Book Co., 1976. Chapters 2 and 3. (Paperback)

Pearl, Richard M. *Geology: An Introduction to Principles of Physical and Historical Geology.* 4th rev. ed. New York: Barnes and Noble, 1975. Chapters 9–13, 15, and 17. (Paperback)

Sediments in the Sea

MARINE GEOLOGISTS *sample the sea floor, to learn how sediment is distributed in the ocean and what kinds of deposits are formed there. Visibility in water is limited to 100 m at best. Even with the help of deep-diving submarines and remote-controlled underwater television cameras, people have seen only a few hundred square kilometers of the sea floor. Oceanographers must rely on other methods of study.*

In the spring of 1971 the United States research vessel Melville *cruised in the Bay of Bengal, off the east coast of India, on its maiden scientific voyage. On the ship's bridge was a satellite navigation receiver. It permitted the ship's officers to determine their position within a few meters. On the stern was a compressor that pumped air under high pressure to a small chamber that trailed in the water behind the* Melville. *About once in 15 seconds, the pressure in the small chamber was released suddenly, producing a loud bang. The sound waves traveled through the water and through the sediments on the bottom of the Bay of Bengal. They finally bounced off the solid rock beneath the sediments and back up to the ship. The time required for the sound waves to travel from the "air gun" through the sediments and back was computed in a recorder. That information plus the position from the satellite navigation receiver went into a computer in the scientists' laboratory. From the computer came a record of the thickness and type of sediments on the floor of the Bay of Bengal.*

For 30 days, the Melville *sailed along a zigzag course, from the northernmost part of the Bay to the waters south of the equator. Each day the scientists aboard became more and more excited. Their data showed great thicknesses of sediments on the sea floor.*

When that part of the cruise was complete, they knew that the Bay of Bengal contains the greatest mass of sediments that has been found in the ocean. A submarine fan spreads from the huge delta of the Ganges and Brahmaputra rivers. It covers 3,000,000 square kilometers and reaches thicknesses of 15,000 m. It has taken at least 30 million years to deposit the sediments. Using space-age technology, scientists learned more about this great volume of sediment in 30 days than they had learned in the previous 80 years.

AFTER COMPLETING THIS CHAPTER, YOU SHOULD BE ABLE TO:

1. **construct or describe a model ocean basin.**
2. **contrast the depositional processes that take place near shore, on the continental shelf, and on the continental slope.**
3. **construct a model of turbidity currents.**
4. **contrast sediments deposited on the continental margins and those in deep ocean basins.**
5. **give examples of events that could change sea level.**

Marine Sediments

8-1 Investigating the Deposition of Sediments

MATERIALS
clock (with second hand), plastic tube (with cap with drain hose and clamp), ring stand and clamp, sediment samples (sorted and mixed)

The violent rush of a stream down a mountain may be more interesting than the quiet lake into which it empties. Even the slow Mississippi River seems more dynamic than the Gulf of Mexico into which it flows (Figure 8-1). However, major parts of the rock cycle go on day after day, year after year, hidden from view in the dark solitude of the deep ocean. One example is the deposition on the sea floor of weathered materials from the land.

When you have completed this investigation, you will have observed sediment deposition in water, and determined the relationship between grain size and deposition speed.

PROCEDURE
Set up the equipment as shown in Figure 8-2. Fill the tube almost to the top with water. Drop a small amount of each sediment into the column and record the time it takes to reach the bottom. Make three trials for each grain size. Then, use the average time of three trials to make a graph of settling time versus grain size. Label the horizontal axis in units of grain size and the vertical axis in seconds.

DISCUSSION
1. Is there a place on your graph where the slope of your curve changes

markedly? If so, why do you think this happens?

2. State the relationship between settling time and grain size.

PROCEDURE

Next, drain off enough water so that the column is only half full. Drop in a handful of mixed sediments and observe what happens. Do this several times.

DISCUSSION

1. Describe how these mixed sediments become arranged above the other sediments in the column.
2. Where in nature might you find deposits like those formed in this investigation?

Figure 8-1 *A high-altitude Landsat photograph of the Mississippi delta. The sediments being carried in the water and deposited on the delta show up as a light blue color.*

8-2 Sediments Reach the Sea.

The products of weathering reach the sea in different ways, but most are brought by rivers. As the materials carried by a river reach the coast, they enter a different environment. Where the fresh and salty waters meet at the mouth of a river, the speed of the river decreases. So does its capacity to carry sediments. The coarser sand particles settle quickly to form sand bars and beaches. Sediments that may have been in the rivers for years and carried for thousands of kilometers are dumped. River mouths are settling basins for the sediments.

At the mouths of all rivers, the sediments form deposits called **deltas.** The Nile Delta of Egypt is shown in Figure 8-3. A delta keeps its shape if there is a balance between the amount of sediment brought by the river and the amount carried away by ocean waves. The Nile Delta had been in perfect balance for thousands of years until 1964 when Egyptians built the Soviet-designed Aswan Dam. The dam changed the flow of sediments.

Figure 8-2 *A setup for investigating sediment deposition.*

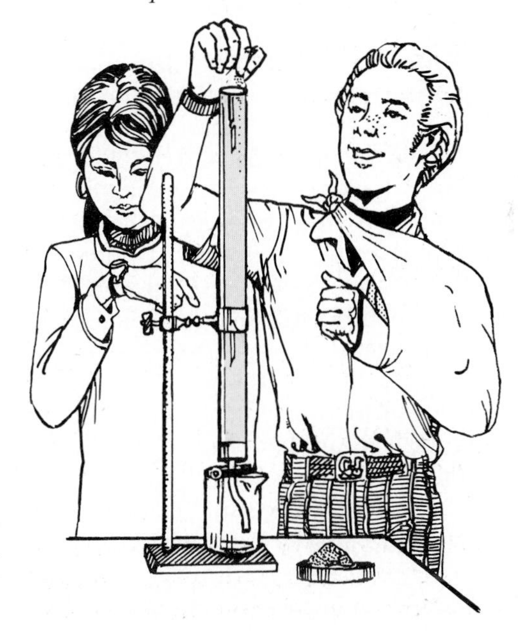

Figure 8-3 *The Nile Delta, Egypt. Compare the sediment flowing from the Mississippi Delta with the clear waters off the mouth of the Nile.*

The delta and its beaches along the Mediterranean shore are now deprived each year of 130 million metric tons of sediments formerly carried by the river. The muds are deposited behind the Aswan Dam. As a result, the Mediterranean waves are eroding the delta. Rich soils are no longer dropped on the Nile Valley and Delta by flooding river waters. For the first time in history, farmers must use commercial fertilizers to raise their crops.

A similar, though less disastrous, case of erosion was created at the Colorado River Delta in Mexico. The cause was a series of dams on the Colorado River, that the United States built during the 1930's, 1950's, and 1960's. Sediments are no longer reaching the Colorado Delta, and it is eroding. There is no human population on this delta, so its change is hardly noticed.

Just the opposite is true of the Mississippi Delta. It has grown 15 km into the Gulf of Mexico since the Civil War. Here is one place along the coast of the United States where one can see new land added during a lifetime. Such a place is rare, however. It is only at the deltas of the great, unchanged rivers of the world where such growth takes place.

Not all of the material carried to the sea by rivers is deposited near the mouth of the river. Some particles enter the sea without settling to the bottom. (See Figure 8-4.) Other material is moved along the seacoast away from the mouths of rivers by nearshore currents. These currents build most of the beaches we see at the shore.

The sand, silt, and clay particles that reach the ocean settle at different rates, depending mainly on their size, shape, and density. This was evident in Investigation 8-1. At the same time the particles are settling, they are also carried along by ocean currents. They may travel far before reaching the sea floor. How far they go depends on the settling rate of the particle, the speed and turbulence of the current, and the depth of water. Look at the settling rate for the finest particle in Figure 8-4. These particles would be deposited far from the river that brought them to the sea. Some even travel halfway around the world before reaching the bottom of the sea.

8-3 Sediments Accumulate on the Ocean Floor.

The products of erosion eventually fall to the ocean floor. The processes that move sediment are most active near the land. So the greatest deposits form on the adjacent sea floor. The **continental shelves** are submerged parts of the continents bordering the ocean. (See Figure 8-5.) The depths of the shelves beneath

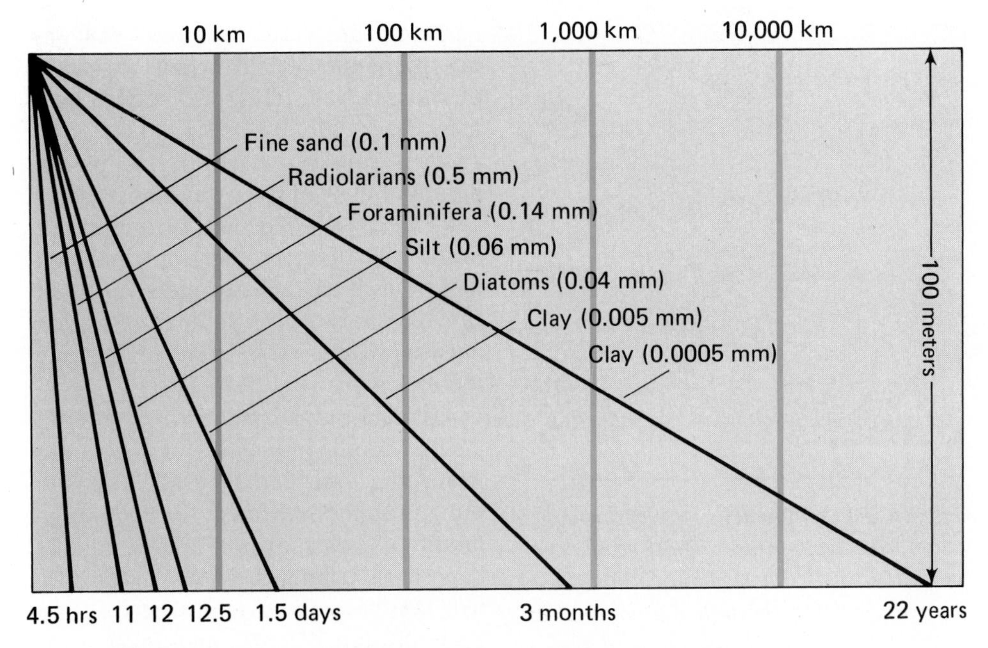

Figure 8-4 *In moving water, particles are carried different distances before they settle. In this example, the current is flowing at 10 cm per second. How does the size of the particle affect the settling rate?*

the surface of the sea vary from one coast to another, and even along the same coast. Off eastern North America, for example, shelf depths range from 50 to 150 m. Beyond the continental shelf the sea floor dips more steeply. This region is called the **continental slope.** Even though the slope becomes steeper, the incline is only about the same as an aisle in a theater. If Figure 8-5 were drawn to scale, it would show a nearly straight line, not a sharp plunge. At the base of most continental slopes, there is an apron of sediments that have moved down the slope and come to rest in deep water. This apron is called the **continental rise.**

Shelves, slopes, and rises border all the continents. Their origins have been debated for many years. However, it is clear that the shelves and slopes are shaped by both erosion and deposition. Both processes are important. During parts of the Ice Age when sea level was lower, erosion could smooth the surface of the shelves. Now that the shelves are submerged, deposition is more active. In the Gulf of Mexico, for example, the shelf and slope features are the result of deposition. In some places along the east

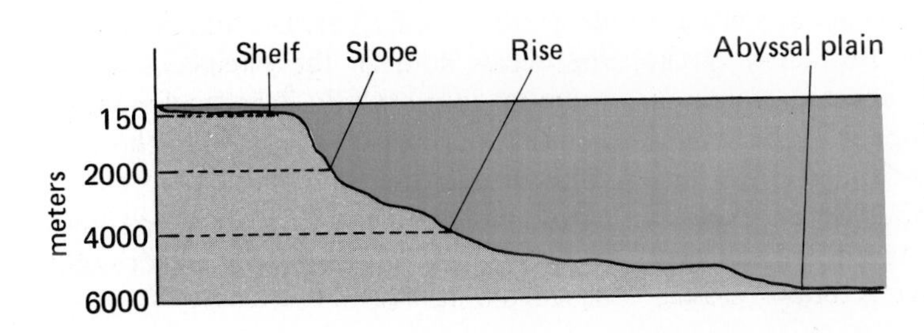

Figure 8-5 *A cross section of a typical continental margin.*

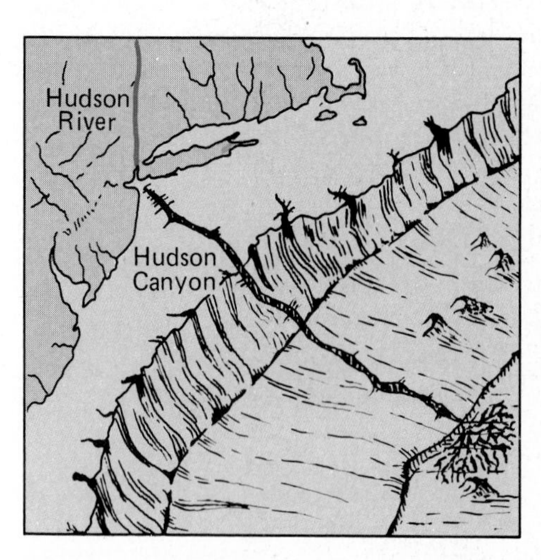

Figure 8-6 *Sediments moving through a submarine canyon are deposited at its mouth. This is the Hudson Submarine Canyon, cutting through the continental margin off the northeast coast of the United States.*

coast of North America, the two processes continue to act together to shape the submerged features. The origin of a continental rise never varies, however. No matter where it is studied it is found to be caused by deposition.

Many deep submarine canyons cut across the continental slopes and into the continental shelves. (See Figure 8-6.) In some ways they are like canyons on land. Some are enormous, extending hundreds of kilometers from near shore into the deep-sea basins. In southern California, some canyons come so close to shore that fishermen can drop their fishing lines from piers into the 60-m-deep water at the head of a canyon.

The canyons off the coast of southern California have been studied in great detail. They reach depths of about 500 m, and many appear to be extensions of canyons on land. It was first thought that the canyons were carved by rivers during the Ice Age, when sea level was 150 m lower than it is today. In other parts of the world, however, canyons dive more than 3000 m below the surface of the sea. The Hudson Canyon near New York City is one example. The greatest canyon of all begins in the mouth of the Congo River in west Africa and extends to 6000 m below the surface of the sea. It is clear, therefore, that submarine canyons are not the remains of canyons cut by rivers. Submarine erosion must play the major role in forming these canyons (Figure 8-7).

Although submarine canyons differ in length, depth, and kinds of rocks into which they are cut, they all have one thing in common. On the sea floor at the mouth of each canyon is a large, fanlike deposit of sediments. These deposits are like fans formed on land, but they are generally much larger and cover great areas of the sea floor. The Bengal Fan in the Indian Ocean is the best example currently known. It is not unusual to dredge up shallow-water shells and twigs from the fan. So we know the sediments come down the canyons, eroding and carving through the continental shelf as they go.

Beyond the continental slope, at depths greater than 3000 m, lie wide, slightly rolling plains. They cover 63 percent of the sea floor. That is nearly half of the earth's surface. (See Figure 8-8.) These plains are interrupted by many mountains and deep trenches. In the Atlantic Ocean the plains are hundreds of kilometers wide. They begin at the continental slopes and reach to the mid-Atlantic ridge. In the Pacific Ocean basin they begin at the great island chains instead of at the continental slopes. In Figure 8-9 you can see that deep trenches lie between the land and the ocean floor in the Pacific. A good example is the Chile Trench along the South American coast. Thus, in the Pacific Ocean, the sediments that move across the continental rise are deposited in deep-sea trenches instead of forming broad plains.

Figure 8-7 *A "sandfall" about 10 m high occurred in a submarine canyon off Baja California.*

8-4 Investigating Turbidity Currents

MATERIALS
clock (with second hand), marking crayon, plastic tube (with cap or stopper), ring stand, clamp, slurry material, two or three large test tubes

Sedimentary deposits at the mouths of submarine canyons blend into the deep-sea plains or fall into trenches. The materials of the fans and the plains both come from the land. Masses of stirred-up sediment flow down the canyons and slopes and spill out onto the plains. Such flows are called **turbidity** (*tur BIHD uh tee*) **currents.**

Coarse fragments are found at the mouths of submarine canyons. They arrived there by streaming across the canyon floor. So far no one has been on the

Figure 8-8 *The percentages of the earth's surface above and below sea level.*

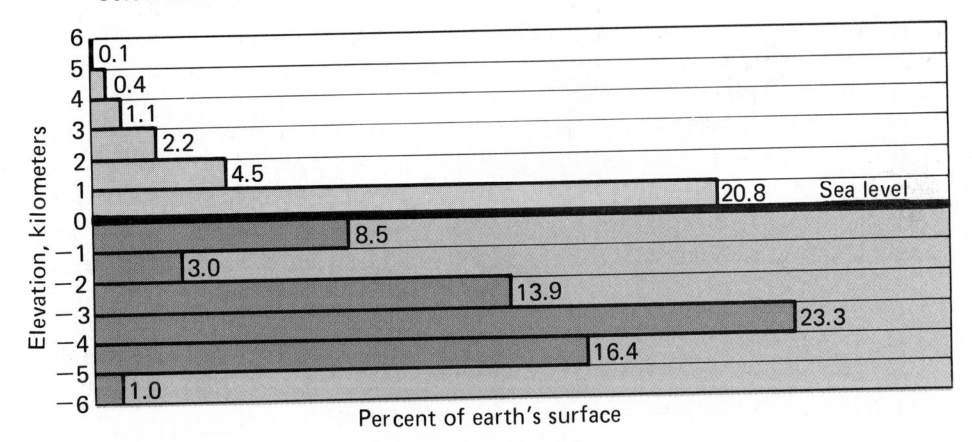

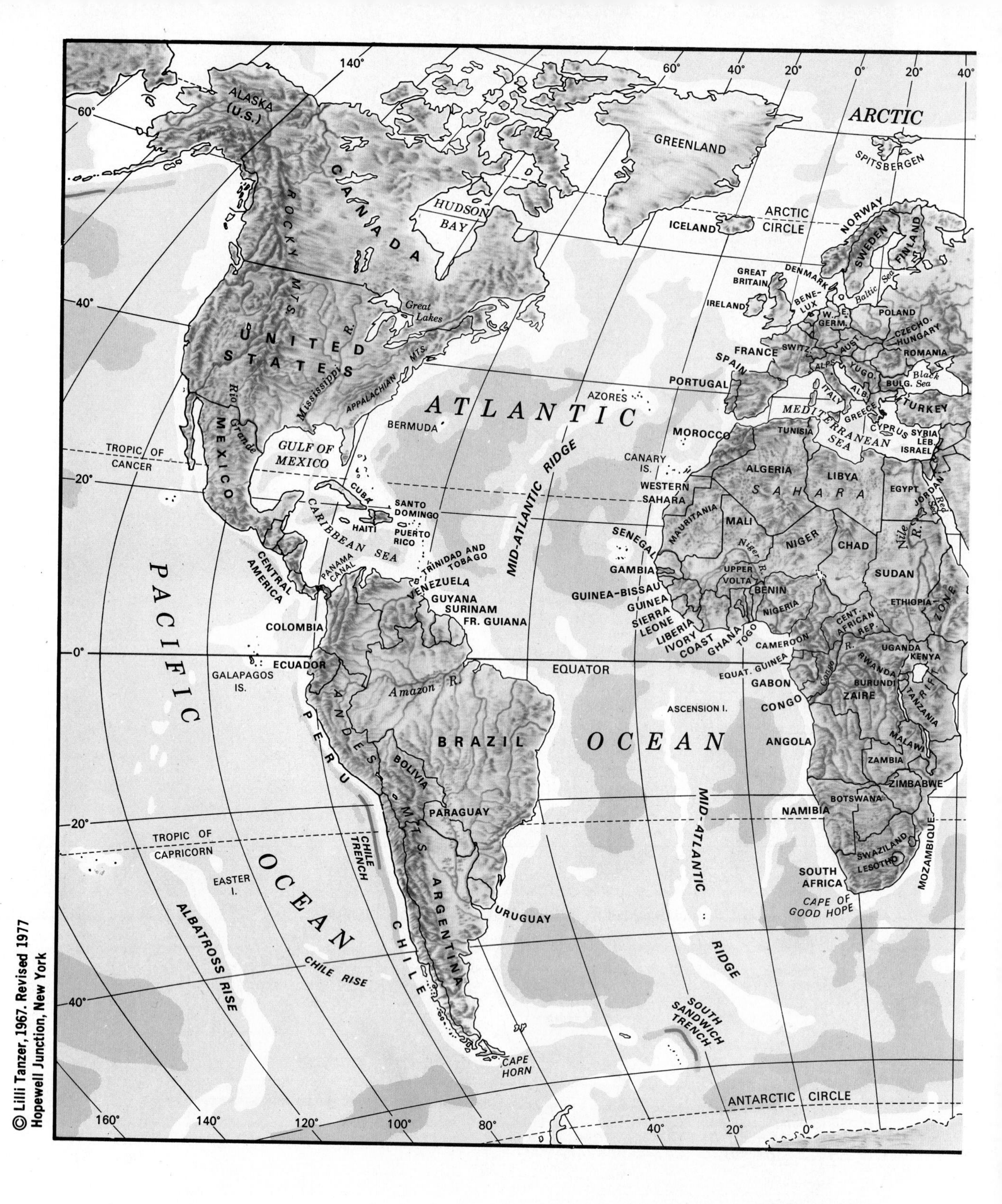

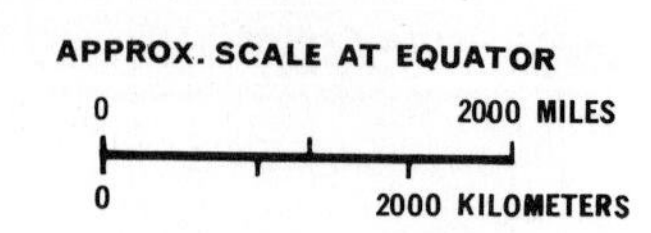

Figure 8-9 *A map of the world.*

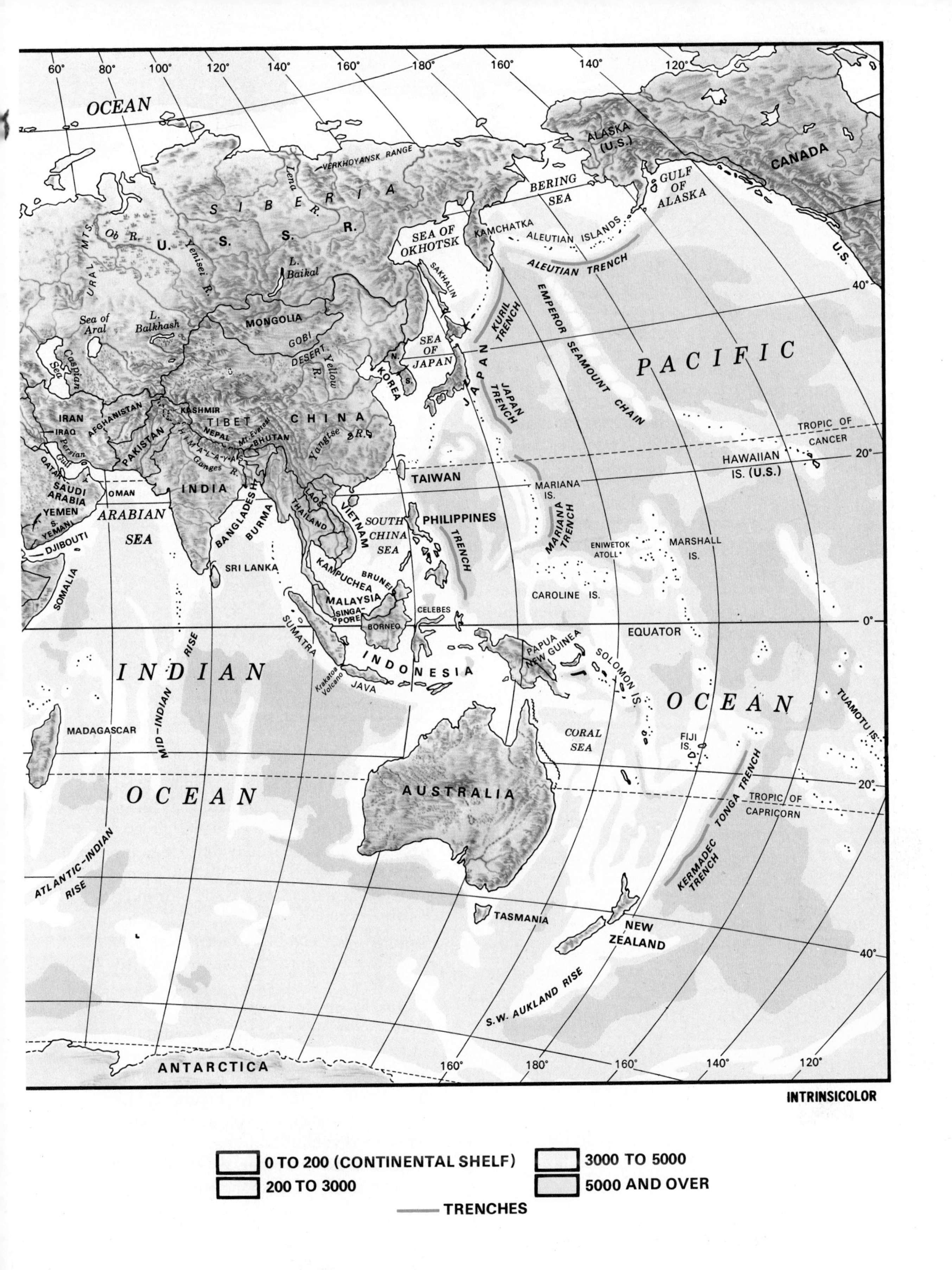
OCEAN
VERKHOYANSK RANGE
Lena
SIBERIA
U.S.S.R.
Ob R.
URAL MTS.
Yenisei R.
L. Baikal
Sea of Aral
L. Balkhash
Caspian Sea
MONGOLIA
GOBI DESERT
Yellow R.
CHINA
TIBET
KASHMIR
NEPAL
Mt. Everest
BHUTAN
HIMALAYAS
Yangtse R.
IRAN
IRAQ
AFGHANISTAN
PAKISTAN
Persian Gulf
QATAR
SAUDI ARABIA
OMAN
YEMEN
S. YEMEN
DJIBOUTI
ARABIAN SEA
SOMALIA
INDIA
Ganges R.
BANGLADESH
BURMA
SRI LANKA
THAILAND
LAOS
VIETNAM
KAMPUCHEA
MALAYSIA
SINGAPORE
BRUNEI
SOUTH CHINA SEA
PHILIPPINES
PHILIPPINES TRENCH
TAIWAN
N. KOREA
S. KOREA
SEA OF JAPAN
JAPAN
JAPAN TRENCH
SEA OF OKHOTSK
SAKHALIN
KAMCHATKA
KURIL TRENCH
BERING SEA
ALASKA (U.S.)
GULF OF ALASKA
CANADA
U.S.
ALEUTIAN ISLANDS
ALEUTIAN TRENCH
EMPEROR SEAMOUNT CHAIN
PACIFIC OCEAN
TROPIC OF CANCER
HAWAIIAN IS. (U.S.)
MARIANA IS.
MARIANA TRENCH
ENIWETOK ATOLL
MARSHALL IS.
CAROLINE IS.
EQUATOR
CELEBES
BORNEO
SUMATRA
INDONESIA
JAVA
Krakatoa Volcano
PAPUA NEW GUINEA
SOLOMON IS.
CORAL SEA
FIJI IS.
TUAMOTU IS.
TONGA TRENCH
TROPIC OF CAPRICORN
KERMADEC TRENCH
INDIAN OCEAN
MID-INDIAN RISE
MADAGASCAR
AUSTRALIA
TASMANIA
NEW ZEALAND
ATLANTIC-INDIAN RISE
S.W. AUKLAND RISE
ANTARCTICA
60°
80°
100°
120°
140°
160°
180°
40°
20°
0°
INTRINSICOLOR
0 TO 200 (CONTINENTAL SHELF)
200 TO 3000
3000 TO 5000
5000 AND OVER
TRENCHES

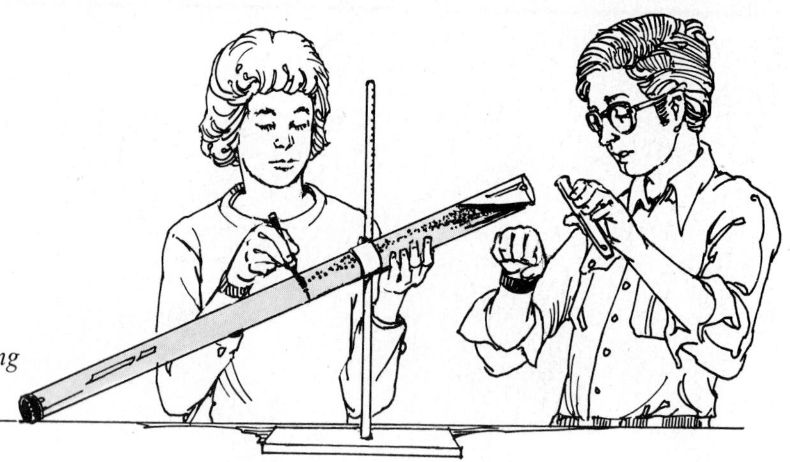

***Figure 8-10** A setup for investigating turbidity currents.*

sea floor at the precise moment when a turbidity current has moved down a canyon. Therefore, we don't know exactly how they look.

When you have completed this investigation, you will have studied turbidity currents by using a laboratory model. You will have seen how sediments can move down a submarine canyon.

PROCEDURE

Set up the equipment shown in Figure 8-10. Mix a slurry of soil and tap water. What do you think will happen when you pour the slurry into the sloping column of water? Test your prediction. Time the movement of the slurry from the water surface to the bottom of the tube. Mark where the front of the slurry is at 5-sec intervals. Pour 10 or 12 more slurries of the same material into the column. Let every second or third one settle. Compare the rate of movement of later slurries with earlier ones. Remove overflow water carefully so you don't disturb the settled sediment.

DISCUSSION

1. Did the results of your investigation agree with your prediction?
2. Describe the speed and motion of the material as it travels down through the plastic column.
3. How do you suppose turbidity currents similar to the ones you produced are caused in nature?
4. How can turbidity currents carry coarse sediments far out into the ocean?
5. Coarse continental sediments are found throughout the basins in the Atlantic. Why are they not found in the Pacific basins?

8-5 Some Sediments Form in the Sea.

If you examined a sample of mud from the sea floor, you would see that all of the material in it did not come directly from the land. You would notice remains of marine organisms and tiny, sharp mineral crystals. These could not have lasted through a long land-to-sea journey. The shells and tiny mineral crystals are chemically formed from ions carried into the sea by rivers.

Marine plants need certain ions to grow, especially the nitrates, silicates, and phosphates. The most familiar marine plants are seaweeds and grasses that grow near shore. (See Figure 8-11.) However, these do not compare in number or total volume with the vast quantities of microscopic plants that live in the open ocean. There may be tens of thou-

Figure 8-11 *Seaweed flourishes along rocky coasts.*

sands in a single liter of water! The animals are also numerous. Larger organisms, such as young fish, eat the tiny ones as part of the food chain.

The number of animals and plants in the surface water is partly controlled by the amounts of nutrient ions and the rate of feeding. The plants can rapidly use up the nutrient ions in a still body of water such as a lake. Plant and animal life are most abundant, therefore, near mouths of rivers where nutrients come from the land, or in ocean areas where nutrients are brought up from the depths of the sea to the surface.

Waters rise to the surface along the borders of the many great currents flowing through the ocean (Figure 8-12). (See also Figure 5-16.) Here the nutrient-rich, rising waters support an enormous pop-

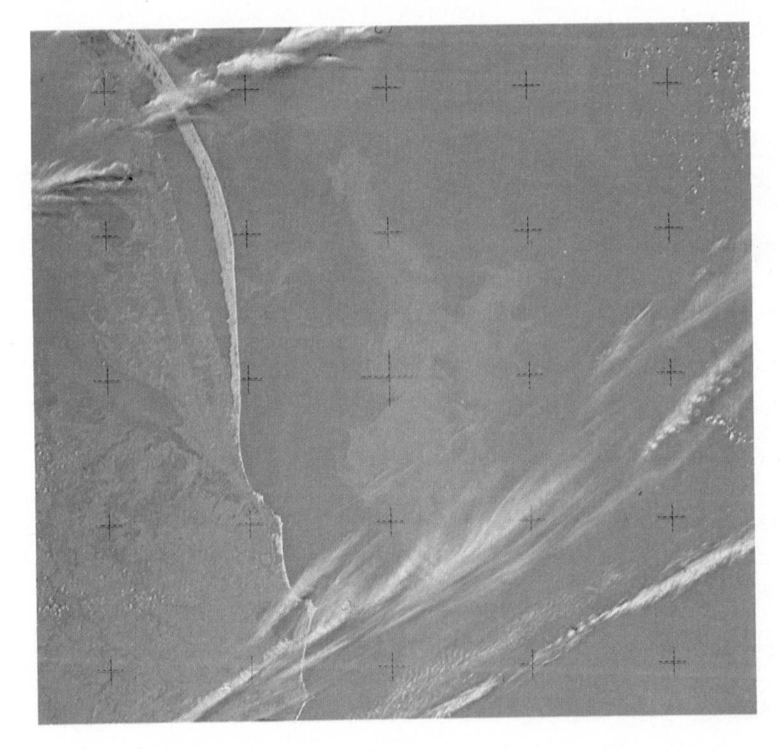

Figure 8-12 *The irregular light patches in the center of this photograph are masses of plankton. The patches extend for hundreds of kilometers along the coast of Brazil.*

Figure 8-13 *The delicate shells of microscopic marine algae called diatoms.*

Figure 8-14 *Remains of foraminifera are found in deep-sea sediments.*

Figure 8-15 *Phillipsite crystals from the* Challenger Reports.

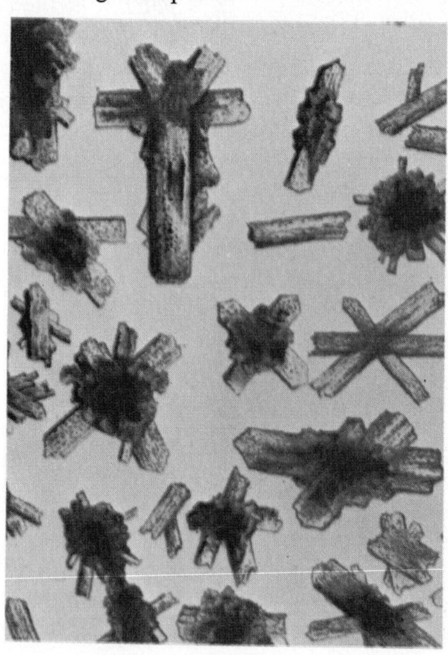

ulation of **diatoms** (*DY uh tahmz*), microscopic plantlike organisms with silica shells (Figure 8-13). The sediments on the nearby sea floor are made up almost entirely of diatom shells.

The most common organic remains in marine sediments are the carbonate skeletons of microscopic organisms called **foraminifera** (*fawr uh mih NIHF uh ruh*), shown in Figure 8-14. Thousands of kinds of foraminifera and billions of individual organisms live in all of the surface waters of the sea, feeding on organisms such as diatoms. Most marine sediments contain remains of foraminifera.

8-6 Minerals Form in the Sea.

When sediments were analyzed from the *Challenger* Expedition, it was learned that the mineral phillipsite was abundant in the deep-sea deposits of the Pacific Ocean (Figure 8-15). Phillipsite is interesting because it is not found in rocks on continents. The crystals have been collected only from deep-sea sediments. Furthermore, phillipsite occurs in sediments as isolated crystals, rarely touching one another. The *Challenger* scientists reasoned that the mineral had crystallized on the surface of deep-sea muds. This was the first real evidence that the deep-sea environment differed greatly from the shallow seas near continents. Minerals that could not form anywhere else on the earth formed in the deeps.

Soon it was learned that many deposits on the sea floor have formed from seawater. Widespread deposits of manganese lumps formed in this way. These, too, were first discovered during the *Challenger* Expedition. Exploration during the International Geophysical Year (1957–58) disclosed thousands of square kilometers, especially in the South Pacific Ocean, covered with manganese deposits. In 1974, scientists on vessels from the United States and West Germany examined vast deposits around the Hawaiian Islands.

The manganese, with small amounts of cobalt and iron, occurs as small grains, larger lumpy nodules, slabs, and coatings on rocks. (See Figure 8-16.) Most of the nodules are about five centimeters in diameter. No one knows how fast they form. They must grow faster than the rate of deposition of particles and organic debris. Otherwise, the first tiny grains of manganese formed would be covered with sediment and not grow any larger. Consequently, manganese nodules form only in areas where there is little deposition of sediment.

Phillipsite and manganese are deep-sea minerals. Under some conditions, different minerals can form in shallow water. Calcium carbonate is the most common. It is soluble in seawater. The amount dissolved depends on the amount of carbon dioxide in the water. Seawater is normally saturated with calcium carbonate.

Figure 8-16 *Manganese nodules. The cross-section shows how nodules grow in layers.*

ACTION

MATERIALS
salt, small glass, filter paper, magnifying glass

Keep adding salt to a small glass of hot water until no more will dissolve. Filter the water and allow the clear filtrate to stand undisturbed for one week while crystals form. Blot the crystals dry with filter paper. If possible, use a magnifying glass to study their shape.

When cold water from deep in the sea rises over a shallow bank, the water warms. The warmth drives off some of the dissolved carbon dioxide in the water. (Warm water can't hold as much carbon dioxide as cold water.) On such shallow banks there is usually a thick growth of sea plants. They carry on photosynthesis, which further reduces the amount of carbon dioxide in the shallow water. As a result, calcium carbonate forms on the banks, as in Figure 8-17.

***Figure** 8-17 Carbonate deposits form in the warm, shallow tropical waters off the Florida Keys.*

Thought and Discussion

1. Do all particles eroded from land stop at the edge of the sea?
2. Where do sediments at the mouths of submarine canyons come from?
3. Where do the materials come from that sea organisms use to make their shells?
4. How do manganese nodules develop on the ocean floor?

The Continental Margins

8-7 The Shorelines Have Moved.

Evidence of ancient beaches comes mainly from marine sedimentary rocks. Today marine sedimentary rocks of many kinds cover about three-fourths of the continents. Many of the rocks contain fossils that show that the rocks were formed in shallow seas. Certain places on the continents have no sedimentary rocks. In these places erosion has stripped them away. Evidently, the continents were at various times covered by shallow seas like those now covering the continental shelves. As the ancient seas advanced or retreated across the land, beaches formed at their edges.

At times sea level was lower than it is today. During the Ice Age, water evaporated from the sea and dropped upon the land as snow, forming glaciers. The glaciers covered much of the land in the Northern Hemisphere. Removing water from the sea to make up the glaciers lowered the level of the ocean. One kind of evidence of this is teeth from prehistoric elephants, which have been recovered from the continental shelf off the east coast of the United States. They were in 60 m of water along with shells of animals that live only in shallow water or mud flats.

As glaciers grew or melted back, they alternately held and released great volumes of water. The level of the sea rose and fell by as much as 150 m (Figure 8-18). Beaches, sand dunes, and mud flats that must have been formed at low levels of the sea have been changed or removed by erosion and deposition. The result is the level shelf surface of today.

From Investigations 8-1 and 8-4, you know that coarse sedimentary material is deposited rapidly. Finer fragments are usually carried far out to sea. This pattern of sediments is obvious on the continental shelf and slope off eastern North America. In some cases, however, layers of sand and mud form one on top of the other. As a delta grows out into the sea, the coarse sand that formed on the beach gradually pushes out onto a bed of fine clay that formed on the outer edge of the delta. In contrast, a retreating shoreline would deposit mud on sand. As the shoreline advances and retreats, layers of sand and mud alternate.

Beaches are easily eroded because the sediments are loose. If sea level falls, stream erosion wears away the beach deposits. The fragments are scattered into the lowered sea and along the new shore. If sea level rises, waves breaking over the old beach soon destroy it. Ancient beaches are rarely preserved within the layers that make up the shelf. However, there are remnants from old beaches, such as pebbles, shells, and sand.

8-8 The Thickening Continental Margins

For millions of years a river system in the United States carried sediments into an ocean basin. This basin, now filled, is the Mississippi Valley. The delta is still growing seaward, just as deltas have in this region for millions of years. In fact, the first delta deposits were laid down near Cairo (*KAIR oh*), Illinois.

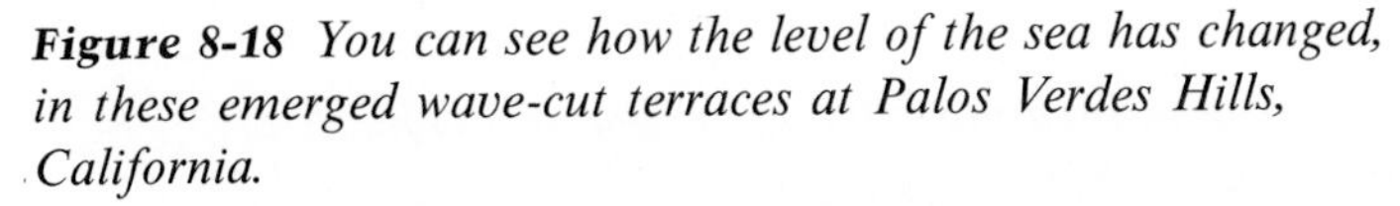

Figure 8-18 *You can see how the level of the sea has changed, in these emerged wave-cut terraces at Palos Verdes Hills, California.*

Each day the Mississippi carries two million metric tons of sediment to the Gulf of Mexico. This sediment is eroded from 41 percent of the land area of the United States. Some of the particles are deposited on the delta and in the shallow water along its edges. Some are swept along the coast into bays and marshes. Others travel down the delta onto the submerged fan in the Gulf of Mexico.

For tens of millions of years, the river has been carrying soils and sediments to the sea. As you would expect, the Gulf of Mexico is gradually filling up. (Most of the sediments come from the Mississippi, but other rivers also enter the Gulf.) Five other coastal regions of the world receive similar volumes of sediment: West Africa off the Congo River, South America off the Amazon and Orinoco, China off the Yangtze, India off the Ganges, and Siberia off the Lena River. In each case, the filling has been going on for millions of years, and a broad, flat coastal plain has been formed. (See Figure 8-19.)

Geologists exploring for oil have learned that the sedimentary layers in the lower Mississippi Valley and along the coast are 15,000 m thick and bend down into the earth's crust. Far out in the Gulf, sediments are only a few hundred meters thick. These layers along the north coast of the Gulf of Mexico have all been deposited by the Mississippi River.

Deep wells drilled along the east coast of the United States have shown that the sediments there are nearly as thick as those along the Gulf Coast. The deposits have come from the rivers draining the Appalachian Mountains. In contrast to these thick deposits near shore, detailed studies made in many of the ocean basins of the world have shown that sediments in the deep sea are only from 300 to 500 m thick. Compare those with the thickness of sedimentary layers along the Gulf of Mexico and the east coast. It seems that most thick sedimentary layers are continental margins.

All of the sediments in the layers that make up the continental shelves have shallow-water features. So all the layers deposited over millions of years must have been deposited in shallow water. That means the shelves must be sinking. Otherwise, the sediments would have long ago built up to sea level. If you assume that the bottom of the shelf has been sinking and the top has been re-

Figure 8-19 *The colored areas bordering continents mark the continental shelves.*

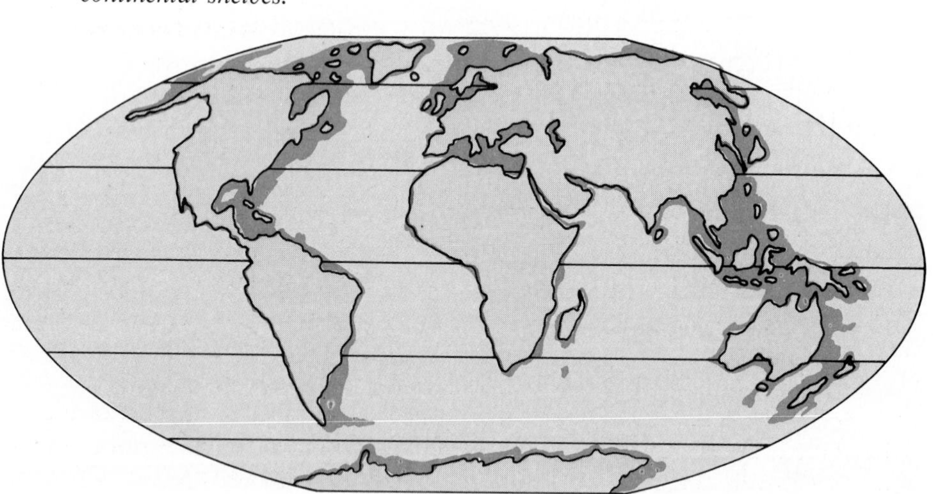

ceiving sediments for millions of years, the great thickness of sediments is explained. The shelves are in very slow motion, then, gradually sinking while sediments are continually added by great rivers (Figure 8-20).

At the same time, masses of material spread farther and farther from the original shore. The shelf and slope are pushed out into the ocean basin. And a great deal of material slides down the continental slope and through the submarine canyons to form the thick layers of the continental rise. If the rate of deposition keeps pace with the rate of sinking, or subsidence, there is no change in water depth or in the position of the shoreline. What happens when deposition is faster or slower than subsidence?

Careful measurements over the last century have shown that the high tide mark has changed differently at some shorelines than at others. These variations in the changes could not be caused by glaciation in the Ice Age. The growth and retreat of glaciers would cause all shores to be alternately covered and then exposed at the same rate. The continental shelf is sinking today where great thicknesses of sediments are still forming (along the east and Gulf coasts of North America). Where such continental margins are absent (west of North America), there is no evidence of sinking.

Thought and Discussion

1. How would you prove that the shorelines of the ocean are not stationary?
2. What do great thicknesses of sediment such as those found along the Gulf of Mexico indicate?
3. How do glaciers influence sea level?
4. Compare the continental shelf deposits of the Gulf of Mexico and the east coast of North America.

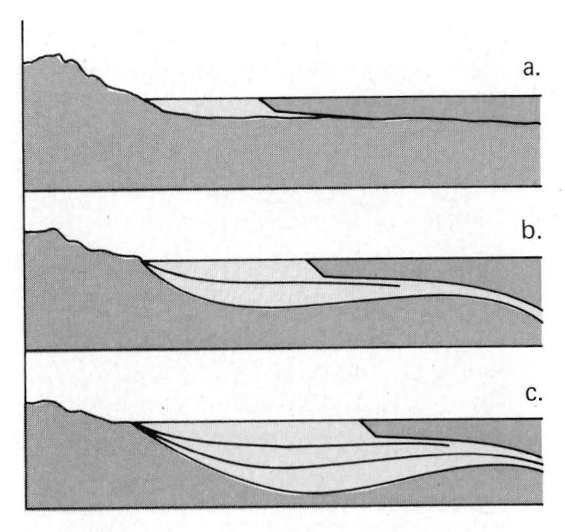

Figure 8-20 *The continental shelf changes slowly as a delta grows seaward. How do the middle and bottom frames differ?*

Unsolved Problems

In 1970, the greatest unsolved problem regarding ocean basins was their origin. By the 1980's, the development of the present basins was well established. Now the question is, "How many ocean basins have there been throughout Earth's history?" Clearly there have been more than we know today, but the puzzle requires many detailed studies of the ocean's rocks.

Have any deep-sea sediments ever formed into rocks that now occur on land? Certainly ancient rocks must have formed from such sediments, and marine geologists would like to examine them. They have not come from cores drilled in the sea floor. A geologist from the University of Hawaii believes she sampled some in the Solomon Islands in 1980. Her analysis is interesting, yet still incomplete.

Certain areas of the sea floor, far away from sources of sediment, have few deposits. Rocks are not covered with sediment, or are close to the surface of the sea floor. In 1981 oceanographers completed a study of these parts of the sea floor, and learned that they are as quiet as had been supposed.

GEOLOGIST

Geology, the science of the earth, consists of three areas: earth materials, earth processes, and earth history. Within each area there are many fields of specialization.

Geologists (*jee AHL uh jihsts*) interested in earth materials may be economic geologists, who look for minerals and coal, or petroleum geologists, who look for usable sources of oil and natural gas. Engineering geologists determine whether a location is appropriate for construction of a dam, tunnel, or buildings. Mineralogists (*mihn ur AHL uh jihsts*) study the structure and composition of minerals, and geochemists study the chemical composition of minerals and rocks.

Geologists who study earth processes are interested in the forces that create surface landscapes. Geomorphologists (*jee oh mawr FAHL uh jists*) study landforms and the effects of different agents of erosion. Volcanologists (*vuhl kuhn AHL uh jihsts*) study volcanoes and their causes.

Geologists interested in earth history may study plant and animal fossils to trace their changes through time, or they may study rock layers and match them up with layers at other locations.

More than half of all geologists work in private industry, most of them for petroleum producers. Public geological surveys and bureaus of mines, land, and resource management employ many geologists. Many geologists work and teach at colleges and universities. An advanced degree is recommended, but not absolutely required, for most research, teaching, and exploration jobs.

Geologists should have curiosity and enjoy solving complicated problems with less than complete data. Exploration geologists should like the outdoors. Field work is sometimes strenuous, but a person of average strength in good health should have few problems. Geologists generally work in teams, so the ability to get along with others is a definite plus.

Chapter Summary

The oceans of the world are immense. They cover 71 percent of the earth's surface and are deep. If all of the land masses were scraped into the oceans, the entire earth would still be covered with 200 m of water. In fact, the lands are being scraped into the oceans by erosion. The United States is being eroded at a rate of about five centimeters every thousand years. Rocks and soils are worn away and deposited in the ocean.

Sediments are moved and deposited by the processes you investigated in this chapter. These sediments are not only deposited in deltas and continental margins, but they are spread over great areas of the sea floor and out into the deep-sea plains. There are some minerals that form in the ocean.

Ancient beach sediments show that seas advanced and retreated across the North American continent in the geologic past. During the Ice Age, sea level was lowered and raised several times, as glaciers grew and melted back. The beaches that formed when sea level was down have been destroyed by erosion.

The shelf, slope, and rise of the continental margins of most of the major land areas of the world are depositional features. They make up tremendously thick deposits along the borders of the continents and ocean basins. The Atlantic and Gulf coasts of North America are depositional areas that are slowly sinking.

Questions and Problems

A

1. What happens to the settling speed of a particle of volcanic ash as it falls and crosses the air-sea interface?

2. Some submarine landscape features such as volcanoes and canyons show sharper outlines than their counterparts on land. Explain this.

3. What determines the rate and quantity of microorganism production and deposition on the sea floor?

B

1. How do turbidity currents move and distribute sediment on the sea floor?

2. What are some minerals that originate in the sea and how do they form?

3. What besides glaciers causes changes in shorelines?

C

1. The average thickness of sedimentary rocks in the earth's crust is about 0.74 km. Assume a uniform rate of deposition of 40 mm per 50,000 years. How long did it take to accumulate the sedimentary rocks of the crust? Do you think your answer equals the total time elapsed since the deposition of the first sedimentary rocks? Why or why not?

2. Some submarine canyons may have been carved by rivers that flowed across continents. How could this happen? What evidence is there that not all of them formed in this way?

Suggested Readings

Emery, K. O. *A Coastal Pond.* New York: American Elsevier Publishing Company, Inc., 1969.

Erickson, D. B., and Wollin, G. *The Ever-changing Sea.* New York: Alfred A. Knopf, 1967.

Shepard, F. P. *Geological Oceanography: Evolution of Coasts, Continental Margins and the Deep Sea Floor.* New York: Crane, Russak, and Co., 1977.

Shepard, F. P. *The Earth Beneath the Sea.* Baltimore: Johns Hopkins Press, 1967.

9 Mountains and the Sea

EARLY IN THIS CENTURY *an extraordinary fossil discovery was made high in the Canadian Rockies. A geologist's pack horse stepped on a loose piece of black shale and turned it over. There, embedded in the dark rock, were the remains of animals that had once lived in the sea! The soft parts of delicate marine organisms, such as jellyfish and marine worms, had been beautifully preserved. This fossil discovery showed that the rocks making up that part of the Rocky Mountains came from the sea.*

More than five hundred million years ago the fossils in the rocks were living animals in a shallow, warm sea. The animals sank into the mud after their deaths. As more and more sediment was deposited, the animal remains were compressed by deposits several kilometers thick and eventually became fossils.

Only a few decades before this discovery, scientists had argued about the origin of fossils. Some said that the remains of animals embedded in rocks had all been left by a worldwide flood. Other scientists pointed out that fossils occurred in deposits that lay one on top of another. These scientists reasoned that the fossils of the bottom layers must be older than the fossils in the upper layers. They decided it must be possible to determine the times when the animals lived and died.

The idea that different kinds of fossils came from different periods of geologic time was accepted before 1900. But geologists still did not know how great mountain ranges form, or how ocean bottoms could be raised thousands of meters above sea level. They knew there must be tremendous forces in the earth to push the rocks upward. But there was no certain understanding of the origin of the great, thick layers of sediments that had been deposited on the bottoms of the ancient oceans.

AFTER COMPLETING THIS CHAPTER, YOU SHOULD BE ABLE TO:

1. **describe the characteristics that are common to geosynclinal sediments.**
2. **use these characteristics (identified in #1) to construct a model of a geosyncline.**
3. **identify mountain ranges on the earth's surface that have apparently developed from ancient geosynclines.**
4. **distinguish between geosynclinal and nongeosynclinal sediments and locate areas on the earth's surface where geosynclines may be in the process of forming today.**
5. **explain how solid rocks are folded and bent without being destroyed.**
6. **describe the development of mountains through the stages of deposition, deformation, and uplift of geosynclines.**
7. **locate belts of crustal mobility on the earth's surface and describe probable reasons why they occur where they do.**

Mountains from the Sea

MATERIALS
graph paper

9-1 Investigating Ancient Marine Sediments

About 1837 James Hall, a geologist on the staff of the New York State Geological Survey, became puzzled by the great layers of sedimentary rocks in western New York. He was intrigued by variations in the thickness of layers and puzzled by the evidence that each layer had been deposited in shallow water. Also, some layers that must have been deposited horizontally were tilted.

When you have completed this investigation, you will have used data gained from observations of rocks to reconstruct past events in the earth's geologic history.

PROCEDURE
You can retrace some of James Hall's studies, using the same kind of information available to him. The data come from the area between Buffalo, New York, and western Massachusetts (Figure 9-1). Study the photographs in Figure 9-2 and note what each one shows you about the past.

Draw a cross section on graph paper along the route shown in Figure 9-1. You will have to decide what horizontal and

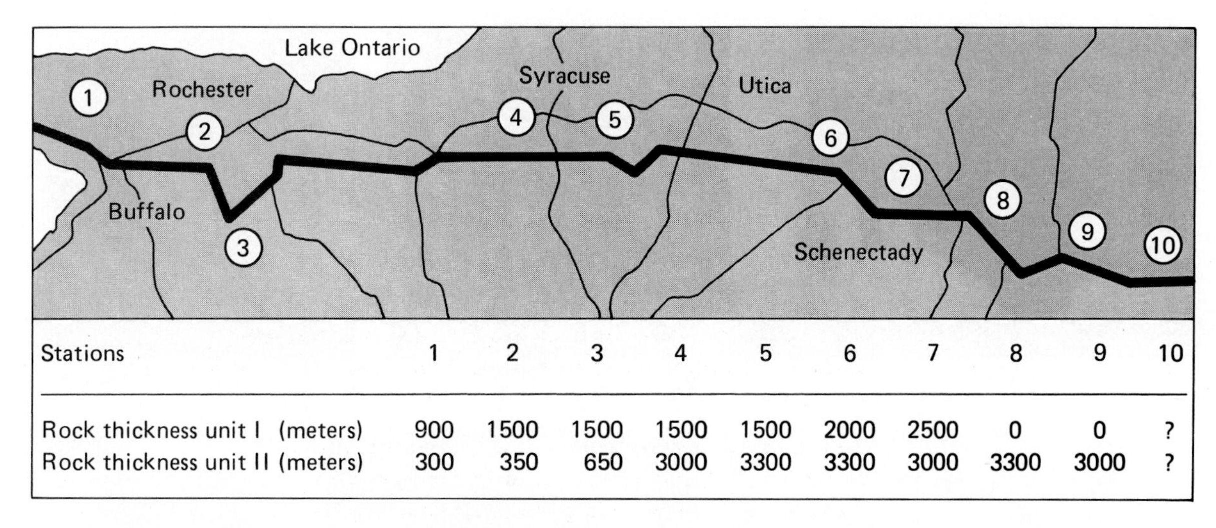

Stations	1	2	3	4	5	6	7	8	9	10
Rock thickness unit I (meters)	900	1500	1500	1500	1500	2000	2500	0	0	?
Rock thickness unit II (meters)	300	350	650	3000	3300	3300	3000	3300	3000	?

Figure 9-1 *The stations in Investigation 9-1.*

vertical scales to use. At each numbered place (called a "station") make sure the rock layers are the proper thickness, as listed at the bottom of the Figure. (Rock Unit I is on top of Rock Unit II. The thickness of Unit II extends from the bottom of Unit I downward.) Remember that sediments are deposited in horizontal layers beneath the ocean's surface. Thus, the layers of sediment at the top of a basin must be flat.

DISCUSSION

1. What evidence is there from the photographs that some of the rocks are marine sediments?
2. Describe the general shape of the basin shown by the cross section.
3. How can you explain the rock types at station 8, 9, and 10 as compared with those found at the other stations?

9-2 The Layers of Geosynclines

James Hall's field trips led him to form the concept of a **geosyncline** (*jee oh SIHN klyn*). He reasoned that great thicknesses of marine sediments containing shallow-water fossils could be formed only in a wide, shallow sea. The basin must have sunk slowly as it was being filled. (Hall did not call this sinking basin of deposition a geosyncline. The name was applied later.)

Hall noted that the sediments became finer-grained toward the midpoint between Buffalo and Massachusetts. On either end there were coarse sands and, in some places, pebbles and larger rocks that formed conglomerates. He considered these coarse sediments as evidence of nearby land masses. He reasoned that there must have been one on the west (the North American mainland) and one on the east ("Appalachia"). Many scientists since Hall's time have tried to locate geologic features that might be the remains of Appalachia. But no evidence of any vanished large land area has ever turned up.

Think back to Chapter 8. What is the natural process that could deposit coarse sediments of shallow-water materials on the outer edge of a great continental margin? Can you suggest a reason why James Hall did not consider this possibility?

Figure 9-2 *(1) The Niagara River flows over sedimentary rock at Niagara Falls. (2) Horizontal sedimentary rock near Batavia. (3) Rock layers at the Genesee River Gorge. (4) and (5) Rocks containing fossil coral. (6) Fossil shells near Canajoharie. (7) Erosion has exposed gently warped layers at the Helderberg Escarpment, near Albany. (8) Contorted and shattered rocks in the New York Berkshires. (9) and (10) Evidence of metamorphism and igneous activity in the Massachusetts Berkshires.*

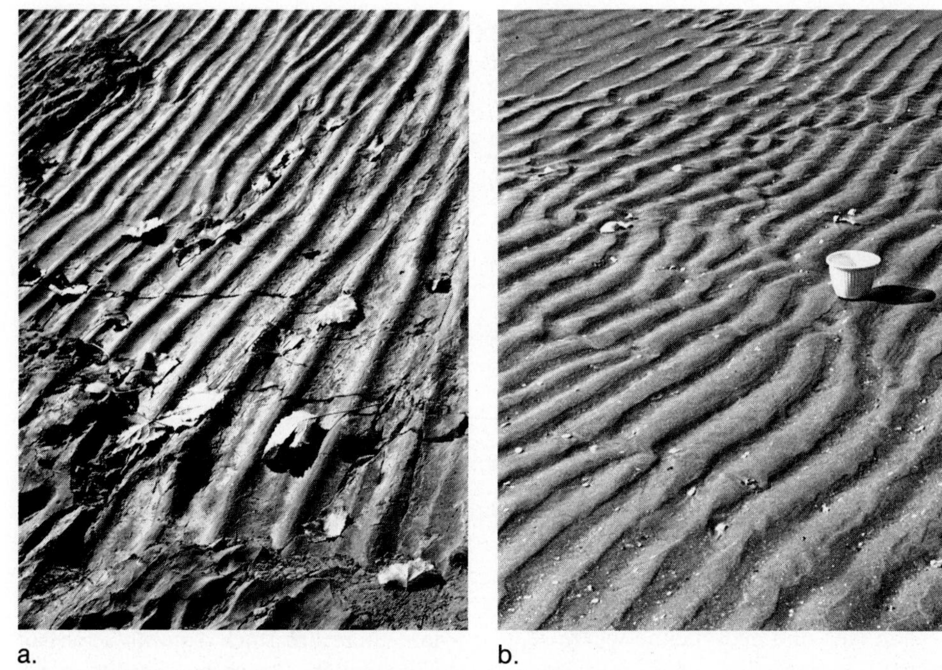

a.
b.

Figure 9-3 a. *Fossil ripple marks formed on the bottom of an ancient sea.* **b.** *Waves form ripple marks on beaches today. Where else might you find ripple marks?*

Suppose you extended Investigation 9-1 to northwestern Pennsylvania and Ohio. You would learn that the rocks there are only one-tenth as thick as the 14,000 m of layers in New York, although all the rocks were deposited through the same period of time. In many mountainous parts of the world, sedimentary rocks are 10,000 to 15,000 m thick. But rocks in the adjoining plains representing the same geologic time span are only a fraction as thick.

It is evident from all these rocks that the sediments were deposited in shallow water. Many rocks contain fossils of organisms similar to those that now live in seas less than 300 m deep. They also contain remains of **ripple marks,** the small ridges that are formed by wave action on sediments in shallow water (Figure 9-3).

Such great thicknesses of sediment could have originated in shallow water in two ways. Deposition may have taken place while the sea floor was sinking slowly. The other possibility is that sea level rose gradually as the sediment layers formed. Sea level cannot rise in one ocean basin, however, without rising in all of the oceans around the world. And the thickness of the layers would require a rise in sea level of 14,000 m. There is no evidence of a rise in sea level of this amount, or anything close to it. Even the great variations in sea level during the Ice Age were only one one-hundredth as much. So the sinking of the sea floor is the more logical explanation for the great thicknesses of shallow water deposits.

About 100 years after James Hall first studied the geology of New York, his ideas were "dusted off" and put into modern terms, and a new concept was born. This was the concept of the geosyncline, or sinking ocean basin.

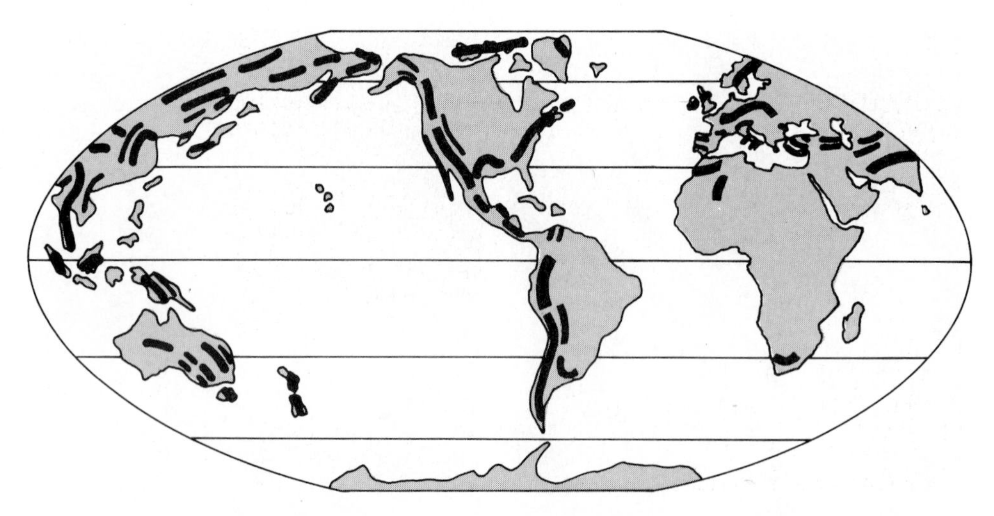

Figure 9-4 *Continental regions where geosynclinal mountains have developed in the past.*

The idea of a geosyncline where sediments accumulated over millions of years has been a useful geologic tool. With some modifications, Hall's explanations have been used to identify and interpret ancient and modern geosynclines all over the earth. You can see in Figure 9-4 the areas of the earth's crust where there are thick accumulations of shallow-water, marine sedimentary rocks. Clearly, these are the mountain ranges of our present landscape. They must also be sites of ancient geosynclines.

The questions now to be considered are:

Figure 9-5 a. *The vegetation on the coastal plain of North Carolina covers sand hills.* **b.** *Padre Island off the coastal plain of the Gulf of Mexico. How were these large sand deposits formed?*

a.

b.

Figure 9-6 Rip currents carry sediments through the surf and out into shelf waters.

1. How do sediments form in a sinking basin at the edge of a continent?
2. How do the basins rise to form long mountain ranges?

9-3 Modern Geosynclines

Great thicknesses of sediment have been deposited on the continental shelves, slopes, and rises of the Gulf of Mexico and the east coast of North America (Figure 9-5). The sediments of both coasts are similar to the sedimentary rocks in the inland portion of James Hall's geosyncline in New York. Let us see then how these modern sediments have formed and how they relate to the rocks of the Appalachian geosyncline.

Marine sediments with the features in which we are interested are mainly in depths no greater than 20 m, the depth at which waves first begin to "feel bottom." Most of the coarse sediments remain shoreward of this depth. Any fine material in suspension is swept across the shelf and out onto the continental slope and rise (see Figure 9-6). Along the east coast of the United States there is not much sediment on the continental slope. This is because turbidity currents and mud slides carry the material deep onto the continental rise.

The apron of sediments laid down on the continental rise by turbidity currents gradually builds up to enormous proportions. As the rise grows, it backs farther and farther up along the continental slope. In response to this growing load, the earth's crust begins to sink along the outer edge of the continental slope. Therefore, the depth of water over the sea floor is constantly changing. It gets shallower where there is deposition and deeper where the sea floor sinks. But the sediments always have features that indicate shallow water. Why is this the case with sediments on the continental rise?

The sediments that form the continental rise have a lower density than the rocks from which they are weathered. They are also only about two-thirds as dense as the underlying rock of the earth's crust and mantle. (The density is 2.3 g/cm^3 for sedimentary rocks of the upper crust compared with 3.7 g/cm^3 for the upper mantle.) Consequently, for every three meters of sediment added to the continental rise, the underlying crust will sink two meters.

Looking at the sequence of deposition, we see that the sediments near the ancient shore are thin (Figure 9-7). These grade into thousands of meters of deposits at the rise. It is easy to understand, then, why the entire margin tilts down

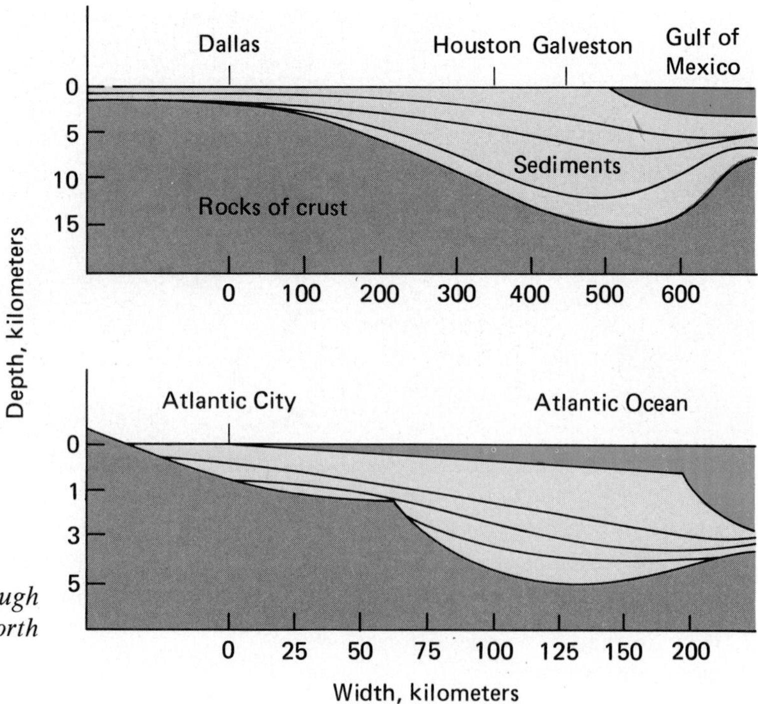

Figure 9-7 *Cross sections through the Gulf and East coasts of North America.*

toward the sea. Although the tilt is greatest under the continental rise, there will be a slight subsidence along the shore (a few millimeters per year). This makes the beach slowly migrate inland.

The wedge of shallow-water sediments grows from the erosion of the nearby land mass. If erosion is rapid, the coastal waters are loaded with suspended mud that is carried across the shelf to the continental rise. This makes the rise grow rapidly and causes more downward tilting of the shelf. As a result the shoreline migrates farther inland. Then a broad and low coastal plain is formed, as there is now along the east and Gulf coasts of the United States. As you know, that is an area of reduced erosion. So early rapid erosion leads in time to less deposition and less subsidence.

MATERIALS
graph paper

ACTION

What does a geosyncline look like drawn to scale? Draw an area 500 km across with 20,000 m of sediments at its thickest. This would be a big geosyncline. Keep this scale drawing in mind when you are considering the movements of the earth's crust during geosynclinal development.

9-4 Deformation Within Geosynclines

The layers of sedimentary rocks in the center of a geosyncline are thicker than those near the borders. They also differ in other ways. On the field trip from Buffalo to western Massachusetts, you saw that the rocks east of Schenectady were bent, broken, and squeezed much more than those at the western edge. These features are much more obvious than the fact that the sediment layers are thicker there than they are near Rochester. It is to Hall's credit that he was able to even imagine the overall shape of the geosyncline. He knew, though, that when the layers are folded, more can fit in a smaller space than when they lie flat.

a.

b.

c.

Figure 9-8 a. *Tilted sedimentary layers.* **b.** *Folded sedimentary layers.* **c.** *A fault in sedimentary rocks.*

Rocks that are bent, broken, squeezed, or stretched are **deformed.** To identify the kind of deformation that has taken place, you must have a good idea of what the rocks looked like before. An inflated basketball that someone is sitting on can be described as slightly flattened because you know what the basketball looked like when no one was sitting on it. You would not say that a football was a deformed basketball. In each case you use your knowledge of the object before it was deformed to detect and describe its changes.

Deformed sedimentary rocks are illustrated in Figure 9-8. The photographs show the way deformed layers of sedimentary rocks most often appear in road cuts or natural outcrops. Figure 9-9a is a diagram of a series of folded rock layers. It represents in a simplified form what is shown in the photograph in Figure 9-8b. As you might guess, the tilted layers in Figure 9-8a could be from one part of a large fold.

The tilted layers in Figure 9-8a might also be tilted because of another type of deformation. Look at Figure 9-9b. The slanted line in the diagram represents a **fault,** or a crack in the crust of the earth along which rocks have moved. The arrow indicates that the rocks on one side of the fault zone have moved upward relative to the rocks on the other side of the fault. Thus, the tilted layers in Figure 9-8a could be the result of movement along a fault.

The other diagrams in Figure 9-9 show other kinds of deformation of the earth's crust. Sometimes blocks of the crust tilt or move up or down. They are called **fault blocks.** The mountains in Nevada, for example, are fault blocks. In some instances great blocks of the crust have actually slid on top of other blocks (**thrust faulting**). This has happened in the Northern Rockies.

The folds, faults, and other fractures were caused by pressures in the earth's crust that acted upon the layers of sedimentary rocks. Of course, igneous and metamorphic rocks are also deformed by the same pressures. Rocks in all of the mountain ranges noted in Figure 9-4 are

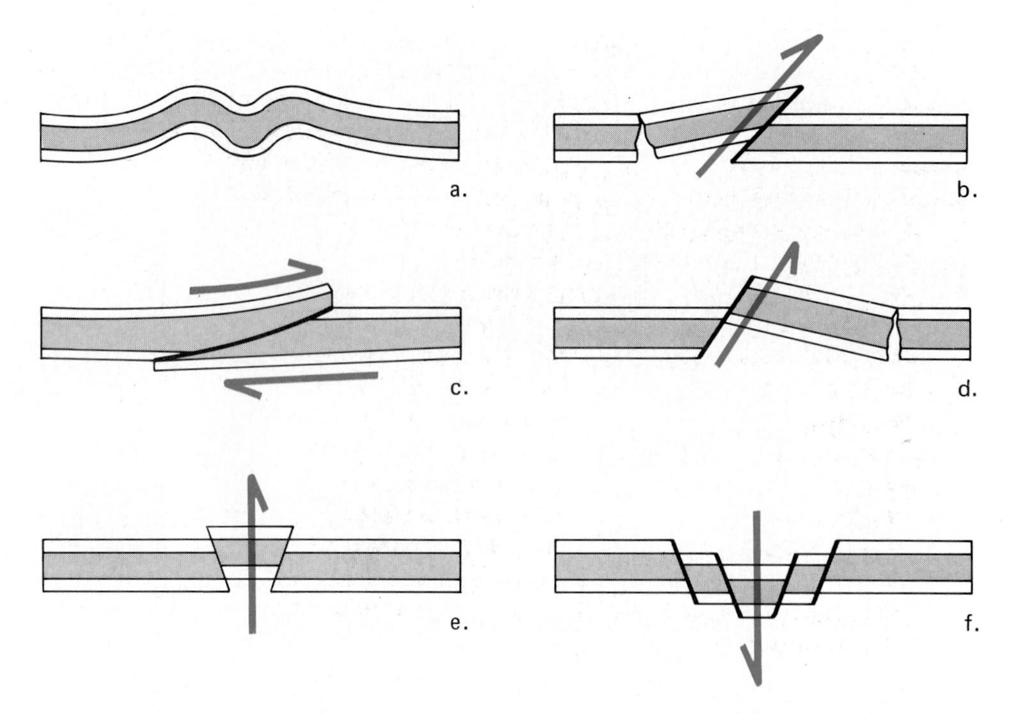

Figure 9-9 *Some of the ways in which rocks are deformed.* **a.** *Folds.* **b.** *A simple fault (reverse fault).* **c.** *A thrust fault.* **d.** *A simple fault (normal fault).* **e.** *An upthrust fault block.* **f.** *Down-dropped fault blocks. Which kinds of deformation show compression of the rocks? Which kinds show tension, or pulling apart of the rocks?*

bent, folded, and fractured. All this evidence means that at some stage in the history of a geosyncline, when the sediments have reached thicknesses of 10,000 m or more, the basin stops sinking and tilting. Great forces begin to squeeze and push the massive deposits. Rocks are slowly deformed from their original horizontal position and pushed up into mountains of broken and folded sedimentary rocks.

The same question occurred to Hall and to every other geologist who has studied mountain formations: What is the source of the squeezing forces that deformed the flat, smooth, geosynclinal sediments?

We cannot see the folding, bending, and squeezing of solid rock happening. It requires hundreds of thousands of years to deform sedimentary rocks into folded layers. It is actually difficult to imagine that a solid rock can be folded or bent without being smashed to bits.

Materials that resist twisting are said to be **rigid.** Solids are rigid, liquids are not. The resistance to flow in liquids is called **viscosity** (*vihs KAHS uh tee*). Butter is more viscous than the cream from which it is made. Molasses is more viscous cold than warm.

Solids may act like liquids for very short periods of time. For example, when wet sands are exposed to strong vibrations (from an earthquake or a dynamite blast), the deposit acts like a liquid for several seconds. If the sand is on a slope, there may be a landslide.

ACTION

Solids sometimes don't act as if they are "solid." If you take an ice cube and strike it with a hammer, it will shatter. Yet, a penny or a quarter laid on the ice cube will slowly sink through it as the ice refreezes behind the coin. Is ice solid or liquid?

Rock salt is solid too. You can easily smash it with a hammer. But when salt crystals are exposed to moist air, they will clump and flow together. (This won't happen with iodized salt.) Is salt a liquid or a solid?

All solid materials will flow, bend, and squeeze if forces are applied over a long period of time. The rate of deformation may be slow, only a few millimeters per thousands of years. But the final product of the squeezing of geosynclinal deposits is a long mountain range with rocks greatly changed from their old sedimentary appearance.

Actually, many substances we consider solids may flow or change under normal forces such as gravity. A sudden landslide is an obvious example. Try to decide which of the following are solid and which are liquid:

MATERIALS
ice cube, hammer, coin, rock salt, iodized salt

limestone	tar
pure water at −50°C	rock candy
pure water at 50°C	modeling clay
gelatin dessert	steel spring
glass	silly putty
fresh paint	caramel candy
rubber	ice cream

Thought and Discussion

1. What evidence did James Hall have for his idea of a shallow depositional basin?
2. Why are the sediments in geosynclines so thick?
3. Why do continental margins tilt toward the ocean basins?
4. What happens to rock layers when they are squeezed?

Mountains in the Sea

9-5 Island Arcs and Volcanoes

There are volcanoes, faults, and earthquakes on the ocean floor as well as on land. Oceanic volcanoes are common. Iceland is a volcanic island. There are isolated volcanoes out in the Atlantic, Pacific, and Indian oceans, far from large land masses. Curved chains of volcanic islands—the Aleutians, the Philippines, the East Indies (Indonesia) and others—form great arcs along the western margin of the Pacific Basin (Figure 9-10). The Antilles in the Caribbean Sea form another island arc. The continental volcanoes seem to be aligned, just like the chains of volcanic islands. Many of the volcanoes are still active and occasionally spew out masses of lava, gases, and vast clouds of ash.

Early on May 18, 1980, Mount St. Helens, Washington, erupted with a blast that blew off 400 meters of the peak, created a crater nearly two kilometers wide, and threw one and a half cubic kilometers of debris into the atmosphere. Fir trees up to 30 km away were stripped and laid flat. An ash cloud covered three states to the east, and the finest ash was

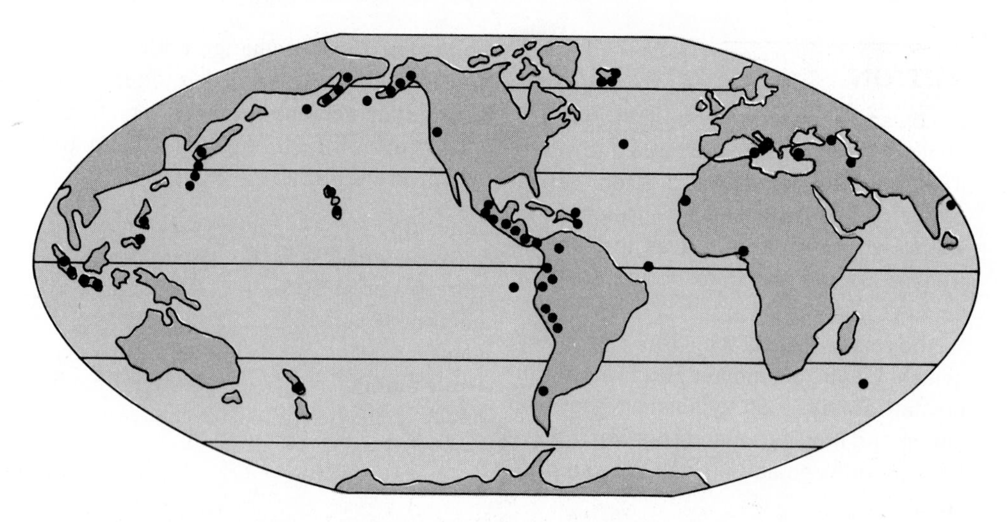

Figure 9-10 *The locations of active and recently extinct volcanoes.*

blown into the stratosphere. Sixty-three people were killed, most by gas asphyxiation within the 300-square kilometer devastation area. Mount St. Helens continues to produce local earthquakes and small gas and steam explosions, showing that the Cascade Range is as active today as it has been for the past one million years. (See Figure 9-11.)

Other volcanoes have created even more havoc. One Sunday afternoon in August 1883, a series of earthquakes rocked the volcanic island of Krakatoa (*krak uh TOH uh*) in the Sunda Strait between Java and Sumatra. The next morning, a violent eruption ripped the cone from the volcano and blasted nearly 5 cubic kilometers of rock into the air. Dust, gas, and ash rose more than 20 km into the atmosphere. The finest pieces of the dust were carried around and around the earth for the next two years before they finally settled. The island of Krakatoa was blown to bits by the eruption. Though there were few people living on the island, uncounted thousands were killed on nearby islands from the sea wave or **tsunami** (*tsoo NAH mee*) created by the eruption.

Nineteen years later, on the Caribbean island of Martinique, Mt. Pelée (*puh LAY*) began to spew out gas and boiling water during the early springtime. This minor activity went on for several weeks. The inhabitants of the city of St. Pierre took little notice, for they knew the volcano had been active but not dangerous in the past. In late April of 1902 some ash began to explode from the main vent, but city officials assured the citizens there was no danger. The activity quieted and everyone breathed sighs of relief. But about 8:00 on the morning of May 9 the volcano exploded without warning in four great blasts. A huge sulfurous cloud roared down the mountainside and engulfed the city. The gases and dust of the cloud suffocated almost the entire population (some 30,000 people) within minutes.

The most startling of recent eruptions was the eruption of El Chichón, Mexico, in 1982. El Chichón sent debris vertically into the atmosphere, causing the largest volcanic cloud since 1912 and affecting temperatures around the world.

These spectacular eruptions tell us that volcanoes all around the world are active. And, even though one volcano erupts and then becomes quiet, others in

***Figure 9-11** Mount St. Helens, erupting in 1980.*

the active regions erupt to take its place. Steam forms a constant cloud around the head of Italy's Mt. Vesuvius. Rumblings are a common occurrence on the slopes of Japan's Fujiyama. Hawaii's Mauna Loa and Kilauea once again sent out fountains of fiery lava in 1983.

Long, deep trenches lie in the sea floor between the volcanic island arcs and the flat deep-sea plains (Figure 9-12). On the eastern border of the Pacific basin, the trenches are adjacent to the continent. In the west Pacific they are thousands of kilometers from the mainland.

The trenches have depths that are as great as the thickness of the sedimentary rocks we've measured in the Appalachian Mountains. The Philippine, Mariana, and Japan trenches are 10,000 m or more deep. One might theorize that the trenches near continents are places where geosynclines are forming.

One problem arises, however, in considering the trenches as future geosyn-

***Figure 9-12** The world's island arcs and deep-sea trenches.*

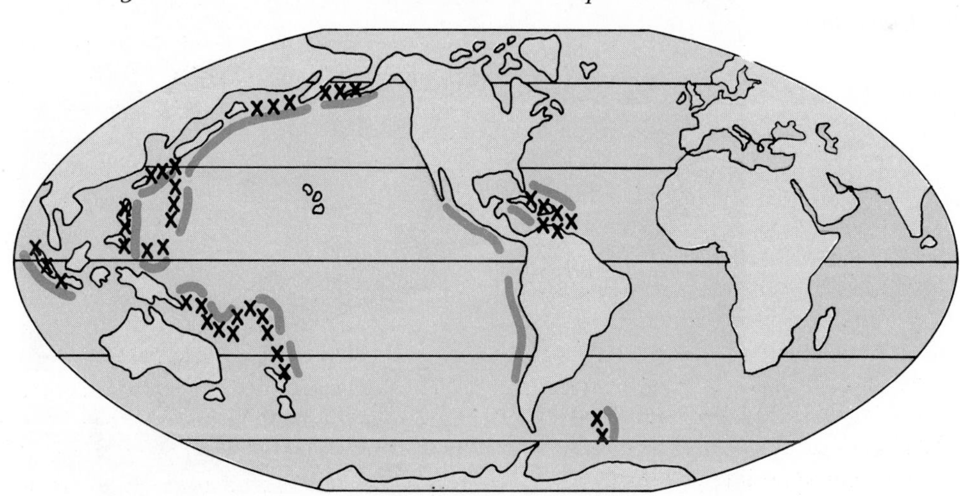

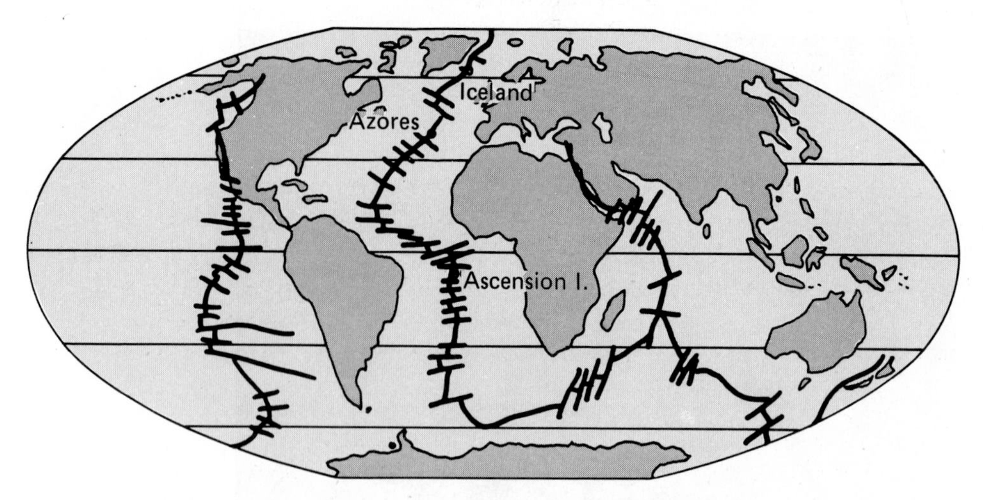

Figure 9-13 The worldwide distribution of oceanic ridges.

clines. They are already deep. No shallow-water sediments can form in them to make the rocks with which we are now so familiar. Look for a possible explanation of their origin in a later chapter.

9-6 Mid-Ocean Ridges

The great mid-ocean ridges are the longest and tallest mountain ranges on earth (Figure 9-13). These submerged ridges wind across the floor of the ocean, following a path that is roughly midway between the continents. (This is best noted in the mid-Atlantic Ridge.) In a few places, such as Iceland, the Azores, and Ascension Island, parts of the ridges rise above sea level (Figure 9-14). The Atlantic mid-ocean ridge is as much as 1500 km wide and the crest is up to 3 km above the sea floor; yet the tops of the mountains are submerged.

The mid-ocean ridges are volcanic. Vast volumes of lava have flowed through giant cracks or fractures and

Figure 9-14 *São Miguel Island in the Azores.*

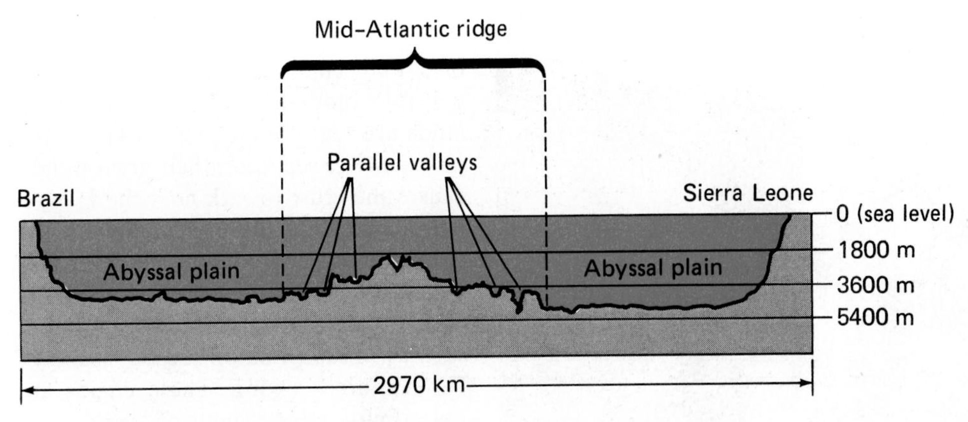

Figure 9-15 *A profile of the ocean floor across the South Atlantic Ocean. The vertical scale is exaggerated about 70 times. How does this profile compare with that in Figure 1-23?*

piled up on the sea floor. The eruption of the volcano Helgafell near Iceland in 1973 shows continuing crustal unrest in the Atlantic mid-ocean ridge.

Ridges have been surveyed in the Atlantic, Pacific, and Indian Ocean basins. Soviet and American scientists working from drifting ice floes, and Americans in nuclear-powered submarines, have mapped a mid-ocean ridge in the Arctic Basin. All of these mid-ocean ridges, except for the isolated Arctic ridge, form a continuous mountain chain that is 65,000 km long.

In some places the ridges are broken by a series of cross faults that give them the look of a crude staircase (Figure 9-13). Notice that the movements along the fault zones have been horizontal. In the other faults you have studied, the movement was more or less vertical. The ocean floor in such areas appears to be composed of blocks.

In 1953 a deep valley was discovered running alongside the mid-Atlantic Ridge (Figure 9-15). Since then, similar valleys or sets of parallel valleys have been discovered and mapped elsewhere in the mid-ocean mountain system.

Using the deep-diving submersible *Alvin,* American scientists in 1975 took a close look at the central valley in the mid-Atlantic Ridge. Through the thick glass of the small porthole, as *Alvin* worked its way through narrow canyons 2000 meters below the sea surface, scientists were able to see and photograph the first signs of faulting in the ridge (Figure 9-16). This was a clear indication that the ridge continues to move and grow.

Mid-ocean ridges, geosynclinal mountains and island arcs suggest that the crust of the earth moves. Geosynclinal mountains and island arcs are created by squeezing (compression). But mid-ocean ridges and their parallel valleys are caused by stretching (tension) in the earth's crust. The tensional and compressional forces, whatever causes them, should equal each other—at least over long periods of time. We know this because there is no evidence that the whole earth is either expanding or contracting. The earth seems to have been the same size since its formation.

9-7 The Sinking Sea Floor

In the ocean basins there are many peaks that are not part of island arcs or mid-ocean ridges. They fit no easily recognized pattern of earth movements. The Hawaiian Islands, for example, are part

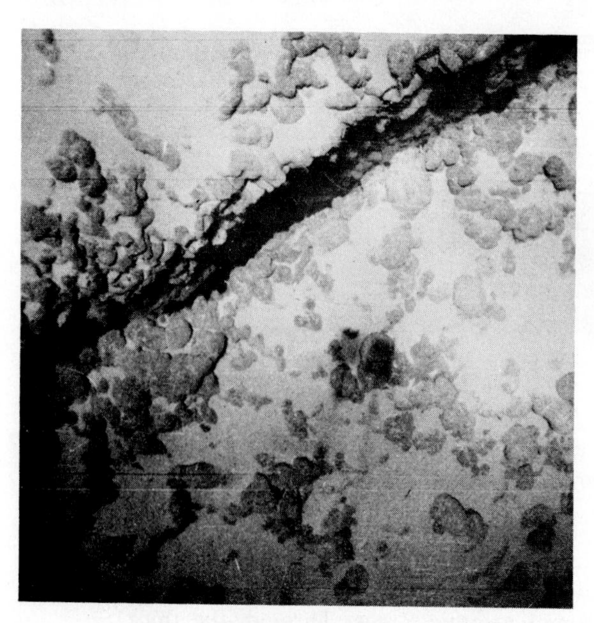

Figure 9-16 *From* Alvin, *scientists were able to see signs of faulting on the mid-Atlantic Ridge—a 1.5 meter-wide crack in the rocks.*

of a long chain that is not near other volcanic mountains. Because these islands are really huge piles of lava lying on top of the sea floor, their great weight causes the crust to sink near the Hawaiian chain. Here there is motion in the crust without any compressional or tensional forces being active. The sinking goes on while huge volcanic mountains, such as Mauna Loa on the big island of Hawaii, are growing. These chains are not at mid-ocean regions of tension, but they do form at zones of weakness.

In addition to great volcanic chains such as the Hawaiian Islands, there are thousands of isolated volcanic mountains and **seamounts** (underwater mountains) scattered across the deep-sea floor. Some have pointed peaks (Figure 9-17). Other seamounts probably had their tops eroded flat by ocean waves. However, their present depth is far too great to be explained by wave erosion, even during the Ice Age when sea level was 150 m lower than it is today. These flat-topped seamounts must have been closer to the sea surface at some time in the past than they are now.

Figure 9-17 *Geologists believe that as the Aleutian Trench sank, the seamounts disappeared below the ocean surface. The trench today is about 8000 m deep.*

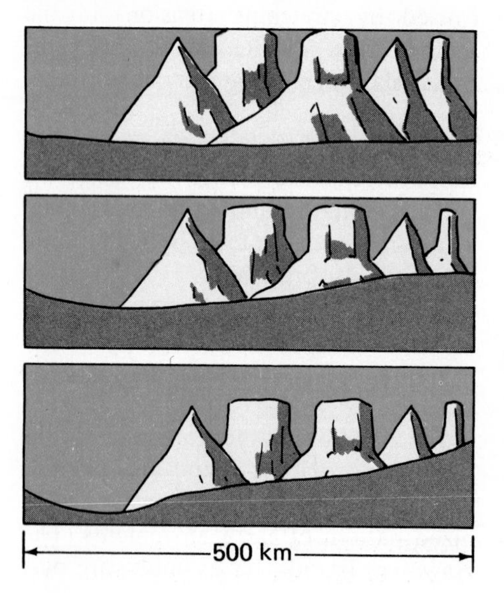

A series of flat and pointed volcanic peaks near the Aleutian Trench off Alaska provides a clue to the formation of some seamounts. Before the formation of the Aleutian Trench, the sea floor was flat and not pulled down into a trench. Some of the volcanic seamounts probably reached the sea surface then. As the trench formed, the sea floor sank, not only in the trench but also for a considerable distance around it. This sinking of the sea floor would have lowered the eroded volcanic peaks in that area. The sinking was greatest near the trench, as you can see in Figure 9-17. This supports the idea that crustal sinking accounts for the present depths of those seamounts.

Coral reefs provide other examples of vertical movement in the crust of the Pacific Ocean. The coral reefs at Eniwetok Atoll (*ehn ih WEE tawk A tawl*) in Micronesia were drilled to a depth of

1400 m. Drill samples showed the entire reef was made of coral skeletons. Corals are small animals that live only in shallow, warm ocean waters. (See Figure 9-18.) Most grow in water from about 5 to 75 m deep, although some are found living 300 m below the surface. The question is how 1400 m of coral rock (calcite) could be deposited by animals that live only in shallow water!

Charles Darwin, the great 19th Century British naturalist, visited coral atolls in the Pacific when he was 22 years old. He was the first person to propose that the sea floor around the islands had slowly subsided. He suggested that the corals grew upward, keeping pace with the lowering of the sea floor. Darwin identified three types of coral reefs, which he interpreted as three stages of development (Figure 9-19). He believed that atolls were old shoreline reefs that had reached the last stage of sinking.

The publication of Darwin's ideas on atolls in 1842 started scientists arguing for a hundred years about his theory. The borings on Eniwetok in the 1950's reached volcanic rock after passing through the 1400 m of coral. These results supported Darwin's coral reef theory.

The earth's crust is moving in all parts of the ocean basins. Earthquakes, faults, and volcanoes are common near the edges and on the great mid-ocean ridge. But, even in other parts of the ocean, volcanic islands and coral reefs show that the crust is mobile. Obviously some great pressures are at work in the earth's crust.

Figure 9-18 *Coral animals grow together in huge colonies. Their skeletons form coral reefs.*

Thought and Discussion

1. How do volcanoes help us describe the active parts of the crust?
2. How are volcanoes related to deep ocean trenches?
3. What is the difference between mid-ocean ridges and geosynclinal mountains?

Unsolved Problems

Scientists aboard HMS *Challenger* used long ropes with lead weights to measure the depths of the oceans. They learned that central parts of the Atlantic Ocean were shallower than the waters nearer the continents. It took another 75 years before marine geologists learned that mid-ocean ridges extended through all ocean basins. We now know much about these submerged mountain ranges. Yet it is only in the Atlantic and Indian Ocean basins that the ridges are truly mid-ocean. In the Indian Ocean basin, a young branch of the ridge extends into the continent through the Red Sea, just as the Pacific ridge cuts through the Gulf of California and across the state of California as the San Andreas Fault. We

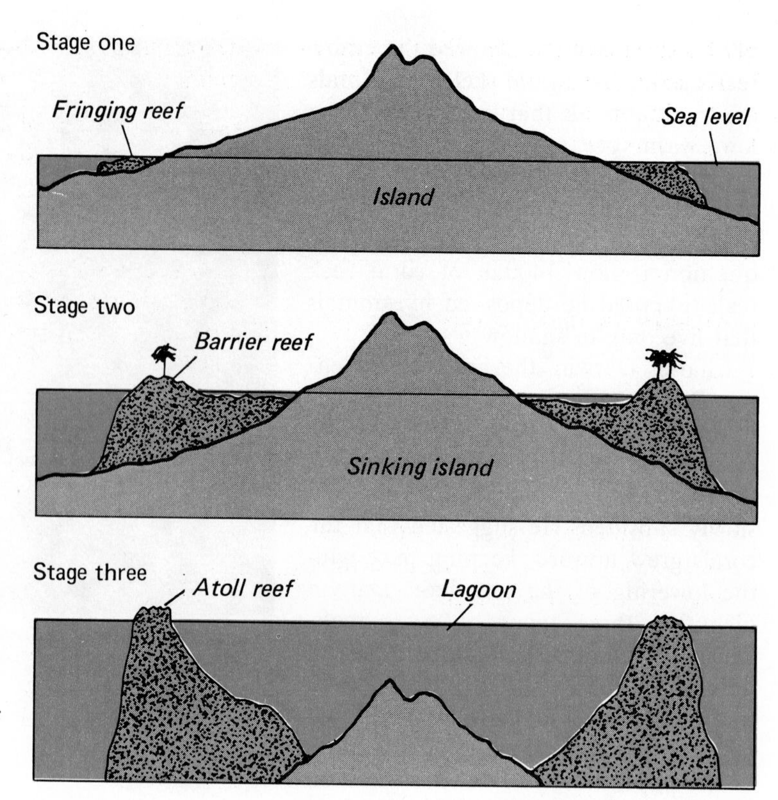

***Figure 9-19** Stages in the development of coral atolls, based on drawings by Charles Darwin.*

must gain a better understanding of mid-ocean ridges to understand these apparent inconsistencies.

When we consider the locations of geosynclinal mountains, as shown in Figure 9-4, some seem to be in long, almost continuous belts. A good example is the belt of mountains that begins in Spain and stretches across southern Europe. It continues through the middle eastern countries, India, and southeast Asia. Mountains along the west coasts of North, Central, and South America also make up a nearly continuous chain. But, what of the Appalachian Mountains, or those in Scandinavia, the Urals in the U.S.S.R. and the great ranges in eastern Australia? Why are they so separate?

Chapter Summary

Rocks in the major mountain ranges were originally deposited in geosynclines. The layers of sediments show evidence of originating in shallow water. Modern geosynclines form along the borders of continents. They are composed of the continental shelf, the slope, and the rise. The thickest deposits are on the rise, and their great weight tilts the continental margin toward the ocean basin.

Rocks within the geosyncline are bent, broken, and deformed into mountains by pressures within the earth's crust.

Volcanoes are evidence of great pressures and movements deep within the earth's crust and mantle. They are most active along ocean borders and on mid-ocean ridges. The ridges make up the longest and tallest mountain ranges on the earth.

Questions and Problems

A

1. What evidence suggests that geosynclinal sediments have been deposited in shallow water?
2. What features of the Gulf of Mexico Basin lead some scientists to believe that it is an active geosyncline?
3. Describe the development of a "typical" geosyncline.
4. How do you know that rocks have been deformed?
5. Describe the features of a folded geosynclinal mountain range.
6. Where on the continents are geosynclinal mountains generally located? Where in the oceans are island arcs generally located?
7. Where is a mid-ocean ridge not a mid-ocean ridge?
8. How do coral reefs form? Explain how they can be used as indicators of crustal movement.

B

1. The shape of the geosyncline that you studied in the James Hall field trip is typical. Therefore, where would most of the rock deformation take place in a geosynclinal basin?
2. What should happen to the crust of the earth where volcanic islands are growing?
3. Give a possible explanation for the

devel

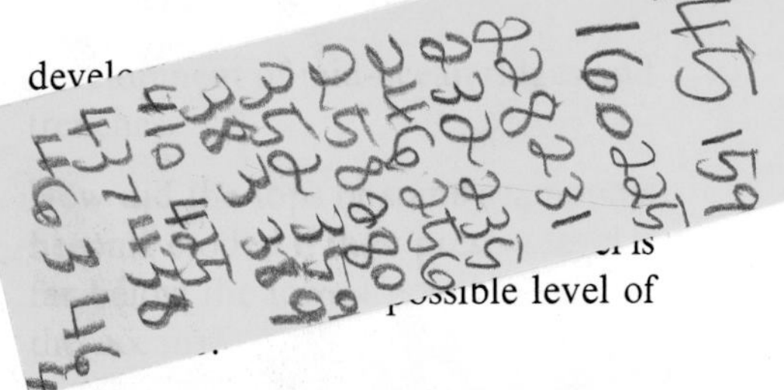

r is
ssible level of

5. How is it possible to find coral skeletons at a depth of 1400 m when coral does not form below a depth of 300 m?
6. How do the water cycle and the rock cycle work together in the formation of geosynclines?

C

1. As sediments move from high areas into geosynclinal basins they lose potential energy. So does the water that moves them. What is the source of the potential energy of the sediments and the water?
2. Under what conditions would the remains of animals that had lived on land be found in sea floor deposits?
3. Describe the processes that could someday make the East Indies and Japan a part of the Asian mainland.
4. Explain how "solid" rock can give evidence of having flowed or been bent and deformed without having broken or melted.

Suggested Readings

Bates, R. L., and Sweet, W. D. *Geology: An Introduction,* 2nd ed. Boston: D. C. Heath & Co., 1973.

Darwin, Charles. *The Voyage of the Beagle.* New York: E. P. Dutton & Co., Inc., 1976. (Paperback)

Heller, R. L. *Geology and Earth Sciences Sourcebook.* New York: Holt, Rinehart & Winston, Inc., 1970. Chapters 2, 3 and 4.

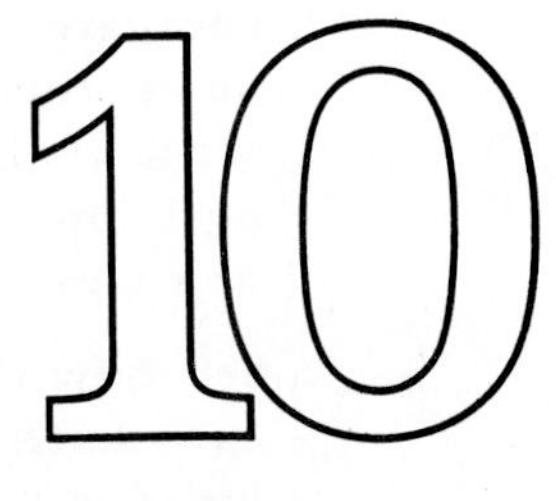

10 Rocks Within Mountains

THE GREAT MOUNTAIN RANGES *on the continents came from the sea. Where mountain ranges now rise there were once shallow seas—geosynclines in which sediments were deposited and then folded and faulted into mountain ranges. This is briefly the account of mountains from the sea presented in Chapter 9. It suggests that if you examine the rocks exposed along the canyons and ridges of high mountains, you will find only sedimentary rocks.*

This, however, is not true. Half Dome, shown on the facing page, is part of the Sierra Nevada in California. It is granite. In any great mountain range you can see an abundance of igneous and metamorphic rocks. In fact, as we shall see, the great bulk of all kinds of rocks on the continents had their origins in long narrow regions that become mountain ranges.

First the sedimentary rocks slowly accumulate to a thickness of 10 or 20 km as the crust beneath them bends downward. Next the sedimentary rocks are folded and faulted, and molten rock material from deep in the crust rises into the deformed rocks. Then there is a period of quiet but vigorous erosion, after which great volcanoes rise to dominate the mountain scenery. Finally the volcanoes become extinct. Erosion proceeds with the task of removing the mountains grain by grain, to be deposited in yet other long narrow regions of sinking crust.

AFTER COMPLETING THIS CHAPTER, YOU SHOULD BE ABLE TO:

1. **compare and contrast the plutonic and volcanic igneous rocks from three points of view: where they form, their mineral composition, and their relative position in the sequence of events in the history of a given mountain range.**
2. **tell how the conditions for the formation of rocks called schist and gneiss suggest a possible non-igneous origin of granite.**
3. **account for the close relationship, in time and area, between regional metamorphism and the formation of plutonic rocks.**
4. **explain why the sequence of formation of the four types of rocks in mountain ranges can be considered the natural rock cycle.**
5. **explain the common occurrence of mineral resources such as coal, oil, and metallic ores in mobile belts.**

Plutonic Rocks

10-1 A Young Mountain Region

The Rocky Mountains are part of a belt of high, rugged mountains like those in Figure 10-1. This belt extends from Alaska to Cape Horn at the southern tip of South America. Within this region there are many mountain ranges, each with its own sequence of rocks. In general, however, the whole region has had a common history of geologic processes and events for a very long time.

There are hundreds of peaks between 4000 and 6500 m in altitude (above sea level). The peaks and ridges commonly rise from 1000 to 3000 m above the valley floors. Glaciers, once abundant throughout the region, still exist around many peaks even near the equator. The sedimentary, igneous and metamorphic rocks exposed are similar to those found in young mountain ranges everywhere. In older mountain regions the landscape is less rugged and the variety of rocks may be smaller.

The development of this great mountain system involves a sequence of events and processes extending over a very long period of time. From the bending of the crust beneath the geosyncline through the deformation of the sedimentary rocks, the formation and rise of magma and the great uplift of the region, movement is the rule. (See Figure 10-2.) For this reason mountain systems are said to develop in **mobile belts** of the crust. This movement or mobility of the crust requires large amounts of force and great quantities of heat energy. In this chapter

Figure 10-1 *High mountain scenery in the Rocky Mountains of Colorado.*

you will investigate the results of these processes.

Of all the sedimentary rocks that have formed, about 25 percent have been deposited outside of geosynclines. The rest formed in geosynclines, and have become parts of mountain ranges. There are, however, two varieties of rocks that have their origin *only* within the mountain ranges of mobile belts. These are plutonic igneous rocks and metamorphic rocks.

10-2 Plutons

Magma, which is molten rock, is formed in the deeper sedimentary rocks or rocks below them during the final stages of folding and faulting. The heat energy may come from the great depth (35 to 45 km) to which these rocks have been forced (see Figures 10-2 and 10-8). Once it is formed, the magma rises to higher levels within the deformed sedimentary rocks. This is because it is less dense than solid rock and is being squeezed upward by the forces of folding. Masses of this magma rarely reach the earth's surface. They cool and solidify below the surface to form large bodies of rock called **plutons** (*PLOO tahnz*). (This name is derived from Pluto, the Roman god of the underworld.) No one has ever seen a pluton form. All of the evidence we have indicates that these masses form below the surface within growing mountain ranges. They are not known to develop in any other environment. So rock bodies of this type are common in folded mountain systems. Plutons cut across the folds and faults in the sedimentary rocks that were produced before the magma rose to its final level. (See Figure 10-2c.)

The thickness of the layers of older rocks that eroded from the top of an exposed pluton is usually unknown (perhaps unknowable). In some instances it can be estimated. Estimates vary from as little as 1 to as much as 8 or 10 km.

Most of the larger plutonic bodies in North America are from 4000 to 50,000 square kilometers in area. One of the largest in the world is in the Coast Range of British Columbia. It is 2000 km long with an average width of about 120 km.

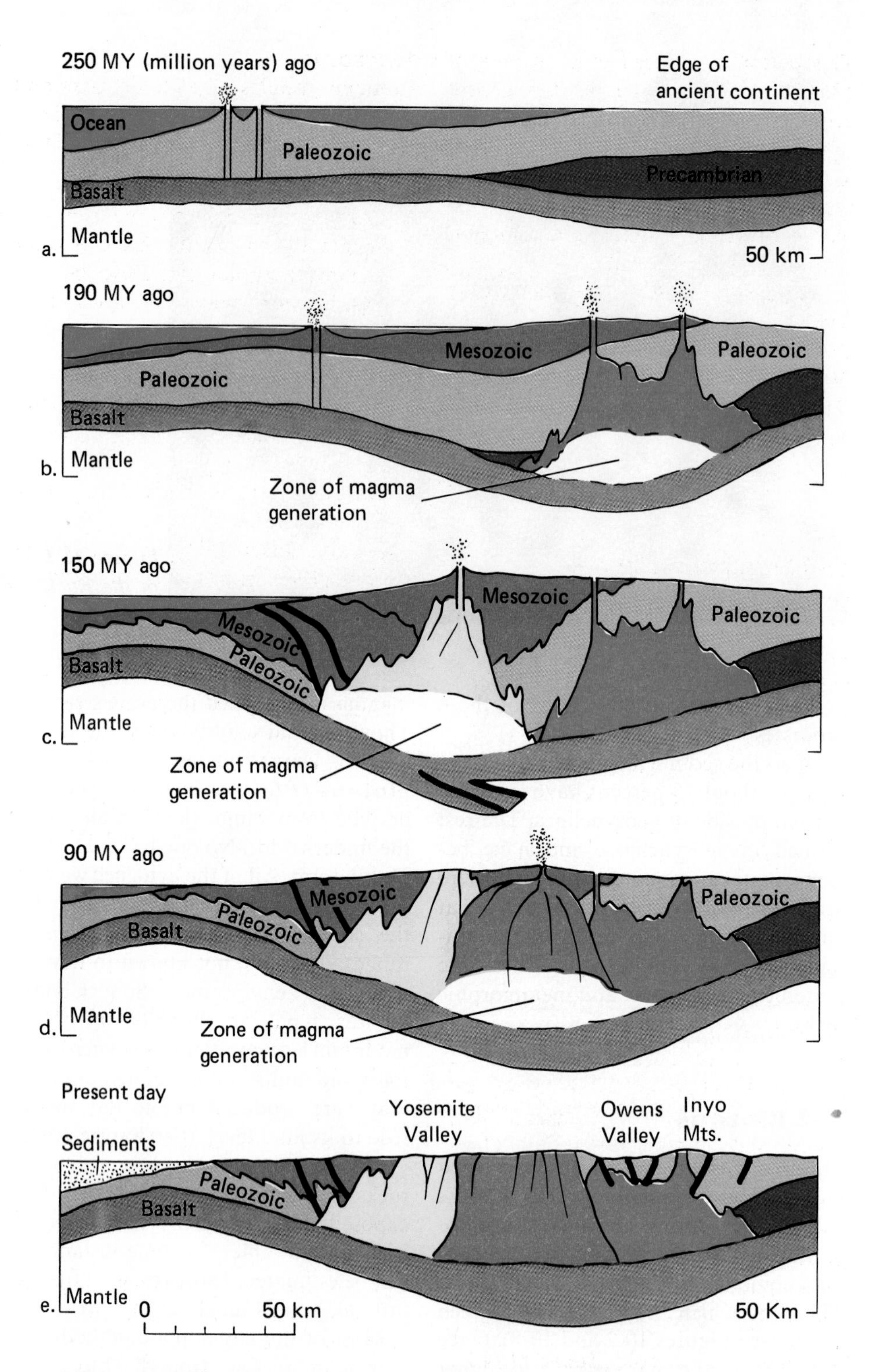

Figure 10-2 *The history of the Sierra Nevada.* **a.** *A geosyncline forms.* **b.** *Sedimentary rocks thicken. The crust bends and folds. Magma is generated and metamorphism starts.* **c.** *Faults form and folding continues. Magma generation and metamorphism continue. Mountains are uplifted.* **d.** *Final pluton development. Mountains are eroded.* **e.** *Eroded mountains are tilted westward.*

This pluton is exposed over an area of 186,000 square kilometers. It is almost certainly not as deep as it is wide.

Most plutonic masses extend parallel to the mountain systems they occur in. A typical large plutonic mass, such as the one in British Columbia, is a combination of smaller plutons.

Magmas formed by the melting or partial melting of older rocks have the chemical composition of those materials. When such magmas solidify they consist mainly of the most abundant minerals, feldspar and quartz.

10-3 Investigating Plutonic Rocks

The rocks of the plutons are, not surprisingly, called **plutonic rocks.** They are a kind of igneous rock that forms below the earth's surface. They are more precisely known as **intrusive igneous rocks.** Plutonic rocks vary in chemical and mineral composition. Most are coarse-grained mixtures of feldspar, quartz, and one or more black minerals. Such a rock is called granite. Granite is generally light-colored (shades of pink and gray).

About 60 percent of a granitic rock is feldspar, the most abundant mineral in the earth's crust. Quartz is next in abundance and amounts to 30 to 35 percent. The remainder consists of iron and magnesium silicates, which are usually black.

There are rare plutonic rocks that are not granitic in chemical and mineral composition. These basaltic plutonic rocks are dark-gray or black. They consist mainly of gray feldspar and other silicates (pyroxenes and amphiboles).

When you have completed this investigation, you should be able to distinguish the common plutonic rocks on the basis of mineral composition, texture and color.

Procedure

Examine samples of plutonic rocks and separate them into two groups: light- and dark-colored. Dark colors are dark-gray and black only.

1. Examine the light-colored samples. Separate them on the basis of color and texture (grain size). Describe each rock in your own words, using variations in color and texture.
2. Look at the light-colored rocks with a low-power lens. Try to identify feldspar and quartz in each of them. Feldspar will be pink, white or light gray, or a combination of these colors. It will show flat shiny faces (cleavage planes). Quartz will usually appear as gray glassy grains without cleavage planes.
3. Examine the black grains with the lens. If any flakes come off when you press a pin or needle against them, the black grains are mica. The others are probably amphibole.
4. Now, turn your attention to the darker rocks. Separate and describe these as you did the light-colored rocks.

MATERIALS
magnifier, rock specimens, teasing needle

Discussion

1. What are the approximate percentages of feldspar and quartz in each of the light rocks?
2. What is the approximate percentage of black grains in the light rocks?
3. Can you easily distinguish the feldspar from the black minerals in the dark rocks?
4. What are the approximate percentages of feldspar and black minerals in the dark rocks?

When you finish, ask your teacher to give you the names of the rocks in each group, or look them up in available books.

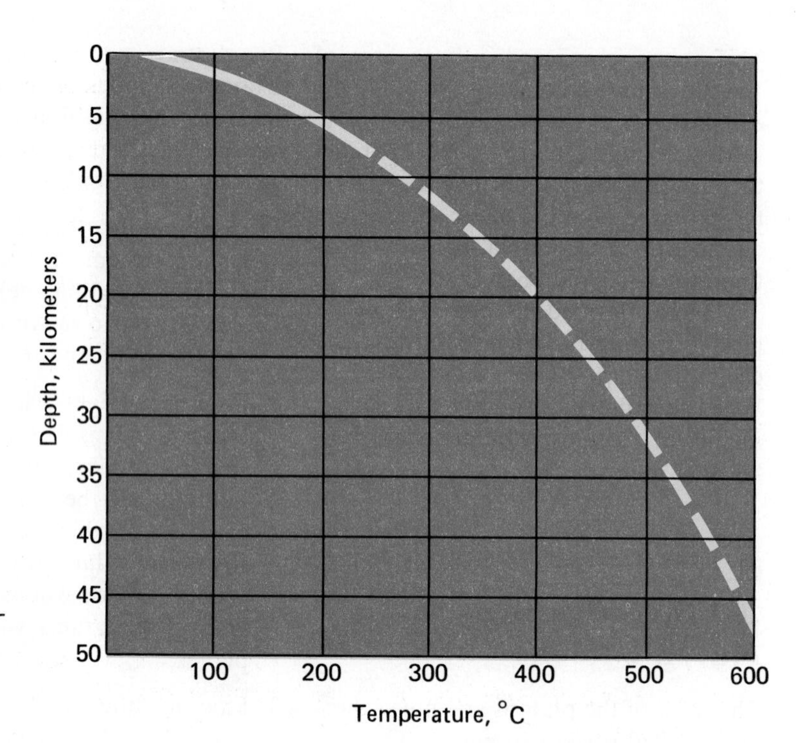

***Figure 10-3** Temperature increases with depth within the earth. The temperatures along the dashed line are estimated.*

Thought and Discussion

1. What is the vertical exaggeration, if any, in the diagrams in Figure 10-2?
2. Assuming that deposition of sediments began about 550 million years ago, what span of time is represented by diagram **a** (Figure 10-2)? What is the thickness of the sedimentary rocks labeled "Paleozoic" shown in **a**?
3. What is the total thickness of sedimentary rocks that was present (in **b**) when folding and magma generation began?
4. How would you describe the changes involved in the growth and expansion of plutons from **b** through **d**?
5. What is the evidence in these diagrams that the continental crust (rocks labeled "Precambrian" plus basalt) has been extended westward by the sequence of events?

Metamorphism

10-4 Metamorphic Rocks

Metamorphic rocks also form within mountain ranges. Typical metamorphic rocks do not form in other environments. In this respect they are similar to the plutonic rocks, which are closely related to them in time and place of origin.

Remember that metamorphic rocks are rocks that have been changed into new forms by high temperatures and pressures. This process occurs deep in the crust of the earth. The temperatures and pressures are higher than those on the surface, but not high enough to melt the rocks.

The graph in Figure 10-3 shows that the temperature of the earth increases with depth. The temperatures represented by the solid line on this graph are averages of measurements made in deep drill holes. The dashed line gives a prediction of temperatures at greater depths.

Pressure also increases with depth. Imagine yourself down several kilometers in a mine. How many tons of rock do you think might be over your head? Besides the thickness of the rock above you, what else would you need to know in order to calculate the pressure of the overlying rock (measured in kilograms per square centimeter)? Suppose you put a rock sample under great pressure on all sides at once. How could the rock change in order to occupy less space?

Look closely at the two rocks in Figure 10-4. Both rocks are made of quartz. One is a sedimentary rock called sandstone. The other is a metamorphic rock called quartzite, which was originally sandstone. What differences can you see between them? Which is the metamorphic rock and which the sedimentary rock? What features of the rocks influenced your choice? The kind of change shown in this figure occurs in almost all metamorphic processes.

Another kind of change is shown in Figure 10-5. Which of these pictures is the sedimentary rock and which the metamorphic? What kinds of changes have occurred? Have new minerals been formed? How do these examples differ from the rocks in Figure 10-4?

If you heated a piece of rock in a laboratory oven to a temperature just below its melting point, changes would begin to occur. Changes in grain size and the growth of new minerals take place

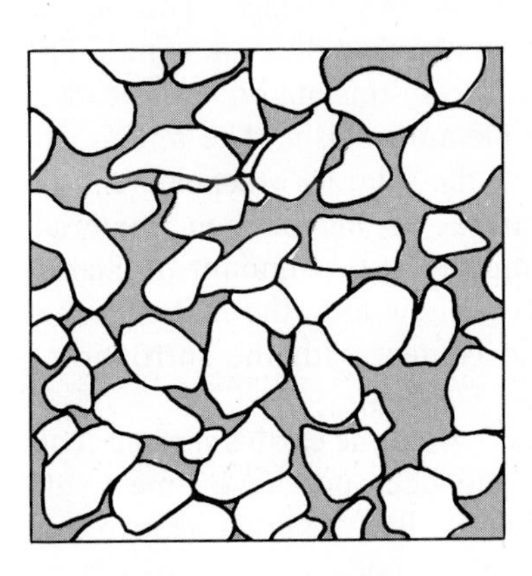
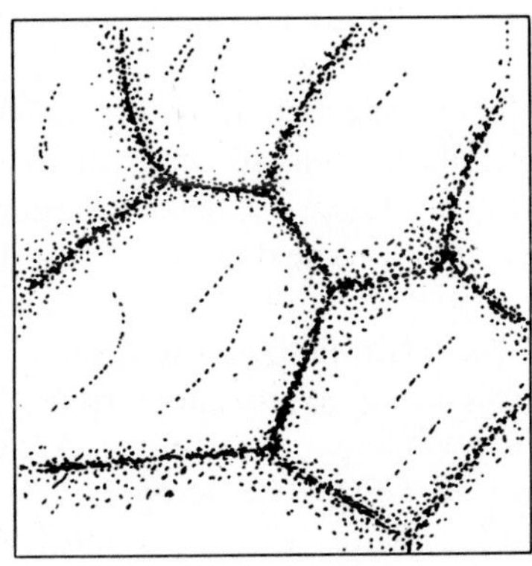

Figure 10-4 *Diagrams of two pieces of rock, magnified about 50 times. Which one is metamorphic?*

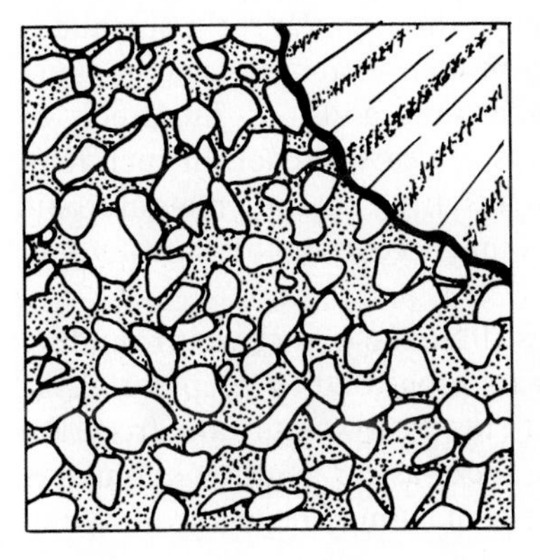
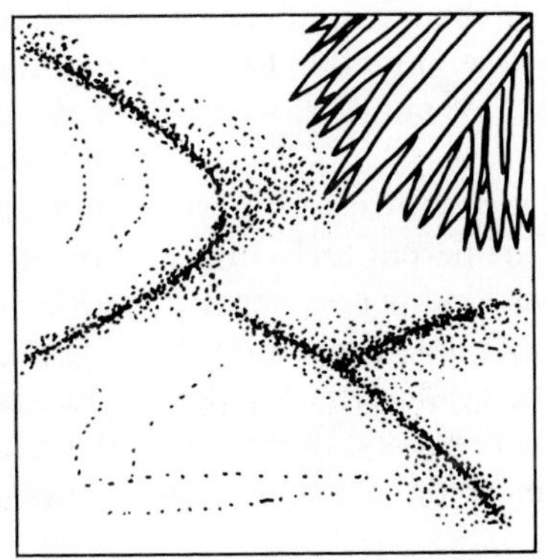

Figure 10-5 *The rock on the left consists of quartz grains (SiO_2) and sea shells ($CaCO_3$). The rock on the right consists of the mineral wollastonite ($CaSiO_3$) and quartz. Both are magnified about 15 times. Which rock is metamorphic and which is sedimentary?*

slowly, however. If you cooled the sample after a short time, you would see little or no effect of heating. Probably you would have to leave the rock in the oven for many months or even years to notice a change. A faster method would be to seal the rock in an airtight container and squeeze it intensely on all sides while you heated it. If there were a small amount of water in the sealed container, the modification of mineral grains and the growth of new minerals would be even faster.

Although high temperature and pressure are important in causing metamorphism, some rocks have been buried several kilometers below the surface without being changed. Remember that minerals contain ions. During metamorphism, some ions break loose from the minerals. For ions to move far in solid rock, small amounts of water or other fluids must be present. Therefore, rocks that have been deeply buried for a long time undergo less metamorphism if little or no water is present.

During metamorphism, ions rearrange themselves to occupy less space then before. A sandstone containing fossil sea shells made up of the mineral calcite ($CaCO_3$) can change and become denser. In the metamorphic rock shown in Figure 10-5, some of the silica from the sand (SiO_2) joined with calcium (Ca) and carbonate (CO_3) ions to form a new mineral and also carbon dioxide (CO_2). The new mineral is a calcium silicate ($CaSiO_3$) that is called wollastonite (*WUL uh stuh nyt*). Wollastonite is denser than the original calcite. The change involves the disappearance of one mineral, calcite, and the appearance of another, wollastonite.

The two sandstones shown in Figures 10-4 and 10-5 are the parent rocks of the two quartzites. Each kind of sedimentary or igneous rock may become a metamorphic rock. The chemical elements that are present in a parent rock determine the kinds of metamorphic rocks that can form from it.

10-5 Metamorphism Varies with Rock Environments.

Different kinds of metamorphic rocks are formed under different conditions. Each combination of heat and pressure can produce another kind of metamorphic rock. The three most important factors in metamorphic environments are temperature, pressure, and the amount of water. Metamorphism can take place at the earth's surface. Imagine a red-hot stream of lava spilling out of a volcano. As lava flows over the top of other rocks, it may bake them. Baking of this type is called **contact metamorphism** because it occurs when molten rock touches another rock. The minerals in the cooler rock change in composition and structure. Zones of contact metamorphism can be quite small, as shown in Figure 10-6.

Some bodies of molten rock exist deep below the earth's surface. This is the molten material called magma. The great heat from this material also causes contact metamorphism. The amount of change in the solid rock depends on the temperature of the molten material. More important, the amount of change depends on how long the molten material is in contact with the surrounding rocks.

A lava flow on the earth's surface cools rapidly and does not have time to do much baking. But the great heat from a deeply buried body of molten magma cannot escape to the air. This heat can only escape through the rocks surrounding the molten material. The magma may take thousands or hundreds of thousands of years to cool. In such long periods of time many metamorphic changes may occur in the surrounding rocks.

The sedimentary rocks surrounding plutons are preheated by deep burial. As the heat slowly flows outward from the plutons, rocks as far as three kilometers from the contact may be changed. When

Figure 10-6 *A lighter-colored layer of basalt on top of a layer of metamorphosed sedimentary rock called hornfels.*

erosion of the mountain range reaches the level of the plutons, metamorphic rocks are seen over hundreds or thousands of square kilometers. This long-distance change is called **regional metamorphism,** in contrast to the more local effects of contact metamorphism.

If the magma contains water, solutions of ions can be forced into the surrounding rocks. Near the pluton these solutions may deposit ores of gold, silver, copper, lead, zinc, and other metals. This is the environment in which most ores form. Farther from the pluton, the solutions may provide water, which will speed up the regional metamorphism. (Remember that water is needed for the ions to rearrange themselves into new minerals.)

Metamorphism can also take place without magma. As the thick pile of sediment in a geosyncline accumulates, the area sinks. Sediments in the lower part are slowly carried deeper and deeper into the crust. Temperature and pressure continue to increase.

The changes that would occur in a layer of clay in a sinking geosyncline are shown by the curve in Figure 10-7. According to the diagram, what might happen to the rock if it were pushed down

Figure 10-7 *As a layer of clay sinks deeper in a geosyncline, its temperature rises as the temperature of the crust rises. The curve shows how the clay changes from one rock type to another as metamorphism increases.*

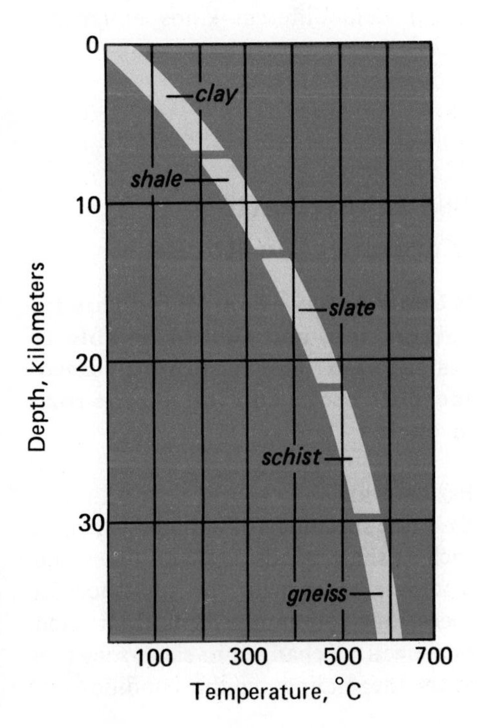

much below 35 km? What would happen if the temperature at a depth of 30 km were 150 or 200 degrees hotter than the temperatures shown by the curve?

Rocks become denser as their environment becomes hotter and as pressure increases. The density of the various rock types shown in Figure 10-7 increases from clay through shale, slate, schist (*shihst*), and gneiss (*nys*). In rock environments where schist and gneiss form, the temperatures and pressures are great enough for denser minerals such as garnet to begin to grow.

The temperatures and pressures shown in Figure 10-7 are based on laboratory experiments. If the experiment is like the conditions in nature, it should be possible to work backwards. That is, we should be able to look at a metamorphic rock and gain some idea of the environment in which it formed. At greater depths and temperatures, if you continue the curve downward, the common rocks in the crust can no longer exist as solids. They begin to melt. Look back to Figure 10-2. Once rock material has melted partially or completely, it can later solidify to form different kinds of igneous rocks.

MATERIALS
magnifier, rock specimens

10-6 Investigating Metamorphic Rocks

When you have completed this investigation, you should be able to distinguish the common metamorphic rocks and the parent rock of each.

PROCEDURE

Examine specimens of sedimentary rocks such as shale, sandstone, and limestone. Compare these with the metamorphic rocks that commonly form from each. Shale will be changed to slate, schist, or gneiss depending on the conditions of metamorphism. Sandstone will be changed to quartzite, and limestone to marble. All of the typical metamorphic rocks will show variations. These variations will depend on impurities in the parent rock and the conditions (temperature, pressure and solutions) of metamorphism. Describe each metamorphic rock in your own words.

DISCUSSION

1. How do the metamorphic rocks differ in texture and mineral composition from the sedimentary rocks?
2. If you have more than one specimen of any of the metamorphic rocks, how do they differ from each other? How would you explain these differences?
3. Plutonic rocks may also be changed by metamorphism. Do you have any metamorphic rocks you think may have originally been granite? If so, what evidence indicates this origin?

10-7 The Origin of Granite

The results of heating granite containing a little water (and most do) are shown in Figure 10-8. The melting temperature depends on the depth at which it is buried in the earth. According to the graph, all the granite does not melt at exactly the same temperature. (Can you think of a reason why not?) For instance, at ground level, you would have to heat the granite to about 900°C before it would begin to melt. As you heated it, more and more would melt, until the temperature got to about 1100°C. At that temperature, all the granite would be melted. At what temperature would granite begin to melt at 12 km depth? 24 km depth?

Probably small amounts of water are present everywhere in the crust. If this is true, is deep magma likely to remain melted at lower temperatures than surface lava? Why?

When the boundary of a granite body cuts sharply across the rocks around it, geologists generally agree that it was probably squeezed in as magma. Most granitic magma comes from the melting of sedimentary and older granitic rocks deep in the roots of mountain chains. Once the magma has formed, it may be squeezed up to higher levels where it cools and solidifies. Other bodies of granitic magma may remain where they are formed and cool more slowly.

Suppose that a group of sedimentary rocks sinks deeper into a geosyncline. Temperature and pressure gradually increase, and the amount of metamorphism of the rocks also increases. Compare Figure 10-8 with Figure 10-7. Notice that the two curves approach each other. You can see that the same minerals could form by metamorphism of sedimentary rock or by solidifying of granitic magma at only slightly higher temperatures. So, it has been proposed that some granite forms by very intense metamorphism just short of much, if any, melting.

Many large granite bodies merge gradually into intensely metamorphosed sedimentary rocks. Their interface is difficult to define. The texture and mineral content of the rock may be just like the texture and mineral content of a granite formed by igneous processes. Therefore, it is not always possible to determine whether a rock that formed deep below the surface is igneous or metamorphic.

If you know the exact make-up of the granite and its relationship to other rocks in the area, you can make a reasonable guess about its origin. But two geologists can study the same body of granite, observe the same relationships, and still disagree on whether the granite formed when a melt solidified or whether it formed by intense metamorphism. It is important to distinguish between what you actually see in rocks and what you infer about their origin.

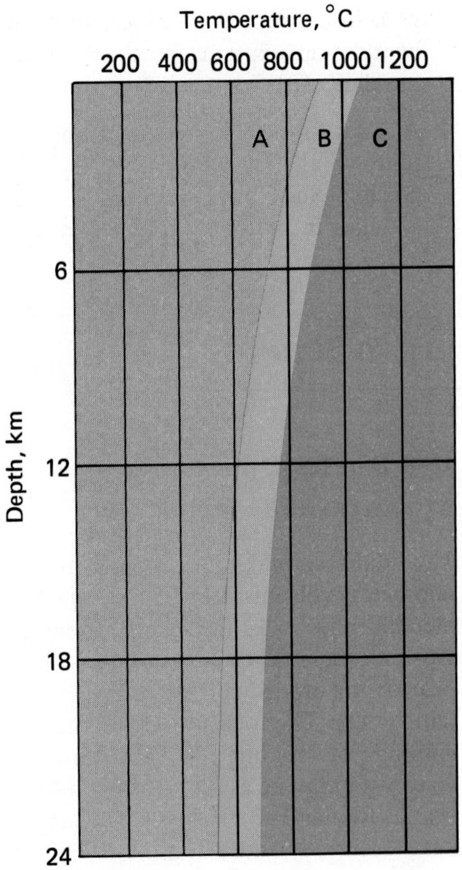

Figure 10-8 *The temperature at which granite with water melts varies with depth. In the shaded area (A) it is solid. In (B) it is partly melted, and in (C) it is completely melted.*

Thought and Discussion

1. Find out how ordinary bricks are made. In what ways does the manufacture of bricks resemble metamorphism?
2. At one time weathering was considered to be a special kind of metamorphism. Why do you suppose people thought that way? Why do they no longer believe this?
3. Do your observations suggest a gradual change in texture in the metamor-

phic series from shale to gneiss? How would you explain the presence or absence of such change?

4. A schist formed at great depth near a pluton may be exposed very slowly by erosion. Such rocks commonly show changes in mineral composition and texture. Would you consider these changes to be metamorphism due to a slow release of pressure?

5. The discussion of metamorphism emphasized temperature, pressure, and solutions as important factors. Can you suggest another factor?

Volcanic Rocks

10-8 Rocks Upon Mountains

The final rock-making event in a mountain belt is volcanism. First plutonic and metamorphic rocks form, and extensive erosion occurs. Then volcanoes may appear on the eroded surface of the mountain system. The total time from the beginning of folding to the end of volcanism may be as much as 30 million years. Remember that great changes have occurred, and 30 million years is not long in geologic time.

The volcanism in mountain belts includes the formation of great volcanic cones and, in some cases, the construction of broad lava plateaus. Volcanoes form over and around the openings of pipes that carry lava to the surface. The volcanic cones are composed of erupted materials. A volcano may be 4000 m high with a base covering tens of square kilometers.

Lava plateaus develop where lava reaches the surface through many fractures. The lava spreads out in great sheets, first filling the valleys and then covering everything except the highest hills. The transfer of great volumes of lava to the surface may cause the crust to sink.

In southern Idaho and westward into Oregon and Washington there is an extensive lava plateau known as the Columbia Plateau. Figure 10-9 is a map of the Columbia Plateau and the Cascade Range to the west of it. The plateau covers over 260,000 square kilometers. Some individual flows are only a few meters thick. Others may be as much as 50 or 60 meters thick. The oldest flows in the plateau occurred about 50 million years ago and the youngest perhaps only a thousand years ago.

The Cascade Range borders the Columbia Plateau on the west and extends southward into California. It consists of broad lava fields that are generally younger than the lavas of the Columbia Plateau. Rising above the lava fields are many spectacular volcanic cones (see Figure 10-10).

The largest individual eruption in the Cascade Range was the eruption that formed Crater Lake about 4600 B.C. Three of the volcanoes in the Cascade Range have been active in the last two centuries, and one of these, Mount St. Helens, is active now. Figure 10-9 shows the dates of the most recent eruptions.

▣

10-9 Investigating Volcanic Rocks

MATERIALS
magnifier, rock specimens

Most plutonic rocks are granitic. Most volcanic rocks are basaltic.

When you have completed this investigation, you will see how volcanic rocks differ from plutonic

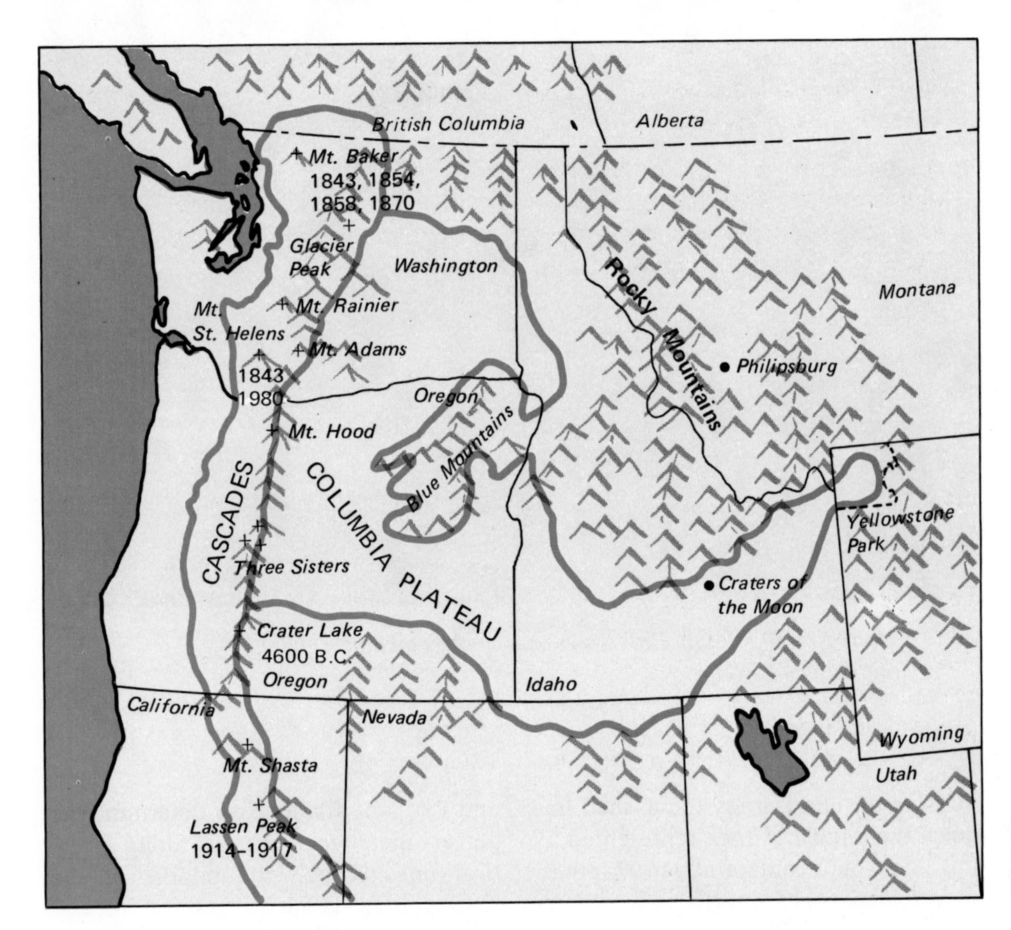

Figure 10-9 The blue lines mark the boundaries of the Columbia Plateau and the Cascade Range.

rocks, and how basaltic rocks differ from granitic rocks in mineral composition and in chemical composition.

Procedure

Examine the specimens of volcanic rocks. Separate the specimens into two groups, light-colored and dark-colored (dark gray or black).

Discussion

1. What is the main difference between all the volcanic rocks and plutonic rocks?
2. Assuming that these rocks have about the same chemical composition as plutonic rocks, how would you explain the physical difference?
3. Which color group do you think is granitic and which is basaltic?
4. Ask your teacher to provide rock names for the specimens, or identify them from available books. If you have a piece of obsidian, in which color group did you place it?
5. Obsidian has the same chemical composition as granite. Can you suggest why some obsidian is black?

Procedure

Figure 10-11 shows the minerals in the two types of igneous rocks. For practice

Figure 10-10 *Mount Rainier is one of the volcanic cones of the Cascade Range.*

in using the chart, study the dashed line down the middle. A rock represented by this line would contain about 9% potassium feldspar, 11% quartz, 54% sodium-calcium feldspar, 2% pyroxene, 6% mica, and 18% amphibole. Now determine the percentage of each mineral along a vertical line through the middle of the shaded area labeled "granitic rocks." Record the results. Do the same for a

Figure 10-11 *The characteristic minerals of igneous rocks.*

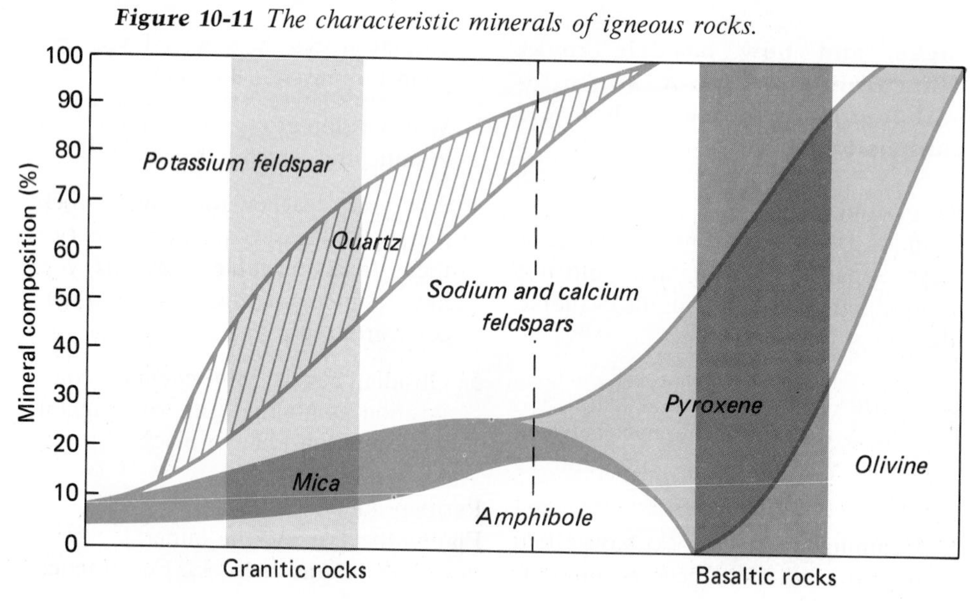

line through the middle of the shaded area labeled "basaltic rocks." As you would imagine, there are rocks that have the mineral combinations found in the unshaded areas of the chart. Your teacher can give you their names if you would like to know them.

DISCUSSION

1. What are the main mineral differences between granitic and basaltic rocks?

Thought and Discussion

1. Nearly all obsidian is geologically young. There are no obsidians among volcanic rocks more than a few million years old. Think of at least two possible reasons for this situation.
2. How could you distinguish between a basaltic lava flow covered by a layer of younger sedimentary rock and a layer of basaltic rock forced between two layers of older sedimentary rock?
3. There are great volumes of granitic rock in plutons. Why are rocks of this composition so rare among the volcanic rocks?

Unsolved Problems

You have studied a sequence of events in the development of a mobile belt at the margin of a continent. That sequence of events is acceptable to most geologists. This is about as far as agreement goes.

What, for example, is the origin of plutonic rocks? Remember that granitic plutonic rocks form below the surface, surrounded by metamorphic rocks. We have said that the intensity of metamorphism decreases away from the plutons in all directions. It could also be said that metamorphism increases in intensity toward the plutons. This suggests to some geologists that the plutonic rocks are actually the results of metamorphism!

Metamorphism might have been so intense that partial melting occurred and produced a molten mass of rock material. This could move upward and intrude into solid rocks above. Do plutons represent the hot spot in the center of the metamorphic mass?

Whatever the answers may be, it seems clear that the processes involve dynamics affecting the whole earth's crust and probably deeper layers of the earth's interior.

Chapter Summary

Distinctive rocks form within and upon mountain ranges. The typical sequence of events is: 1. Sedimentary rocks are deposited during the geosynclinal phase. 2. The rocks are deformed by folding and faulting. 3. Late in the deformation phase but commonly before its completion, plutons of granitic rock are formed. This is accompanied by metamorphism and the development of metamorphic rocks. 4. There is much uplift and erosion, producing rugged mountain scenery and in some cases exposing the plutonic and metamorphic rocks. 5. Basaltic lava pours out from vents and fissures within the mountain belt. 6. There is additional uplift and erosion until the mountain system is destroyed. The deeper levels of plutonic and metamorphic processes are exposed. 7. The eroded material is deposited in a new geosyncline. This is the rock cycle.

Some plutonic rocks are light colored, others are dark. The light, or granitic,

TRANSLATOR OF SCIENTIFIC DOCUMENTS

The number of scientists in the world is increasing every year. As new discoveries are made, it is important that information about them be readily available. This frequently involves translating information from one language to another.

The language problem in science exists on a large scale. For example, in recent years there have been many scientific advances made in the U.S.S.R. Since only a tiny fraction of English-speaking scientists can read Russian, it is impossible for most of them to read about Soviet discoveries without the work of translators.

Translators are employed by many international agencies. Some of these are the World Health Organization, the World Meteorological Organization, and the International Atomic Energy Agency. Some of the international scientific societies also employ translators. And scientific journals are increasing their need for translators.

There is no standard required training for work as a translator. However, since scientific translating deals with technical subjects, a degree in science is usual. Fluency in two or three languages is necessary. This may be obtained through formal course work or experience in speaking the languages. There are a few universities with special programs in interpreting and translating.

rocks contain light feldspar, quartz, and black femic silicates. The dark, or basaltic, rocks contain dark feldspar and dark silicates.

Metamorphic rocks have formed from pre-existing rocks without melting, but at high temperatures and pressures. The degree of metamorphism of a rock reflects the temperature and pressure the rock was subjected to. During metamorphism, some minerals change into others.

Igneous rocks are believed to form from magma. Magma is molten rock material containing small amounts of water vapor and other gases. Igneous rocks can probably originate in at least three ways. Some granitic rocks may have developed from an original basaltic parent magma. Others have crystallized from magma formed during melting of sedimentary rocks. Rocks that appear to be igneous may also form during mountain building by a kind of metamorphism.

Volcanic rocks differ from plutonic rocks in texture, but their mineral compositions are similar. There are light granitic and dark basaltic rocks. The volcanic rocks form from hot, mobile rock material that reaches the surface. It probably comes from below the crust.

Questions and Problems

A

1. Why are environmental conditions within the crust different from those on the surface of the earth?
2. Why are some igneous rocks fine-grained and some coarse-grained?
3. What is likely to happen to sandy clay sediment buried for a long time at a depth of 20 km? How might such rock be changed into granite?

B

1. Estimate the difference between the approximate temperature when rocks begin to melt, and the average crustal temperature at a depth of 20 km. Assume the rock composition is granitic. (Refer to Figures 10-7 and 10-8.) What is the significance of this temperature difference?
2. Are all granites igneous in origin? In what other way may granites form?

C

1. By melting granite in an open container, early investigators tried to discover the lowest temperature at which a magma of this composition could exist. How meaningful were such experiments? What other factors should be controlled to make such experiments more meaningful?
2. Certain ancient rocks exposed in mountains are called volcanic rocks. Since their formation was not seen, how do you think it can be determined that they are volcanic?
3. Suppose that the earth had no atmosphere. Would this have any effect on the amounts and kinds of rocks that compose the crust?
4. What evidence indicates that plutonic and volcanic rocks are not two parts of a single event in a mobile belt?

Suggested Readings

Carr, Michael H. "The Volcanoes of Mars." *Scientific American,* January 1976, pp. 33–43.

Dietrich, Richard V., and Brian J. Skinner. *Rocks and Minerals.* New York: John Wiley & Sons, 1979.

Price, Larry W. *Mountains and Man.* Berkeley, California: University of California Press, 1981.

Romey, William D. *Field Guide to Plutonic and Metamorphic Rocks.* Boston: Houghton Mifflin Co. (ESCP Pamphlet Series), 1971.

Shelton, John S. *Geology Illustrated.* San Francisco: W. H. Freeman and Co., 1966.

11 Evolution of Landscapes

SOME YEARS AGO *many students had small plastic rulers that carried this message: "Study nature, not books." This is an excellent motto for anyone who would like to understand landscapes.*

While you study this chapter, observe your local landscape. Keep a written record of your observations. If there are hills and valleys, describe their size, shape, and arrangement. Try to find maps that show details of the local landforms. If there are exposures of rock in road cuts, quarries, or parks, try to identify the kinds of rock you find there. You may not be able to see much of a natural landscape where you live if it is covered with buildings and highways. In that case, try to take field trips into the countryside. And always notice backgrounds in photographs, movies, or on T.V. See how many different kinds of landscapes you can identify.

The casual observer of the landscape sees only scenery; the informed observer reads the storybook of geologic change. The landscape on the opposite page can be admired simply for its sculptured shape. It can also be read as a single, beautiful episode in an endless tale of deposition, uplift, and erosion.

AFTER COMPLETING THIS CHAPTER, YOU SHOULD BE ABLE TO:

1. **describe the relative importance of external and internal earth processes in the development of a given landscape.**
2. **explain how mountains, plains, and plateaus represent differences in the degree of balance between erosion, deposition, and uplift.**
3. **describe the important differences in landscapes developed on similar rocks in humid and in dry climates.**
4. **recognize types of stream drainage patterns and how they develop.**
5. **recognize evidence of buried landscapes.**
6. **compare and contrast the landscape in your region with that in other regions of North America.**

Landscapes in Perspective

11-1 Growth versus Breakdown

The shape of the land depends mainly on three factors. The first is *the kind of rock* at the earth's surface and how well it resists weathering and erosion. It makes a difference, for example, whether the bedrock is a uniform mass of granite, horizontal layers of sedimentary rocks, or folded sedimentary rocks.

The second factor is *the movement of the earth's crust.* This includes uplift, sinking, tilting, bending, breaking, and volcanic eruptions. As we have seen, these processes unlevel the land.

The third factor is a group of *external processes* including weathering, downslope movements such as creep, and erosion. These processes are powered by the force of gravity and by energy from the sun. They tend to level the land.

In some places the tug-of-war between uplifting and downcutting can be studied in detail. For example, in Figure 11-1 you are looking northward across a low ridge. It descends gently to the east (right). The land drains toward the north (the top of the picture). Most of the stream channels join to flow through the lowest notch in the ridge. The large notch to the west is now merely a pass. Both of these notches have been cut by streams flowing north. Why has the stream ceased to flow through the larger notch?

Examine the ridge more closely. The two notches divide the ridge into three parts. Notice that the western part is cut by many large gullies, but the central part has only a few smaller gullies on its slopes. The low eastern part at the far right has no gullies at all. Can you think of an explanation for this?

The most likely explanation for these different conditions is that the western part of the ridge was uplifted first and has been eroding the longest. The parts toward the east were uplifted more recently. These events occurred during the past one or two million years. If the ridge keeps rising, all the drainage may be forced eastward once again. The streams would cross the ridge through a new channel farther to the east.

Now look at Figure 11-2. Like Wheeler Ridge, this is an **anticline** (*AN tih klyn*), an upward fold of sedimentary rocks. A stream crosses it in a narrow canyon. In contrast to Wheeler Ridge, the canyon is near the widest and highest part of the ridge.

Sometime after the folding that produced the anticline at Sheep Mountain, the area was eroded and then buried under a thick deposit of sedimentary rocks. A river system developed on the surface of these younger rocks. Eventually the river and its tributaries cut down to the older folded rocks. Continued erosion by this river cut the canyon, and erosion by the smaller tributaries left the anticline standing above its surroundings.

At Wheeler Ridge each successive stream channel across the anticline is

Figure 11-1 *Wheeler Ridge, California is slowly rising above the surrounding plain. The diagrams show how the rising ridge affects the major stream in the area.*

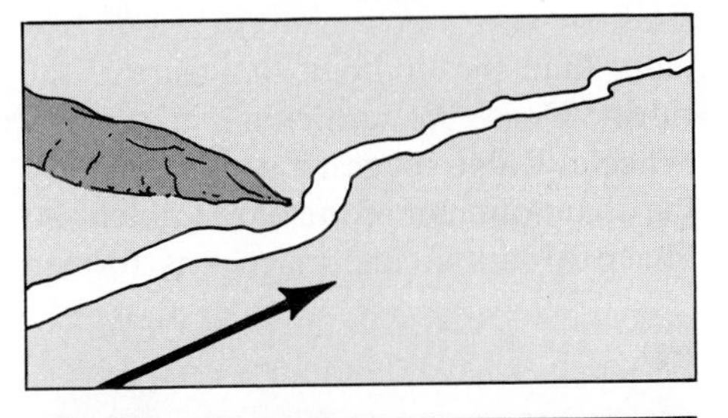

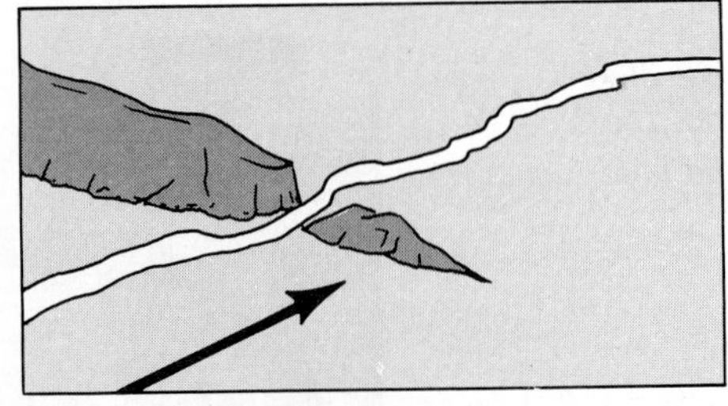

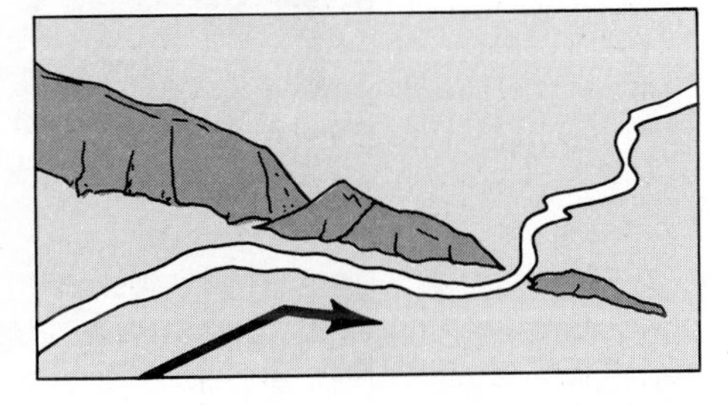

Figure 11-2 *Sheep Mountain, Wyoming. How does this ridge compare to the one in Figure 11-1?*

older than the uplift of that part of the ridge. Does the anticlinal nature of Wheeler Ridge suggest a good reason for the abandonment of the oldest notch? At Sheep Mountain the ridge was formed by the erosion of less resistant rocks surrounding it, rather than by uplift.

Crustal movements have been accurately measured many times over the years. Near Cajon Pass in California measurements begun in 1906 show that the land is being uplifted by nearly one centimeter a year.

The Serapeo at Naples, Italy, displays evidence of both subsidence and uplift since Roman times. The average movement here has also been about one centimeter a year (see Figure 11-3).

The anticline in Figure 11-4 has been rising at a slower rate. You can compute the rate from the information in the caption. By comparison with volcanic activity, these rates are all slow. A single volcanic outpouring can suddenly add several meters of lava to the surrounding area.

11-2 Mountains, Plains, and Plateaus

The great variety of landscapes can be divided into three main groups: mountains, plains, and plateaus. The lofty peaks of the Sierra Nevada in California and the Andes Mountains in western

Figure 11-3 *The Serapeo in Italy. What is the evidence that the level of the sea in relation to the land has changed?*

Figure 11-4 *The youngest rocks in this upfolding at Kettleman Hills, California are 1 to 3 million years old. If the rocks have been arched about 3000 m, what has been the average rate of uplift?*

South America are places where, for the time being, the internal processes are far ahead of the external ones. These areas have lifted faster than erosion could cut them down. Great internal forces still struggle with each other in these high mountain ranges and produce faulting, earthquakes, and volcanism.

The Appalachian Mountains, by contrast, appear to be the worn down remains of a large mountain range uplifted long before the Andes and Sierra Nevada. Much of this region contains folded sedimentary rocks and has a distinctive landscape of alternating ridges and valleys (see Figure 11-5).

Plains are large areas that are flat and either horizontal or gently inclined. They may occur near sea level or at elevations of many hundreds of meters. Some plains are formed from deposition, others are created by erosion. The plains along the Atlantic and Gulf coasts of the United States and the Great Plains east of the Rocky Mountains in Canada and the United States owe their flatness to

Figure 11-5 *An aerial view of the Appalachian Mountains near Harrisburg, Pennsylvania.*

Figure 11-6 *What evidence is there that this stream channel has shifted? This is the Mudjatik River, Saskatchewan.*

Figure 11-7 *The Grand Canyon was formed by the Colorado River and its tributaries. The canyon averages about 1.5 km in depth and 16 km in width at the top. The river is only 50 m wide.*

deposits of sediments by streams. In both instances the stream deposits are underlain by older sedimentary rocks. The older rocks were deposited mainly in the sea and are now uplifted and tilted.

Most of eastern Canada is a low-level erosion plain. Much of this plain is underlain by igneous and metamorphic rocks, the remnants of mountain ranges older than the Appalachians.

Smaller plains of deposition (flood plains) are found along many large streams. Flood waters and a shifting stream channel result in a plain underlain by sediment (Figure 11-6).

A **plateau** (*pla TOH*) is a large elevated tableland. Some plateaus are produced by the piling up of lava flows. Other plateaus form when a wide area is gently uplifted. There is little folding or faulting, and the layers of rock remain nearly horizontal. A plateau may be cut by deep canyons like the Grand Canyon in the Colorado Plateau of northern Arizona (Figure 11-7).

Figure 11-8 *Can you divide this landscape into two kinds of terrain?*

Thought and Discussion

1. Look again at the ridge in Figure 11-1. Suppose that uplift began at the same time for all parts of the ridge but was more rapid toward the west end. Would this account for the present landscape?
2. If the processes of weathering and erosion have been acting on the land areas for billions of years, why do we find high mountain ranges today?

Analyzing Landscapes

11-3 The Parts of a Landscape

Look at the landscape in Figure 11-8. Two kinds of surfaces appear in this scene. One, mostly at the top of the scene, is rough and cut by several gullies and one major valley or canyon. This landscape is being eroded. The second kind of surface, mostly at the bottom of the photograph, is relatively smooth. It includes the deposits on the floor of the canyon and the fan beyond the canyon's end. The fan, in turn, seems to be spread over a broader area of sediment coming from outside the area. Apparently this second area is not eroding but is receiving sediment instead.

Place a piece of tracing paper over Figure 11-8 and lightly draw a line around the borders of the picture. Now carefully trace a line across it that will separate the two kinds of terrain: land eroding and land receiving sediment. What are the major contrasts between the landscapes on the two sides of your line? How will these landscapes change?

In Figure 11-9 you see the effects of thousands of years of erosion and deposition. A road runs along the edge of the fan. Notice the patch of light-colored gravel that has been deposited on the steep fan at the end of the canyon. This is the only important addition of material to the fan in the last few decades. More sediment in this area will build it up until the stream shifts its course to another part of the fan. The gravel makes up about one thousandth of the whole deposit. If it took 30 years to add this amount of material, how long did it take to build the fan to its present size? What factors can you think of that would affect the accuracy of your answer?

Again, it is possible to draw a line separating areas of erosion and deposition. Although these processes are sometimes interrupted and go on very slowly by human standards, it is clear that they are changing the landscape.

Figure 11-9 Compare this view of the steep east wall of Death Valley, California with the landscape in Figure 11-8.

How are the two landscapes in Figures 11-8 and 11-9 different? Would it surprise you to learn that the first shows an area only about one meter square? The "pole" in the photograph is a pencil. This miniature landscape with its canyon and fan was sculptured in loose soil by a single rainstorm. Thus, the two scenes differ because the second is 1000 times larger than the first and took much longer to form. Yet these two scenes have the same basic characteristics.

You can divide any landscape into areas losing material and areas gaining material. In fact, a good first step in analyzing any landscape is to distinguish between the areas of erosion and deposition.

ACTION

Observe some landscapes (large and small) in your locality. Can you identify areas that are eroding? areas that are receiving sediments?

Try to visit a system of valleys or gullies near your school or home. Can you identify a main valley or gulley with tributaries? Sketch a map showing the relation of the tributaries to the main valley. Does the main valley have a flat bottom that is many times wider than the channel? Do the tributary channels meet the main channel at a common level (without rapids or waterfalls)? Prepare a brief description of the valleys you have observed with sketches to show sizes, shapes, and arrangements.

11-4 Investigating Areas of Erosion and Deposition

When you have completed this investigation, you will have observed model landscapes formed by erosion and deposition in a stream table.

MATERIALS
bucket, soil, support to raise one end of tray, tray (with drain at one end), tubing (with clamp), water

PROCEDURE

Set up a stream table and place soil or sand in a flattened mound in the higher part. Leave the lower end of the tray empty. Then, run water through a tube onto the mound and observe what happens. Concentrate on the boundary line between erosion and deposition.

DISCUSSION

1. What causes the boundary between erosion and deposition to move?
2. Predict how your landscape will change as the stream flows through it.

11-5 Rates of Change

When you try to read the life story of a natural landscape from the features around you, it soon becomes obvious how slowly the features change. This suggests that in addition to the structure of rocks and the internal and external processes acting on them, you must add a fourth factor that influences landscapes. The fourth factor is the length of time involved.

Most landscapes do not change rapidly enough for you to watch them. Other methods must be found to interpret them. One method is to examine miniature landscapes outdoors or in a stream table. You can assume that many of the same processes operate in miniature landscapes and major ones. The rates of change, however, are faster in a miniature landscape.

Another method is to compare landscapes in different areas that have formed under similar conditions. This can help you predict how a particular structure will wear away and change. Look at the landscapes in Figure 11-10. If you assume that the rocks and the climates are similar in each case, which plateau has been weathering and eroding longer?

11-6 Landscapes and Climate

Even in deserts, where it rarely rains, streams are the most important cause of erosion and deposition. Many tourists take pictures of the Painted Desert near Holbrook, Arizona to show their friends the colorful "sand dunes" there (Figure 11-11). The "dunes" are hills produced by stream erosion of brightly colored, horizontal sedimentary rocks.

Figure 11-10 *Two views of the Colorado Plateau area. Could the lower landscape develop into the upper one?*

Figure 11-11 *Stream erosion in a dry climate carved out the Painted Desert in Arizona.*

Streams are even more important land movers than glaciers. Glaciers in the high mountains of Alaska occupy valleys originally formed by streams. The glaciers merely deepen and reshape the valleys.

In humid climates (50 cm or more of precipitation annually) landscapes are usually covered with vegetation. The plant roots tend to hold the weathered rock in place. This favors deep and complete weathering of the bedrock and the development of soils. The landscape develops smoothly rounded slopes (Figure 11-12a).

There is less weathering and soil formation in dry regions. The landscapes are angular, and valleys tend to be narrow canyons between vertical walls. Most of us live in the more humid areas and find the landscapes of dry regions fascinating and spectacular (Figure 11-12b). If the bedrock is colorful, so much the better. Our familiar landscapes may appear dull and uninteresting by contrast.

Figure 11-12 *Typical mature landscapes can be found in both* **a.** *humid climate areas and* **b.** *dry climate areas.*

a.

b.

a.

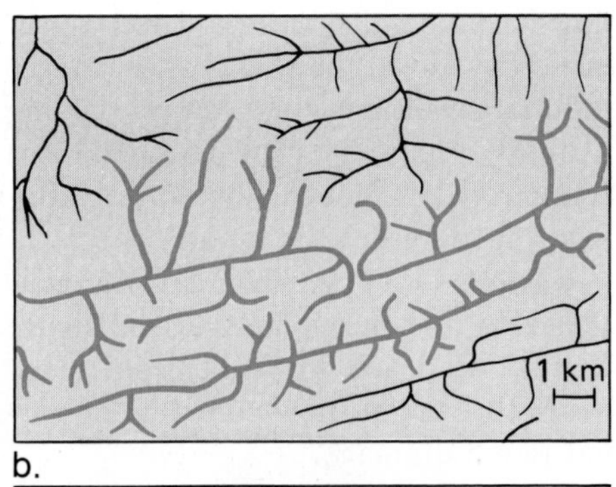

b.

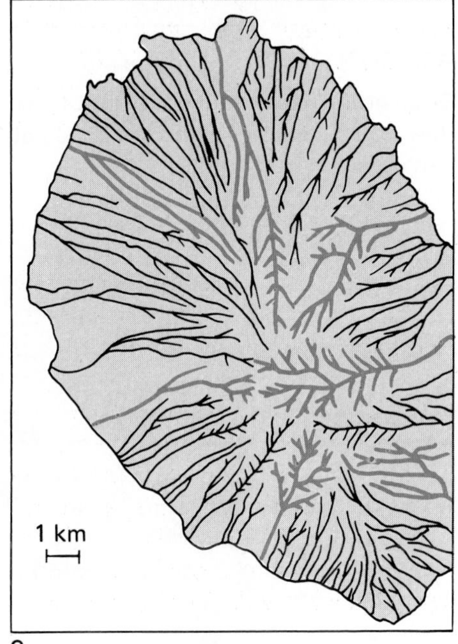

c.

Figure 11-13 *Compare the three kinds of drainage patterns:* **a.** *dendritic* **b.** *trellised* **c.** *radial.*

A typical landscape consists of many valleys. These valleys form different patterns depending on how water drains into and across the land. The **drainage pattern** is usually determined by the kind of bedrock in the area. In the case of sedimentary rocks, the structure of the rocks also affects the drainage pattern. There are three kinds of patterns.

Dendritic (*dehn DRIHT ihk*) **drainage** forms a tree-like pattern. It develops in areas where the bedrock has uniform resistance to erosion (Figure 11-13a). This would be true in an area of horizontal sedimentary rocks. One layer may be more or less resistant than the ones above and below. However, each layer has uniform resistance over wide areas. Plutonic rocks also have uniform resistance in any horizontal plane. Broad areas of plutonic rocks usually show dendritic drainage. Most landscapes have a more or less dendritic drainage pattern, but there are some exceptions.

Trellised (*TREHL ihsd*) **drainage** is typical of some areas where there are folded sedimentary rocks. Such landscapes consist of parallel ridges and valleys. The ridges are resistant to erosion, and the valleys are less resistant. The major streams have valleys across the parallel ridges. Their largest tributaries flow in the parallel valleys. Small tributaries flow down the slopes of the ridges to form a right-angle pattern resembling a trellis (Figure 11-13b). (A trellis is a framework of wooden slats for a climbing vine or bush to grow on.)

Radial (*RAY dee uhl*) **drainage** is typical of volcanic mountains (see Figure 11-13c). Can you think of other landforms that might show complete or partial radial drainage?

ACTION

Examine the photographs in Figure 11-14. How would you classify the drainage patterns?

11-7 How Low Can a Landscape Get?

If the leveling processes were the only ones at work on the earth, they would eventually remove all the high land. All irregularities in the surface would be worn down. Finally there would be a smooth landscape at the lowest possible elevation—one on which no particle could fall or roll anywhere. Is there such an area on the earth?

As long as a landscape remains above sea level, streams and rivers can move parts of the land to lower elevations and eventually into the ocean basins. Unless sea level itself changes greatly, the sediment will become a part of the conti-

a.

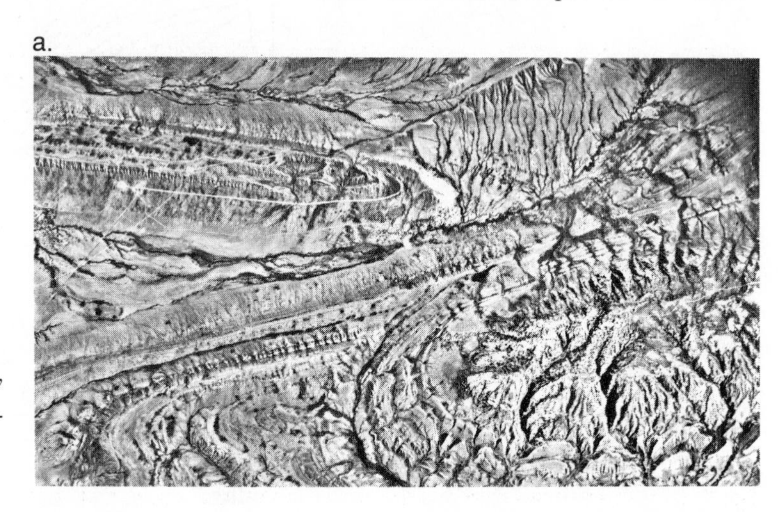

Figure 11-14 *How would you classify these drainage patterns from* **a.** *Wyoming* **b.** *Arizona* **c.** *New Mexico?*

b.

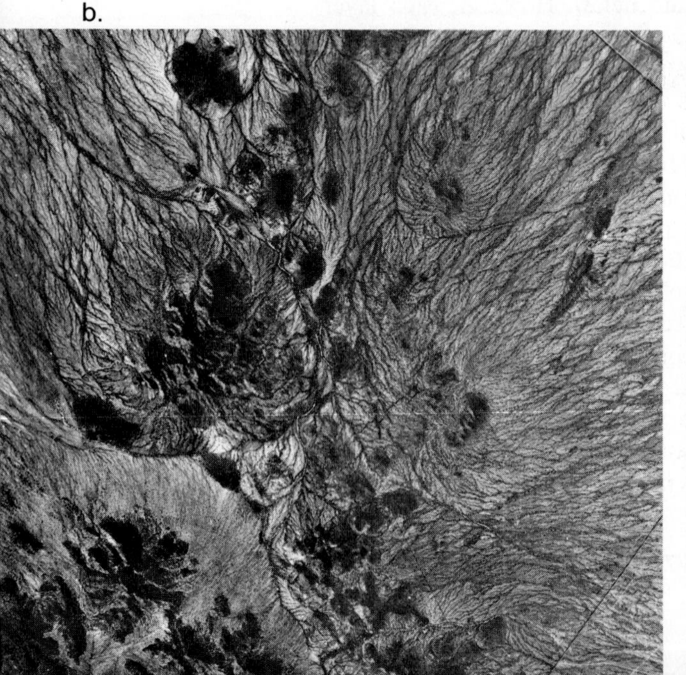

c.

Figure 11-15 *Rocks formed beneath the surface were lifted into high mountains and later eroded to this low plain.*

nents again through uplift of the ocean floor during part of the rock cycle.

The lowest level to which a land area can be eroded is usually considered to be sea level. This does not mean that there is no erosion below sea level. Materials underwater may be eroded by waves and currents. You saw in Chapter 9 that the ocean floor has a strange and varied landscape. Would you expect the landforms on the sea floor to show the same patterns, shapes, and sizes as those on land?

On the land the dominant leveling agent is running water. Near the shore on the continental shelf, wave action is most important. Glaciers entering the sea can also gouge valleys below sea level. But since ice floats, this gouging cannot continue very far from land. Over most of the deep ocean bottom the only important leveling agents are turbidity currents.

No continent today is so eroded that its entire surface is nearly at sea level. However, many large areas have been almost leveled by erosion. From Washington, D.C. to Atlanta, Georgia there is a broad flat area between the Appalachian Mountains and the coastal plain. The region is underlain by igneous and metamorphic rocks (see Figure 11-15). The existence of these rocks at the earth's surface is proof that this area was once deep within a high mountain range. The thick cover under which these rocks formed has been removed by erosion.

11-8 Buried Landscapes

A landscape of low relief such as the one in Figure 11-16 may later be covered with younger sedimentary rocks. These younger rocks rest on the eroded surface of older rocks. This buried erosion surface is called an **unconformity.** There are three major types of unconformity. The younger rocks may rest on eroded plutonic and metamorphic rocks, on tilted or folded sedimentary rocks, or on horizontal sedimentary rocks.

The photograph in Figure 11-16 is a picture of the north wall of the Grand Canyon. The upper horizontal layers of sedimentary rocks rest across the eroded edges of older tilted sedimentary rocks. These tilted sedimentary rocks rest, in turn, on an older complex of granites and schists to the left of this view. The sketches in Figure 11-6 illustrate the se-

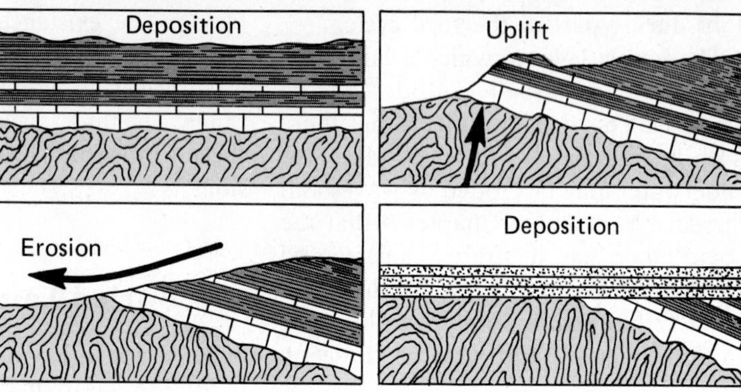

Figure 11-16 *A photograph of the north wall of the Grand Canyon, showing the angular unconformity at the base of thick, horizontal sedimentary rocks. The four diagrams show how this unconformity was formed and its relationship to an older one.*

quence of events in the formation of the two unconformities.

On the walls of the Grand Canyon you can follow the edge of the younger unconformity for more than 300 km. For most of this distance the unconformity is almost horizontal. In places this surface cuts across metamorphic, plutonic, and folded sedimentary rocks. This means there must have been a great thickness of rocks removed before the sediments of the overlying rocks were deposited. The rocks just above the unconformity are about 500 million years old. To produce such a large, nearly smooth plain must have required a long period of erosion.

ACTION

Imagine that a mountain range and the nearby areas have been eroded to a flat surface. Horizontal layers of sedimentary rocks are later deposited on this surface. Make a sketch to show the positions of 1) the younger sedimentary rocks, 2) the plutonic and metamorphic rocks of the interior of the old mountain range, 3) the folded sedimentary rocks adjoining these, and 4) the undisturbed older sedimentary rocks beyond the area of deformation. Can you identify the three types of unconformities shown in your sketch?

11-9 Other Landscapes

There are unusual landscape features that you may know about. One that is of special interest is **impact craters** formed by meteoroids. Over 20 impact craters have been found on the surface of the earth. These landforms do not owe their shape or origin to internal movements of the crust.

Craters that occur in moist climates usually contain lakes. Some craters look recent, like the one pictured in Figure 11-17. It may be 2000 or 3000 years old. Others have been so modified by weathering and erosion that they look like depressions or pits that could have formed in other ways. Proof of their explosive origin rests entirely on clues within the shattered rock.

Interest in the geology of the moon has led to intensive study of impact craters on the earth. Drilling in Meteor Crater in Arizona failed to reveal any huge chunk of a meteorite beneath its floor. From sampling the surrounding area, geologists estimate that thousands of tons of very small meteorite particles are mixed with the soil.

Sandstone exposed below the rim of the crater is shattered and some of it has been fused to glass. Some of the quartz in the sandstone has also been converted to rare high-density forms of silica. From laboratory experiments we know that these forms of silica can be produced only by great amounts of heat and pressure. These conditions occur naturally at well over 100 km below the earth's surface. A logical conclusion is that the necessary heat and pressure were supplied by the impact of a meteoroid.

Meteor Crater is nearly 180 m deep and almost 1.6 km in diameter. But it has been calculated that it could have been caused by a meteoroid only 25 m in diameter traveling 50,000 kilometers per hour! If it had traveled faster, as many meteoroids do, it might have been even smaller. In any case it was largely destroyed by the impact, which tossed out the lumpy deposits of light-colored rock debris.

11-10 Investigating Regional Landscapes

MATERIALS
clear plastic sheet, grease pencil, landform map of North America, stereo atlas, stereo viewer, topographic map

By now you should be able to examine a landscape and say something about the nature of the surface rocks, the effect of the internal forces, and the kinds of external forces at work there. You may also

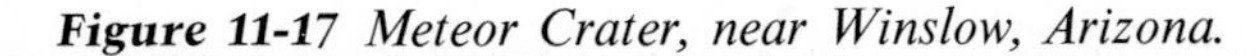

Figure 11-17 *Meteor Crater, near Winslow, Arizona.*

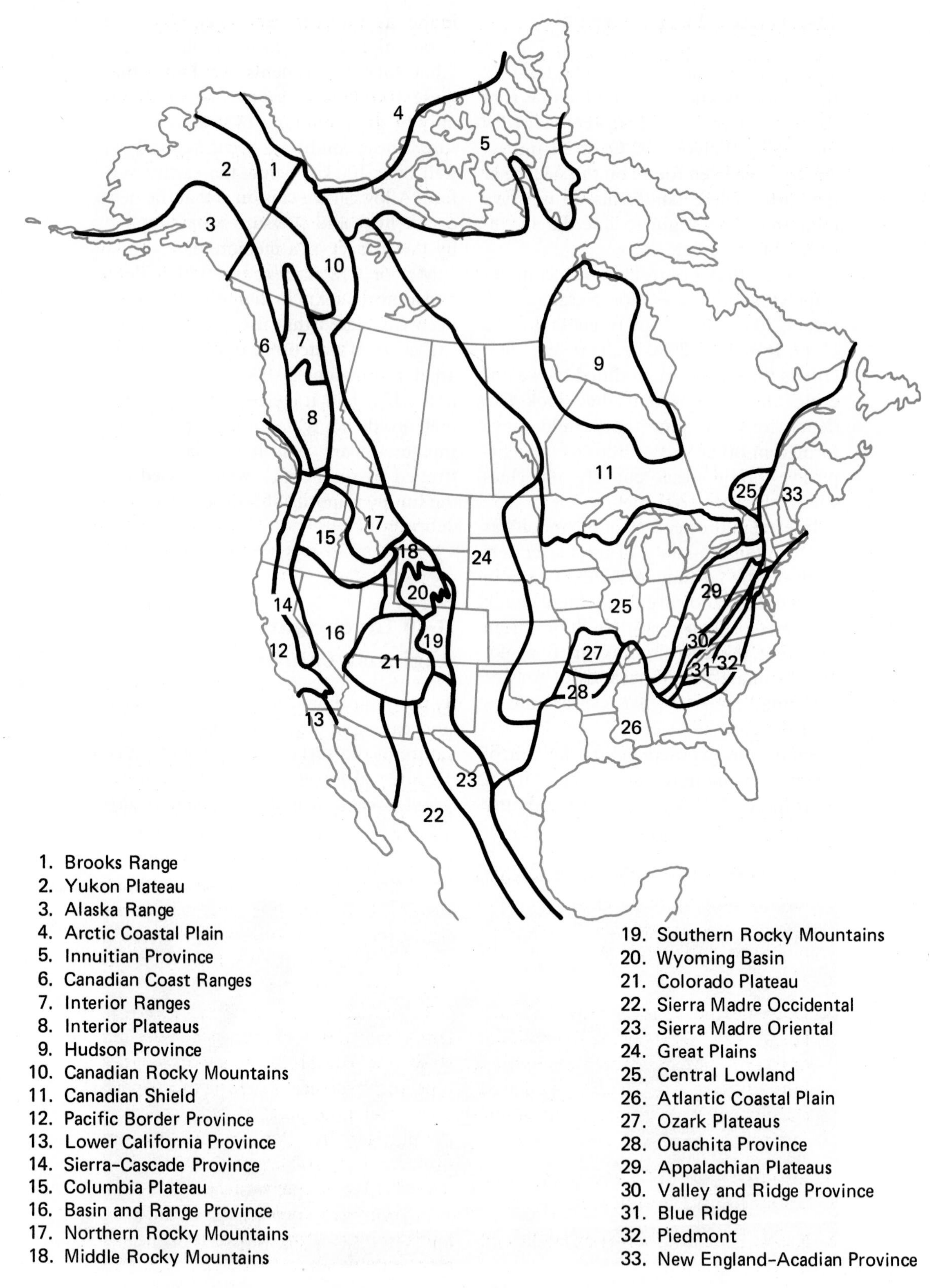

Figure 11-18 *The physiographic (landscape) provinces of North America.*

be able to say whether internal or external forces dominate the landscape at present.

When you have completed this investigation you will have observed, described, and interpreted typical landscapes found in the mountains, plains, and plateaus of North America.

PROCEDURE

Figure 11-18 shows the landscape regions (provinces) of North America. In which region do you live?

This investigation should start with an analysis of your local landscape. From this you will proceed to investigate landscapes typical of some of the other regions. You will examine each landscape by means of three-dimensional aerial photographs and topographic maps. In each instance compare and contrast the landscape of your locality with the one being observed. The materials you are to use and the necessary background material will be provided by your teacher.

Analysis and comparison of landscapes by observing maps and photographs should include the following steps.

1. Determine the region or province in which the landscape occurs.
2. Carefully consider the scale of each map or photo you observe. The same landscape will appear quite different on large and small scale maps or photos.
3. Describe the characteristics of each landscape. This should include: drainage pattern, relief (mountains, hills, rolling, flat), and land use (farming, urban, mining, recreation, forests).
4. Describe the differences in landscapes located in different regions.
5. Record specific observations indicated by your teacher.

Thought and Discussion

1. In what ways are turbidity currents on the sea floor a special case of erosion and deposition by streams?
2. What is the role of gravity in shaping landscapes?
3. In previous chapters you have studied features due to deposition or erosion by some of the leveling agents other than streams. Which of these were not influenced in size, shape, or location by earlier deposition or erosion by streams?
4. Which leveling agent is most important in shaping landforms in your locality?
5. Under what circumstances could there be radial drainage with streams flowing toward rather than away from a common point?

Unsolved Problems

One of the major difficulties in trying to understand landscapes is the way many factors interact to form them. Various specialists have examined certain aspects of the landscape with great care. The geologist knows a lot about how rocks weather and erode. The soil scientist has studied soil formation, and the biologist has learned much about soil and plant relationships. But these specialists have really only begun to study the complex ways in which all these and other elements work together.

To understand many of the details of landscape development, we need to study the combined influence of rock type and structure, soil formation, moisture, insolation, plant growth, and even human activity. Perhaps the greatest problem is to establish the actual time it takes for landscapes to develop through youth, maturity, and old age on different kinds of rocks. How long does it take for a mountain range to become a plain?

SURVEYOR

Before any construction project can begin, the planners need exact measurements of the site, information about the land features, and accurate location of the boundaries. Making these kinds of measurements is the work of the surveyor. Surveyors usually work in groups of three to six. The party chief plans the work, records the information, and prepares the maps and reports. Assisting the party chief are other workers who operate the surveying instruments and who measure heights above sea level (elevations), distance, and directions. Some surveyors specialize in highway surveys. Others perform land surveys and provide information needed for legal documents such as deeds. Topographic surveyors measure elevations and describe the physical features of the land so that contour maps can be made.

About a third of all surveyors are employed by government agencies. Others work for construction companies, engineering firms, and petroleum companies. Many surveyors have their own businesses. Training is usually gained through a combination of on-the-job training and courses after high school. Completion of a two-year program is good preparation. A state license is required for surveyors responsible for locating boundaries. Requirements are 4-8 years experience, formal training, and passing an examination.

Surveying usually requires that the person like the outdoors. Good health is needed, since surveying involves a lot of walking and carrying equipment over all types of land surfaces. Other qualifications for success include the ability to make accurate mathematical calculations, and leadership ability for advancement to party chief.

Chapter Summary

Two groups of external processes continually shape the surface of the earth. One is the weathering and erosion of solid rock, the other is the transportation and deposition of sediment. Both are closely related to the water cycle, and both depend on gravity. Together these processes make the land more nearly level.

But the exact shape of the land at any particular place and time is not determined by leveling processes alone. The shape of the land also depends on events within the earth's crust, which determine the kinds of rocks exposed at the surface, how they are arranged, and whether they are being elevated.

To best interpret the shape of the land you must recognize what is going on now. Then you work back through time looking for evidence of different conditions in the past. In this way you can often reconstruct a step-by-step sequence of events in the struggle between internal and external processes.

Impact craters are a special kind of landscape. While not created by normal rock cycle processes, they are destroyed by weathering, erosion, and deposition.

Questions and Problems

A

1. How can areas of erosion be recognized? Describe such an area.
2. How can areas of deposition be recognized? Describe such an area.
3. What are the leveling processes?
4. What is the lowest elevation to which leveling processes could possibly erode the land?
5. What do unconformities indicate?

B

1. Describe the leveling processes in terms of energy.
2. Why is it unlikely that the leveling processes will ever lower all the land to sea level?
3. Once sediments have reached the ocean, how can they still be eroded?
4. Describe a region in which internal forces dominate external forces.
5. Describe a region in which external forces dominate internal forces.
6. Cite evidence that Meteor Crater in Arizona was created by a meteoroid.

C

1. How does the kind of rock help to shape the landscape in a region?
2. How do events within the crust help shape the landscape of a region?
3. Can you think of conditions under which rocks in a region would not be physically weathered?
4. Why might physical weathering stop long before chemical weathering?

Suggested Readings

Price, Larry W. *Mountains and Man.* Berkeley, California: University of California Press, 1981.

Schultz, Peter H. *Moon Morphology.* Austin: University of Texas at Austin Press, 1976.

Sheffield, Charles. *Earth Watch: A Survey of the World from Space.* New York: The Macmillan Company, 1981.

Short, Nicholas M. *Planetary Geology.* Englewood Cliffs, New Jersey: Prentice-Hall, 1975.

Tuttle, Sherwood D. *Landscapes and Landforms,* 2nd ed. Dubuque, Iowa: William C. Brown Publishers, 1975.

unit IV
Earth's Biography

12 The Record in the Rocks

IN *1893 the captain of the sailing ship* Jason *set foot on an Antarctic island largely covered by ice. Imagine his surprise when he found petrified ferns, snails, and trees in the rocks there!*

Equally surprising discoveries were made by members of Admiral Richard E. Byrd's expedition to the Antarctic in 1935. They found fossilized tree trunks about a half meter in diameter and layers of coal that were separated by beds of shale and limestone. Since then, coal has also been found on the Arctic island of Spitsbergen. And in 1969 and 1970 the fossilized remains of reptiles were found in Antarctica. Most modern relatives of these fossilized plants and animals live only in tropical and subtropical climates. Climates must have been quite different when these earlier organisms lived.

You may have read that rivers have cut long deep trenches, such as the Grand Canyon, in the earth's surface. But do you know that seas covered the Grand Canyon area millions of years ago? Great thicknesses of sediment were deposited on the floors of these ancient seas. The rocks formed from these sediments can be seen in the walls of the Grand Canyon today. These layered rocks are like "pages" in the history of the earth, for they contain the geologic history of the Grand Canyon region.

Geologists reconstruct the geologic history of an area from clues found in the rocks themselves. Like detectives solving a crime, geologists must gather clues, assemble them, and finally determine what they mean. In this chapter you will have a chance to become a "geo-detective." You will become familiar with the various types of evidence found in the rock record and will use this information to reconstruct certain events that took place in the geologic past.

AFTER COMPLETING THIS CHAPTER, YOU SHOULD BE ABLE TO:

1. **infer the origin of various sedimentary rocks from their physical characteristics.**
2. **distinguish differences in the origins of various layered rocks from samples or descriptions, and group them by rock type.**
3. **describe how it is possible to distinguish the top of a rock layer from its bottom.**
4. **show how rock layers can be "matched up" by means of fossil content, rock type, or other means.**
5. **construct a cross-section or geologic column from geologic data, and reconstruct a sequence of events in the geologic history of that area.**
6. **interpret the geologic history of an area from a geologic map or cross-section, and explain the sequence of key geologic events.**
7. **explain how ancient climates can be inferred from fossil content and other clues in the rocks.**

Learning to Read the Record

12-1 Clues to the Origins of Rocks

Imagine that you are looking at a rock outcrop. Figure 12-1a gives an overall view of the outcrop and the surrounding rocks. If you move closer (Figure 12-1b) you can learn more about the nature of the rocks. An even closer examination (Figure 12-1c) shows their texture and provides additional information about their history.

Trace minerals (minerals that are usually present in small quantities) often can be used to help determine the origins of sedimentary rocks. Assume, for example, that grains of the mineral garnet are found in sandstone. If the only other garnets in the area occur in an outcrop of metamorphic rock 40 km to the north, it is likely that the metamorphic rock was the source of at least some of the sediment. This interpretation would be strengthened if you discovered that the percentage of garnet increased as you got closer to the metamorphic rock.

Most sedimentary rocks are made up of fragments of other rocks. The sizes of the sedimentary particles may be a clue to the speed and volume of the water current that deposited them. Coarse sediments, such as the pebbles and gravelly sands that form conglomerates, are carried by larger, fast-flowing streams. Fine-grained sand and the smaller particles that form shales are usually carried by more slowly moving bodies of water. The shapes of the particles can also indicate something about their past. Rock fragments may be rounded and smoothed as they are carried by water, wind, or ice. If a sedimentary particle is rough and angular, it probably was not carried too far.

Certain sedimentary rocks may contain fossils. Some **fossiliferous** (fossil-bearing) rocks consist almost entirely of the broken shells of marine organisms. Ancient coral reef deposits, for example, are built almost entirely of fossils. We assume that these ancient organisms lived in warm saltwater environments like their present-day relatives. Thus, these fossils tell us about environments that existed in the geologic past.

The texture of an igneous rock depends on how it formed. A coarse-grained crystalline texture like that of granite suggests that the rock cooled slowly. In most cases, it cooled deep beneath the earth's surface. In contrast, a fine-grained igneous rock like basalt cooled quickly, at or near the earth's surface. Metamorphic rocks may also contain clues to their origin. Most metamorphic rocks have been subjected to high pressure. This sometimes causes flaky, crystalline textures, as you can see in schist.

ACTION

Carefully examine the materials in the rock in Figure 12-2a. Can you tell where the sediments that formed it came from? What do the sizes and shapes of the sedimentary particles tell about the rock in Figure 12-2b? Is one rock made up of sedimentary particles that were probably moved a long distance? Which rock contains fragments that have undergone little erosion and transportation?

a.

b.

c.

Figure 12-1 *Zeroing in on a rock outcrop in Los Angeles, California.*

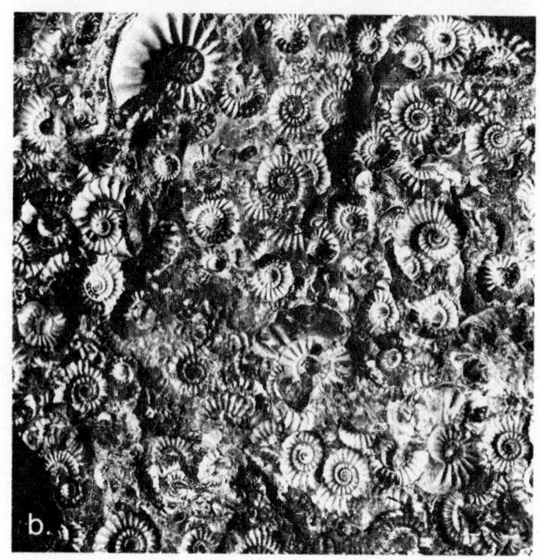

Figure 12-2 *Where did each of these rocks probably originate?*

12-2 Layered Rocks Are Like Pages of Earth History.

The most common feature of sedimentary rocks is layering or bedding. (See Figure 12-3.) You might wish to review the conclusions you came to in Investigations 8-1 and 8-4. Each rock layer gives clues to the conditions under which it was originally deposited as sediments. Some of these clues reveal how—and perhaps when—the rocks were formed. Figure 12-4a shows how sediments might accumulate in still water. However, currents are present in most bodies of water, including lakes and oceans. Most sedimentary particles settle under conditions such as shown in Figure 12-4b–d. (Which particles have greater potential energy, the ones that settle on top of the small humps or those that settle in between?)

Layers of sediment are usually deposited in a horizontal position. Tilted rock layers (as in Figure 12-1a) show that the rocks have been disturbed since they were formed, and provide additional clues to the geologic history of an area.

The boundary between layers of sediment can be sharp or gradual. The surfaces between layers are called **bedding planes.** Bedding planes form between layers when the conditions of deposition change. What changes would you expect to find in a sediment layer if the speed of a depositing current suddenly decreased?

Let us assume that a stream that usually deposits large amounts of sand in a basin suddenly dries up. After that stream stops flowing, another stream from a different area begins to empty into the basin. It deposits sand with different characteristics. The bedding plane

Figure 12-3 *Typical layered sedimentary rocks. Sandstone and shale in the Capitol Reef National Monument, Utah.*

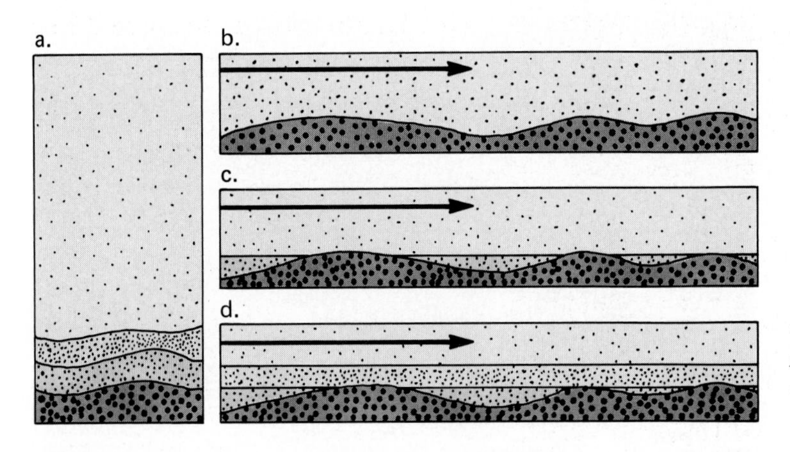

Figure 12-4 *Compare the settling of particles in* **a.** *still water and* **b.-d.** *moving water.*

between the different layers of sand would be distinct. Many times, however, changes in the type of sediments take place slowly. Slow changes cause indistinct bedding planes that are not always easy to recognize.

A single layer of sediment that covers several square kilometers may vary considerably from place to place. In one area the single layer may consist of coarse sand and gravel. These relatively heavy sediments were probably deposited in places where the stream current was strong. Coarse sediments are also usually deposited close to shore. In other places, the same single sedimentary layer may be made up of clay and shale. These lighter, finer sediments were probably carried greater distances, and the particle size gradually decreases away from shore. Study Figure 12-5. Note the places where clay extends into several of the sand layers. What might this indicate about conditions during deposition?

Many sediments are deposited on the bottoms of oceans. Sea shells have been found as fossils in sedimentary rocks in central Kansas, on the tops of mountains in British Columbia, and in a desert in California. You can be certain that these rocks formed from sediment deposited in an ancient ocean. The recognition of marine sedimentary rocks is particularly important to petroleum geologists. Most oil and natural gas occur in this type of rock. Other sedimentary rocks appear to have formed on land. If you find the remains of a fossil horse in a sedimentary rock, it is likely that the rock formed on the land.

Deposits of loose sand can become packed and cemented together to form

Figure 12-5 *Horizontal variation in marine sediments.*

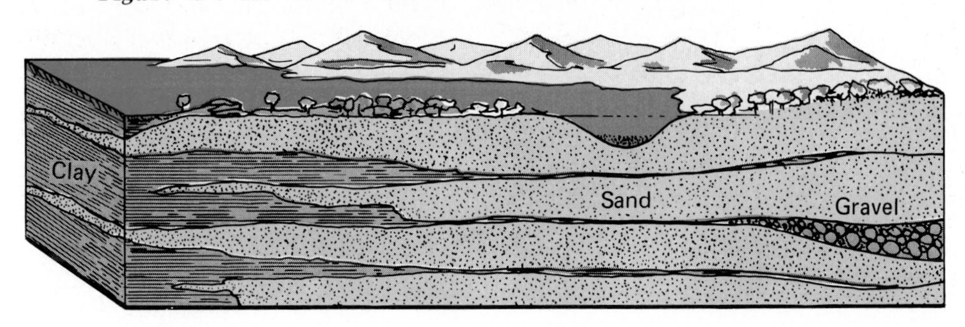

Figure 12-6 *Layers of lava at the Palouse River Falls, Washington.*

solid rock. So can soil. The cement that holds the sediments together usually comes from the weathering of other rocks. Ground water that flows through the pores of rocks contains ions of calcium, silicon, and iron. These ions can combine with oxygen and other elements in the water to form a cement of calcite ($CaCO_3$), quartz (SiO_2), or various iron minerals.

Not all layered rocks are sedimentary. There are thick piles of layered lava on the Columbia Plateau in the northwestern United States. (See Figure 12-6.) Layers also form on the sides of volcanic cones when flows of lava pour down the sides and harden. If ashes and dust are blasted from the crater, they may also settle in layers on the cone. Because they are composed of igneous rocks, layers of lava and volcanic debris can usually be easily distinguished from sedimentary rocks.

Some volcanic flows may include different-colored layers that are wrinkled like an untidy tablecloth. The rocks

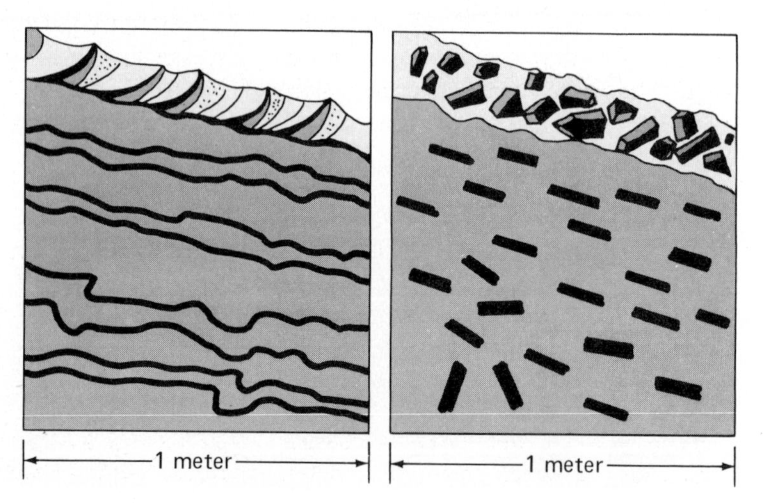

Figure 12-7 *In what direction did the lava flow in these diagrams?*

within certain lava flows contain long sliver-shaped crystals that line up in the direction of flow, like logs floating in a river. The arrangement of these crystals can sometimes be used to determine the source of the lava. Lava flows may also develop surfaces that look like waves on water. How could these wavy surfaces, plus the way the crystals are arranged, help determine the origin and direction of the flow? (See Figure 12-7.)

Certain metamorphic rocks also show evidence of layering and may resemble sedimentary rocks. For instance, the layering in marble and quartzite was present in the original sedimentary rocks. What looks like layering in other metamorphic rocks probably developed during metamorphism. The layered or banded appearance of most schists is thought to be due to the great heat and pressure that formed them. The soft or molten minerals gathered together into bands of different colors. In some cases, the appearance of layering might be caused by weathering.

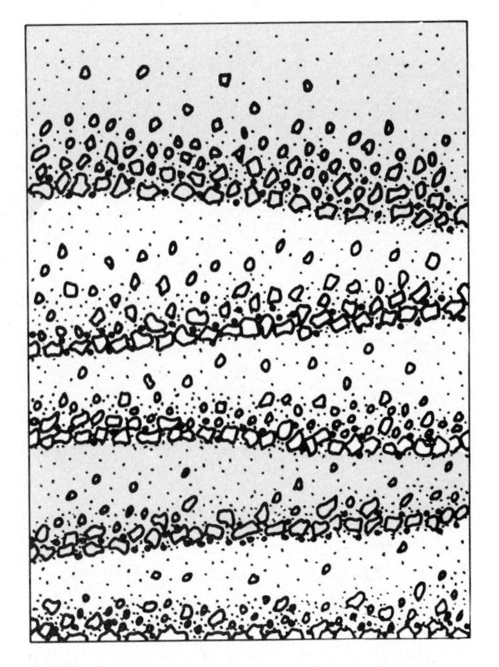

Figure 12-8 *Layers of sediment show a gradation in the size of the particles. Is this sketch right side up or upside down?*

12-3 Cross-beds and Ripple Marks As Clues

Before you can determine the history of most layered rocks, you must be able to tell the bottoms of the layers from the tops. In many areas, rock layers have been tilted or even turned upside down. Fortunately, these rocks may contain clues to the upper surfaces of the layers. Examine Figure 12-8. How can you tell top from bottom?

Some sedimentary layers contain **cross-beds.** Cross-beds are thin rock layers that lie at an angle to the larger layer that contains them. The outcrop of cross-bedded sandstone shown in Figure 12-9 was formed from an ancient dune of windblown sand.

The cross-beds shown in Figure 12-10 were formed in water. They are characteristic of the development of a delta.

Figure 12-9 *Cross-bedded sandstone in Oak Creek Canyon, Arizona.*

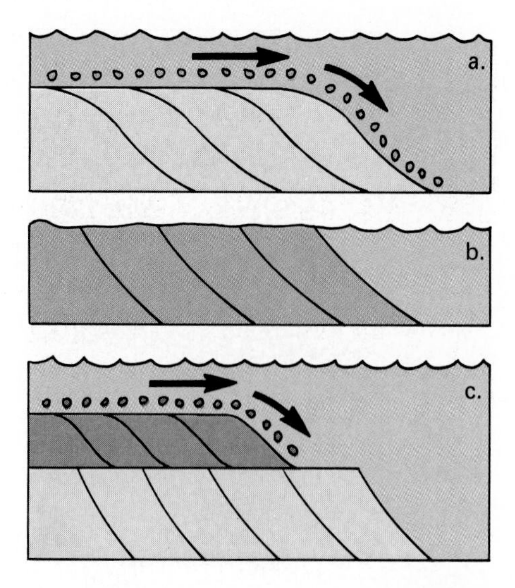

Figure 12-10 *Cross-bedded sediments form when deposition and erosion take turns.*

The cross-beds form where the river dumps its load of sediment upon entering a larger body of water such as the ocean. Reconstruct the series of events represented in the diagram. From the shapes of cross-beds shown in the diagram, how could you use the cross-beds to tell the top layers from the bottom? Would they provide clues as to the direction of the current at the time of deposition?

MATERIALS
small tray, mud

Ripple marks (Figure 12-11 and Figure 9-3) are another feature occasionally found in sedimentary rocks. You may have seen them formed by currents in the shallow waters of a lake or ocean. Ripple marks may also be found on sand dunes, on the bottoms of streams, on snowdrifts, or even in puddles after a rainstorm. Ripple marks that are formed by currents moving in a single direction have a definite shape. The down-current side of each ripple mark is steeper.

Ripple marks whose sides are symmetrical develop in water that constantly moves back and forth. The crests of symmetrical ripple marks point upward, so they can be used to tell whether a layer has been overturned.

Fossils can also be used to determine whether a rock layer is upside down. When empty clam shells come to rest on a beach, waves or currents usually turn them over so that the hollow side is down. (See Figure 12-12.) The relative ages of fossils may be another clue to whether layers have been overturned. For instance, you might find an outcrop in which the fossils in an upper layer belong to a period of geologic time that is millions of years older than the period of the fossils in a lower layer. Could these layers have been overturned? What other explanation is there for such a situation?

ACTION

Place a small tray of wet mud beside a radiator, in a sunny window, or under a heat lamp for several days. Examine the cracks that form on the surface. Imagine similar cracks on a mud layer that has been changed to rock. How could you use these cracks to recognize the top of the layer?

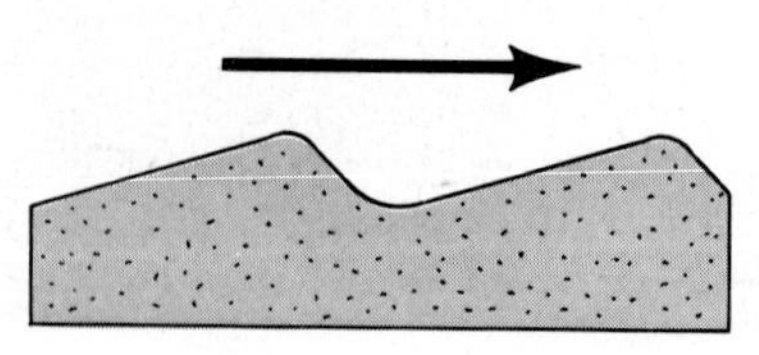

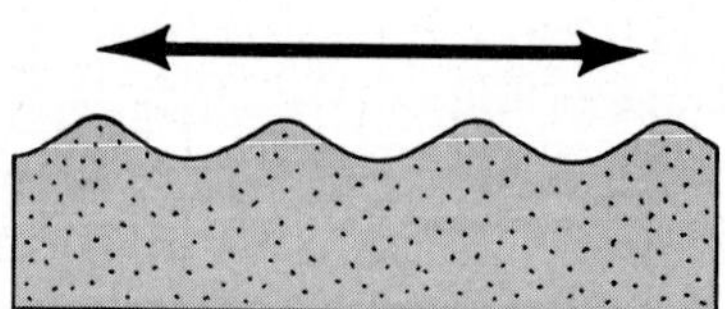

Figure 12-11 *Current and oscillation ripple marks in sand. The arrows show the directions of water flow. Compare these diagrams with the ripple marks in Figure 9-3.*

12-4 Investigating an Ancient Stream Channel

Suppose that you are a geologist who has found ripple-marked, cross-bedded sandstone exposed at the surface. You suspect that the sandstone was deposited by an ancient stream.

When you have completed this investigation, you will have traced a portion of the channel of the ancient stream.

MATERIALS
ruler, tracing paper (or clear plastic sheet)

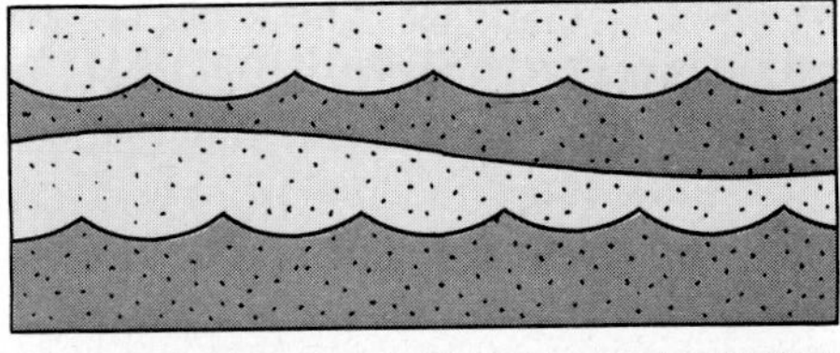

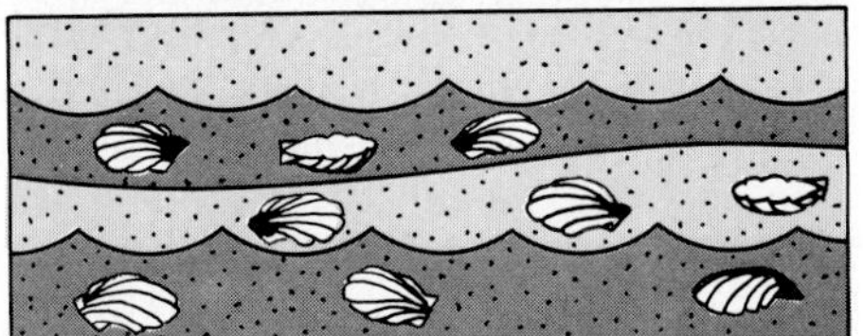

Figure 12-12 *Symmetrical ripple marks. Is either of these examples upside down?*

PROCEDURE

Assume that you have already done the field work for this problem. The map you have prepared (Figure 12-13) shows where you measured the thickness of the rocks beneath the surface. This was done with drill holes. Numbers (in parentheses) are located next to many of the drill-hole markers. They give the thickness (in meters) of a sandstone that looks as though it might have been deposited

Figure 12-13 *Data for investigating an ancient stream channel.*

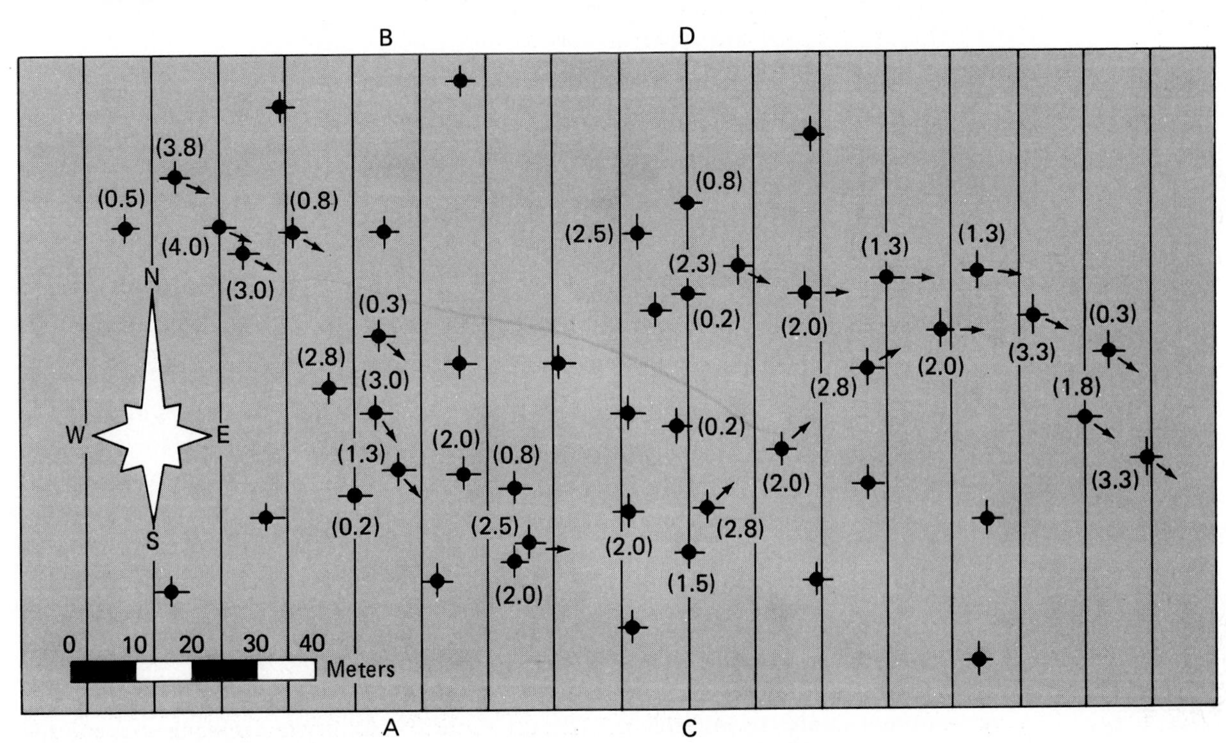

by an ancient stream. The markers with no numbers represent drill holes in which no sandstone was found. The arrows show the direction of flow in the ancient stream as determined by cross-beds and ripple marks.

Put a piece of tracing paper or clear plastic over the map and use a soft pencil or crayon to draw the shape of the ancient stream channel.

DISCUSSION

1. What evidence can be used to locate the buried stream channel?
2. How do you know where to draw the lines showing the ancient stream channel?
3. Did the ancient stream meander in loops, or did it flow in a straight line?
4. Did the stream have any branches? Can you be certain from the data?
5. Make a sketch showing what this sandstone might look like if you could dig trenches along one of the lines (marked A-B and C-D) in Figure 12-13.

Thought and Discussion

1. How can the characteristics of sedimentary rocks be used to determine the environmental conditions under which they formed? Give specific examples.
2. Think about this statement: All layered rocks are sedimentary, and all sedimentary rocks are layered. Is the statement true or false? What are your reasons?
3. Describe several ways to determine which is the upper surface of a sedimentary layer.
4. Where would you expect to find the most coarse-grained texture in an igneous intrusion? Why?

Putting the Pieces Together

12-5 Investigating Puzzles in the Earth's Crust

Most of us enjoy putting together the pieces of a jigsaw puzzle. But occasionally pieces of the puzzle are lost, and we find blank spots in the final picture. The clues geologists use to reconstruct earth history are like the pieces of a giant, rocky jigsaw puzzle. In many areas the missing pieces far outnumber those that are available for study.

When you have completed this investigation, you will have used the "clues" present in rock outcrops to solve a geologic puzzle.

PROCEDURE

Look at the block diagrams of the rock layers in Figure 12-14. These exposed rock layers are found considerable distances from each other. Now study Figure 12-15, which illustrates fossils from several periods in the earth's history. The numbers in the rock layers correspond to the fossil numbers. For example, a layer in outcrop C has the numbers 10 and 16. That means that fossils 10 (*Mucrospirifer*) and 16 (*Phacops*) are found in the layer. Geologists often use fossils to help in identifying rock layers from one place to another.

Use the information provided in Figure 12-14 to try to decide the *relative* age

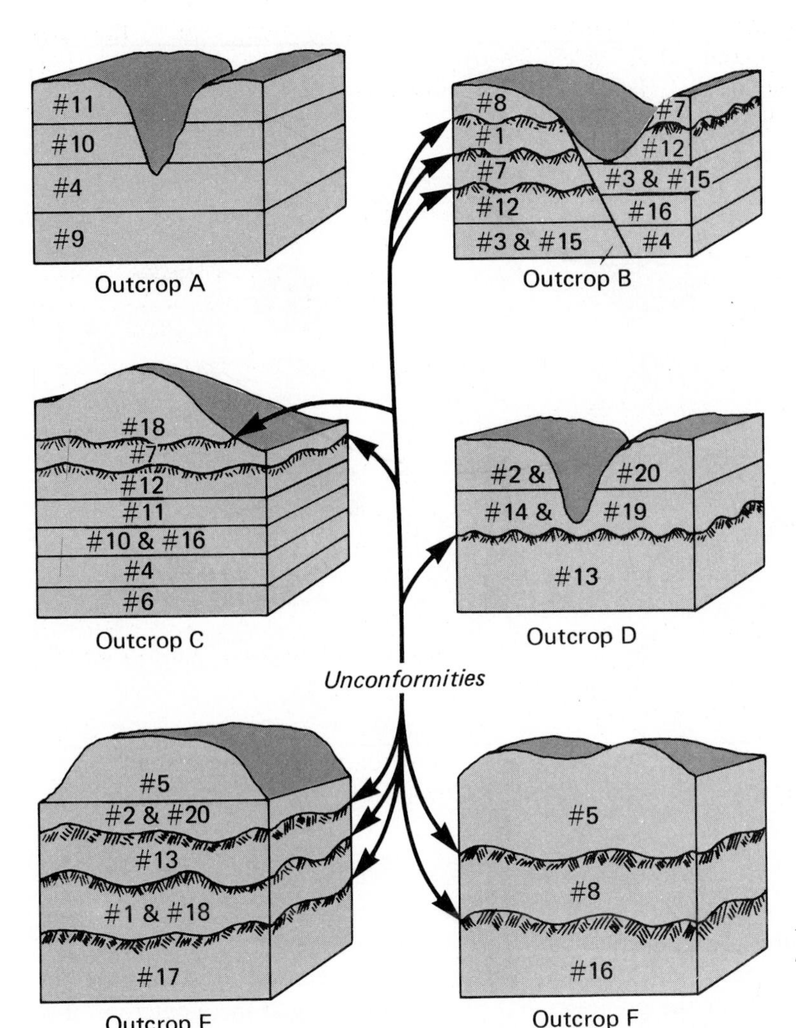

Figure 12-14 *Sections of six widely scattered rock outcrops.*

of each of the layers in all six outcrops (which is the youngest, next oldest, and so on, to the oldest layer). Since some layers are found in more than one outcrop, you can compare outcrop to outcrop until you think you know the relative ages of all the layers in the area. Then you will also have some ideas about which fossils are older than others.

As additional information becomes available to you, during the remainder of this chapter and in Chapter 13, you may wish to make some changes in your sequence. At the end of Chapter 13 (see Section 13-11) you will be asked to determine the exact order in which these rock layers formed. You may wish to refer to the questions in Section 13-11 to see what answers to look for between here and there.

12-6 Correlating Rock Layers

In Investigation 12-5 you attempted to match up, or correlate, the layers of rock according to the order in which they formed. Geologists also try to correlate rocks to find out which ones were formed at the same time.

In some places, correlation is simple.

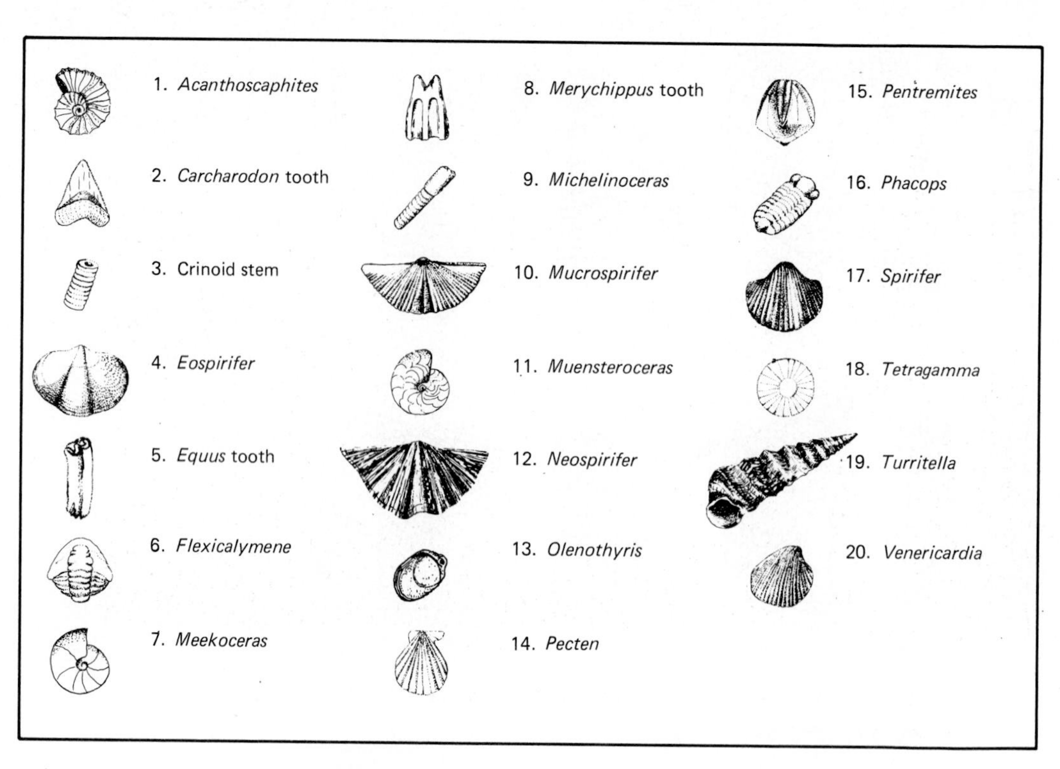

Figure 12-15 *Fossils found in the outcrops in Figure 12-14.*

Examine the photograph of the Grand Canyon in Figure 12-16. Correlate the different layers exposed in the canyon wall on the left side of the photograph with layers on the right side. Can you see any evidence that the layers on the right and left are the same? Trace several distinct layers across the picture. Would this be more difficult if you were standing down in the canyon rather than up on the rim? Why?

Sedimentary rock layers that have a distinctive color, texture, or set of fossils can easily be traced short distances. But the farther apart the outcrops are, the harder it is to correlate them. Nearly two hundred years ago William Smith discovered that physical characteristics and fossils could be used to correlate rock layers many kilometers apart. This English surveyor's discovery was one of the "kilometerstones" in our understanding of earth history.

An example of a correlation problem is shown in Figure 12-17. The diagram shows rock outcrops in four different places. In looking at these outcrops, it is possible to recognize several different layers. The geologist wants to know whether the rocks at section a are related to the rocks at section d, which is nine kilometers away. It is known that the white limestone in sections b, c, and d contains fossils and is covered by conglomerate. The conglomerate in turn is covered by basalt. Because the sequence of layers is the same, it is possible to correlate the conglomerate layers in the three sections.

The basalt in sections a and c is covered by green sandstone. So, section a can be matched with section c. This reveals how sections a and d are related to each other, even though they have no layers in common. The geologist concludes that the red sandstone in section

Figure 12-16 *Sedimentary layers exposed in the Grand Canyon.*

Figure 12-17 *Outcrops a and d do not have any rock layers in common. But the rocks in the outcrops can still be correlated.*

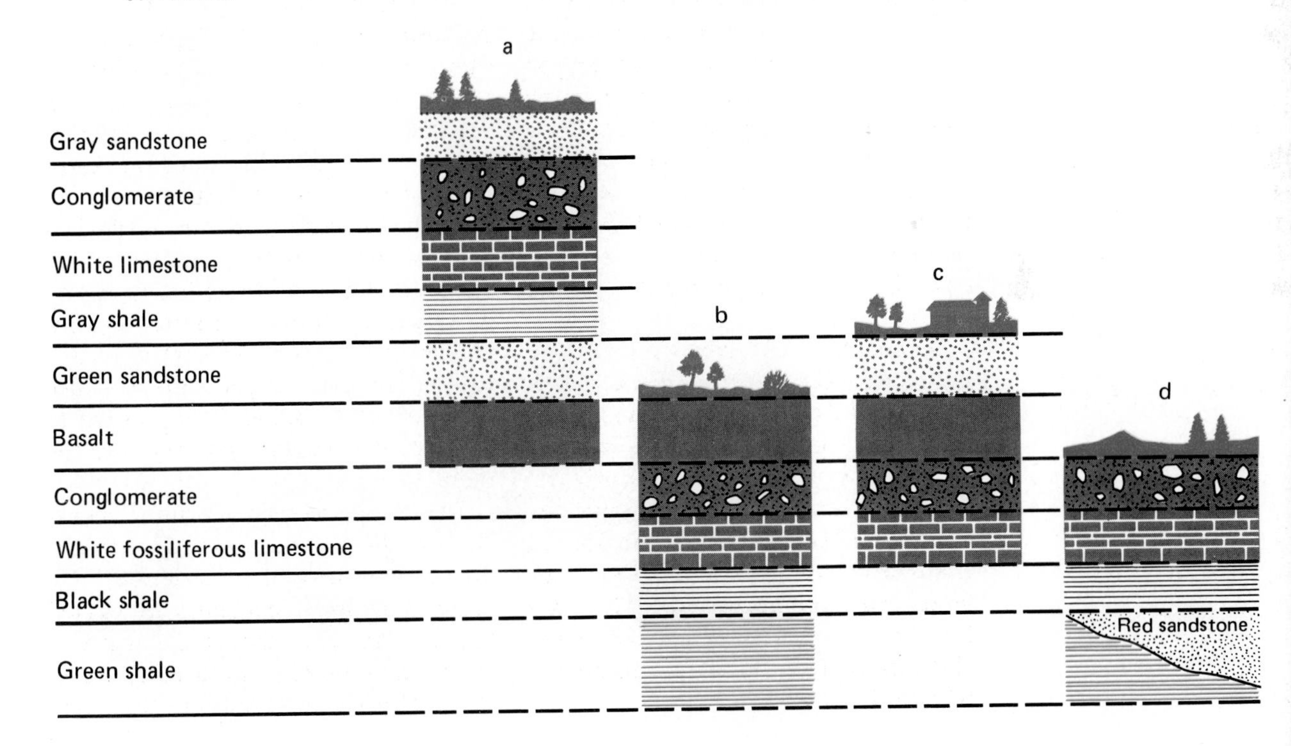

d gradually thins out and does not extend into location b. Is this a reasonable conclusion? Carefully examine the sections for other evidence that might support this correlation.

It is much more difficult to correlate rock layers in widely separated areas, such as on different continents. During the past, conditions in different areas varied just as they do today. Deposition was going on in some places, erosion in others. There are no individual rock layers that span the entire earth or even an entire continent. Nevertheless, earth scientists can correlate rocks in different continents.

One of the best ways to correlate rocks in widely separated areas is by using fossils. For example, similar species of fossils are found in Africa, France, and North America. Although widely separated, these creatures all lived at the same geologic time. Scientists assume the rocks that contain these fossils consist of sediments deposited during the same part of geologic history.

12-7 Outcrops Reveal a Sequence of Events.

Sedimentary layers are deposited horizontally. However, the layers may later be deformed by folding, faulting, or erosion. Suppose you conclude from fossils in two different layers of rock that limestone 100 million years old lies directly on top of sandstone that is 600 million years old. How can you account for the missing layers?

This situation would be similar to discovering that the middle 100 pages of a detective novel had been torn out. What would you do to find out about the full story? You might be able to guess at the events that occurred in the middle of the

Figure 12-18 *Find the unconformity in this outcrop. Explain how this unconformity might have developed.*

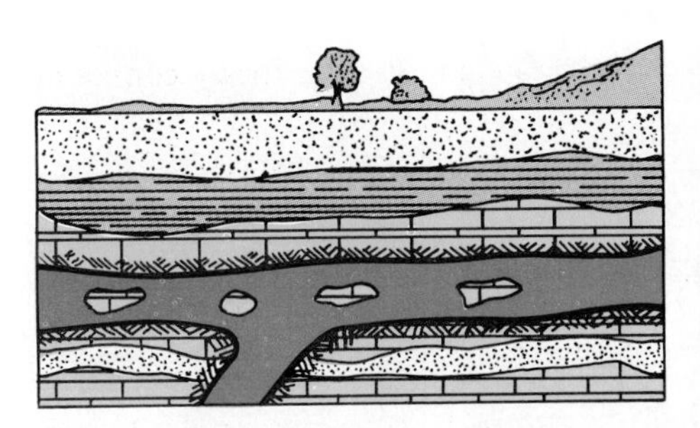
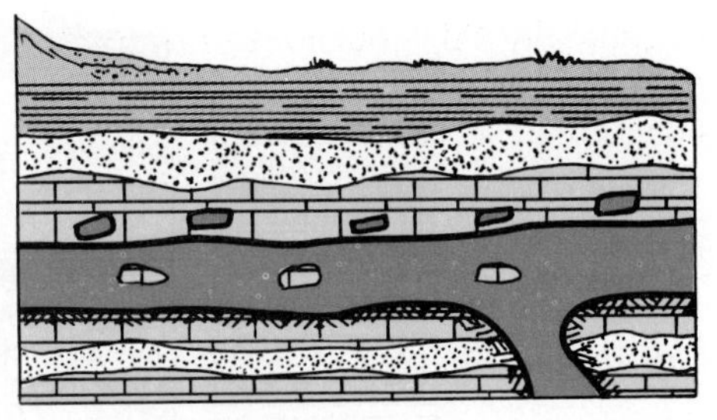

Figure 12-19 *Which diagram shows an intrusion and which shows a buried lava flow? Upon what evidence do you base your answer?*

book if you had read other detective novels written by the same author. In studying rocks, you might not be able to tell the complete story from a single outcrop. But by correlating the rocks in different outcrops you might be able to find out what happened during the period of the missing layers.

Where layers are missing from a sequence, the upper surface of the older rocks may be an old erosion surface or **unconformity.** Unconformities are easily recognized when folded or tilted rocks are eroded and horizontal layers are deposited on top of them. (See Figure 12-18.)

The geologist must also be able to determine the relative ages of igneous and sedimentary rocks in the same outcrop. Suppose that you found a layer of igneous rock in the middle of a thick pile of sedimentary rocks. The igneous rock could be a lava flow that was buried by sediments. Or it could be an intrusion squeezed in between the layers of sedimentary rocks already deposited. There is a good way to tell the difference. An igneous intrusion will cut through and bake the rocks around it or carry rock fragments along with it.

Examine Figure 12-19 carefully. Can you tell which diagram shows a buried lava flow and which shows an intrusion? What is your evidence? What are the relative ages of the igneous rock and the top rock layer in each sketch?

ACTION

A series of additional geologic cross sections is given in Figure 12-20. Examine each of these cross sections and describe the sequence of events that occurred in each area. Begin with the oldest and end with the most recent event.

12-8 Interpreting a Chapter in Earth History

MATERIALS

knife or cheese cutter, modeling clay (or other modeling material), powder (or flour or plastic film), wooden roller

If you examine a rock outcrop carefully, you can usually piece together the sequence of events that formed it.

When you have completed this investigation you will have made and interpreted geological models made of clay.

PROCEDURE

Figure 12-21 shows how you can make geological models out of clay. Your group should make up its own outcrop. Be sure to keep a list showing the exact sequence you used to make the model. When the models have been completed, exchange them with another group of students in the class. Now try to determine the steps used to construct the new model. The group you exchanged with

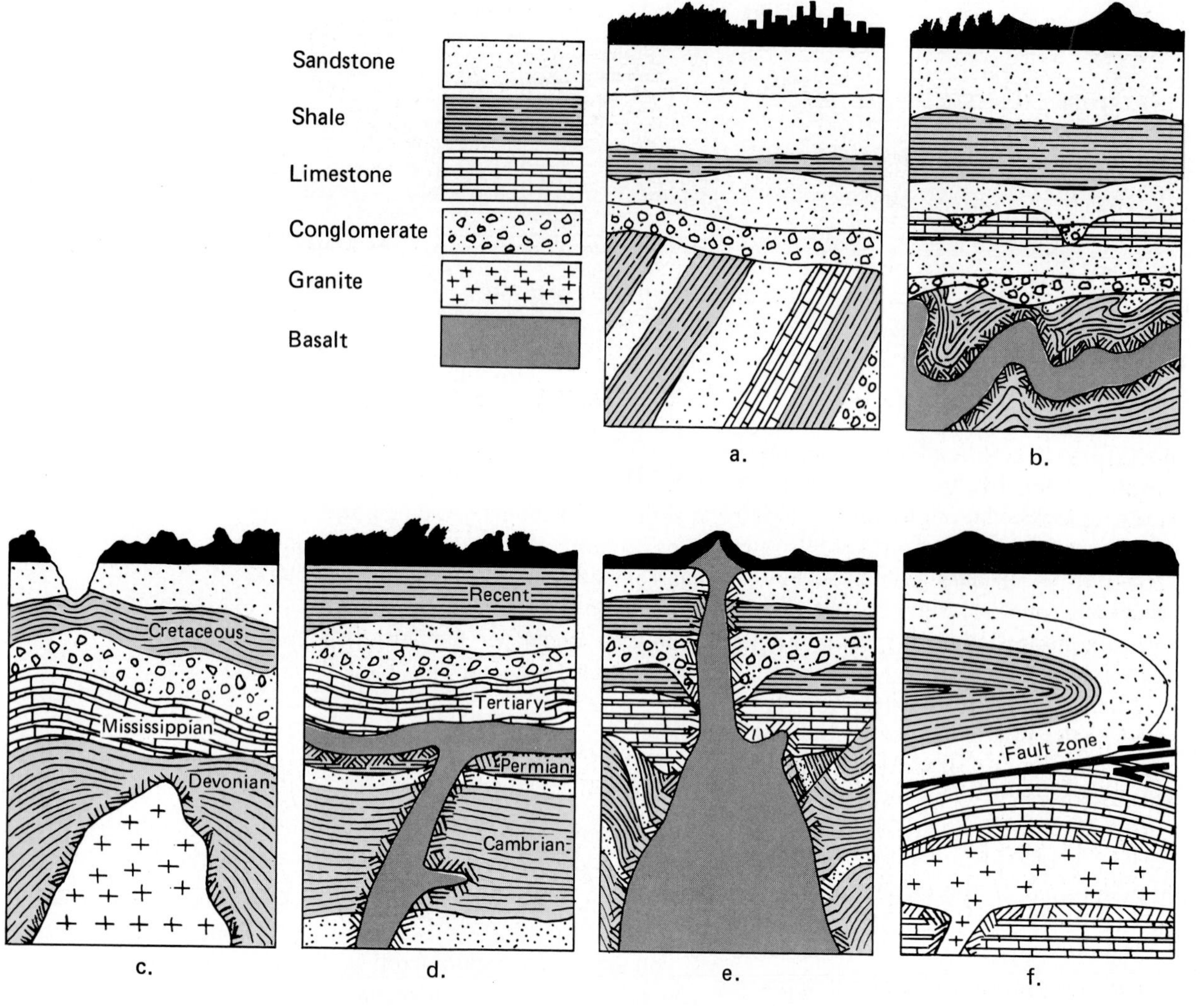

Figure 12-20 *Describe the sequence of geologic events revealed in each of the above cross-sections.*

will do the same with the model you made.

Discussion

As you examine the model, try to answer these questions:

1. Which layer is the oldest? How do you know?
2. What are the relative ages of the rest of the layers?
3. If your model contains folded layers, when did folding occur? Can you determine from the shape of the folds how they were formed?
4. Is there evidence that any of the layers have been eroded? How do you recognize a former erosion surface?
5. How would you explain a layer of one color cutting through layers of other colors?

6. Have any layers been disrupted by faults?

If you have difficulty answering any of these questions, what additional information do you need to help you interpret the model?

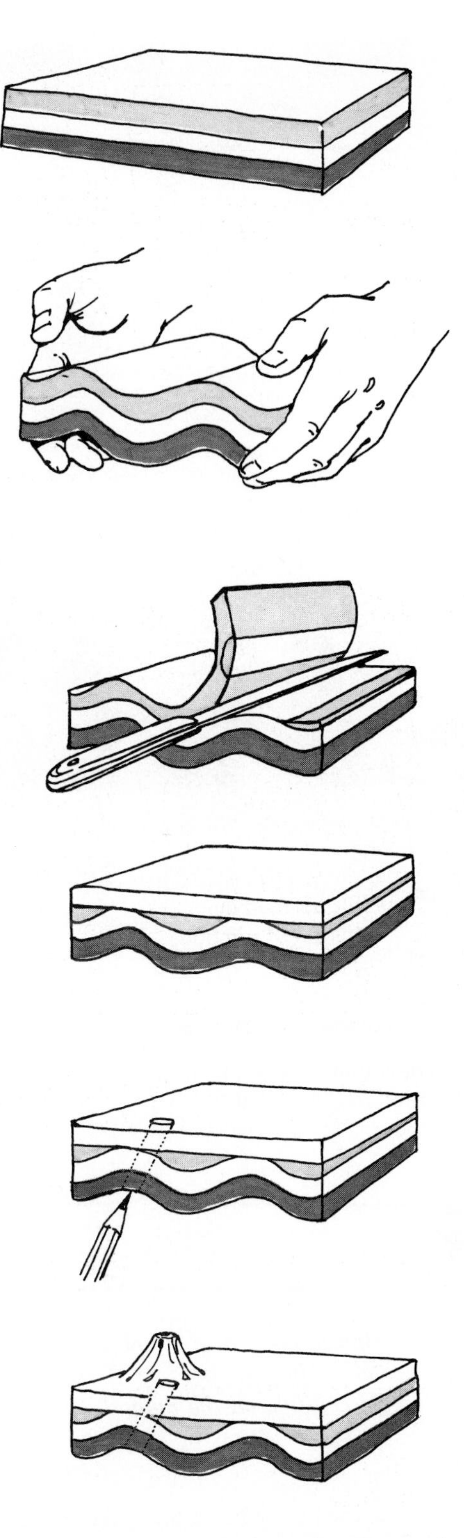

Figure 12-21 *One possible sequence of steps in making a geologic model.*

12-9 Investigating the Record in the Rocks with a Geologic Map

Western Montana is an area of high, rugged mountains. Along the Continental Divide there are many peaks higher than 3000 m. The peaks and ridges rise 750 to 1500 m above the valley floors. Glaciers once nestled in valleys around many of the peaks. The rocks exposed here are typical of those found in young, complex mountain ranges. In old mountain ranges the sedimentary rocks are gone, and only igneous and metamorphic rocks remain.

In this investigation, you will use a geologic map to "read" a part of the rock record in this part of Montana.

The Philipsburg area is about halfway between Butte and Missoula, Montana. The geologic map in Figure 12-22 shows the distribution of rocks at the surface or under the soil cover. The types of rocks and their relative ages are shown in the legend. The oldest rocks are at the bottom of the legend and the youngest are at the top.

PROCEDURE

Examine the map. How would you describe the pattern of distribution of the sedimentary rocks? "Confused" might be the first word that comes to mind, but can you do better in one or several words? How can layered rocks occur as bands?

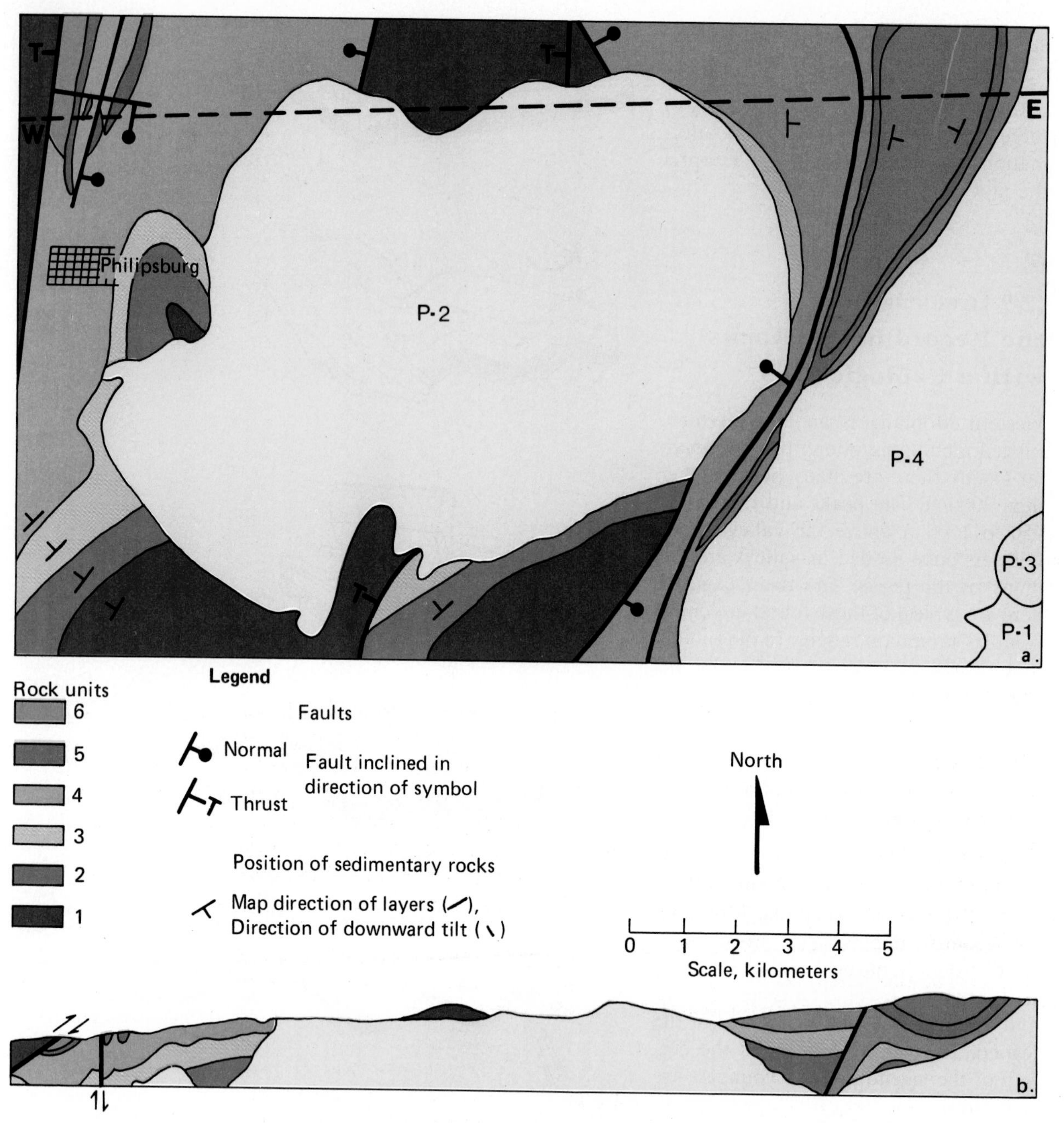

***Figure 12-22* a.** *A simplified geologic map of the Philipsburg, Montana area.* **b.** *A cross section of the area along the dashed line on the map.*

Notice the heavy black lines like those just north of Philipsburg. These lines mark the faults at the surface of the earth. The faults extend below the surface for a considerable, but usually unknown, distance. Most fault surfaces are not really flat, but are gently warped and inclined from the horizontal at steep angles. The map symbols point in the direction that the fault tilts downward from the surface.

The rock masses labeled P-1 through P-4 in Figure 12-22a are plutons. (Plutons were described in Section 10-3.) Rock bodies of this type are common in folded mountains. All these rock masses have the same relation to the sedimentary rocks and the faults. They are numbered differently to indicate different mineral compositions. Figure 12-22b is a vertical cross section to a depth of 2750 meters along the line W–E (west to east) on the map. Notice that the sides of Pluton 2 are inclined steeply. The bottom of the figure is not the bottom of the pluton. It extends deep into the earth.

No one has ever seen a pluton form. All of the evidence we have indicates that these masses form below the surface within a growing mountain range. They are not known to develop in any other environment, except under unusual circumstances.

Discussion

1. If rocks are displaced by faults, might faults displace faults?
2. Look at the fault lines north of Philipsburg. Two run almost north–south, one runs nearly east–west. Which do you think is younger?
3. Some of the faults shown shift the boundary lines between various sedimentary rock layers. Does this mean that the faults are older or younger than the rocks?
4. Look at pluton P-2 just east of Philipsburg. Does this rock cut across or parallel the units of sedimentary rock?
5. Would you conclude that this rock is older or younger than the sedimentary rocks it touches at the surface?
6. Would you conclude that P-2 is older or younger than the faults?
7. Notice that P-2 is not shown in the legend of the map. Where would you put it in the legend?

The thickness of the older rocks that were eroded from the top of the pluton is usually unknown (perhaps unknowable). In some instances it can be estimated. Estimates vary from as little as 1 to as much as 8 or 10 kilometers.

The Philipsburg pluton is modest in size compared to many. Most of the larger plutonic bodies in North America range from 4000 to 50,000 square kilometers in area.

12-10 Reconstructing Ancient Climates

Certain rocks contain evidence that climates have radically changed during the past. Rocks that contain fossils are especially helpful indicators of ancient climates.

Living organisms, especially plants, grow best in certain climates. For example, palm trees do not grow in Alaska, nor does reindeer moss grow in the tropics. Fossilized plants and animals probably lived in the same climates as their present-day relatives. Fossils of subtropical plants such as magnolias are found in rocks on the Arctic island of Spitsbergen. These fossils suggest that the climate there was once much warmer than it is now.

Some fossil plants have features that suggest they grew in a warm climate. Large cells or a lack of annual growth rings are examples. Dinosaurs were cold-blooded, like the modern-day reptiles. Yet, these extinct reptiles were once abundant in what is now the United States and southern Canada. The climate there was probably mild during this time, because reptiles must hibernate when the temperature drops toward freezing. A dinosaur 20 meters long and weighing several metric tons would have had trouble finding a cave large enough to sleep in during cold winters.

The number of fossils can also be a clue to an ancient climate. More species of plants grow in the tropics than in the higher latitudes. The situation was probably the same in the past. Therefore, finding a large variety of fossil plants is good evidence that an area once had a tropical or semitropical climate.

You can study the worldwide distribution of certain fossils during a single geologic period. You may then discover that each species is found in one latitude zone. This indicates that it could only live under particular climate conditions. The maps in Figure 12-23 describe probable climate zones from three different periods. Note the subtropical zone in the map on the left. Rocks have been found there containing fossils whose modern relatives live in subtropical areas. Do you think that such maps would be reliable for much earlier periods?

Reef-building corals are especially useful for identifying ancient environmental conditions. Most present-day species need warm water in order to digest their food and to carry on other life processes. Corals live only in salt water and prefer temperatures between 25 and 29°C. In addition, most corals prefer depths of less than 75 m.

Certain minerals can also indicate ancient climates. Gypsum ($CaSO_4 \cdot 2H_2O$) and rock salt ($NaCl$) are examples of such minerals. When climates are hot and dry there is much evaporation of water from lakes or soil, and salts may crystallize and form layers. Salt layers formed this way are now being mined in Michigan, Kansas, and Germany. How would these salt deposits indicate climates during ancient times?

ACTION

Imagine that you have been transported several million years into the future. What might you find in the rock record representing the 1980's in the area where you now live? What objects might survive? Which ones might not? What could a future geologist learn about our civilization from common objects?

Figure 12-23 *Climate changes have occurred in North and Central America since Early Cenozoic time.*

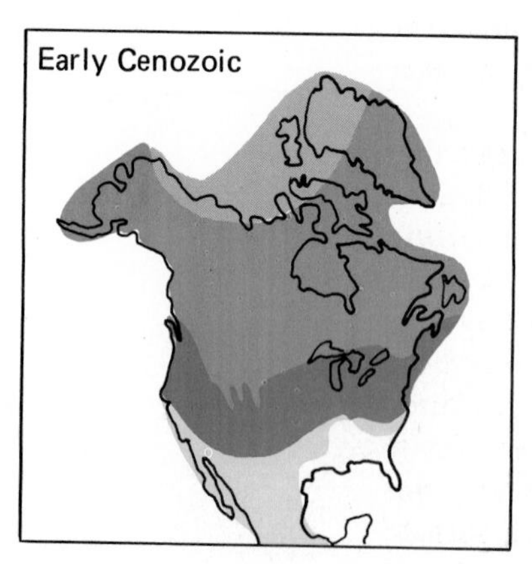

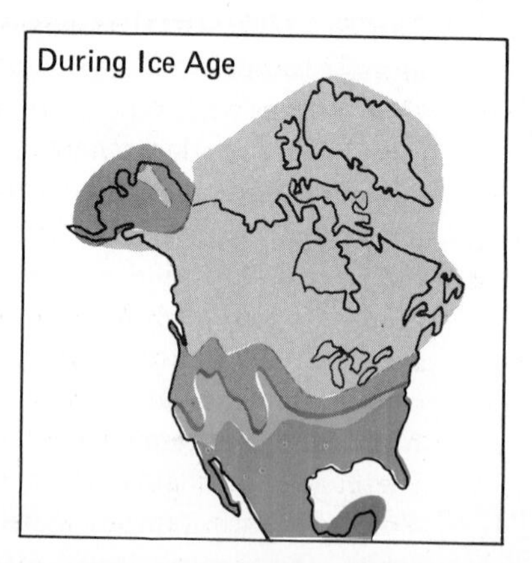

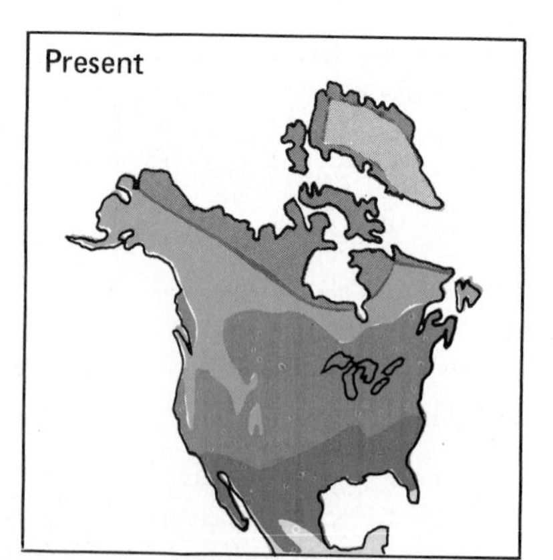

Thought and Discussion

1. How could you recognize a buried erosion surface?
2. How could you tell whether an igneous rock formed from a lava flow or an intrusion?
3. What types of evidence about ancient climates are found in rocks?
4. How are fossils used in correlating rocks separated by great distances?

Unsolved Problems

Many areas of the earth's surface have not been carefully examined, and some parts have not been studied at all. As a result, our knowledge of earth history is far from complete. Even areas that have been studied in great detail provide new information when improved methods are used to study them. The geologic history of your area probably has large gaps in it, too, and some of these gaps may never be filled. Find out what you can about your local geology and determine how much information is missing from the geologic record.

Are sedimentary, igneous, or metamorphic rocks forming or eroding in your area right now? You might contact local conservation groups, travel to mines, quarries, or deep road cuts to collect your information.

Chapter Summary

Rocks tell many things about past events on the earth. The layering in sedimentary rocks can show the conditions under which the rocks formed and the types of sediment involved. Fossils may reveal whether sedimentary rocks originated in the ocean or on land. Layering in igneous and metamorphic rocks may also give clues to the conditions of their origin.

In sedimentary rocks, cross-beds and ripple marks help geologists distinguish the tops of layers and the directions of ancient currents. The position of fossils and the sizes of sedimentary particles also may show if layers have tilted or been overturned.

The texture and mineral content of rocks are important clues to their origin. Rounded or coarse particles, shell fragments, and trace minerals in sedimentary rocks all provide information. The texture of igneous rocks reveals how fast they cooled.

Rocks in different outcrops can be correlated. The types of rocks and their fossils can tell whether they are related. Correlation helps fill in missing parts of the geologic record. Certain layers missing from the sequence at one location may be found at another location. Unconformities, or buried erosion surfaces, often account for the missing layers.

The relative ages of rocks that cut through each other can also be determined. For example, igneous rock that has baked surrounding sedimentary rock and that contains sedimentary fragments is younger than the rock it has intruded.

Rocks may also be useful in revealing ancient climates. Fossils of tropical plants and animals have been found in areas that are now arctic or subarctic. They suggest that the climate in these regions was once much warmer. There are abundant deposits of minerals such as salt or gypsum in some temperate areas of the world. They probably formed at a time when these areas had different climates.

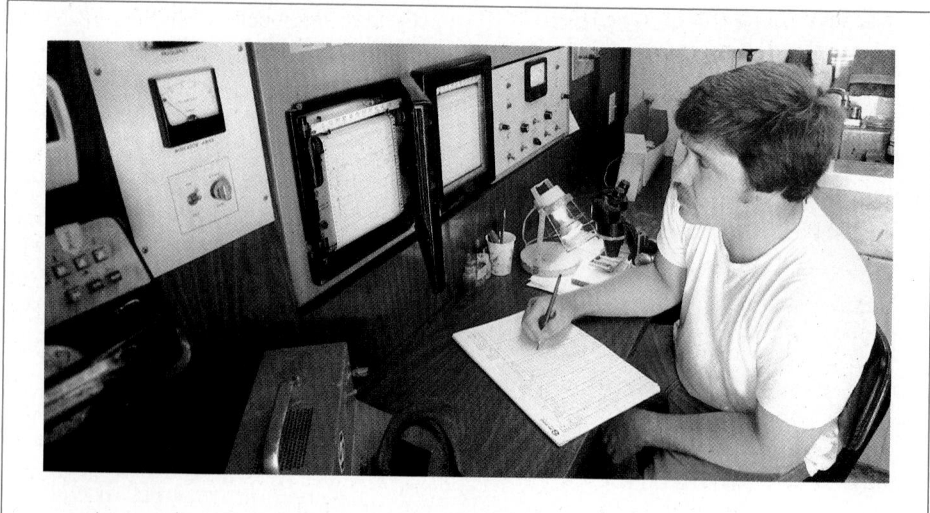

MUDLOGGER

Mudloggers are people who watch the functions of an oil drilling rig and keep a running record of their observations. The mud is the fluid that lubricates the cutting of the drill bit. The log is the record, like a ship's log.

A mudlogger watches carefully for gas, both to make sure the rig is safe and to discover gas that can be sold. He or she analyzes the rock chips from the bottom of the hole. That helps the geologist and driller know what rock is being drilled through. Loggers detect evidence of oil, if there is any.

Loggers keep a 24-hour check on how the drilling machinery is working. They inspect, maintain, and repair the sensors and recording instruments that give them the information. During and at the end of a working shift, the logger adds to the log by drawing and typing any information not recorded by the instruments.

Mudloggers are hired by drilling companies or by well-logging companies. With experience and suitable academic training, a mudlogger may advance to be a well-site geologist or subsurface geologist. A mudlogger should have a college degree or equivalent knowledge. Most have majors in geology or another physical science.

Questions and Problems

A

1. In section 12-2 there are examples of a single layer of sediment varying from one place to another. Describe some other circumstances in which horizontal variation might occur within a body of rock.
2. How would you explain a sedimentary layer with large grains on top and smaller grains on the bottom?
3. How could a sandstone form that is composed almost entirely of well-rounded sand grains?
4. What features do sedimentary rocks have that igneous and metamorphic rocks don't?
5. Suppose you find two sedimentary rocks. One of these, composed of rounded fragments of calcite, is relatively soft. The other rock, composed of fairly angular fragments of quartz, is relatively hard. Can you tell

whether the calcite grains have been transported farther than the quartz grains? Why or why not?

B

1. Quartz is probably the most common mineral in sandstone throughout the world. Yet on many beaches in Florida most sand grains are composed of calcite. Suggest reasons for this situation.

2. Some rocks consist entirely of volcanic ash and fragments of rock that have fallen into a pile at the foot of a volcano. Would rock formed from this material be sedimentary or igneous? Why?

3. Refer to Figure 12-1, at the beginning of this chapter. If "plutonic" rock is very common in this area, would it be found to the left or to the right of where this picture was taken? What kind of rock appears in the photograph? Would it be younger or older than the "pluton" that exists near-by? Explain how you arrived at your answer.

4. Three rock outcrops are located about 1 kilometer apart from each other. Outcrop A, exposed at the surface, is a brown sandstone, resting on a green shale. It is 11 cm thick and contains fossils X and Y. Outcrop B is also exposed at the surface. It is a brown sandstone resting on green shale. It is 9 cm thick and contains fossils W and Z. Outcrop C is a light tan sandstone found 2 m below the surface of the ground. It is 1 cm in thickness and contains only fossils of the W type. It is resting on limestone. Geologists have determined that two of these outcrops are the same and one is different. Which two could most likely be "correlated" with each other? What factors are most important in determining which of the rock outcrops does not belong with the other two?

C

1. Construct a geologic cross section, given the following sequence of events (use a pencil, so that you can change your mind and erase, if necessary): sandstone, then shale, then sandstone deposited; all rocks folded; extensive erosion occurs in the area, removing some, but not all, of the layers; further deposition of limestone and shale; earthquakes, as a result of local faulting, occur; a stream erodes through the shale to the limestone; conglomerate, then sandstone, then shale are deposited; further erosion takes place; a mass of molten rock intrudes into the existing rock, with a small section of it reaching the surface to form a volcano.

2. From the sequence of events provided in the previous question (C-1), how many times does it appear that the area described was *not* covered by a large body of water?

3. What basic assumption must be made in order to determine the changes in climate that have occurred in an area from the kinds of fossils that are found there?

Suggested Readings

Freeman, Tom. *Field Guide to Layered Rocks.* Boston: Houghton Mifflin Company (ESCP Pamphlet Series), 1971. (Paperback)

Matthews, William H., III. *Geology Made Simple.* Rev. ed. New York: Doubleday and Co., 1982. Chapters 18 and 19. (Paperback)

Romey, William D. *Field Guide to Plutonic and Metamorphic Rocks.* Boston: Houghton Mifflin Company (ESCP Pamphlet Series), 1971. (Paperback)

13
Time and How It Is Measured

"WHAT DAY IS IT?" *"What time will this class be over?" These questions, and many like them, show how closely our lives are regulated by calendars and clocks. In fact, everything we do is related to time.*

Have you ever missed a bus or part of a television program simply because you were a few minutes late? Were you late because you did not allow yourself enough time, or perhaps because your watch was not running properly? Stop for a minute and make a mental or written list of the things you do each day that depend on the measurement of time.

For many centuries people wondered about the age of the earth and whether or not our planet was formed at the same time as the rest of the universe. Finally, during the last 200 years, scientists have been able to answer some of these time-related questions about the history of our planet Earth.

But what about future time? Earth scientists are always looking for clues that might tell something about the future. For example, they want to know how long the sun will give off enough radiant energy to keep Earth's plants and animals alive, when the next earthquake might occur, and what climate changes lie ahead. Earth scientists can already make some predictions about the future because they have studied time, past and present. In the following chapters you will learn about these predictions and how they are made.

AFTER COMPLETING THIS CHAPTER, YOU SHOULD BE ABLE TO:

1. **explain the difference between relative time and measured time.**
2. **describe the process of radioactive decay and explain how radioactive elements are used to date the rocks that contain them.**
3. **determine the approximate age of a sample, when you are given the half-life of a radioactive substance and the amount of that substance present in the sample.**
4. **identify methods used to measure the occurrence of events and the amount of time between them.**
5. **construct and analyze a time-line for the earth that indicates both the relative and exact ages of events.**
6. **explain how the Geologic Time Scale can be used to compare the ages and lengths of various segments of geologic time.**
7. **identify the geologic age of rocks when you know of guide fossils in certain rock layers.**
8. **compare human existence on the earth with the entire range of geologic time.**

Measuring Time

13-1 What Is Time?

Time affects the things you do every day. But can you think of a way to define or describe time?

ACTION

Use any method you can think of, except your watch, to determine the length of a five-minute period. Cover all clocks and watches in the room, and choose one student to be timekeeper. When the timekeeper makes a mark on the chalkboard, start to measure a five-minute period of time. When you think five minutes have passed, signal the timekeeper. The timekeeper will make a mark each time someone signals.

Did everyone signal at the same time? What do the marks on the chalkboard tell you? How did you decide when five minutes had passed? What other methods did your classmates use? What is time?

We are usually aware of time because something has changed. Change affects not only the earth but the plants and

Figure 13-1 *Buttes in Monument Valley, Utah. Can you match any rock layers in the pinnacles?*

animals that live on it. Changes not only make us aware of the time but also provide a way to measure time.

Can you imagine what it would be like to live in a totally dark, air-conditioned, soundproof room without any time-measuring devices? You would not know whether it was night or day, summer or winter, or hot or cold outside. You could not tell the difference between day sounds and night sounds. In fact, you would be in an unchanging environment in which time seemed to stand still. (Do you think that you could *feel* the passage of time?)

ACTION

List four events of your past life. Put the most recent event at the top of your list. Now add to your list the events that one or two of your classmates listed. Try to put all of these events in the order in which they happened.

Did you have difficulty in deciding whether a certain event occurred before or after other events? How long did it take for each of these events to occur? Was the amount of time between events the same? Did most events listed occur recently or when you were very young?

13-2 Relative Time—Measured Time

Time is remembered by certain events. You can also mark the passage of time by relating it to a series of events. However, if you want to construct an exact history of past events, you must know the length of time between them.

Earth scientists are interested in events that took place long before humans were present on Earth. The rocks of the earth's crust contain evidence of these events. A geologist can reconstruct the geologic history of an area by studying the rocks there. Figure 13-1 shows some sedimentary rocks. Which lettered rock layer is the oldest, a, b, or c? How can you tell?

When you list events in the order they happened, you make a time sequence. This series of events is a **relative time scale,** which is simply a "before-or-after" scale for a series of events. A relative time scale does not provide information about the *amount* of time involved, but it does show whether one event happened before or after another.

The rock layers in Figure 13-1 have not been disturbed since they were originally deposited. This means that the top layer (c) is younger than the layers beneath it. But how *much* younger is the top layer: ten million years? a thousand years? fifteen years? To answer this question, the geologist must gather more information from the rocks.

Suppose that it took a million years to deposit each rock layer and that five million years passed before the formation of each new layer. You can now figure out how much older the bottom layer (a) is than the top layer. You can say that the second layer (b) is five million years older than the first. However, you still do not know when layer c formed. In other words, you can determine "how long" but not "how long ago."

In order to determine how long ago, you would have to measure backward from some known event in time. When you relate these ages to the present, you have made a measured time scale. A **measured time scale** tells you how long ago an event took place.

Events in history, whether the history of people or the history of the earth, should be dated in relation to the present. The events can then be arranged in order. For example, knowing that dinosaurs became extinct before humans appeared on earth is not as useful in constructing the history of life as knowing that the dinosaurs became extinct about 70 million years ago. This event can be related to the earliest history of humans, which dates back more than two million years ago. Now you can say how many years separated human life from the dinosaurs.

13-3 Calendars and Clocks

We normally use clocks and calendars to set events in their proper place in history. These time-keepers can be used to keep track of years, seasons, months, days, hours, minutes, and seconds. Three of these units, the year, the season, and the day, are based on natural events. Other time units are fractions of these. An hour is $\frac{1}{24}$ of a day, a minute is $\frac{1}{60}$ of an hour, and a second is $\frac{1}{60}$ of a minute or $\frac{1}{86,400}$ of a day.

We know that calendars and clocks work well for recent happenings. But what about events that occurred thousands or millions of years ago?

Mechanical clocks have not been around very long. Thus scientists must use natural clocks to date events that took place millions of years ago.

The day, the seasons, and the years are natural units of time resulting from motions of the earth. It is reasonable to assume that the earth has been rotating on its axis and revolving around the sun since the solar system formed. It is possible, therefore, that evidence of seasonal change might be found in the rock and fossil records.

ACTION

Look at Figure 13-2. See if you can find signs of changes that have happened in the recent past. Figure 13-2a is a magnified view of tree rings. Which wood cells represent spring growth and which wood cells represent summer growth?

Could the two trees shown in Figure 13-2b and 13-2c have lived at the same time during any part of their lives? What evidence is there for seasonal changes in the shell growth and the sedimentary layers of Figure 13-2d and 13-2e?

In the preceding ACTION you examined plants, animals, and rocks that have recorded changes at various times in the past. These natural clocks give information about climate conditions over short periods of time. However, their recording period is much too short to be used in establishing events that took place on the earth millions of years ago.

Thought and Discussion

1. In a time-ordered sequence of events, event A happened before event B, which in turn happened before event C. Event D, however, happened before event B, but after event A. Can you represent these events in their proper order? Place the most recent event at the top of your list.
2. Name an event that does not involve change.
3. Why is it important that earth scientists be able to determine relative and measured geologic time?
4. How would you define time? Compare your definition with your classmates'.

a.

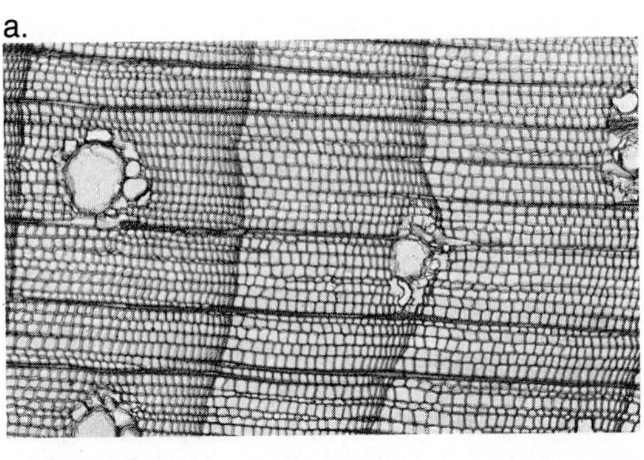

b.

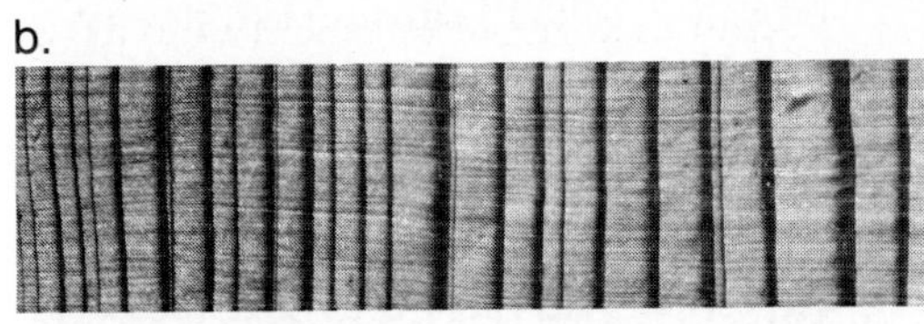

c.

d.

Figure 13-2 a. *A magnified section of wood showing cells. How many years of growth show?* **b.,c.** *Growth rings from two different trees. Do they match at any points?* **d.** *Growth rings on a clam shell.* **e.** *Sedimentary layers.*

e.

13-4 Radioactive Elements and Atomic Clocks

MATERIALS
uranium ore or mineral containing uranium, cloud chamber, unexposed film

ACTION

Obtain a small amount of uranium ore or a specimen of uranium-bearing mineral from your teacher and place it in a cloud chamber. Observe the emission of particles from your specimen. Next, place the uranium specimen in a drawer next to an unexposed piece of cut film or a roll of film. Be sure that the film is tightly wrapped so light cannot seep in. After the uranium and the film have been next to each other for a few days, have the film developed.

In the preceding ACTION, you have recreated one of the greatest discoveries of science. In 1896 Henri Becquerel (*ahn REE beh KREHL*), a French physicist, placed an unexposed photographic plate next to a sample containing uranium in his darkroom. When he later decided to use the photographic plate, he found that it had been partially exposed, and an image was already on it. This puzzled the physicist, for the plate had been carefully protected from light in the darkroom. Becquerel later realized that he had made an important discovery: certain natural substances, such as uranium, release energy that could expose a photographic plate. See Figure 13-3. This scientific breakthrough led others to study this mysterious property, which we now know as **radioactivity.**

Figure 13-3 *These radioactive tomato seeds took their own picture by exposing nearby film.*

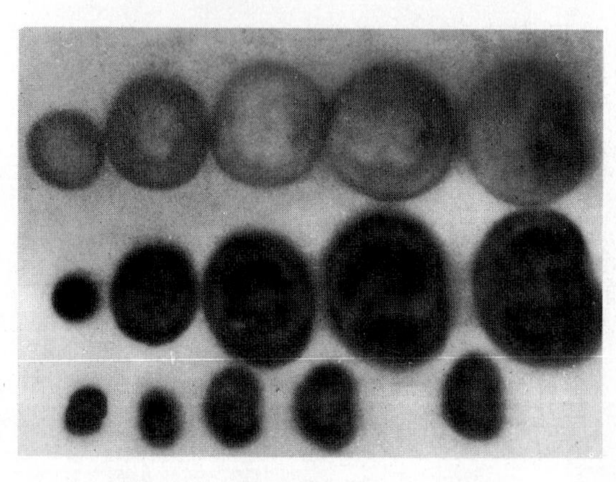

The nucleus of an atom is composed of protons and neutrons. The positively charged protons of the nucleus repel each other, and the uncharged neutrons add mass to the nucleus. The forces that hold the nucleus together are very strong and are not fully understood. The balance between repelling and binding forces in the nuclei of some kinds of atoms is often disturbed. As a result, the nuclei split apart, and energy and charged particles are given off. This process is called **radioactive decay.** A radioactive element decays, or loses some energy and charged particles, to form another, more stable element.

Certain minerals contain radioactive elements. The rates at which these elements decay have been determined. Therefore, radioactive minerals in rocks can be used to date events in the earth's past.

13-5 Investigating Rates of Radioactive Decay

The nuclei of all radioactive atoms must eventually decay. However, all of the atoms do not decay at the same time. The decay process involves chance, and it is impossible to know exactly when any particular nucleus will decay.

Because even a very small sample of a radioactive element contains billions of atoms, the *average rate* of decay can be determined. Once this average rate is found, calculations can be made to find out the amount of time it would take for 50 percent of the atoms to decay. This time is called the **half-life.** The half-life of a radioactive substance may be as little as a fraction of a second or as much as billions of years.

When you have completed this investigation, you will be able to explain the role of probability in radioactive decay.

PROCEDURE

Place the objects you are going to use as markers into a square or rectangular box that has one side marked. Shake the box vigorously (Figure 13-4). Remove the markers that point to the marked side of the box and assume that these "atoms" have decayed. Record the number of markers remaining in the box as Trial 1. Repeat the procedure for Trials 2, 3, and so on. Continue until the box is empty. Make a graph by plotting the number of markers left in the box after each trial versus the trial number. Put the number of markers left on the vertical axis and the trial numbers on the horizontal axis.

Next, do the same thing with *two* sides of the box marked, and again with *three* sides marked. Graph these data, too.

DISCUSSION

Use your three graphs to answer the following questions. Assume that each trial represents 100 years.

1. What was the half-life for each model?
2. How did you change the half-life in the models?
3. What difference would it make in your results if a classmate added more markers to the box during your investigation? Try it!

Figure 13-4 *Making a model of radioactive decay.*

MATERIALS
graph paper, box (with lid), pointed markers

4. How many "years" does it take for the "decaying material" in each model to reach 25% (¼) of the original amount?

13-6 Early Attempts to Measure Geologic Time

Most 18th and 19th century earth scientists believed that the earth was very old, but proving its great age was not a simple matter. One of the earliest attempts to date the earth was made in 1715 by an English astronomer, Edmund Halley. He assumed that the sea was originally fresh water and that it gradually became saltier as it became older. Halley knew that the salts in the sea had been dissolved from rocks on land and later carried to the sea by streams. This led him to believe that the total amount of salt in the sea might be a clue to the age of the oceans. In turn, the age of the oceans would give an estimate of the age of the earth.

Halley was unable to try his salinity method of dating because he did not have the necessary data. But in 1899 John Joly, an Irish scientist, believed that he had assembled enough information to make a reasonable estimate of the age of the ocean. His calculations suggested that it had taken from 80 to 90 million years for the oceans to reach their present salinity. And since the earth formed before the sea, our planet had to be more than 80 million years old.

During this same period, other geologists were trying to find out how long it took to form all the sedimentary rocks in the earth's crust. They studied the rates of accumulation of various sediments. They then estimated how much time was required to deposit the sediment needed to form one meter of sandstone, limestone, shale, and other rocks. Geologists examined exposed rocks all over the world, and tried to determine the maximum thickness of rock formed during each period of geologic time. They added together the thicknesses of all the rock beds. Finally, they calculated the time needed to deposit all of the sediments. Based on this method, estimates of the earth's age ranged from less than 100 million to more than 400 million years.

Unfortunately, these results were far from accurate, for there were too many factors involved. For one thing, sediments accumulate at different rates in different environments. Also, it is impossible to determine the amount of time that is missing in the gaps in rock layers caused by erosion. Old, buried erosion surfaces may account for tens of millions of years. They indicate missing parts of the geologic record.

Most early physicists were convinced that they could prove mathematically that the earth could be no more than 20 to 40 million years old. Their method was based on the idea that the earth had cooled from a very hot liquid. They determined the temperature of rocks of the earth's crust and an approximate rate of cooling for the earth. Using these figures, they then calculated the earth's age. As additional support, the physicists pointed out that there was no known source of energy that could keep the sun hot for more than 20 million years. It was not logical to assume the earth was older than the sun.

13-7 How Atomic Clocks Measure Geologic Time

The discovery of radioactivity by Becquerel provided geologists with a key to the age of the rocks that contain radioactive elements. Dates obtained from radioactive minerals helped convince the physicists that the earth was much older than they had originally thought.

The discovery of the property of radioactivity soon led to other discoveries about radioactive substances. These included rates of decay, the amount of energy generated, and products of the decay process. In 1907 the American chemist and physicist B. B. Boltwood discovered that uranium decays, forming lead as the final product. Boltwood concluded that the age of a particular mineral can be found if you measure the amount of the **parent material** (uranium) that it contains and compare this with the amount of the **decay product** (lead). You would also have to calculate the rate of decay of the parent material.

One popular radioactive dating method is based on the breakdown of uranium-238 (^{238}U). Uranium-238 decays through a series of 14 steps; the end product is lead-206 (^{206}Pb). As the breakdown from uranium continues, the amount of lead increases. The rates of decay of ^{238}U and other elements used for dating have been precisely deter-

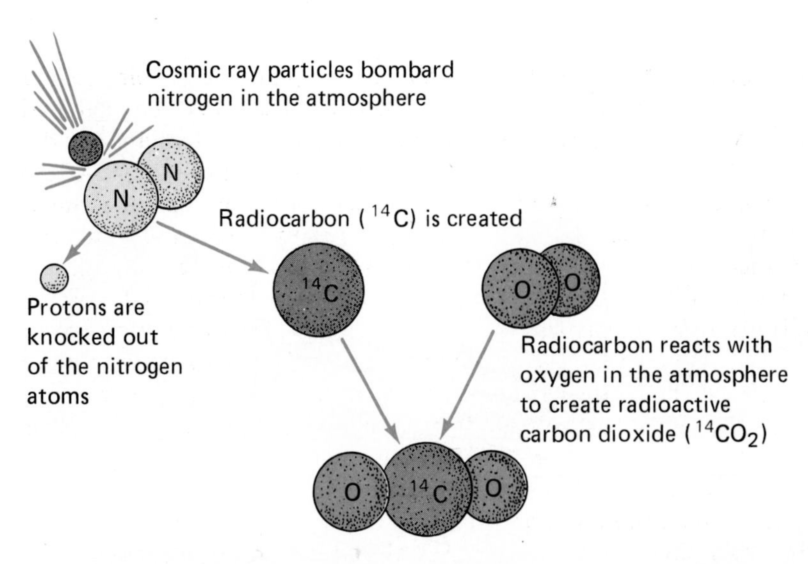

Figure 13-5 *Radiocarbon atoms form in the upper atmosphere.*

mined and have been found to be constant throughout geologic time.

After a mineral containing uranium atoms is formed, the products of uranium decay begin to collect in the mineral. The age of the mineral is found by determining the ratio of the parent material (^{238}U) to the end product (^{206}Pb). Special equipment must be used for determining uranium-to-lead ratios. In using this method, it is assumed that none of the lead escapes from the mineral, that no outside lead is added, and that no lead from a non-radioactive source was present to begin with. If any of these conditions have affected the sample being tested, the results will not be accurate. Can you suggest how the ages obtained in dating three samples might be affected if each of the samples was altered in one of the above ways?

The half-life of ^{238}U is incredibly long: 4.51 billion years. Therefore, ^{238}U is used only to date very old rocks in the earth's crust. Certain rocks found in southwestern Greenland, dated by this method, were found to be around 4 billion years old. These are the oldest reliably dated earth rocks known. To date events that have occurred during the past 40,000 to 50,000 years, a radioactive element having a much shorter half-life must be used. Because carbon-14 has a half-life of about 5700 years, it has been used widely to date relatively recent events.

Radiocarbon (^{14}C) is continuously forming in the earth's upper atmosphere. This happens naturally as nitrogen atoms (^{14}N) are hit by high-energy cosmic rays (Figure 13-5). Cosmic rays are streams of fast-moving particles from atoms that reach the earth from space. Once the carbon-14 atoms are formed, they can unite with oxygen to form carbon dioxide. This reaction is part of the carbon cycle.

The radiocarbon dating method has been used to date organic materials such as wood, bones, hair, and even old food. Figure 13-6 summarizes the events that make radiocarbon dating possible. Note also the recently discovered double check on this dating method in Figure 13-7.

Other radioactive dating methods are also used by scientists to learn the ages of rocks. Some of these methods are more useful than others because they use ele-

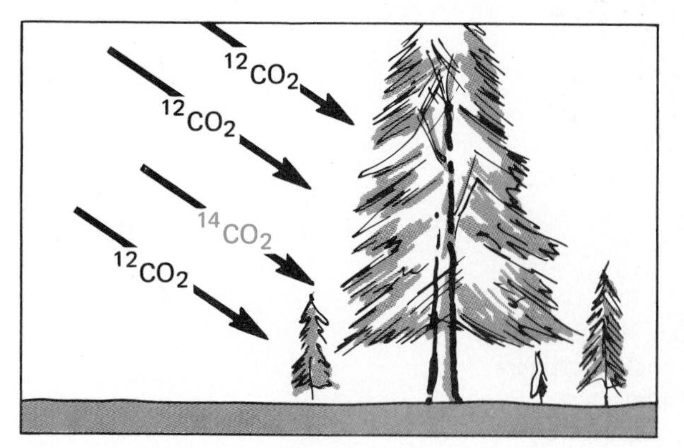

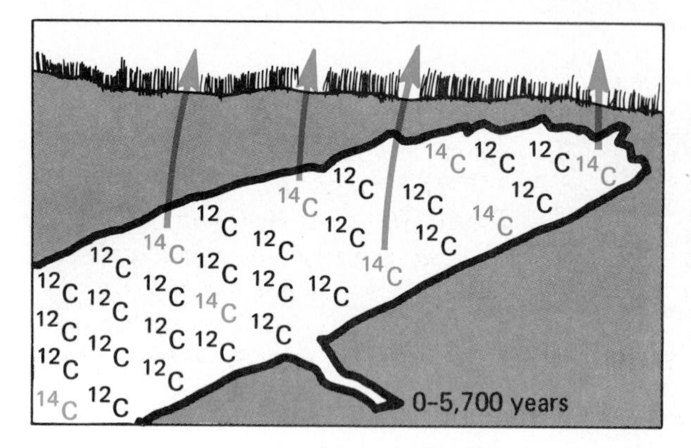

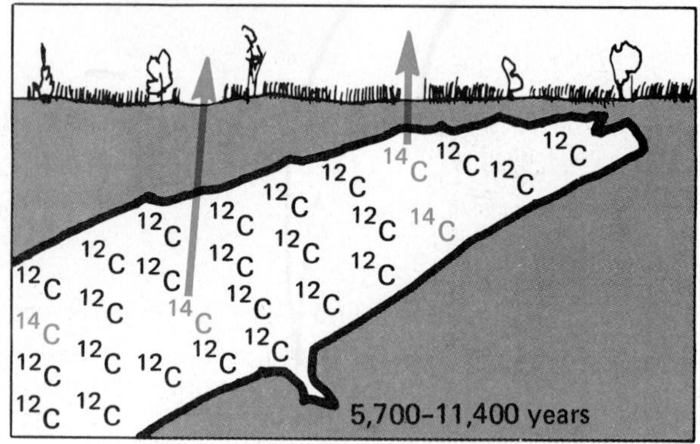

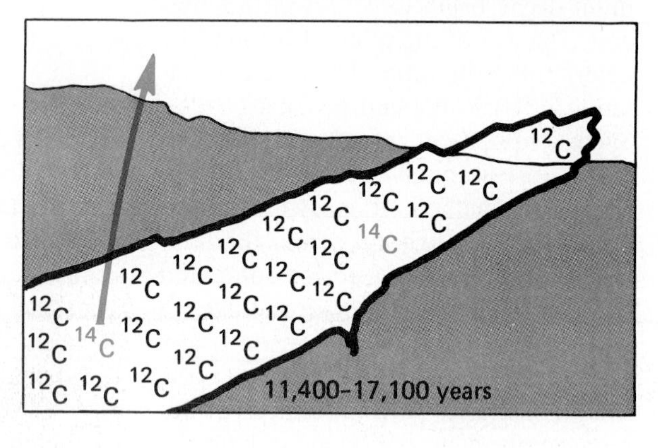

Figure 13-6 *$^{14}CO_2$ enters a living tree in a fixed amount. When the tree dies and decays, the amount of ^{14}C in the wood begins to decrease. How does the ratio of ^{14}C to ^{12}C change in time?*

ments that are more common in most rocks than ^{238}U and ^{14}C. One of these methods uses potassium-40 (^{40}K). This rather common form of potassium is weakly radioactive. The decay of ^{40}K is very complex, but one of its decay products is argon-40 (^{40}Ar). Potassium is common in many igneous rocks. In addition, ^{40}K has a relatively short half-life of 1.31 billion years. The common occurrence of potassium, plus the short half-life of ^{40}K, has made the potassium-argon method the most frequently used atomic clock. If you have not al-

Figure 13-7 *Bristlecone pines are among the oldest living things on the earth. One in the Humboldt National Forest, Nevada is 4900 years old. Their growth rings show that radiocarbon formation has not been constant during the earth's recent history. Many* ^{14}C *dates have had to be revised.*

ready done so, answer the question in Figure 13-8.

Figure 13-8 *The Dead Sea scrolls were discovered by archaeologists in a cave in Jordan. Which dating method was probably used to show that they are about 2000 years old?*

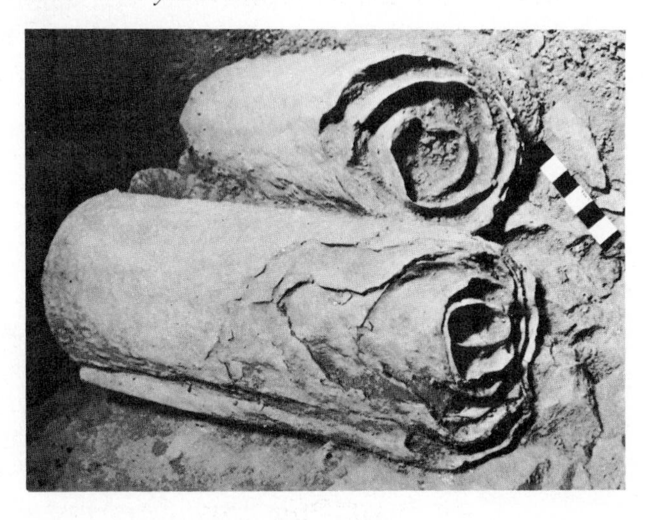

Thought and Discussion

1. What information besides a time sequence can be obtained from tree rings?
2. Define the term "half-life." Do all radioactive elements have the same half-life?
3. What effect does the radioactive decay of unstable elements in the earth's crust have on the rocks surrounding them?
4. Which method of radioactive dating is used for relatively recent events? Why?

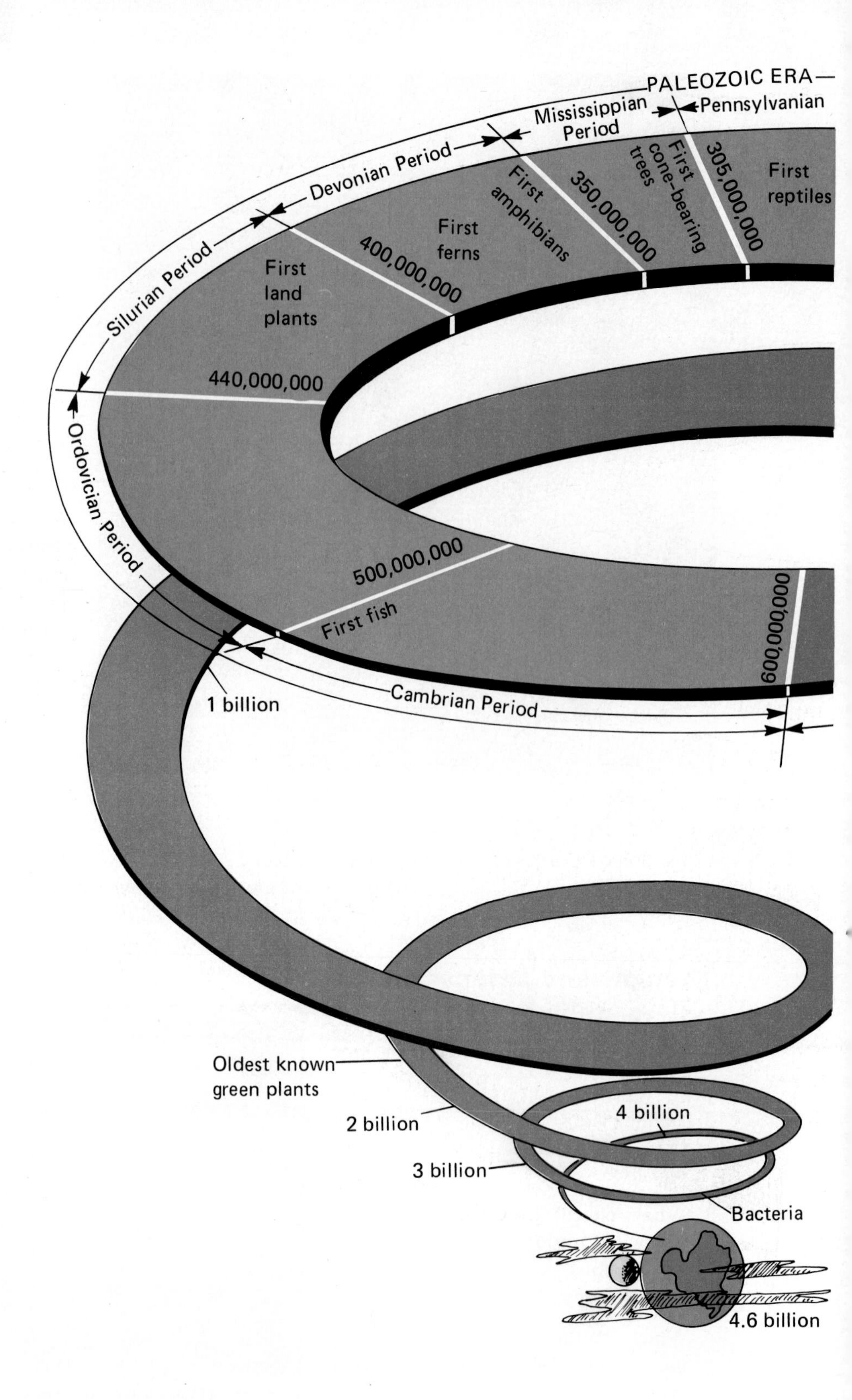

Figure 13-9 *The Geologic Time Scale.*

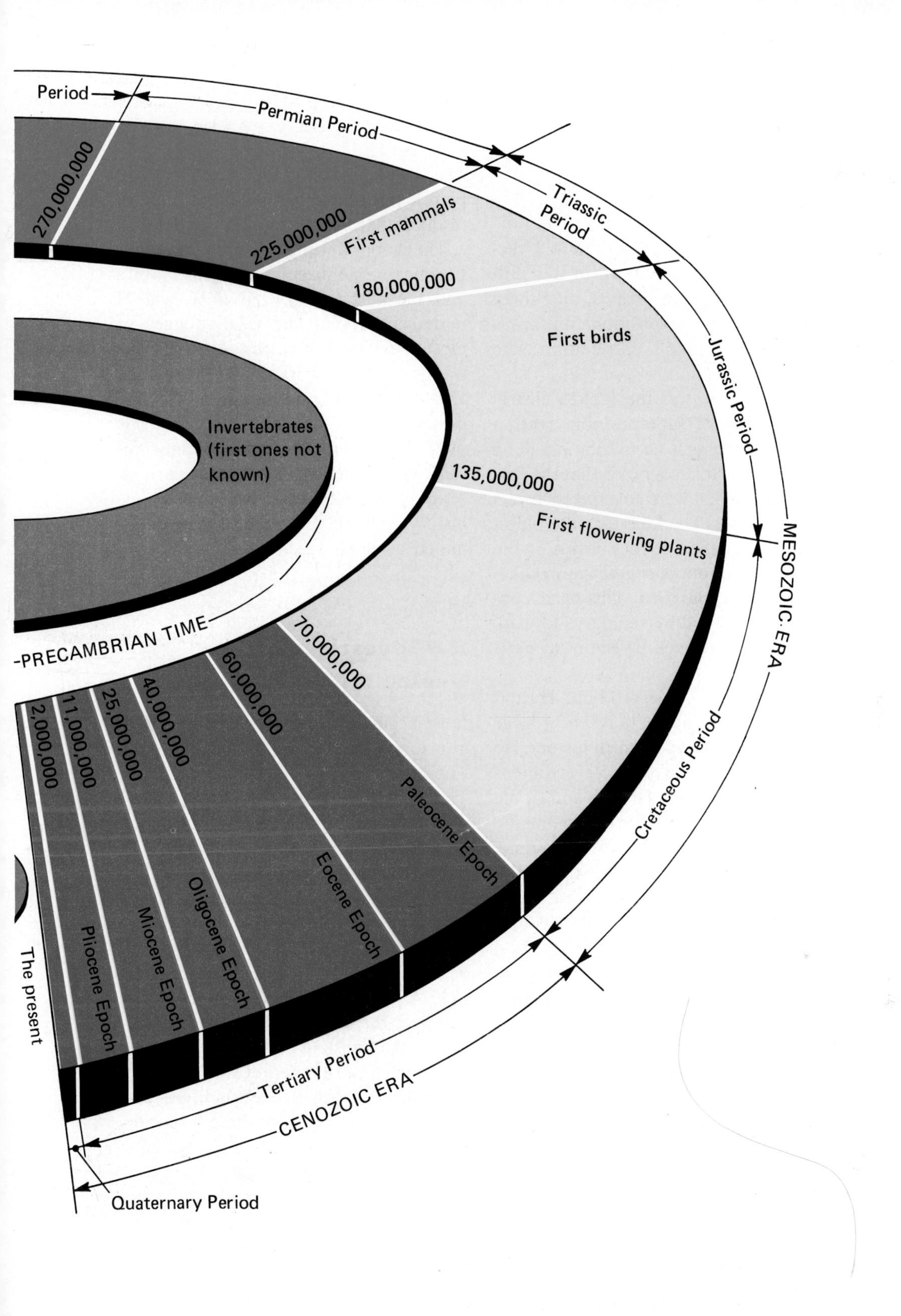
Period
Permian Period
Triassic Period
Jurassic Period
MESOZOIC ERA
Cretaceous Period
270,000,000
225,000,000
First mammals
180,000,000
First birds
Invertebrates (first ones not known)
135,000,000
First flowering plants
PRECAMBRIAN TIME
70,000,000
60,000,000
40,000,000
25,000,000
11,000,000
2,000,000
Paleocene Epoch
Eocene Epoch
Oligocene Epoch
Miocene Epoch
Pliocene Epoch
The present
Tertiary Period
CENOZOIC ERA
Quaternary Period

13-8 Organizing the Rock Record

Real progress in organizing the rock record did not begin until late in the 18th century. During this period, the three basic ideas of superposition, uniformitarianism, and fossil correlation were proposed.

James Hutton was the first to clearly state the idea of superposition. Hutton examined sediments accumulating along the seashore. He recognized that the layers deposited first were covered by layers deposited later. The idea that the oldest bed in a sequence of rock layers is the one on the bottom is called the **principle of superposition.** This basic concept has been used by geologists to work out sequences of rock layers in all parts of the world.

In addition to superposition, Hutton stated another principle that is now basic to our understanding of earth history. He observed that features of old sedimentary rocks, such as mud cracks and ripple marks, were duplicated in sediments he saw being deposited in his own time. From his observations in the field, Hutton concluded that the same processes that affect the earth today also affected the earth in the past. Therefore, the present can be used as a key to the past. This idea is now known as the **principle of uniformitarianism.**

The third fundamental idea is that fossils may differ from one layer of sedimentary rocks to the next. Consequently, these different fossils can be used to identify the beds that contain them. This idea, called **fossil correlation,** was first recognized by William Smith, an English engineer. Smith was primarily concerned with identifying rock beds in different places during construction of roads and canals. Today fossil correlation is widely used in the search for coal, petroleum, and other valuable mineral resources (see Sections 13-11 and 14-7).

The work of Hutton, Smith, and other European scientists during the 1700's and the early 1800's led to a general understanding of the relative ages of most rocks on the earth's surface. This was done by first working out the relative ages of rocks in many areas. It was like fitting together pieces of a huge jig-saw puzzle. The series of ages obtained locally were matched, or **correlated,** with series found in other areas. The geologic history of large regions has been determined using this technique.

▣

13-9 Investigating the Geologic Time Scale

It took more than 200 years for geologists to put together a workable Geologic Time Scale (Figure 13-9). The Geologic Time Scale subdivides geologic history into units of time based on the formation of certain rocks.

The largest of these time units is called an **era.** Each era is divided into **periods,** and each period may be divided into smaller units called **epochs** (*EH puhks*). When placed in proper order, these time units form a geologic calendar. If you compare the divisions of the earth's history to the divisions of your textbook, an era is like a unit, a period is like a chapter, and an epoch is like a section. The chief difference between these is that the time scale divides time, and the book divides information.

The dividing lines between eras, periods, and epochs are based on recognizable changes. These include changes in kinds of plants and animals, and episodes of mountain building. For example, the extinction of dinosaurs separates the Mesozoic (*mehs uh ZOH ihk*) Era from the Cenozoic (*sehn uh ZOH ihk*) Era. The Quaternary Period is divided into an earlier Pleistocene Epoch and a later Holocene Epoch, in which we are now living. The beginning of the Pleistocene is marked by the first advance of American and Eurasian ice caps. This started the great Ice Age of the Pleistocene Epoch. The Holocene Epoch began when the Ice Age glaciers disappeared, 10,000 years ago. (The opener to Chapter 14 and Sections 15-7 and 15-8 have more information on the Pleistocene Ice Age.) However, the dividing lines between divisions of geologic time are never sharp ones. They are more like zones of gradual transition in time.

It is difficult to understand such long periods of geologic time when a human's life span may be only 70 years. You may think of ways to help yourself visualize large numbers of things, such as pebbles. Can you do the same with large numbers of years? One way to help is to set up a demonstration that will compare a few years to millions of years.

When you have completed this investigation, you will have demonstrated the relative lengths of time between events and the scale of geologic time.

PROCEDURE

MATERIALS
paper tape, meterstick

Examine the list of events in Figure 13-10. Then decide how to represent these events in a time-ordered sequence. A roll of paper tape will be provided on which to plot your model (see Figure 13-11).

Figure 13-10 *Ages of events in the past.*

Today.
Last New Year's Day.
First person on the moon, 1969.
First U.S. satellite orbited, 1958.
Mount Vesuvius eruption destroys Pompeii, A.D. 79.
End of the Ice Age, 10,000 years ago.
Beginning of the Pleistocene Ice Age, 1 million years ago.
First humans, about 2.5–3 million years ago.
Beginning of the Pliocene, 11 million years ago.
Beginning of the Miocene, 25 million years ago.
Beginning of the Oligocene and first elephants, 40 million years ago.
Beginning of the Eocene, 60 million years ago.
Beginning of the Paleocene and first primates, 70 million years ago.
Beginning of the Cretaceous, 135 million years ago.
First birds, 160 million years ago.
Beginning of the Jurassic, 180 million years ago.
Beginning of the Triassic, and the first dinosaurs and mammals, 225 million years ago.
Beginning of the Permian, 270 million years ago.
Beginning of the Pennsylvanian and the first reptiles, 305 million years ago.
Beginning of the Mississippian, 350 million years ago.
Beginning of the Devonian and the first amphibians, 400 million years ago.
Beginning of the Silurian and the first land plants, 440 million years ago.
Beginning of the Ordovician, 500 million years ago.
Beginning of the Cambrian, first abundant fossils, and first vertebrate, 600 million years ago.
First known animal (jellyfish), 1.2 billion years ago.
First known plants (algae), 3.2 billion years ago.
Oldest known rocks, 3.9 billion years ago.

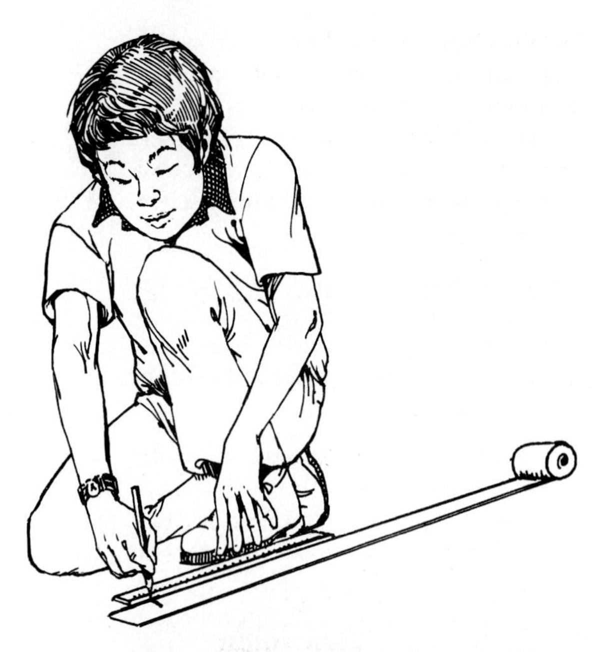

Figure 13-11 *Plot the events in Figure 13-10 on a roll of paper tape.*

DISCUSSION

1. How would you count or measure a million of anything?
2. How long ago did the earliest humans appear? How does this compare to the total span of the earth's history?

13-10 Calibrating the Geologic Time Scale

As more rocks are dated by radioactive methods, earth scientists can determine the actual ages of the events on the Geologic Time Scale. The dates that are generally accepted for the major units of geologic time are the same as those on your tape from Investigation 13-9.

To see how earth scientists calibrate the time scale, let us examine a simplified version of a technique they use. In Figure 13-12 you can tell that the granite is younger than rock unit x because it intrudes into unit x. The rocks in unit x are older than those in unit y, which contains weathered pieces of the granite. You can also tell that the granite is older than some actual age determinations. Assume that tests show that the granite is 150 million years old and unit z is 130 million years old. Now what can you determine about the age relationships of the rock units x, y, and z in the diagram?

The oldest rocks that have been dated are about 4 billion years old. These rocks have been intruded into still older rocks that have not been dated. Evidence of the great age of the earth has also been obtained from meteorites that contain radioactive elements. These meteorites have been dated and appear to be more than 4.5 billion years old. In addition, some rock samples from the moon are approximately 4.6 billion years old. How does this age compare with that of the oldest rocks found in the earth's crust?

13-11 Investigating Fossils As Clues to Correlating Rocks

In the previous chapter (Section 12-5) you began an investigation of rock layer correlation. You were asked to "match up" several rock layers from widely scattered outcrops and attempt to identify which of the rock layers were older or younger than the others. You may or may not have settled on what you think are the correct answers by now. Some additional information, provided here, may help you to either change your mind or confirm your conclusions as correct.

How fossils are formed and their many uses are discussed in Chapter 14. One of the most important uses of fossils is in rock correlation. Some fossils represent plants and animals that only lived a very short time in geologic history. Yet while they were alive, they may have been widely distributed. Certain fossils are so characteristic of certain parts of geologic time that they have been called **guide** or **index fossils.** Guide fossils are especially useful in identifying the rock layers that contain them. It is known, for example, that dinosaurs lived only during the Mesozoic Era. Thus, when dinosaur remains are found, it is usually safe to assume that the rocks containing them are Mesozoic in age. However, paleontologists usually prefer to correlate by using groups of fossils rather than by means of a single kind of fossil animal or plant. Explain why using groups would be more reliable.

Several of the fossils that are found in the rock layers to be correlated in Investigation 12-5 are recognized as guide fossils. For example, trilobites (*TRY luh byts*) only lived during the Paleozoic (*pay lee uh ZOH ihk*) Era. But the particular form of trilobite known as *Phacops* (#16) has only been found in rocks that formed during the Devonian Period. *Phacops,* then, is a guide to the Devonian Period of the Paleozoic Era. The *Spirifer* (#17) and *Pentremites* (#15) were found in rocks of the Mississippian age. *Neospirifer* (#12) is Pennsylvanian in origin. *Olenothyris* (#13) is an early Cenozoic fossil, from the Eocene epoch. The *Merychippus* tooth (#8) is only found in Miocene sediments while *Carcharodon* teeth (#2) can be found in sediments deposited from the Late Miocene to the Holocene. While there are other index fossils present in the rock layers, you have enough information to arrange the rock layers in order of formation and to answer the questions below.

Discussion

1. Which of the rock layers are the oldest? How do you know?
2. Which of the rock layers formed most recently? How do you know?
3. What reason can be given to explain why layers 7 and 12 can be found under 18 and also under 1, but not under the combination of 1 and 18?
4. Notice that some fossils, such as 10 and 16, can be found together only in some cases. Explain why this could happen.
5. Only three of the rock layers in the outcrops were deposited when the dinosaurs were present on the earth. From the information you have been given, identify those rock layers.
6. *Mucrospirifer* (#10) is also recognized as an index fossil. From the information you have been given, identify the age of its origin.
7. Imagine that several million years from now a visitor from another

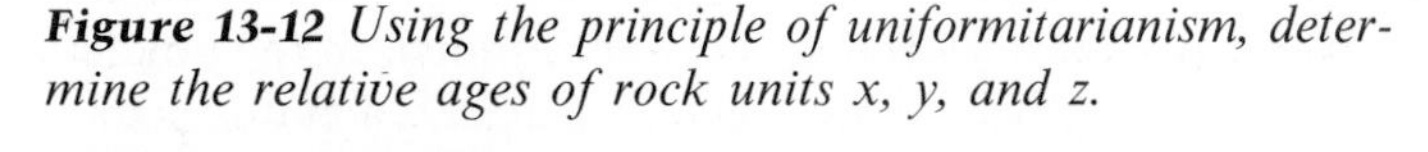
Figure 13-12 *Using the principle of uniformitarianism, determine the relative ages of rock units x, y, and z.*

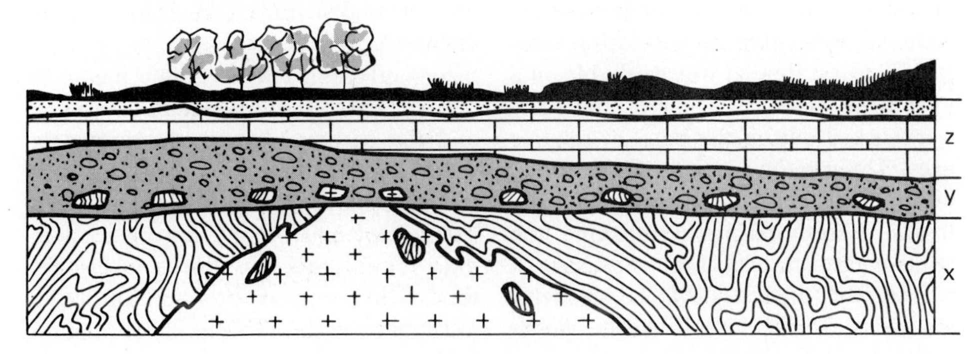

planet is here to study the earth's geologic history. The visitor recognizes the *Equus* tooth as being "index" to a particular 50,000 year span of sediment deposition. In addition to *Equus* teeth, what would most likely be the most common index fossil of that particular geologic age? Why?

8. What is the correct sequence of rock layer formation in the outcrops identified in Investigation 12-5?

Thought and Discussion

1. Were the earliest methods of classifying geologic time relative or measured? Why were such methods used?
2. How does the correlation of fossil species relate to the development of a Geologic Time Scale?
3. How might you develop a Geologic Time Scale for your local area?

Unsolved Problems

Will we ever know the exact age of the earth? Radioactive age determinations of some meteorites and certain moon rocks are about 4.5 billion years old. Is this the age of the earth? Geologists also want to know how long it took for the earth's crust tó form and how old the ocean basins are.

By dating events more accurately, scientists can determine more closely the rates of geologic processes such as uplift and erosion. Were these rates slower or faster in the past? Knowledge of past rates of processes will give us a clearer picture of the development of the earth as we know it now. Such information might also permit speculation as to the future of our planet.

Finally, are the oldest rocks in each continent about the same age? If not, which continent is the oldest? Which continent is the youngest?

Chapter Summary

Time is measured by events. We can consider time in a relative sense (old, older, oldest) or we can consider it as a measure of duration (how long) or age (how long ago). Earth scientists consider time in all these ways. Long before the discovery of radioactivity in 1896, earth scientists were able to develop a Geologic Time Scale that was workable on a worldwide basis. This early time scale was worked out on the basis of relative ages. Three basic ideas—superposition, uniformitarianism, and correlation by fossils—were used in establishing the relative ages of rocks of the earth's crust.

With the discovery of the property known as radioactivity, it became possible to date events that occurred at various times in the distant past. Some minerals that occur in the rocks of the earth's crust contain unstable elements whose atomic nuclei start to decay as soon as they are formed. In the process of decay, energy is released and unstable elements are gradually transformed into stable elements.

Several radioactive dating methods, including the uranium-lead ratio method and the carbon-14 method, are widely used in measuring the ages of rocks of the earth's crust. Carbon-14, which has a half-life of 5700 years, can be used to date young rocks and objects of historic time. Other methods are used for dating much older rocks, some as old as about 4 billion years. Based on the ages of these rocks as well as the ages of certain meteorites and moon rocks, earth scientists now believe the earth to be at least 4.5 billion years old. The span of geologic time is incredibly long, especially when compared to the short time that people have been on Earth.

Questions and Problems

A

1. It is believed that the earth's rate of rotation is slowing down as a result of the moon's gravitational attraction. What effect will this change have on the length of a day? the length of a year?
2. Why is carbon-14 not used for dating rocks of Paleozoic age?
3. Why is carbon-14 more useful in dating certain earth materials than other dating methods?
4. How old are the oldest rocks dated thus far? Do these rocks represent the original crust of the earth?

B

1. Why were earth scientists unable to prove before the year 1907 that the earth was more than 20 to 40 million years old?
2. Upon what basis is "geologic time" subdivided into its various eras and periods?
3. Why is it necessary to study carefully both the rocks and the geology of an area from which a sample for radioactive dating is obtained?
4. If rocks on other continents contain the fossil remains of large dinosaurs, would they be approximately the same age as rocks in the United States containing similar fossils? Explain your answer.

C

1. Some charcoal and charred, broken bones of deer and rabbits were dug from beneath several feet of sand and gravel along the banks of a river. Analysis of the charcoal in a laboratory showed that one-eighth of the ^{14}C remained in the charcoal. How old is the charcoal? Reconstruct the sequence of events that may have taken place at the site of this discovery.
2. What *percentage* of geologic time is represented by the existence of human beings on the earth?

Suggested Readings

Eicher, Don L. *Geologic Time.* 2nd ed. Englewood Cliffs, New Jersey: Prentice-Hall, 1976. (Paperback)

Matthews, William H., III. *Geology Made Simple.* Rev. ed. New York: Doubleday and Co., 1982. Chapters 18 and 19. (Paperback)

Ojakangas, Richard W., and David G. Darby. *The Earth: Past and Present.* New York: McGraw-Hill Book Co., 1976. Chapter 5. (Paperback)

Pearl, Richard M. *Geology: An Introduction to Principles of Physical and Historical Geology.* 4th rev. ed. New York: Barnes and Noble, 1975. Chapter 20. (Paperback)

14

Life: Present, Past, and Future

THE YEAR WAS 1900 *and a weary Russian hunter carefully made his way along the banks of Siberia's Berezovka River. Although he was tracking a wounded deer, he found something quite different. Imagine the hunter's surprise when he came upon the head of a full-grown fur-covered elephant sticking out of the frozen river bank. The discovery of this well-preserved shaggy beast caused great excitement. It was found 100 km inside the Arctic Circle, more than 3250 km north of where elephants normally live today.*

News of the amazing discovery eventually reached scientists in Saint Petersburg (now Leningrad), Russia. An expedition was sent to collect the unusual specimen. Although part of the flesh had been eaten by wild animals and the body was badly decayed, it was easy to tell that this was no ordinary elephant. The creature, which is shown on the opposite page, had very long curved tusks, and its body was covered with thick hair. Beneath the hair was a protective undercoat of woolly fur, and a thick layer of fat was underneath the animal's skin.

The hunter had found one of the world's most famous fossils—the frozen body of a woolly mammoth. These extinct elephant-like creatures lived in Eurasia and North America many thousands of years ago. Since the discovery of the Berezovka mammoth, similar frozen fossils have been found in Siberia and Alaska. We know that elephants do not now live in Siberia and Alaska. These remarkable fossils are reminders of a time when the Arctic region supported life forms entirely different from the ones in that area today.

You have already learned that fossils are clues to ancient climates. Fossils also provide valuable information about the development of life on Earth. You will also learn how the world of living things depends on and interacts with the atmosphere, the hydrosphere, and the lithosphere.

AFTER COMPLETING THIS CHAPTER, YOU SHOULD BE ABLE TO:

1. **list three criteria that distinguish a living organism from a non-living object.**
2. **describe some of the ways that organisms receive, use, and store energy.**
3. **explain the paths of the water and carbon cycles between living and non-living things.**
4. **describe the various ways that fossils can be formed and identify the factors that must be present for fossilization to occur.**
5. **interpret fossil imprints and models of casts and molds as either a record of some event occurring in the geologic past or as preserved "clues" about the organisms or objects that formed them.**
6. **explain how fossils give clues to the climate, geological events, and changes of life forms in earth history.**
7. **observe and describe the variations within a species and explain why species variations are necessary for an explanation of the changes in organisms through time.**
8. **describe the changes in one plant or animal as preserved in the fossil record.**
9. **explain the geometrical increase of numbers associated with population growth and the dangers of overpopulation of the earth.**

Life Today

14-1 What is Life?

Many billions of organisms inhabit the earth. They live on the land and in the water. Microscopic organisms live in the air that you breathe. Organisms range in size from microscopic plants and animals to the giant sequoia tree of California and the great whales of the oceans. There is hardly a place on Earth where some form of life does not exist. But despite the variety and abundance of living forms, life is a most difficult term to define.

ACTION

Consider a frog, a rock, and a plant. What differences between the animal and the rock in Figure 14-1 determine which is alive? How are the plant and the frog similar? How are they different? Was the rock ever alive? Could the rock show any evidence of life?

The chemical makeup of living things is different from that of non-living things. Living organisms consist mostly of **organic** (*awr GAN ihk*) **compounds.** These are compounds containing carbon atoms that join with one another and with the atoms of other elements, especially hydrogen and oxygen. The amount of carbon in living things might suggest that it is one of the more common elements in the earth's crust. Actually, carbon is seventeenth in order of abundance. Crustal rocks contain less than one-tenth of a percent of carbon by mass. Oxygen and silicon are the most abundant elements by mass.

A few non-living or **inorganic compounds** may be produced by plants and animals. Many animals such as clams, snails, and oysters grow shells that contain calcite ($CaCO_3$). Others, like certain

Figure 14-1 *Which of these things are alive? How are living things and inanimate objects different?*

Figure 14-2 *The carbon cycle. What different pathways could one carbon atom follow to go from the atmosphere to living organisms and back again?*

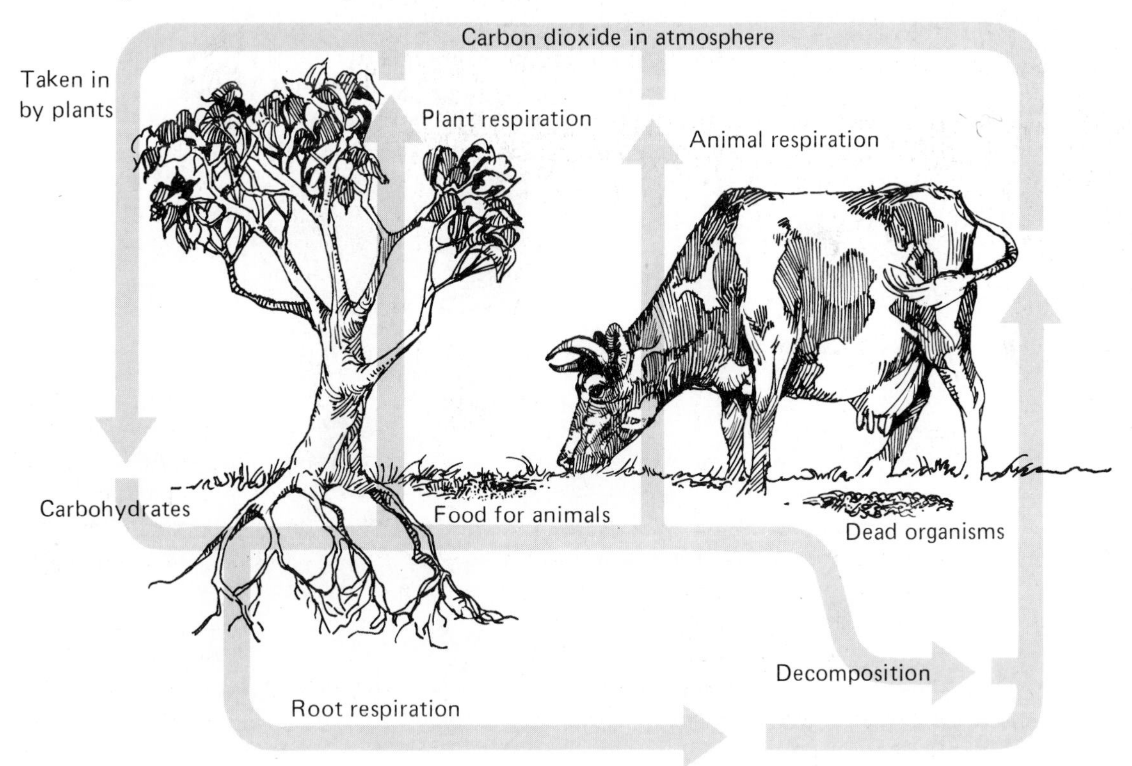

sponges, have hard parts made of silica (SiO_2). But most inorganic substances are not produced by living things.

14-2 Organisms and Chemical Cycles

Chemical elements can move through a cycle from non-living to living objects and back again. An example is the **carbon cycle** (Figure 14-2). The main pathway in the carbon cycle is from the atmosphere into living things and then back again. Living organisms give off carbon-bearing wastes that are returned to the land, air, and water. When living things die, carbon compounds are left in their bodies. The remains of each dead plant or animal provide food for the small organisms called **decomposers.** The decomposers, in turn, release carbon into the atmosphere as carbon dioxide. In this way, decomposers make materials available for reuse by plants. Most decomposers are microscopic. But some decomposers, like the funguses (mushrooms and molds), are much larger.

In some cases carbon takes a different path. For example, certain organisms take carbon into their shells as calcite ($CaCO_3$). When these animals die, the shells are not always broken down by decomposers. Instead, they may be deposited as sediment.

Earlier in the earth's history, great masses of carbon compounds were changed into coal or petroleum. Coal is made up mostly of ancient plant remains that have been enriched in carbon by the removal of other elements. The origin of **petroleum** (oil and natural gas) is not clearly understood. Perhaps decomposers changed the remains of microscopic plants and animals into crude oil. Petroleum and coal are called **fossil fuels** because they are formed from the remains of prehistoric plants and animals. Fossil fuels are a source of solar energy that was trapped by plants and stored in the earth's crust many millions of years ago.

The water cycle is also necessary for life. If rain and snow didn't continuously

Figure 14-3 *Fossil shells weathering from a rock outcrop near Leeds, New York.*

return fresh water to the land, all land would soon become lifeless. Organisms that live on land may pick up water at a number of points in the water cycle. Land plants usually absorb water from the soil, while land animals drink it. The amount of water in different organisms varies. Our bodies are about 66 percent water. But a jellyfish may be more than 95 percent water.

Moisture absorbed by a land plant is carried to its leaves. Most of this water evaporates through openings in the leaves. Animals return moisture to the atmosphere through breathing, perspiration, and waste products.

Thought and Discussion

1. Give your own definition of life.
2. Can you explain the statement: Coal is "petrified sunshine"?
3. Explain, with examples, how the cycles are important natural processes.

Life of the Past

14-3 Fossils As Evidence of Prehistoric Organisms

Chapter 13 explained how the history of the earth can be outlined from evidence found in the rocks. For example, certain rocks in Utah contain fossil bones of the dinosaurs that once lived there. And fossil evidence (leaves and stems) also suggests that great swampy forests once covered parts of what is now Pennsylvania and Illinois and other states, as well as southeastern Canada. Scientists who do the detective work with this kind of fossil evidence are called **paleontologists** (*pay lee uhn TAHL uh jihsts*). The earth scientists use fossils as clues to trace the development of life and to reconstruct the geologic past.

To learn from the fossils of organisms that disappeared from the earth hundreds of millions of years ago, the paleontologist must know as much as possible about present-day organisms. For example, the environments of extinct organisms are not always known. But often a group of fossil organisms closely resembles a living group. Then you can usually assume that the two groups lived under similar conditions. The principle of uniformitarianism is again applied to interpreting the past. The present is used as a clue to the past.

The term **"fossil"** comes from the Latin word *fossilis,* meaning "dug up." But most fossils are not dug from the ground. They are usually uncovered by weathering and erosion (Figure 14-3). There are many different kinds of fossil plants and animals. Animal fossils range in size from dinosaur bones more than two meters long and weighing several hundred kilograms to fossils so tiny that hundreds of them would fit on the head of a pin! These smaller forms are called **microfossils** because they must be studied with a microscope (Figure 14-4).

ACTION

If you live in an area where rocks containing fossils are exposed, take a field trip. Make a collection of fossils if collecting is allowed. Visiting a nearby museum or reading several of the books listed in the Suggested Readings at the end of this chapter will give you some ideas about how fossils can be collected, identified, and displayed. The pamphlet *Field Guide to Fossils* will be very useful in planning your fossil collecting field trip.

Figure 14-4 *This tooth-like object is a microfossil called a conodont. It was found in north-central Utah and is Early Triassic in age. It is magnified forty times in the photograph. Conodonts may be mouth parts of otherwise unknown organisms, possibly worms.*

Many fossils are the actual remains of plants and animals, such as bones, teeth, leaves, and shells. Other fossils are mere traces of organisms. These include trails left by worms, the imprints of leaves, and the footprints of dinosaurs. (See Figure 14-5.) What could you learn about a dinosaur from its tracks?

14-4 Investigating a Footprint Puzzle

Suppose you discovered a set of fossilized tracks like those in Figure 14-6. You would have a chance to determine past events from limited evidence. The only clues would be the footprints.

When you have completed this investigation, you should be able to explain how paleontologists can reconstruct prehistoric events from evidence preserved in the rocks.

PROCEDURE

Look at Figure 14-6. Tracks like these are common in some rocks of Canada, New England, and the southwestern United States.

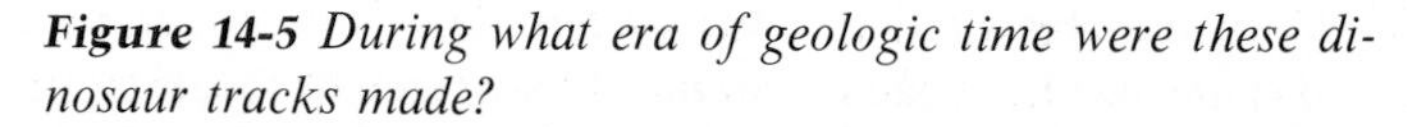

Figure 14-5 *During what era of geologic time were these dinosaur tracks made?*

DISCUSSION

1. What can you tell about the size and kind of the animals from their footprints?
2. Were all the tracks made at the same time?
3. How many animals were involved?
4. Did the animals walk on four legs or two legs? How can you tell?
5. Reconstruct the series of events represented by this set of fossil tracks. Your teacher will show you two or more parts of this footprint puzzle.

Figure 14-6 *What kind of a story do these fossil footprints tell?*

14-5 How Fossils Are Formed

Dead plants and animals decay rapidly, but hard parts such as teeth, shells, and wood are sometimes fossilized. Figures 14-7, 14-8, and 14-9 show some examples. Under some special conditions, organisms with no hard parts, such as jellyfish, have been preserved.

Yet not all organisms with hard parts will be fossilized. You know that you hardly ever see the complete skeleton of a dead animal. It is sometimes hard to find a perfect seashell on the beach. Many shells are broken or badly worn. There are many ways that the remains of organisms are destroyed. After an animal dies, its flesh is attacked by a variety of organisms, such as vultures, coyotes, insects, and bacteria. The sun, wind, and rain help the decay process. The flesh soon disappears, leaving only bones, teeth, or shell. These hard parts may also be destroyed. Bones will weather away, if they do not get buried. Some shells and bones are crushed by the weight of overlying sediments. Others are broken and eroded as they tumble along a stream bed or are tossed about by waves.

Although a number of factors affect fossilization, there are two factors that are most important. First, if the organism has hard body parts, the chances of fossilization are greatly increased. Second, the plant or animal remains must be quickly covered by some sort of protective material. The environment of the organism determines the kind of covering material and the type of fossilization. For example, the remains of marine animals can be preserved because they fall to the ocean floor shortly after death and are buried by soft mud and sand (Figure 14-10). In general, the finer the sediment covering the organisms, the more likely that the remains will be preserved as fossils. How might a land-living organism be buried in order to become fossilized?

Most organisms change greatly during fossilization. But when conditions are just right, the actual plants and animals are sometimes preserved intact. The Berezovka mammoth is an example of a whole animal preserved by freezing. Insects and spiders have been preserved in **amber,** a fossil resin (pitch) that flowed

a.

b.

Figure 14-7 *Some invertebrate fossils.* **a.** *A clam.* **b.** *A brachiopod.* **c.** *A sea urchin.* **d.** *A trilobite.* **e.** *A coral.*

c.

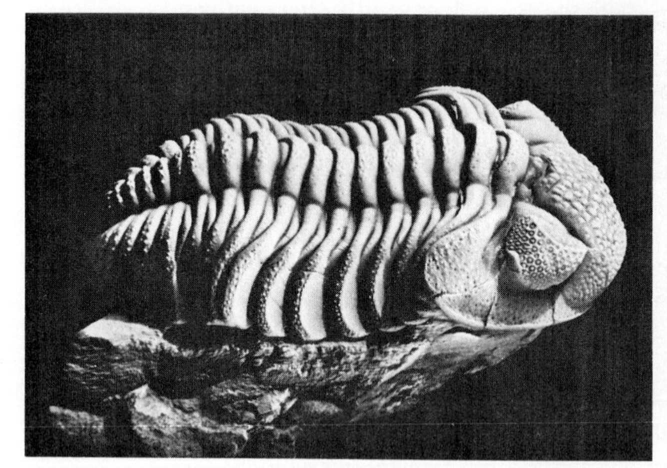

d.

e.

a.
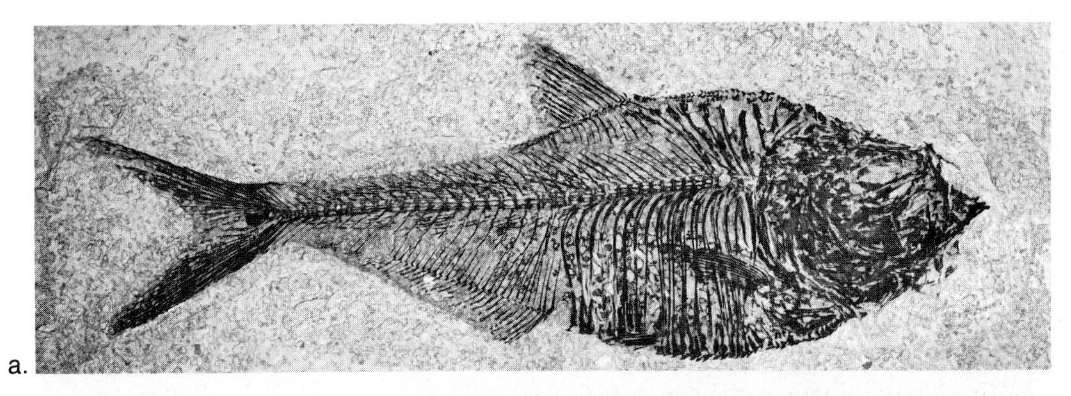

b.

c.

d.
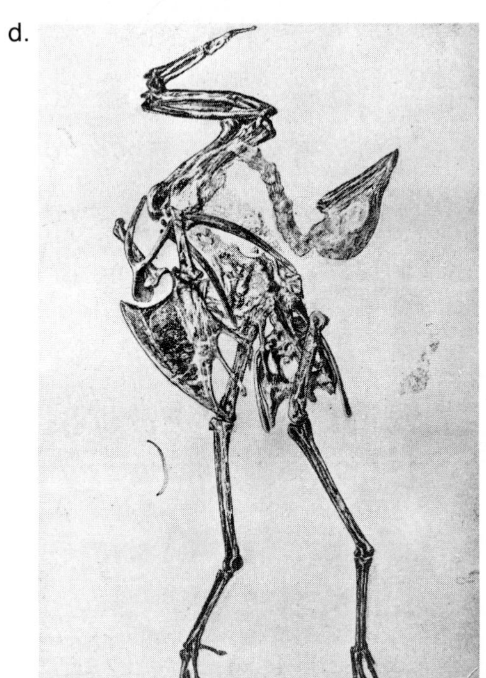

Figure 14-8 *Some vertebrate fossils.* **a.** *A bony fish.* **b.** *An amphibian.* **c.** *A mammal.* **d.** *A bird.*

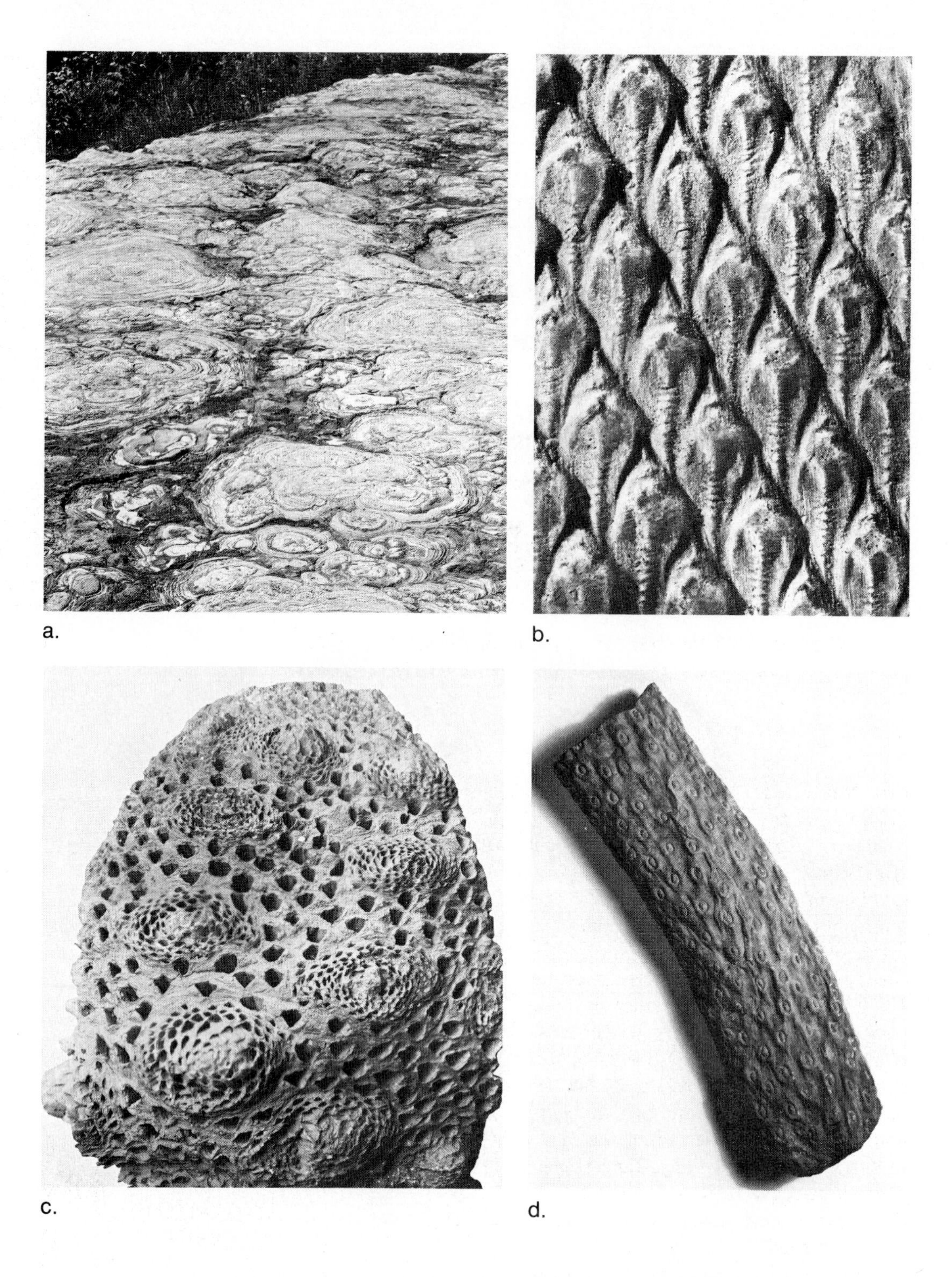

Figure 14-9 *Some plant fossils.* **a.** *A fossil reef formed by secretions of algae.* **b.** *A natural cast of the outer part of a Coal Age tree trunk* (Lepidodendron). **c.** *A petrified cycad stem.* **d.** *A natural cast of the root from a lepidodendron tree.*

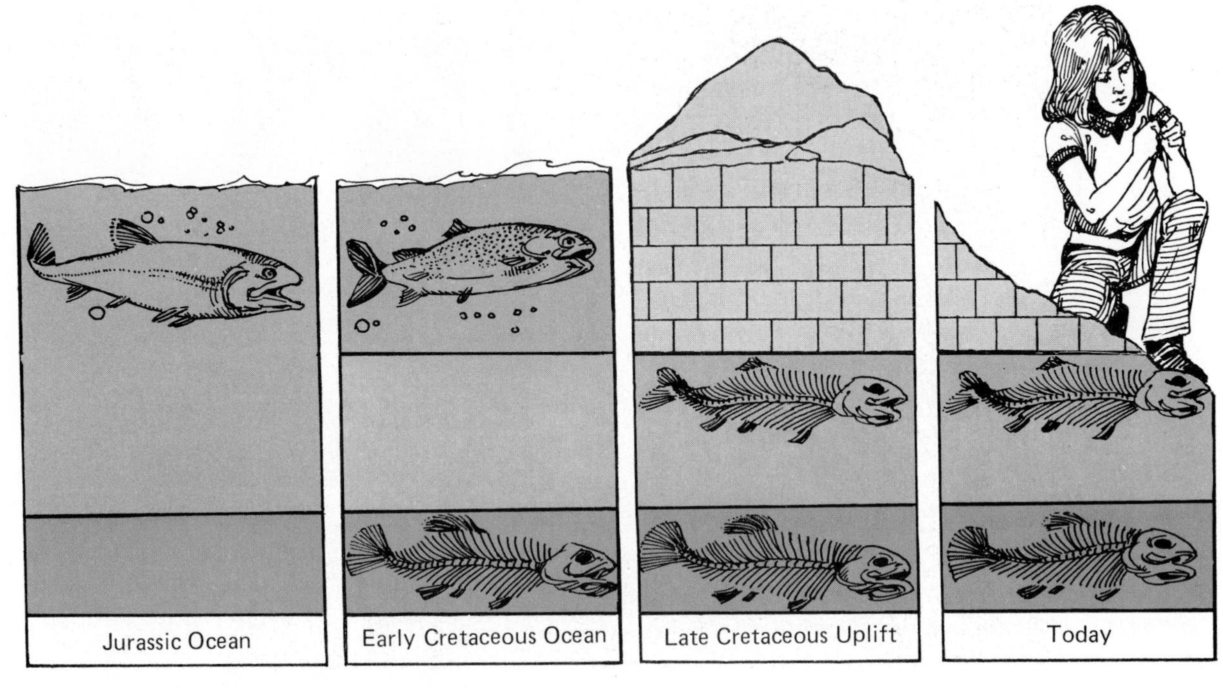

Figure 14-10 *Describe the stages in the fossilization of these fishes, from their life in prehistoric seas until their discovery as fossils.*

from certain cone-bearing trees (Figure 14-11).

Organic hard parts normally change after they are covered by sediment. Mineral-bearing water seeping through the sediment may gradually dissolve mineral matter such as calcium carbonate (calcite) from a shell. The calcium carbonate might later be replaced by silica or some other material in the water. In this way, remains become **petrified,** or "turned to stone." Under other conditions the replacing solution may be rich in compounds of iron, magnesium, or calcium. In fossils that have been formed by **replacement,** even microscopic details of the original hard parts may be beautifully preserved.

Many fossils have been formed through the gradual decay of organic material after burial. During the process of decay, the organic material leaves behind a thin film of carbon that shows a detailed outline of the original organism. (See Figure 14-12.) Fossils of this type are formed by **carbonization.**

Figure 14-11 *An insect preserved in amber. Why could this specimen be called a "fossil within a fossil"?*

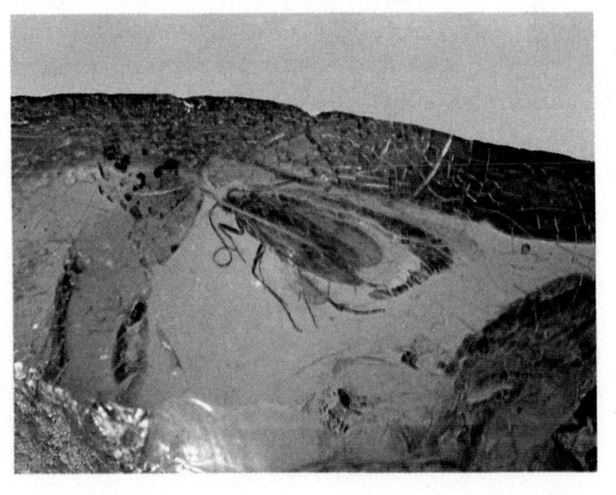

a.

b.

c.

Figure 14-12 *Fossils preserved by carbonization.* **a.** *A fern leaf.* **b.** *The leaf of a flowering plant* (Aralia). **c.** *A fossil fly.*

MATERIALS
aluminum pie plate or milk carton, container for mixing plaster, fast-drying plaster, objects to cast, petroleum jelly or soap solution, stirring device

Large numbers of fossils are found in the forms of **molds** and **casts.** To understand how these are formed, consider a seashell buried in ocean sediments. After the sediments have been changed to rock, ground water may slowly dissolve the shell, leaving an empty space where the shell was. The space is a mold, that preserves the marking of the shell. How would a ridge on a shell appear on this mold? As time passes, the mold may be filled with minerals deposited by ground water. The minerals may form a cast of the original shell. (See Figure 14-13.)

14-6 Investigating Casts and Molds

In this investigation you will make models of fossil casts and molds and try to interpret evidence that gives you clues about the "organisms" you have preserved.

When you have completed this investigation, you will be able to describe how fossil casts and molds are formed, and explain how they can be used to identify the organisms or objects they represent.

PROCEDURE

Prepare both plaster molds and casts of various objects. Then exchange the plaster blocks containing your home-made "fossils" with other members of the class. The following procedures should be used when making the plaster molds and casts.

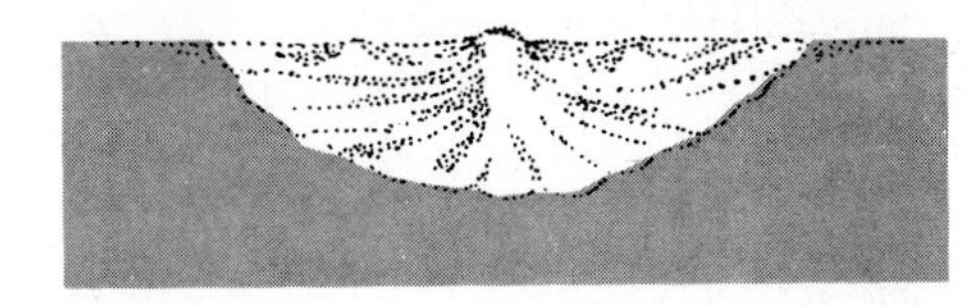

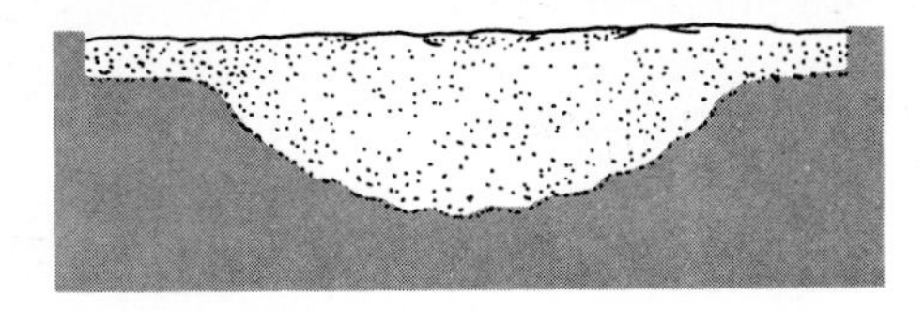

Figure 14-13 *Suppose a shell is buried in sediment. Remove or dissolve it and a mold is left* (*center*). *If the mold is filled in, a cast is formed* (*bottom*).

MAKING A MOLD

1. Cover the selected objects with a thin film of liquid soap or grease.
2. Place a small amount (about 75 ml) of water in the mixing container.
3. Slowly sift plaster through your fingers into the water, mixing the plaster and water as you sift. Add plaster until the mixture has the consistency of thick cream. If the mixture is too thick, add a small amount of water. If it is too thin, add more plaster.
4. Pour this mixture into another container. Tap the container to eliminate air bubbles.
5. Press the lightly coated object *gently* into the plaster. *Do not submerge it.*
6. Allow the plaster to harden completely.
7. To expose the molds, remove the objects from the plaster.

MAKING A CAST

1. Lightly coat the surface of the block containing the molds with grease or liquid soap. For a strong mold, let the plaster dry first.
2. Mix another batch of plaster. Add a little food coloring or ink to the water, if you wish. Cover the original plaster mold block completely with fresh plaster.
3. After the fresh plaster has set, carefully separate the two blocks. The raised areas on the new block are the cast of the objects from which the mold was made. If it is difficult to separate the two blocks, use a screwdriver or chisel to pry them apart. If the blocks break, use them anyway. Many broken fossils are found, but the paleontologist will still attempt to interpret them.

DISCUSSION

1. Can you identify the objects that were used to make the molds or casts that you have been given?
2. What does the evidence tell you about the "organism"?
3. If some objects didn't fossilize well, tell why.

14-7 Fossils Are Clues to Earth History.

As you saw in the previous chapter (Section 13-11), fossils are very useful in correlating rock layers. Scientists have found that fossils also provide other kinds of clues.

Fossils can give evidence of ancient climates. Although fossil reindeer have been found in Arkansas, modern reindeer live in cold climates. This suggests that Arkansas in the geologic past had a climate different from what it is now. (It might also suggest that reindeer once lived in a different climate.) The distribution of marine and land fossils have helped to locate ancient lands and seas. This information is used to draw maps showing the continents and oceans at various times during geologic history.

Fossils also lead geologists to rock formations that may contain valuable deposits of ore, coal, or petroleum. Certain kinds of microfossils are especially useful to the oil geologist. Microfossils are so small that they are not usually broken by the bits used to drill oil wells. They can be brought to the surface almost undamaged and used as markers to provide information about the age of underground formations.

Thought and Discussion

1. Why are soft-bodied animals such as jellyfish seldom fossilized?
2. Name some materials that have been known to replace organic matter.
3. How are fossils useful to human beings?

Changing Patterns of Life

14-8 Plants and Animals Change Through Time.

In 1859 Charles Darwin, an English naturalist, published a book called *On the Origin of Species by Means of Natural Selection.* It contained his theory about how plants and animals have evolved over long periods of time. Darwin's work, based on many years of study of living and fossil organisms, started a controversy that has not yet ended. Some people object to Darwin's theory because they feel it conflicts with religious teachings. Current scientific controversy, however, centers around the mechanisms by which evolution occurs.

What two applications does Darwin's theory of organic evolution have to a study of earth history? First, it assumes that species of living things have changed over long periods of time. Plants and animals now inhabiting the earth are different species from the first organisms on Earth. A **species** is a group of organisms that can produce fertile offspring. Second, it means that many species that were once abundant no longer exist. They have become extinct, and are now known only as fossils.

If a plant or animal species changes, the following generations will be different from their ancestors. Darwin called these natural changes **variations.** This concept of change provided the basis for his theory.

Members of a single species vary from each other. All dogs don't look alike. Some people can run faster than others. The child of a fast-running person may run faster or slower than the parent could run at the same age. Yet these small differences *within* a species do not necessarily form a new species. Darwin's theory says that a steady trend over a series of generations will produce an organism that is recognizably different from its ancestors. Darwin's work suggested that such trends existed, and that they depended on what he termed natural selection. By **natural selection** Darwin meant that the individuals best suited to their environment are most likely to survive and pass life-saving characteristics on to their offspring. Through natural selection over a series of generations, the descendants of the original organisms would be better fitted to survive in their environment.

Darwin saw many examples of such adjustment, or **adaptation,** to new environments. The adjustment of these new species seemed to support his theory of adaptation through natural selection. Most of Darwin's earlier observations were made in the Galápagos (*guh LA puh guhs*) Islands. These volcanic islands are located about 960 km off the western coast of South America. They were an ideal outdoor laboratory.

The rocky islands contain a great variety of unusual plants and animals, but Darwin was especially interested in the birds. Although there were fewer species of birds than on the mainland, he found a large number of finches. The Galápagos finches were remarkably different in the shapes and sizes of their beaks. Darwin saw that the birds had beaks that were especially adapted for the types of food they ate. Some of the finches had heavy beaks that could be used to crack seeds. Others were adapted to catching small insects. Some of the different beaks of the Galápagos finches are shown in Figure 14-14. Which bird had a beak well suited for picking up small insects? Which bird could crack the hardest seeds?

When Darwin had completed his study of finches, he was convinced that the various beak types were the result of adaptation. Each kind of beak had slowly changed and become specialized to peck, or bite, or to dig up a different kind of food. The beaks had developed through natural selection from one original kind of beak. Darwin thought that this probably took place over a period of about one million years.

Figure 14-14 *Darwin's finches. What kinds of food might each of these birds eat? Upon what evidence do you base your answer?*

14-9 Investigating Variation During a Long Period of Time

MATERIALS
metric ruler, small objects (for indicating which fossils have been measured)

Do the fossils in Figure 14-15 show any changes that might have occurred over a long period of time? How could you find out? Questions like those were asked by paleontologists a century ago. The studies and interpretations of Charles Darwin helped give answers. In this investigation you will study the drawings in Figure 14-15 of two slabs of rock containing fossils. You will examine these to determine for yourself what changes might have taken place.

When you have completed this investigation, you should be able to explain one way paleontologists can study fossils to learn if changes have occurred during long periods of time.

PROCEDURE

Examine the reproductions or drawings of the two slabs of rock. How could you describe the differences and similarities of the fossils? Discuss with the rest of the

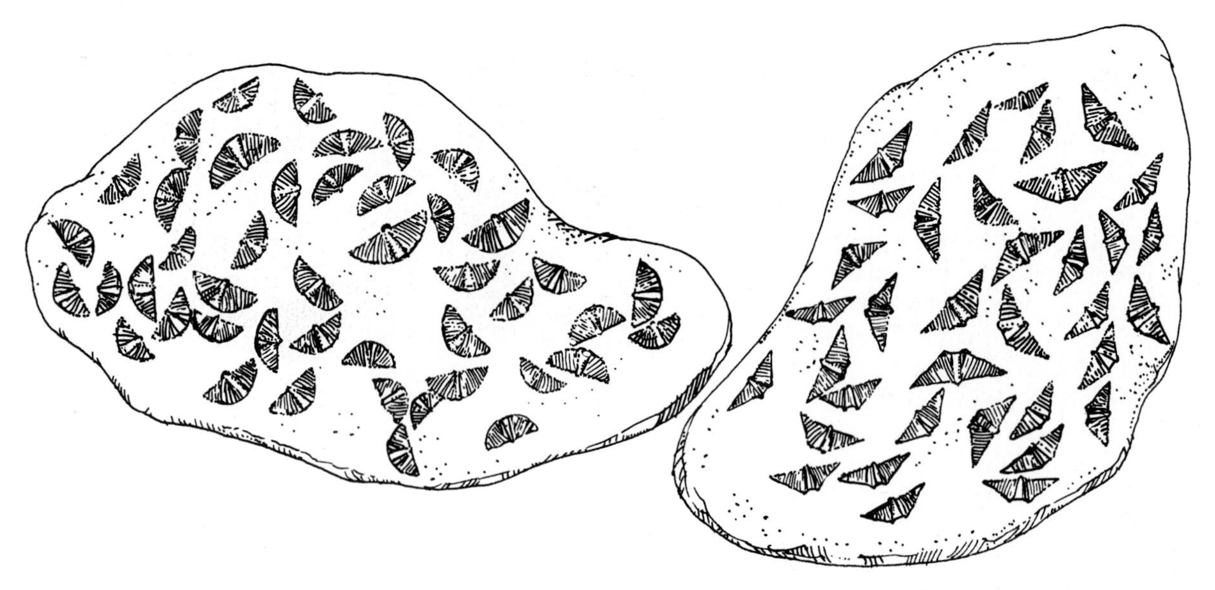

***Figure 14-15** How do the brachiopods on these two slabs of rock differ? In what ways are they alike?*

class the characteristics that can be used, as your teacher directs.

Measure the length and width of each fossil preserved on the slabs. Your teacher will suggest ways of making graphs of your results.

DISCUSSION

1. How do the fossils on each slab differ from each other?
2. What similarities are apparent in each group? What differences exist between the two groups?
3. What changes have taken place between the time of the older group and the time of the younger group?

Thought and Discussion

1. How does the theory of evolution relate to paleontology?
2. How did Darwin explain the process of natural selection?
3. How did the finches of the Galápagos Islands adapt to their environment?

The Parade of Life

14-10 Paleozoic Time: The Age of Invertebrates

No one knows when or where life first appeared on Earth. However, fossils of simple life forms have been found in Precambrian rocks 3.5 billion years old. Certain of these fossil bacteria are illustrated in Figure 14-16.

Not much is known about the earliest life forms because Precambrian organisms left very little evidence of their presence. However, evidence of life is abundant from the beginning of the Cambrian Period to the present. This clearly defined record stretches 600 million years back into time. (See Figure 14-17.)

The fossil record of Paleozoic time shows the successive development of new forms of invertebrates. **Invertebrates** (*ihn VUR tuh brayts*) are animals without backbones. During the Cambrian Period the first known **vertebrates** (animals with backbones) appeared. They were fishlike creatures that lived more than 500 million years ago.

Life was apparently confined to the sea during Cambrian and Ordovician (*awr duh VIHSH uhn*) time. However, the fossil record shows that during the Silurian (*sih LOOR ee uhn*) Period both plants and animals moved onto the land. The earliest land dwellers were probably simple plants. Fossils show that the first land or fresh water animals seem to have appeared after plants had invaded the land.

Near the middle of the Devonian (*dih VOH nee uhn*) Period about 375 million years ago, the specialized fish called the "lobe-fins" appeared. One kind of these fish eventually changed into the first amphibians. The "lobe-fins" had muscular fins that were used as flipperlike paddles. During seasonal dry spells the fish were able to crawl from drying lakes or streams to more permanent bodies of water. How do amphibians differ from fishes?

The early amphibians spent most of their lives in water. They were the dominant animals outside of the ocean until the Pennsylvanian Period, when the first reptiles appeared. Reptiles do not need to live near the water. Their outer skin allows them to live in very dry places. Unlike eggs of amphibians, eggs of reptiles can develop out of water.

Fish, amphibians, and reptiles became common in Paleozoic time, but this era was really most favorable for the development of marine invertebrates. Trilobites (Figure 14-18) were numerous on the bottoms of Early Paleozoic seas. But near the end of Paleozoic time their

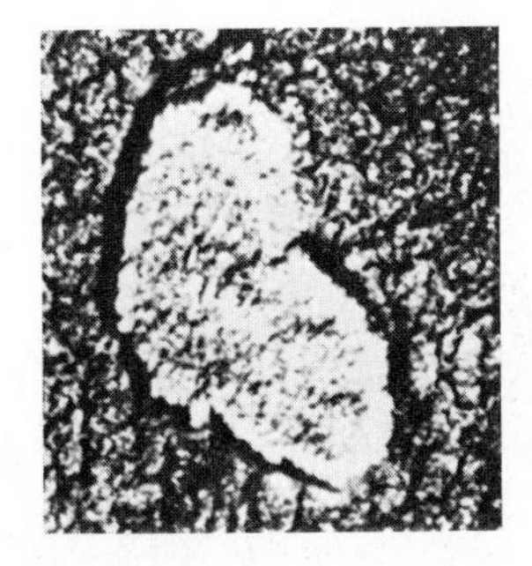
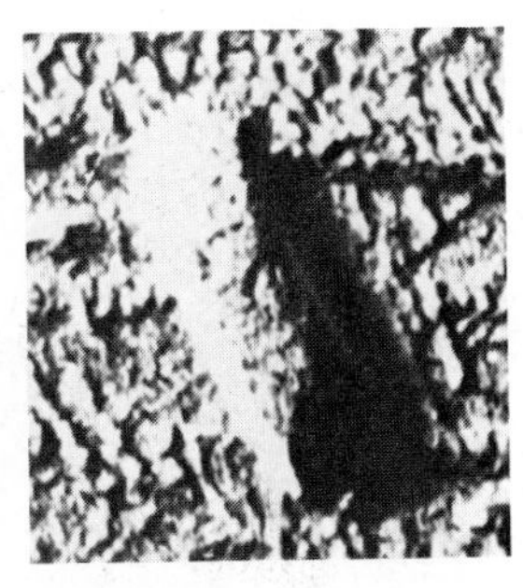

Figure 14-16 *These bacteria-like fossils are less than one ten-thousandth of a centimeter in diameter.*

numbers began to decrease. Finally, at the end of the Permian Period about 230 million years ago, the trilobites became extinct. What might be some reasons for the extinction of this once abundant and successful group of animals?

Brachiopods (*BRAK ee uh pahdz*) are animals with shells somewhat like clam shells. Figures 14-7b and 14-15 show brachiopods. They were among the most abundant creatures on Earth during Middle Paleozoic time. Sedimentary rocks in which their fossils are found suggest that most of these animals lived in shallow water. Throughout geologic time, certain species of brachiopods seem to have been able to adapt to different kinds of marine environments. Some lived on muddy bottoms and others lived on sandy bottoms. Some lived in shallow water and others lived in deep water. Some lived near coral reefs. Perhaps this adaptability is one reason why brachiopods are still found in shallow seas today—600 million years after brachiopod shells were buried and preserved in the mud of a Cambrian sea.

During the last part of the Paleozoic Era, erosion lowered the level of great areas of the continents. Large swamps formed in some of these areas. As partly decayed vegetation slowly built up in the swamps, it formed a mixture of plant fragments called **peat.** Then, layers of

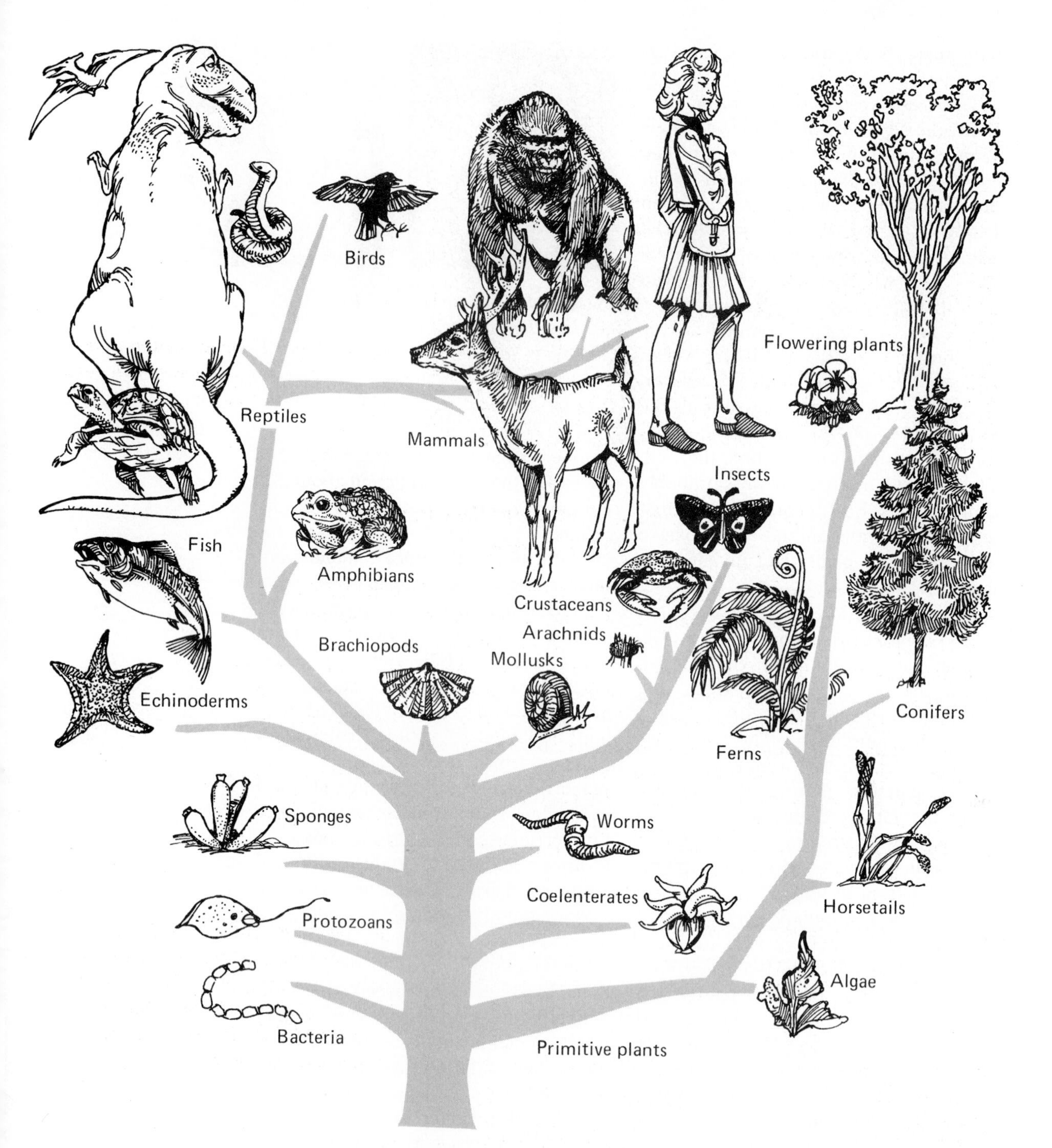

Figure 14-17 *The tree of life.*

Figure 14-18 *A reconstruction of a Cambrian sea floor showing trilobites, jellyfish, primitive sponges, and worms.*

sediments were laid down on top of the peat.

Thousands of years after the peat was buried and compressed beneath heavy layers of sediments, it slowly changed into coal. These thick deposits of Paleozoic coal make up one of the largest concentrations of carbon on earth. It is certainly one of the greatest sources of energy available today. Coal will become even more important in the future as we use up our limited deposits of oil and natural gas. (Coal deposits are limited, too. Then what?)

14-11 Reptiles Rule the Earth.

Beginning with the Triassic (*try AS ihk*) Period, reptiles ruled the land until the end of the Mesozoic Era.

In size, these reptiles were less than a meter to over 30 meters in length. The presence of flying as well as sea and land forms of reptiles shows that they had adapted to life in many different environments. Because of their numbers, the Mesozoic Era is called the Age of Reptiles. (See Figure 14-19.)

Figure 14-19 *Cretaceous dinosaurs. The dinosaurs on the left* (Triceratops) *are staying out of the way of the two* (Tyrannosaurus) *that are fighting over a kill. The trees are cycads and hardwoods.*

a.

b.

Figure 14-20 *The earliest known bird,* Archaeopteryx, *had a skeleton that somewhat resembled a dinosaur, with toothed jaws and claws on its wings.* **a.** *The skeleton as it was found in the rock.* **b.** *A reconstruction of the bird, in the same position as that of the fossil skeleton.*

Among the many kinds of reptiles that developed, dinosaurs were particularly numerous and varied. Dinosaur National Monument in Utah is one of the few places in the world where dinosaur skeletons can be seen in the rock, as in the photograph at the beginning of this Unit. Paleontologists have uncovered the bones by chiseling away the rock, so that you can see them in much the same positions as when the sediment covered them millions of years ago.

The skeleton of the oldest known bird was found in southern Germany. It was found in Jurassic limestone about 140 million years old. If only the skeleton had been found, this primitive bird would probably have been classified as a reptile. However, impressions in the rock showed that it was covered with feathers (Figure 14-20). This was a most remarkable find. Not only is it the oldest known bird, but it also supports the theory that birds developed from reptiles.

At the same time, the land plants increased in number and variety. During the Cretaceous (*krih TAY shuhs*) Period, grasses, fruit trees, and other flowering plants appeared.

Near the end of the Cretaceous Period the dinosaurs, flying reptiles, and most of the marine reptiles mysteriously disappeared. After 140 million years the Age of Reptiles came to an end. Can you think of any reasons why these animals, which had so successfully adapted to such a wide variety of habitats, might have become extinct at what may have been the very peak of their development?

14-12 The Cenozoic Era: Golden Age of Mammals

The first mammals appeared early in the Triassic Period. These early warm-blooded animals were primitive and reptilelike. The fossils of the earliest known mammals are very rare and incomplete. They indicate that these creatures were about the size of a mouse. Although mammals were present and continued to develop during Jurassic (*joo RAS ihk*) and Cretaceous time, they were overshadowed by the reptiles. It was not until the Cenozoic Era that mammals began to adapt to new environments and increase in numbers. Could the extinction of the dinosaurs and other reptiles have had anything to do with the rapid rise of the mammals?

Some later mammals, such as the giant ground sloths, woolly mammoths, and saber-toothed cats of the Ice Age, flourished for hundreds of thousands of years and then became extinct. Some of these animals were preserved in tar (see Figure 14-21).

As time passed, the numbers of grasses and other Cenozoic flowering plants increased. Many of today's trees, such as the poplar, maple, elm, and oak, spread rapidly over the land. What part could these developments in the plant kingdom have played in the rise of mammals?

14-13 Investigating Population Growth

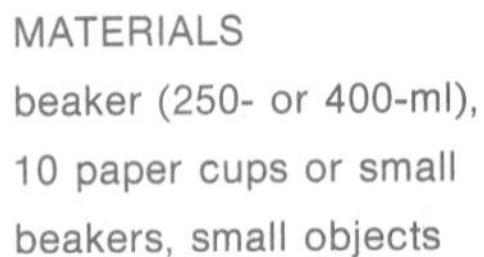
MATERIALS
beaker (250- or 400-ml), 10 paper cups or small beakers, small objects

Within recent years it has become obvious that we are living in an environmental crisis. Whenever environmental problems are investigated, it is usually found that humans have produced the basic causes. The problems are usually the result of advanced technology and increasing population density. In the investigation that follows you are going to investigate the mathematical nature of human population growth.

When you have completed this investigation, you will be able to develop a model that can be used to

Figure 14-21 *A scene at the La Brea (California) tar pits during the Ice Age. The tar pits are now part of Los Angeles.*

find and predict trends in population growth.

PROCEDURE

Place a glass beaker on your desk with two objects in it. This will represent the earth, which will hold only a certain size population.

Put a row of paper cups on your desk (ten should be enough). In the first cup, place two of the objects. In the second cup, place twice as many as in the first cup, or four objects. Write on the outside of the cups the number of objects that have been placed in each cup.

In cups 3 through 10, double the number of objects that are in the previous cup (that is, cup number 3 will contain 8 and cup number 4 will contain 16). Write the amount in each cup on the outside. Determine the height of the beaker with the two objects in it. What is the approximate volume (in percent) of the empty space in the beaker? Record this at 0 time. Make a table to record your data.

In 35 seconds, add the contents of cup 1 (that is, 2 objects) to the beaker and record in the table the total population and the approximate percent of the volume of the beaker that is empty. At 35-second intervals, add the contents of cups 2 through 10. Record your results.

Make a graph of your results, with population on the vertical axis and time on the horizontal axis.

DISCUSSION

1. The human population of the earth is thought to have had a slow start, with early periods of doubling in size as long as 1 million years. The present world population is thought to be doubling every 37 years. How would the mathematical nature of this growth rate compare to your investigation?
2. The present world population is well over 4 billion people. To answer this question, assume that it is 4.5 billion. The earth's radius is about 6400 kilometers and about 7/10 of its surface is covered with water. What is the present density of human population in terms of people per square kilometer of land surface? (Area of a sphere = $4\pi r^2$)
3. Assume that the present population growth rate will continue. What will the density per square kilometer be 37 years from now? 111 years? 1110 years?
4. Is space the only limiting factor in determining maximum human population? If not, describe others.

14-14 What Lies Ahead?

Conditions on our planet have been good for the development of the human species. Earth has also given us the necessary raw materials and energy sources for the development of civilization. Modern technology seems to have made our lives more pleasant. However, progress in technology is also responsible for many of our environmental problems, especially pollution.

As populations expand, we will need increasing amounts of limited resources such as fresh water, petroleum, and metal ores. This will place additional demands on our already strained environment. In time, our ecological problems will become even more critical. The earth is our home and we should be more responsible for its care. How do you think we can help to preserve and repair our natural environment?

Thought and Discussion

1. What can fossils tell us about ancient environmental conditions?
2. Which plant and animal types can adapt to the widest range of environ-

ments? Make a list of the evidence to support your answer.

3. Discuss some of the relationships between plant and animal development during the Cenozoic Era.

Unsolved Problems

Many fossils have been discovered, but some pieces in the puzzle of life history are still missing. For example, when, where, and why did the different animal groups develop hard parts? By the end of the Cambrian Period most of the major groups of invertebrates had developed hard parts. Yet in Precambrian time most organisms had poorly developed hard parts or had none at all.

At various times in geologic history, organisms such as the dinosaurs have become extinct. Although several explanations for this have been proposed, none of these extinction theories can be proved. The world-wide disappearance of a single group of animals cannot be explained from the available evidence. Mountain building, climate changes, and other environmental changes cannot affect all of the individuals in a group. What single or combined events could have caused extinctions?

Scientists do not yet know exactly how plant and animal substances are changed into fossil fuels. Knowledge of how fossil fuels are formed could help geologists locate more of these important energy-producing materials.

Chapter Summary

Although life surrounds us, it is not always easy to distinguish living from non-living objects. In general, however, living objects are characterized by growth, response to outside stimulation, and the ability to reproduce. Living matter is composed of organic compounds, all containing carbon. During the more than three billion years that life has been present on Earth, it has spread to all parts of the globe.

Although the record of past life is incomplete, fossils have provided valuable clues to earth history. In studying fossils, paleontologists assume the plants and animals of the past lived in much the same manner as their modern relatives.

The way an organism becomes fossilized depends somewhat on the original composition of its body and the physical and chemical conditions that surround the animal before and after burial. Some organisms are petrified by the minerals in underground water. Others are preserved as thin films of carbon. More rarely, the organic remains are preserved in their original unchanged condition.

However, most prehistoric organisms have not been preserved. And of those that have been fossilized, most have been destroyed through natural chemical and physical processes. Others will never be found by paleontologists.

Plants and animals have always been restricted in time, space, and environment. Thus, their fossil remains are valuable aids in interpreting earth history.

The rock record suggests that the first organisms to appear on Earth were probably much less complex than those that developed during later geologic periods. The record of Precambrian life is scarce, but the fossil record is relatively clear from the beginning of the Cambrian Period. Although marine plants and animals were the dominant forms of Paleozoic life, vertebrate animals appeared early in the era and the numbers

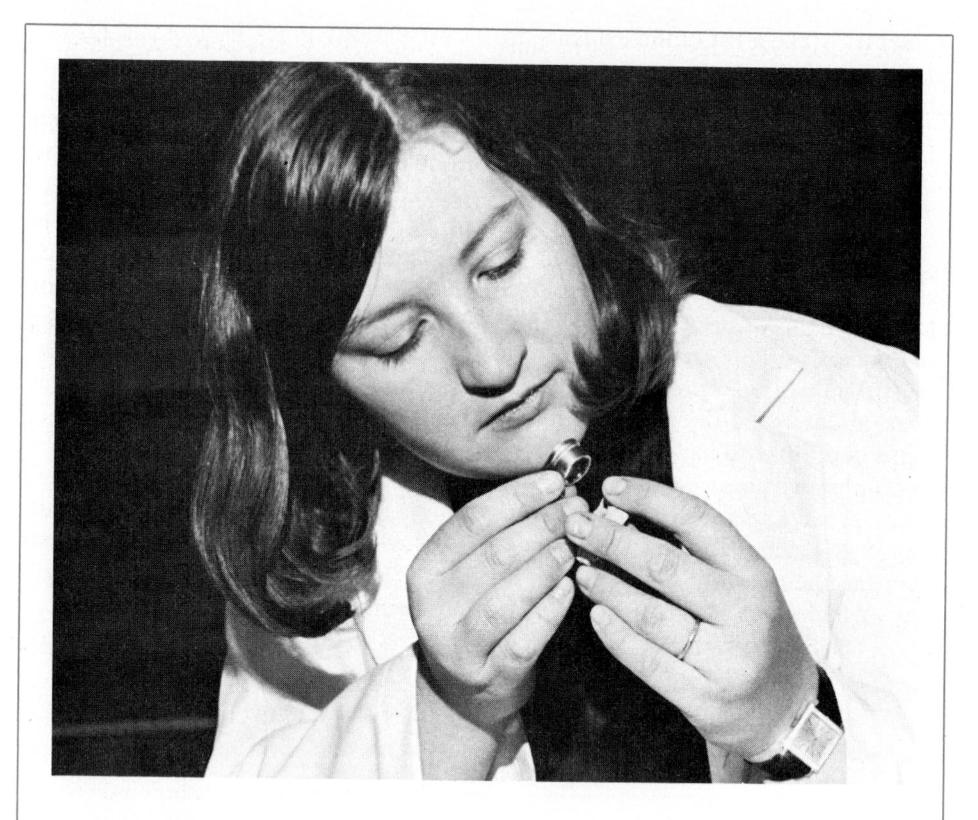

PALEONTOLOGIST

Paleontology is a field of study appropriate for people who are interested in both geology and biology. Paleontologists study plant and animal fossils. Their purpose may be to trace the development of a particular life form. Or, they may study the fossils found in a rock bed to gain more information about the rock itself. You have learned about the importance of fossils in determining the environment of deposition of sediments that may form rock.

The work of paleontologists, like that of other geologists, is divided into three basic parts. First, the paleontologist works in the field, locating and collecting data related to the particular question under study. Some paleontologists spend several months each year doing field work. Then the paleontologist returns to the laboratory to study the collected specimens. They may be sliced and studied under a microscope. Analyzing the specimens requires precise, painstaking work. Lastly, the paleontologist organizes all the data and presents it in a written report. This is a very important task, because only through a well-written report can a paleontologist communicate findings to other researchers.

Many paleontologists are employed by educational institutions such as museums or colleges and universities. The petroleum industry also employs paleontologists. Other paleontologists may work for agencies such as federal, state or provincial, or local geological surveys. The educational requirements are high (see the description under *Geologist*), but the work is highly rewarding.

of them increased rapidly. Fish, amphibians, and reptiles seem to have developed in that order. Following their first appearance in the Silurian Period, land plants evolved rapidly and soon covered much of the earth.

Reptiles reached the height of their development in the warm Mesozoic climates. Dinosaurs, huge swimming reptiles, and flying reptiles also lived during that era. Mammals appeared in the Triassic Period but did not become common and varied until the Cenozoic Era. Modern types of trees, grasses, and other kinds of flowering plants developed at that time.

Despite the successful adaptation of human beings, problems of overpopulation and pollution threaten the human species. People have a great responsibility to recognize and solve these problems.

Questions and Problems

A

1. Describe the source of the energy that is found in fossil fuels such as petroleum and coal.
2. What are fossils?
3. Why are fossils not likely to be found in igneous and metamorphic rocks?
4. Why are microfossils especially useful to the petroleum geologist?
5. From what group of animals did birds and mammals probably develop?

B

1. A computer responds to outside stimulation. Why, then, is it not considered to be "alive"?
2. Distinguish between organic and inorganic compounds.
3. How do fossils support Darwin's theory of natural selection?
4. What can a paleontologist learn from the cast and mold of a prehistoric organism? Can each be considered a true "fossil"?

C

1. Discuss the various ways in which living organisms, the lithosphere, the hydrosphere, and the atmosphere may interact with one another. Describe the interfaces or boundaries at which such interaction may occur.
2. Explain how the carbon and water cycles are necessary for the support and continuation of life.
3. Describe the differences of the Galápagos finches and explain how they supported Darwin's theory of natural selection.
4. Describe the adaptations made by the amphibians as they evolved into reptiles.
5. What are some of the environmental problems caused by overpopulation that are faced by modern humans? How might they be remedied?

Suggested Readings

Lane, Gary L. *Life of the Past.* Columbus, Ohio: Charles E. Merrill Publishing Co., 1978. (Paperback)

Levin, Harold L. *Life Through Time.* Dubuque, Iowa: William C. Brown Co., 1975.

McAlester, A. Lee. *The History of Life.* 2nd ed. Englewood Cliffs, New Jersey: Prentice-Hall, 1977. (Paperback)

Ojakangas, Richard W., and David G. Darby. *The Earth: Past and Present.* New York: McGraw-Hill Book Co., 1976. (Paperback)

15 Development of a Continent

THE DEVELOPMENT OF LIFE, *discussed in the preceding chapter, is like a parade through time. It involves events and their relationships to one another in time. Some of the most difficult questions for earth scientists to answer involve the relationship of events in space. These questions ask* how *and* where *the events occurred.*

Today, camels are found in the desert regions of Africa and Asia. However, evidence indicates that the original camel ancestors lived in North America. How did they migrate to the other continents? Did they swim? Even if the Bering Strait were much warmer in the past than it is now, the shortest swimming distance is about 85 km. Did camels travel on rafts of matted vegetation carried out to sea by rivers and across oceans by currents and winds? Were there times when more of the continents were connected by "bridges" like the Isthmus of Panama that connects North and South America? Or is there another possible explanation? These questions cannot be answered correctly until some problems about the continents have been solved.

For instance, how old are the continents? Is there evidence that can lead to conclusions about their ages and the ways in which they developed? Answers to these questions would help not only to unravel the history of the continents but also to provide information about animal and plant migration over the surface of the earth.

AFTER COMPLETING THIS CHAPTER, YOU SHOULD BE ABLE TO:

1. **explain how the principles of uniformitarianism, superposition, and fossil correlation can be used to understand geologic history.**
2. **demonstrate how cross sections, geologic maps, and other geologic illustrations can be used to reach conclusions about the geology and history of an area.**
3. **briefly outline the general structure and historical development of the major geologic regions of North America.**
4. **discuss a major unsolved problem in the geologic record, such as how continents form or how geosynclines develop.**
5. **briefly describe the effects of the Pleistocene ice sheets on the landscape of North America.**

Early History of North America

15-1 North America— A Sample Continent

A view from space (refer back to Figure 6-1) shows that some parts of our planet look smooth, while others appear to be rough and irregular. The uneven patches on the face of the earth mark the continents, which make up about 40% of the earth's surface. These rocky platforms are made largely of granitic rocks. The continents have an average elevation of almost 5 km above the floors of the surrounding ocean basins.

The surfaces of the continents are quite uneven. There are smooth areas of plains and broad plateaus. There are also very irregular surfaces that consist of high mountain ranges. In elevation these surfaces range from coastal plains that touch the sea to the summit of Mount Everest, which towers almost 8848 m above sea level. But on the average, the continental surface lies only about 805 m above sea level.

Each continent has three basic parts. The **shields** are large areas where Precambrian rocks are exposed. (See Figure 15-1.) Each shield consists of igneous and metamorphic rocks. There is at least one shield on each continent. Each one has been eroded down to near sea level. There are also **covered shields** on each continent. These are parts of the continental shields that are covered by relatively thin layers of flat-lying sedimentary rocks. The third basic part of each continent is its **folded mountains.** These usually occur along the margins of the continents. These mountain belts consist largely of sedimentary rocks that have been squeezed into tight folds during mountain-building movements.

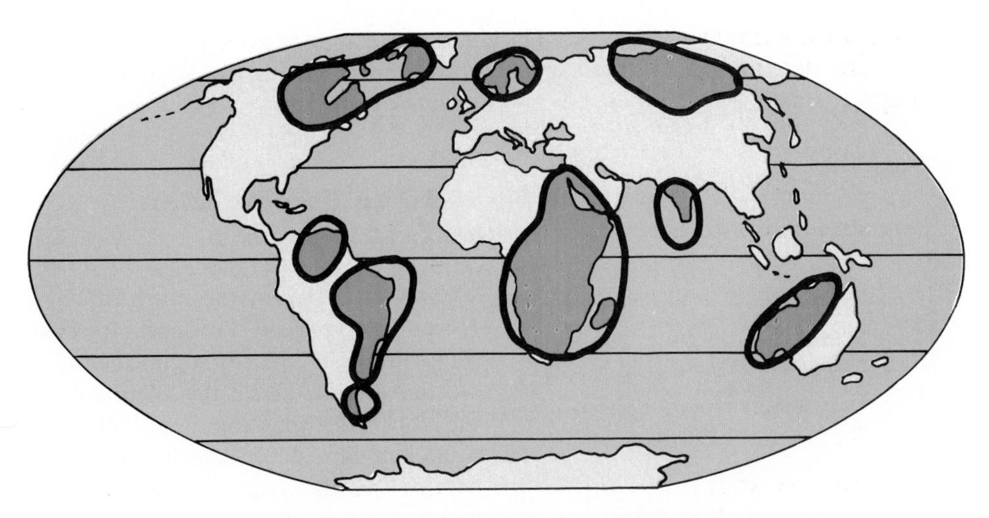

Figure 15-1 *The shield areas on the world's continents.*

Although the continents differ in detail, there are many features that they have in common. In this chapter, North America will serve as a sample in which you can examine problems typical of the geology of all continents. You can begin by looking at the entire continent and then at some of its parts. Suppose you were in a satellite over North America, viewing different parts of the land. What would be the most noticeable features of the continent? Mountain ranges, plains, and lakes, shown in Figure 15-2a, stand out as the most obvious features of the terrain. Large rivers drain the interior of the continent, carrying sediments to the seas that border it.

The distribution of the different kinds of rock in North America is shown in Figure 15-2b. Areas in which igneous, sedimentary, and metamorphic rocks are found together correspond roughly to the mountain regions. Areas with nearly horizontal layers of sedimentary rocks lie within the plateau and plains areas or in some coastal regions. Rocks that make up the plains and plateaus surrounding Hudson Bay are almost all igneous and metamorphic, like those found in mountainous areas. What might this mean?

15-2 Investigating Precambrian Rocks

Figure 15-3 shows three cross sections of Precambrian rocks found on the north shore of Lake Huron in Ontario. Information was gathered from many outcrops and pieced together. Earth scientists think that these cross sections best represent the geology of this area. This sequence of igneous and metamorphic rocks in Ontario is typical of Precambrian rocks at many places in northeastern North America.

When you have completed this investigation, you will have pieced together data from rock outcrops to outline the geologic history of the area.

PROCEDURE

List the rocks shown in each of these cross sections in the order in which you think they were formed. If you are uncertain of the order, make a note of the letters that identify the questionable rocks and explain why you cannot be sure. Your teacher will then give you a list showing the ages of these rocks. The ages were determined by radioactive dating methods.

DISCUSSION

1. Can you now put all the rock units in a relative time sequence?
2. Do the ages of these rocks agree with the order you worked out from the cross sections?

15-3 The Precambrian Record

The rocks found on the north of Lake Huron are typical of Precambrian rocks found at many places in northeastern North America. This is the shield area of the North American continent, called the Canadian Shield because it lies mostly within Canada. It is common to find within the Canadian Shield some older, folded metamorphic rocks under younger Precambrian rocks that are not as highly metamorphosed. In most places, the two kinds of rock are separated by a great unconformity.

Around the borders of the shield and at some places within the region, Cambrian and younger rocks lie on the Precambrian. Figure 15-4 shows one place where we can see evidence of an unconformity or gap in time between two rock units. The overlying horizontal rock layers are Late Cambrian in age. This makes them at least 50 million years younger than the underlying, tilted Precambrian rocks.

Unconformities and other evidence indicate that almost no ocean sediments have been deposited over most of northeastern North America since the early Paleozoic. This suggests that the region has been quite stable at least since the end of the Precambrian, about 600 million years ago. Precambrian metamorphic and igneous rocks are now exposed over thousands of square kilometers of

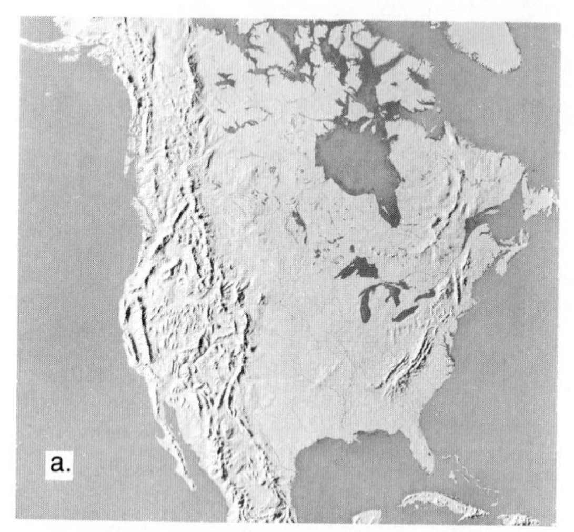

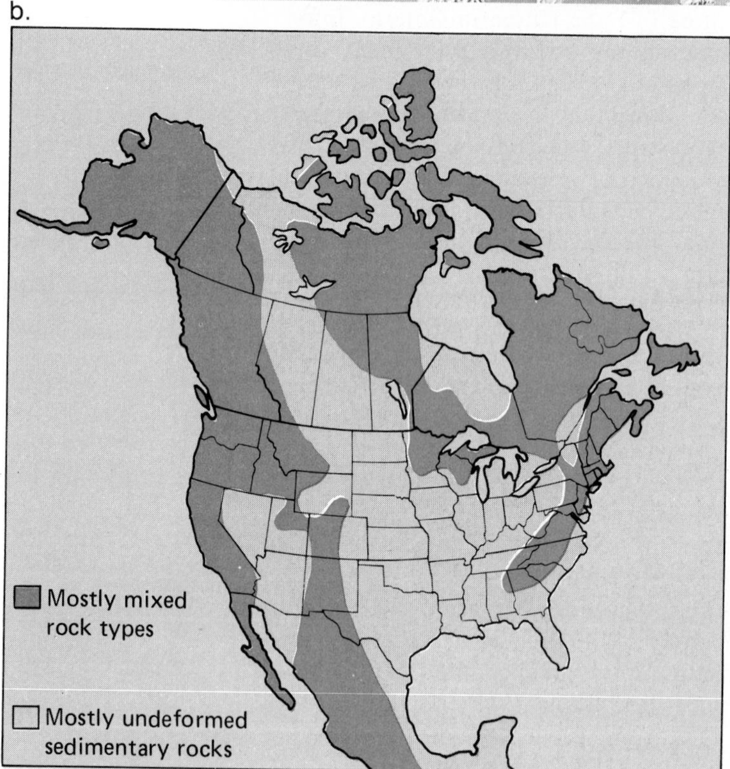

Figure 15-2 **a.** *The relief of North America.* **b.** *The distribution of major rock types. Is there a relationship between the rocks and the relief?*

Figure 15-3 *Cross sections of Precambrian rocks on the north shore of Lake Huron, Ontario.*

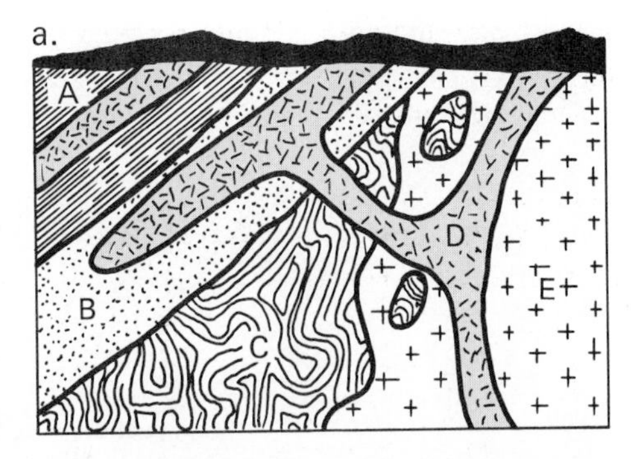

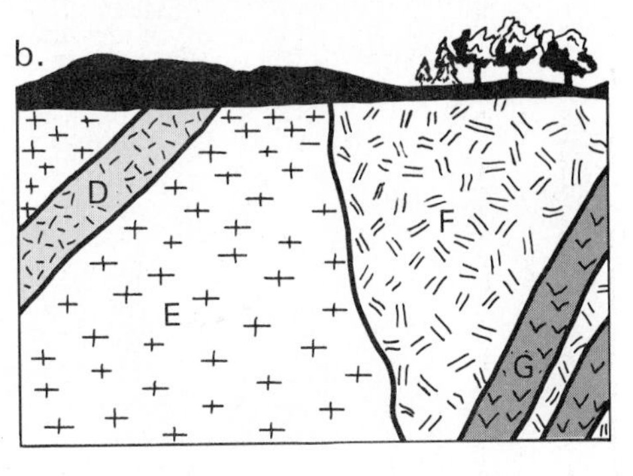

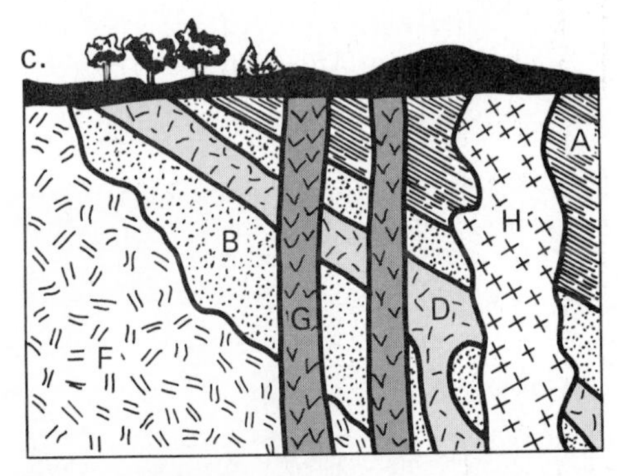

this part of our continent. Such rocks are typical of the shield areas that characterize all continents.

In some places within the Canadian Shield, interlayered lava flows and sedimentary rocks are over 5000 m thick. After they accumulated, these rocks were folded, intruded by igneous rocks, and metamorphosed. They are now exposed in low-lying areas like that shown in Figure 15-5. They appear to be the deeper parts of old mountain ranges formed so long ago that the higher parts have been eroded away. The long narrow belts where they are found are probably sites of former geosynclines. If these conclusions are correct, the Canadian Shield was an active area during Precambrian time. Mountain building occurred several times and at many places in the Canadian Shield.

The map in Figure 15-6 shows the areas in North America where Precambrian rocks are found at the surface. No fossils have been found in these rocks that can be used to correlate the ages of

Figure 15-4 *An unconformity between Precambrian and Cambrian rocks is exposed in Elgin, Ontario.*

Figure 15-5 *Folded and eroded Precambrian rocks of the Canadian Shield in the Belcher Islands, Hudson Bay.*

Figure 15-6 *Areas where Precambrian rocks are exposed at the surface in North America. In many of the places outside the Canadian Shield, they are mixed in with younger rocks.*

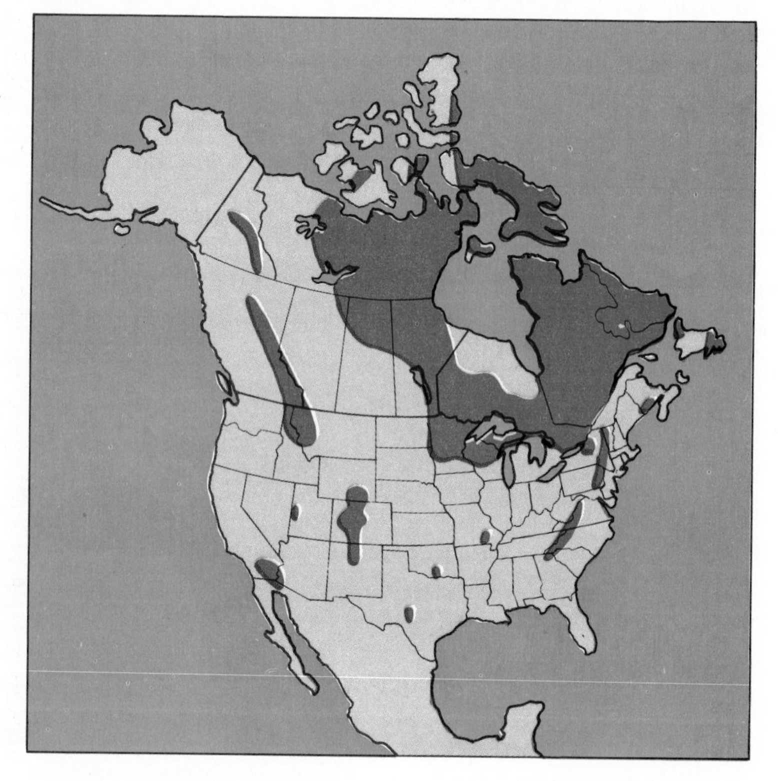

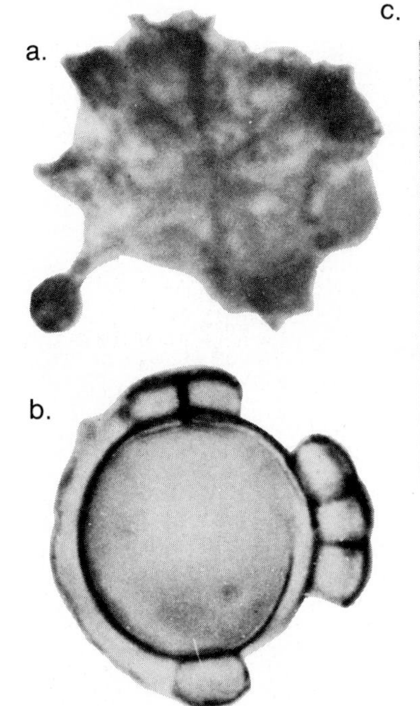

Figure 15-7 *Some fossils found in Precambrian rocks.* **a.** *and* **b.** *Fossil algae from Ontario.* **c.** *A fossil segmented worm from Australia.*

the rocks. There are enough radioactive dates so that geologists can try to correlate some of the Precambrian rocks. The time between the formation of the youngest known Precambrian rocks and the oldest Cambrian rocks is about 500 million years in some places.

Although very rare in Precambrian rocks, more fossils are being found. The oldest known undisputed fossils are bacteria found in western Australia in 1980. They are believed to be about 3.5 billion years old. Precambrian algae from South Africa have been dated at about 3.2 billion years. Carbonized objects believed to be fossils have been found in rocks 3.8 billion years old in Greenland. However, it is not certain that they are of organic origin.

Precambrian fossils have been found also in Canada (see Figure 15-7). Fossils of more complex but still primitive organisms, such as sponges and worms, have been found in Australia in rocks believed to have formed in very late Precambrian time. In the Grand Canyon, fossils of some Precambrian invertebrate animals have been found.

Thought and Discussion

1. Name three basic parts common to all continents.
2. What evidence is there that the Canadian Shield has been stable since the Early Paleozoic?
3. What evidence is there that the Canadian Shield has not always been stable?

The Later History of North America

15-4 The Paleozoic Record

The map in Figure 15-8 tells much about the development of North America. It shows areas that contain Paleozoic rocks as well as locations where there are surface exposures of Precambrian rocks.

The earliest Paleozoic sediments, deposited in a Cambrian Period sea, formed sandstones and shales. They

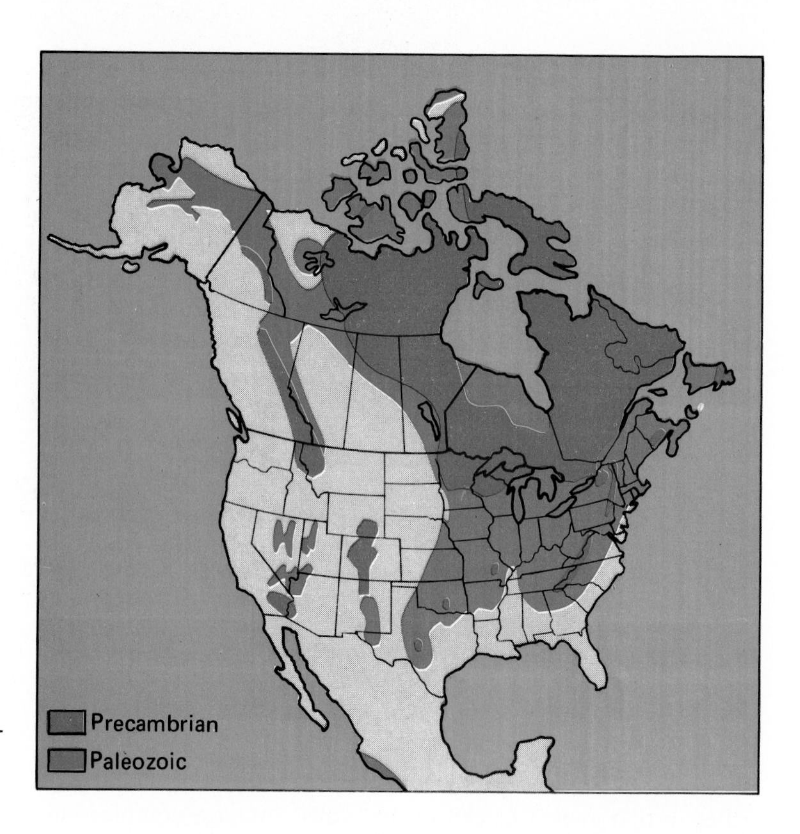

Figure 15-8 *Precambrian and Paleozoic rocks in North America.*

contain fossils of trilobites and brachiopods. A very thick sequence of limestone from dissolved seashells overlies the Cambrian layers. Coral reefs are common in these limestones, extending from the Arctic Circle to Kentucky. Salt deposits, although less abundant than other rock types, extend over thousands of square kilometers in the rock layers of some regions. (See Figure 15-9.) The youngest Paleozoic rocks are mostly sandstones and shales containing thin coal beds (Figure 15-10). This sequence of rock types is characteristic of Paleozoic rocks in the interiors of all continents between their shields and mountains.

In some areas of North America, Paleozoic rocks of the continental margins are eight times thicker than those in the interior. These geosynclinal rocks provide evidence for several episodes of mountain building, extensive volcanic eruptions, and other events. Such events indicate the unstable character of these belts during Paleozoic times. Look at the

Figure 15-9 *A salt mine in Fairport, Ohio.*

map of North America in Figure 15-11. It shows the locations of geosynclines along the borders of the continent during the Paleozoic Era. What information did scientists use to draw this map?

In reconstructing the history of the continent it is important to know not only where geosynclines were and what they were like but also when they existed. Near the boundary between the United States and Canada in the Rocky Mountain Range more than 13,000 m of Precambrian rocks are exposed. They lie beneath thick Paleozoic rocks that formed in the western geosyncline. Evidence in the Appalachian area suggests that a similar thickness of Precambrian rocks is found beneath the Cambrian layer. It seems likely that geosynclines were in existence along the continental margins during Late Precambrian time.

The presence of fossils makes it possible to correlate geosynclinal rocks with those of the continental interior. Figure 15-12 is a cross section of rocks in Nevada and Arizona drawn across the margin of the ancient geosyncline. *Olenellus,* a kind of trilobite, and other fossils associated with it are found only in the very

Figure 15-10 *Late Paleozoic coal interbedded with limestone, near Provo, Utah.*

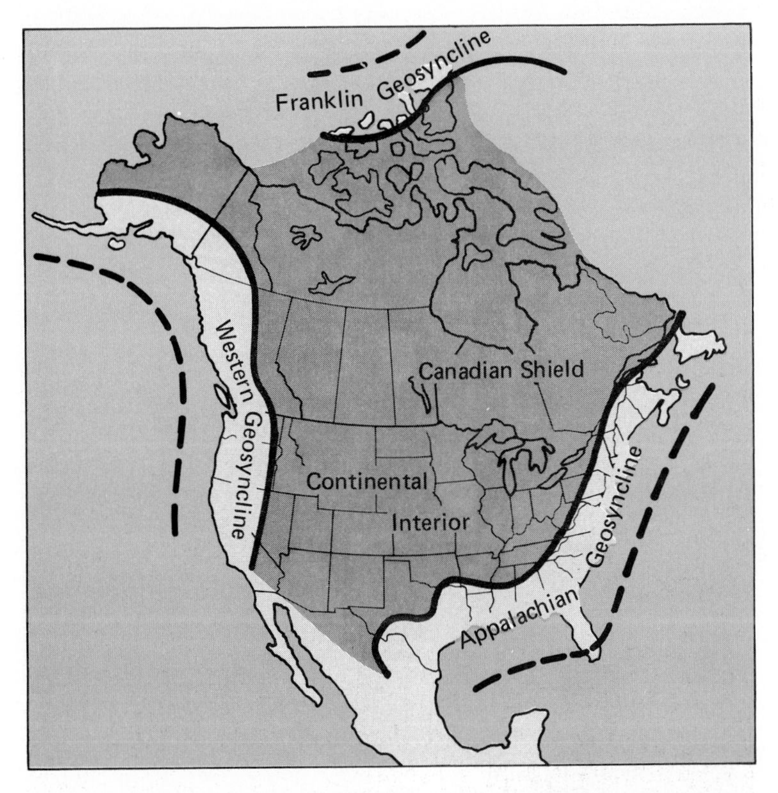

Figure 15-11 *The Paleozoic geosynclines of North America.*

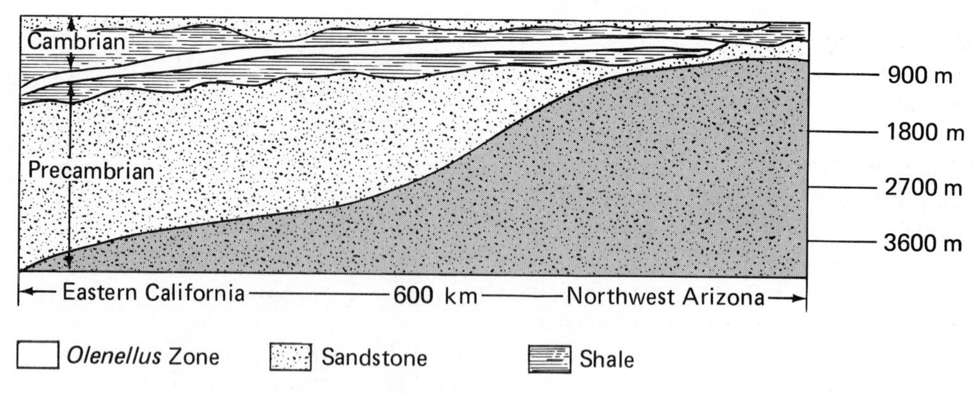

Figure 15-12 *A cross section of Precambrian and Cambrian rocks in Nevada and Arizona.*

oldest marine Cambrian rocks. If *Olenellus* and the other fossils were not present, there would probably be no way to correlate the thick sediments in the geosynclines with thinner sediments in the continental interior.

Fossils are used in the same way in other regions. For instance, some of the thick beds of the Michigan area correlate with much thinner beds in the Cincinnati arch and Ozark dome regions. (See Figure 15-13.) Reefs in one place correlate with sandstones and shales in other places. They also correlate with unconformities in other regions. We assume that all of these features—reefs, sand

Figure 15-13 *Major structural features of the North American continent: (1) Michigan Basin. (2) Cincinnati Arch. (3) Illinois Basin. (4) Ozark Dome. (5) Llano Uplift. (6) West Texas Basins. (7) Central Kansas Uplift. (8) Williston Basin. (9) Colorado Plateau. (10) Columbia Plateau. The bars show the locations of ancient and modern mountains.*

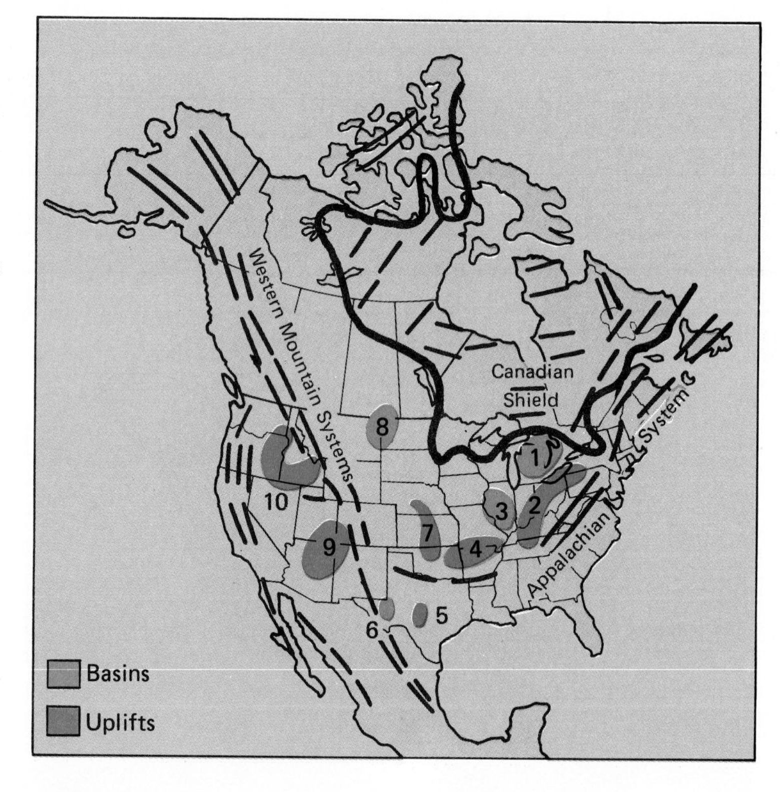

deposits, and shale deposits—were forming at the same time in different locations just as they are today.

The thick sedimentary rocks in the Canadian Arctic and the Appalachian-New England-Maritime Provinces area were folded, faulted, intruded by igneous rocks, and uplifted into mountains. This happened during the Paleozoic Era, followed by uplift in the Early Mesozoic and Cenozoic eras. There was also some Paleozoic mountain building in the western geosyncline. As you will see in the next section, this area was not added to the continent by major mountain-building processes until Late Mesozoic and Early Cenozoic time.

The rock layers in the geosynclinal zones show the mountain-building history of the Paleozoic Era. These zones mark a period of important continental growth that lasted 400 million years. By the end of that era, the North American continent had grown outward from the older and smaller shield area. When mountain building stopped, the Appalachian mountain chain had become a new part of the continent. (See Figure 15-14.)

15-5 The Mesozoic-Cenozoic Record

Figure 15-15 is a generalized geologic map that shows where Precambrian, Paleozoic, and Mesozoic rocks are exposed in North America. A close look at the Mesozoic rocks would show that they consist mostly of rocks that were deposited on land. Stream, lake, and wind deposits are spread across several of the states and provinces. Some areas, such as most of the eastern half of North America, were above sea level. Such areas were being eroded during much of the Mesozoic Era. Marine sediments were deposited along the east and Gulf

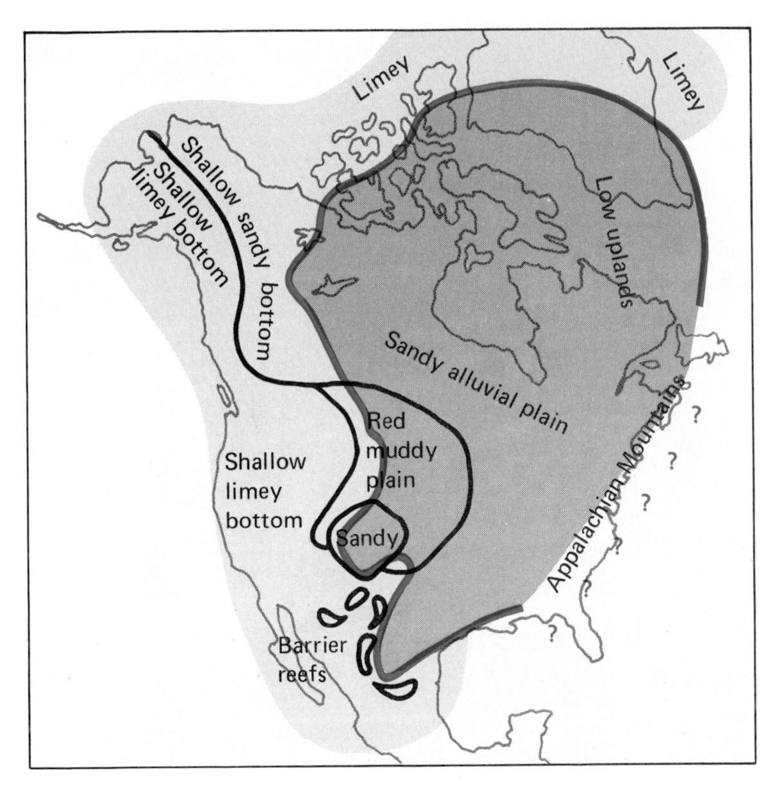

Figure 15-14 *This map indicates what North America may have been like 250 million years ago. After the uplifting of mountain ranges, the sea covered less of the land than it had before.*

Coasts. Other marine sediments were deposited in an arm of the sea reaching across what is now the Rockies.

At some times during the Mesozoic, there were desert-like conditions in the southwestern part of the continent. This resulted in many piles of windblown sand. Toward the middle of the Mesozoic, the western part of the continent sank, and waters from the open ocean covered it. This Mesozoic sea spread out over the Central United States and eventually connected with the Gulf of Mexico. Near the end of Mesozoic time North America once again rose above sea level, and the seas retreated.

During the Jurassic Period, there was mountain-building in the west. Folding and igneous intrusions formed mountains where the Sierra Nevada of California is now. The Mesozoic closed with the uplift of mountains in the Rocky Mountain region. At the same time there was continental uplift in other parts of North America. These Late Cretaceous uplifts were accompanied by great climate change and shifting of the lands and seas. These changes may have been the cause of the extinction of many Mesozoic life forms, such as the dinosaurs.

The geologic map of North America shown in Figure 15-16 indicates areas where rocks of each of the major subdivisions of geologic time are found at the surface. Deposits distributed by ice during the last Ice Age have been omitted. Cenozoic rocks are exposed in the Great Plains and in areas between the western mountain ranges. They contain fossils of horses, camels, cats, dogs, and ancestors of other familiar land animals and plants. The rocks containing these fossils are conglomerates, sandstones, and shales. They formed in lakes and streams and on flood plains of rivers that flowed from the growing mountain ranges.

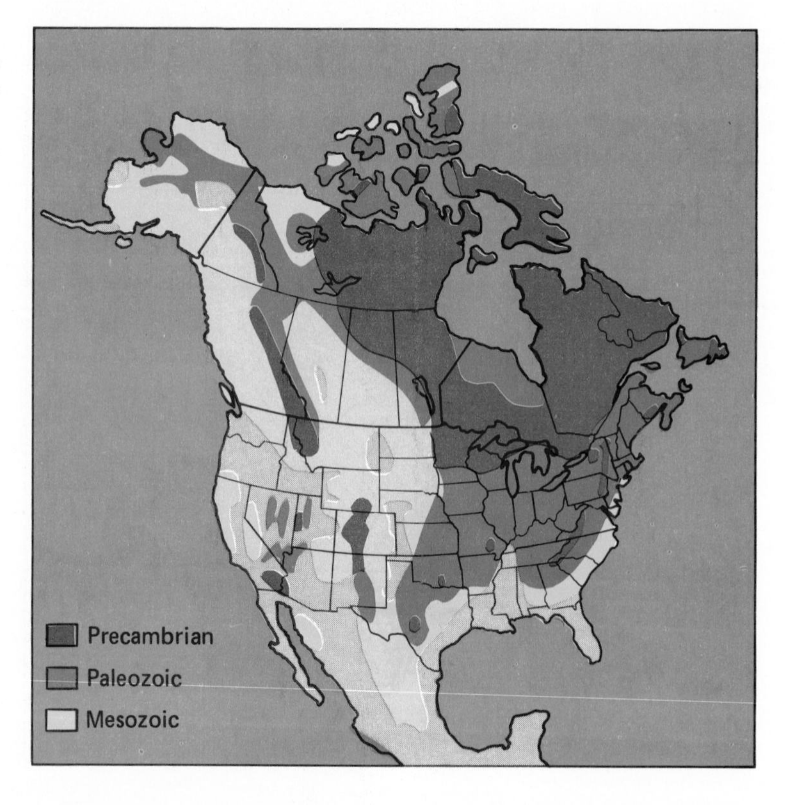

Figure 15-15 *Precambrian, Paleozoic, and Mesozoic rocks exposed in North America. Compare this map with Figures 15-6 and 15-8. What kind of a pattern do you see?*

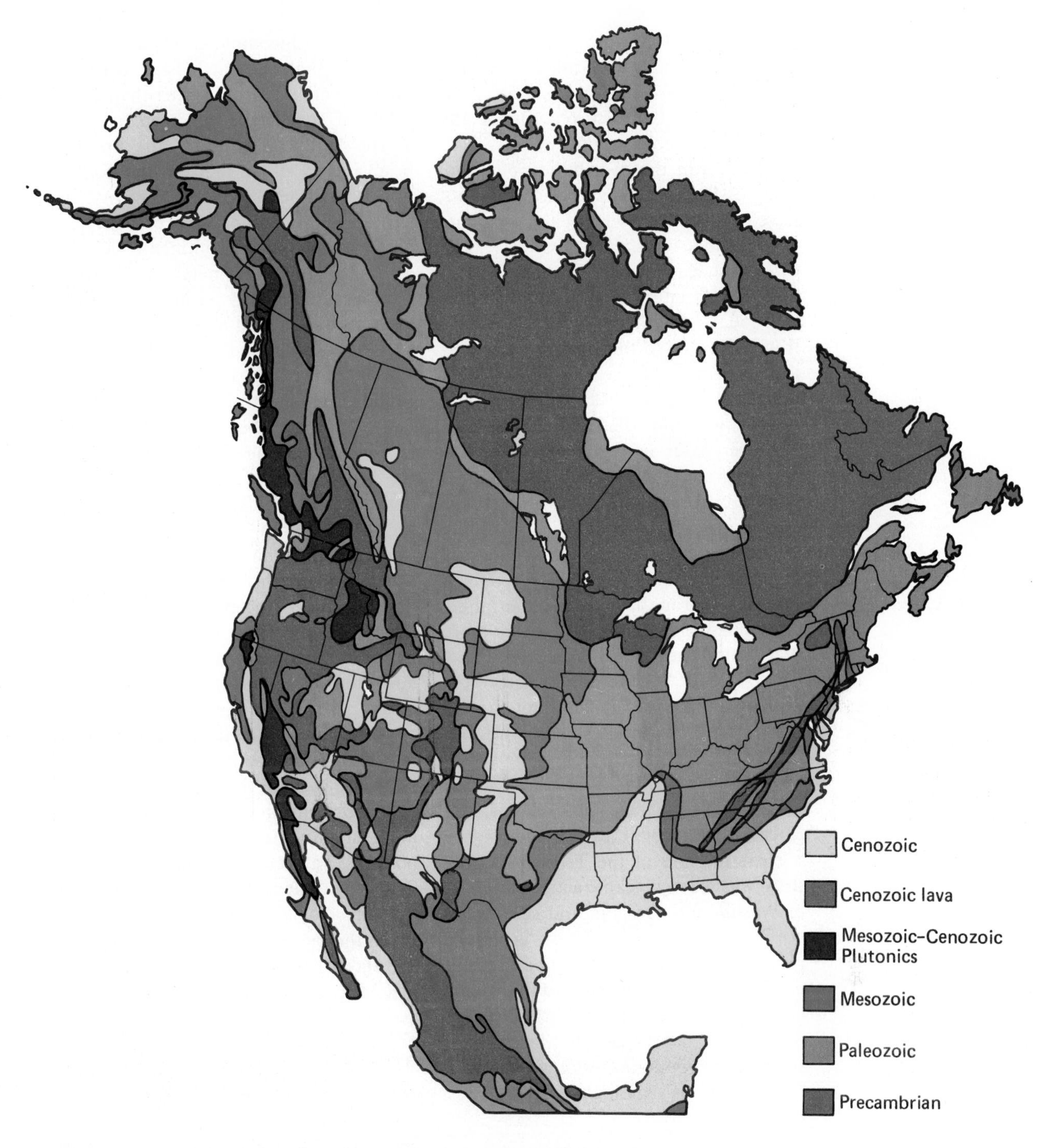

Figure 15-16 *A geologic map of North America.*

MATERIALS
tracing paper

During the Cenozoic Era, marine invertebrate animals were much like those of today. The fossil record is more complete than are the fossil records for any other part of geologic time.

Changes in the distribution of Cenozoic fossils, especially those of plants, have been studied to learn about climate changes that took place. Evidence indicates that early in the Tertiary Period, northern climates were warmer than they are today. Southern North America, from Texas to North Carolina, was tropical and most of the northern United States was subtropical. Canada and Alaska were in the temperate zone! However, later in the Tertiary the climates grew cooler due to regional mountain formation. The warmer belts moved toward the equator.

There was considerable volcanic and mountain-building activity in North America throughout Cenozoic time. Great basalt lava flows covered more than 260,000 square kilometers in Idaho, Oregon, and Washington. There were also many mountain-building uplifts that formed new ranges in the western part of the continent. These same activities further uplifted earlier-formed ranges such as the Appalachians, the Rockies, and the Sierra Nevada.

Near the end of the Tertiary Period, there was much volcanic activity in parts of Washington, Oregon, and southwestern British Columbia. These eruptions produced the Cascade Range and other mountains around the edge of the Pacific Ocean. The series of eruptions of Mount St. Helens that began in March 1980 shows that mountain-building continues in parts of this region today.

15-6 Four Billion Years of History

In investigating the Precambrian record, you have seen that the record for the early history of the North American continent is incomplete. Scientists have evidence that the earth is at least 4.5 billion years old. Rocks older than about 3 billion years have not been found in North America. However, the Precambrian record shows that erosion, formation of sediments, volcanic activity, intrusion, and metamorphism combined to build and shape the continent. The oldest rocks known are metamorphic rocks that were originally sedimentary. The location of the land area that eroded to provide the sediments is not known.

Most of the radioactive dates shown in Figure 15-17 are from igneous rocks believed to have been intruded during mountain building. Dates shown within the area of the Canadian Shield were obtained from exposed Precambrian rocks. Some dates outside the Shield were gotten from samples taken from wells drilled through younger rocks down to the Precambrian layers. The way the dates are grouped suggests that parts of North America may have been added to the continent at different times in the past. In the figure, possible stages of growth are outlined on maps of North America. The hypothesis on which these maps are based is that the North American continent developed from two or more small nuclei. According to this hypothesis, an idea called **accretion,** the first stage of development included rocks that are 2.5 billion years old or older. From this beginning the continent grew. By the early Paleozoic it may have looked as in Figure 15-17d.

ACTION

The process of accretion is easier to see from the study of Paleozoic, Mesozoic, and Cenozoic rocks. Review Figure 15-11. If you were to add maps e and f to Figure 15-17, what would they look like? Your teacher will give you some tracing paper. Trace the maps in Figure 15-17, and add two more.

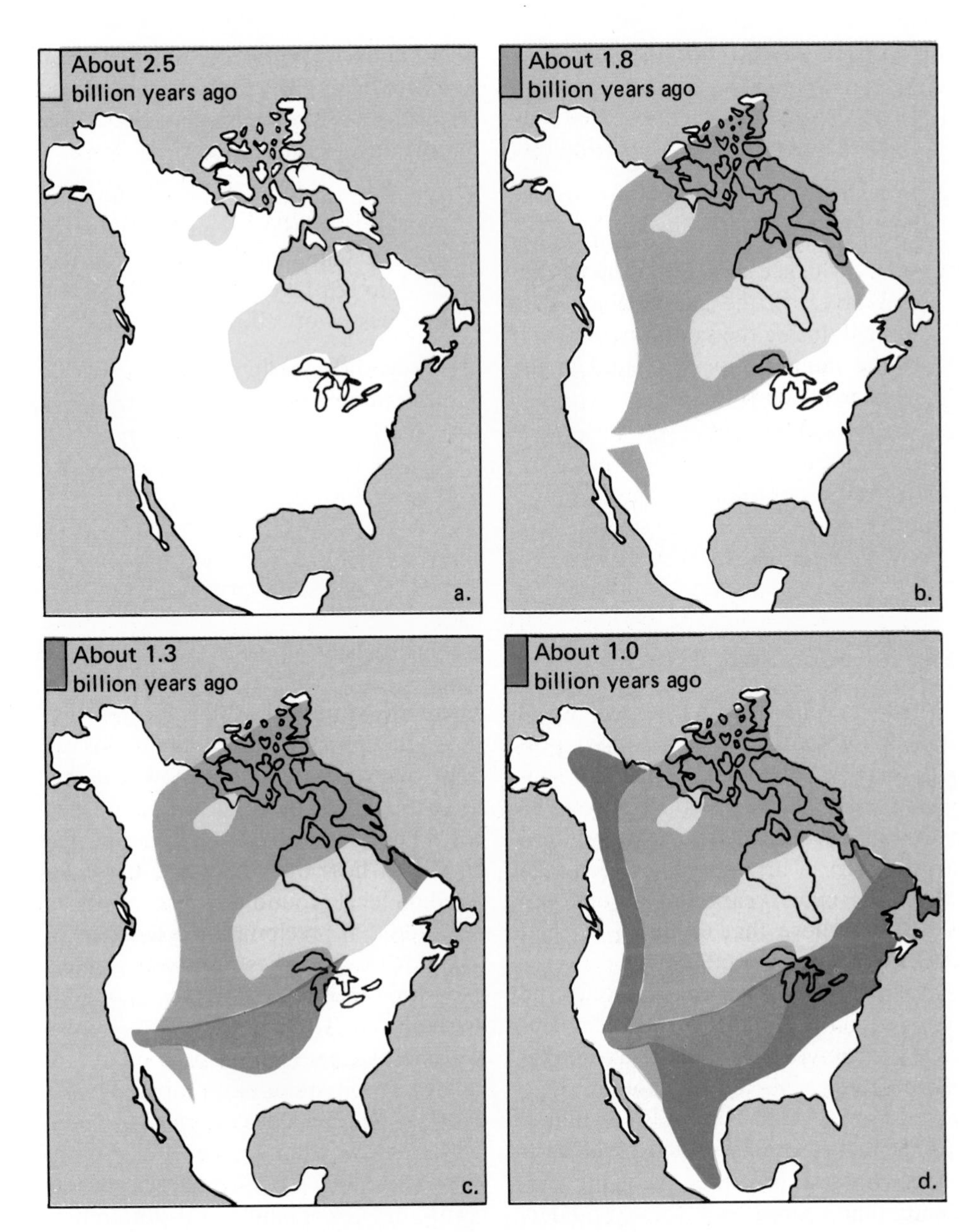

Figure 15-17 *Radioactive dates suggest that North America grew from a central area in Canada over a period of a billion and a half years.*

If growth occurred in this fashion, the presence of geosynclinal rocks in the shield suggests that growth may have taken place at the edges of the continent. Early in each stage geosynclines may have developed at the edges and received thick accumulations of sediments and volcanic rocks. Later, during mountain building, the geosynclinal rocks appear to have been lifted and folded to become part of the continent. Along the newly established margins new geosynclines developed and the mountain-building was repeated.

Thought and Discussion

1. How do geologists know that the Early Cambrian seas were shallow?
2. In what era of geologic time are we now living?
3. What evidence shows that the thicker rock layers in the geosynclines were formed during the same time interval as the thinner deposits found in the interior of the continents?
4. What destroyed the geosynclines that existed along the east and west coasts of North America during the Paleozoic and Mesozoic Eras?
5. What might happen in the future to the Gulf coastal region where the present land surface is nearly flat, close to sea level, and covering great thicknesses of sedimentary rocks?
6. When did North America reach its present size and shape?

The Great Ice Age

15-7 Theory and Proof

In 1836 a young Swiss scientist, Louis Agassiz (*AG uh see*), began studying alpine glaciers. His studies strongly suggested that glaciers had once covered much of the Northern and Southern hemispheres. This idea had been discussed for many years, but no one was ready to believe that so much ice had once been on the continents.

Agassiz pointed out that many of the boulders on the plains of northern Europe are composed of igneous and metamorphic rocks. Yet oddly enough, these boulders are found hundreds of kilometers from the nearest outcrops of such rocks. Agassiz presented geologic evidence that proved that these boulders could have come only from Scandinavia. This led him to propose that the glaciers that had once covered much of northern Europe had carried these boulders with them. Agassiz also suggested that the valley glaciers had been much larger about a million years ago. He said that these ice streams were responsible for carving the sharp peaks and U-shaped valleys seen in the Alps today.

Where did these huge "rivers" of ice come from? Glaciers are huge, slowly moving masses of snow and ice. (See Figures 7-23 and 7-26.) They form where the winter snowfall is greater than summer melting. Snow then builds up from year to year. As the snowfields become thicker, the lower layers are packed down and turn into ice. Glaciers located on slopes flow downhill, and those located on level ground flow outward from the middle. If precipitation is lessened or the climate becomes warmer, glaciers may stop flowing. In time they may even gradually melt away.

Some glaciers form in high mountain valleys or on mountain slopes. Others develop on the sides of ridges, where drifting snow is left by the wind. As you know (Section 7-9), much larger glaciers occur in Greenland and Antarctica. These are usually called ice sheets or ice caps. They are survivors of the Pleistocene Ice Age that Louis Agassiz found evidence of. Most of the ice melted away about 10,000 years ago.

The balance between snowfall and melting determines whether a glacier grows or shrinks. Glaciers on the west coast of Greenland increased in size from 1850, when they were first surveyed, until 1860. Since that time they have been shrinking. Although the average temperature of the earth has fallen since the late 1940's, not all glaciers have been altered by this change. In some regions of the earth they have advanced, but in other places they continue to shrink.

The rate at which glaciers melt depends partly on the type of snow surface. The heat that isn't reflected is absorbed. Fresh snow may reflect more than 80 percent of the sun's radiation. In time, however, the form of the crystals changes, and old snow may reflect less than 50 percent of the sun's energy. If the winter snowfall is *greater* than normal, melting during the following summer can be *less* than normal. The fresh snow reflects more energy, and less is available for melting. (How might the melting of glaciers be artificially controlled?)

Agassiz's theories came from studying glacial deposits in Europe. However, glaciers were also present in North America. There is evidence that thick ice sheets covered more than half of North America at four different times. Grooves and scratches on the bedrock indicate that these continental glaciers originated in Canada. The types of rock in the glacial sediments and the location of glacial deposits are further evidence for a Canadian origin. The glaciers flowed outward from centers near Hudson Bay and in the Canadian Rocky Mountains. (See Figure 15-18.)

Beginning about three million years ago, the glaciers grew and melted at least four times. Each of these ice sheets covered the land for thousands of years. Between each glacial advance, there were much longer periods when the land surface was free of ice. During these periods, the climates were warmer than they are now. Radiocarbon dates of plant fossils found in glacial deposits show that the last ice sheet melted only about 10,000 years ago.

During parts of the Pleistocene Epoch more than 30 percent of the earth's land surface was covered by ice. Today, only about 10 percent of the land is ice-covered. But the Pleistocene was not the

Figure 15-18 *The greatest extent of the ice cap in North America during the Pleistocene Ice Age.*

first icy chapter in earth history. The earliest known glacial period occurred during Precambrian time. There were other ice ages in Ordovician-Silurian time and during the Pennsylvanian and Permian Periods. Today, earth scientists are not sure whether we are living in an ice age. Until more is learned about the causes of glaciation, scientists cannot predict what will happen to the modern glaciers. Will they grow or melt away?

What causes glaciation and ice ages? It is not known for certain, but more than 60 hypotheses have been proposed to explain them. Some earth scientists believe that changes in the land and water masses near to the poles set the stage for the ice ages. Mountains and land-locked seas close to the earth's poles would favor snow and ice. In mountains and highlands, the snows of winter tend to stay through the summers. If polar seas are landlocked, or nearly so, ocean currents cannot move heat into the polar regions.

When ice sheets are formed, they tend to persist and grow. On the average, the earth and its atmosphere reflect about 30 percent of the solar radiation. However, ice reflects about 70 percent of the insolation reaching it. When large areas of snow and ice last through the summers, much more solar energy is lost to space. The summer circulation pattern becomes more like that of winter, and more snow falls to feed the glaciers.

The cold glacial periods may have appeared after a series of mild winters and cool summers. In mild winters, the polar front is located far to the north of its average position in the Northern Hemisphere. Then more snow falls in winter at these latitudes, instead of farther south. In cool summers, there is less melting of the winter snowfall. Thus, after many mild winters and cool summers, the polar ice sheets could spread toward the equator. To explain the reversal of an ice age, scientists usually assume a reversal of the conditions that produced it. When warmer summers returned, the ice would melt.

Astronomers have pointed out that relatively mild winters and cool summers at high latitudes can result from slow changes in the earth's motions around the sun. These changes, every 360,000 years, provide more radiation at high latitudes during winter, and less in summer. Attempts have been made to relate this cycle of changes in insolation to known glacial and interglacial periods. The onset of a glacial period is gradual (about 90,000 years), but the warming is sudden (about 10,000 years).

15-8 Evidence of the Great Ice Age

In Chapter 7 you learned that glaciers can erode, carry, and deposit great amounts of rock debris. The process of mountain glaciation has created some of the world's most beautiful scenery. The huge continental glaciers of the Pleistocene time also had a long-lasting effect on many parts of North America. The Finger Lakes of New York and the Great Lakes were produced as the result of drainage changes caused by continental glaciers.

The rock debris deposited by glaciers is called **drift.** People originally thought the loose glacial sediment was deposited by melting icebergs as they drifted through water. But as a result of Agassiz's work, drift was finally accepted as evidence of glaciation. Although its glacial origin is now universally accepted, the name drift is still used.

There are two main types of drift: till and outwash. **Till** is an unsorted, unlayered jumble of clay, sand, pebbles, and boulders (Figure 7-27). This is the material that settles out of the glacier as the ice melts. Many of the larger till fragments have sharp angles, and the surfaces of others are scratched and grooved. Large accumulations of till are found around each basin of the Great Lakes. These till accumulations mark the limits of the various glacial advances.

When glaciers melt, streams of meltwater pick up some of the rock debris that was locked in the ice. This mixture of clay, sand, and gravel is carried along by the glacial stream and deposited as **outwash.** The running water sorts the sediments and deposits them in layers. The largest rock fragments are deposited first. The finest material, such as clay, remains suspended in the water and is carried farther away.

Because the glacier may pick up more rock fragments one place than another, some parts of the ice will carry a larger load. When this material is deposited, some areas receive more sediment than others. This is one reason that the surfaces of glacial deposits are typically irregular and bumpy.

During parts of the Great Ice Age, the ice over Hudson Bay was more than 3300 m thick. The great weight of the ice sheets caused the earth's crust to sink. Where the ice was thickest, the surface was lowered more than 500 m. Hudson Bay is one of the depressions that was created by the ice and later filled in by the sea.

When the ice melted and the weight decreased, the land surface began to rise to its original position. Some of the land still has not returned to its original level. The area near Niagara Falls is still rising at the rate of about 25 cm per century. At this rate, the slow upward movement will continue for a very long time. Can you name some areas on other continents that might have been depressed by ice sheets?

When the huge ice sheets accumulated on the continents, large amounts of water were temporarily stored as ice. As water was locked into the glaciers, sea level was gradually lowered on at least four different occasions. What would happen if all glaciers of the modern world should melt? Old beach deposits and wave-cut cliffs provide evidence for these changes in sea level. (Can you explain why some old beaches and other coastal features are still above sea level?)

Great continental ice sheets still cover Greenland and Antarctica (Figure 15-19). There are also smaller ice caps, like the ones on Iceland. And thousands

Figure 15-19 *The Antarctic ice sheet.*

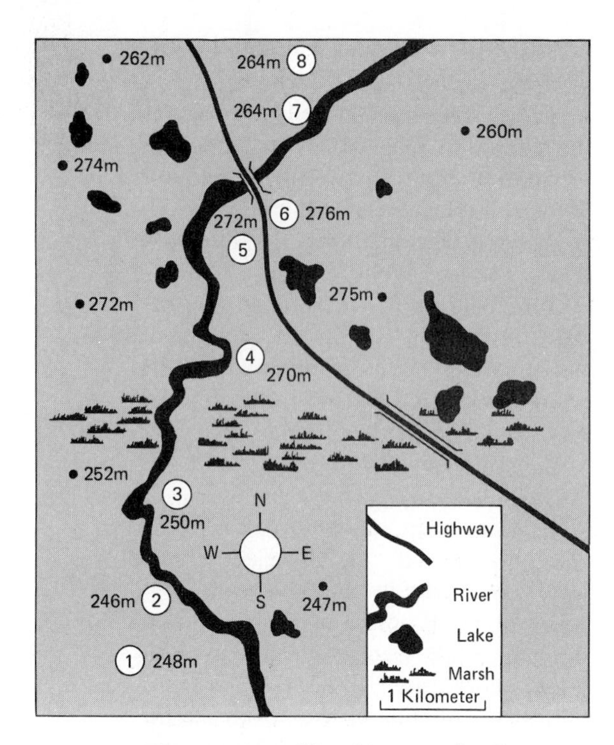

Figure 15-20 *Use this map for Investigation 15-9.*

of valley glaciers can still be found on mountain slopes in many parts of the world. These valley glaciers are continually grinding away the land and reshaping valleys and mountain ranges into distinctive landforms. During the Ice Age, when ice sheets were widespread, valley glaciers were also much more numerous. The typical landforms that are created by valley glaciers can be seen in the Alps and the Rocky Mountains. Such landforms can also be seen in the Sierra Nevada of California and the White Mountains of New Hampshire, as well as on other high mountains on all continents.

MATERIALS
graph paper

15-9 Investigating an Ice Age Puzzle

When you have completed this investigation, you will have examined an area that was once covered by an ice sheet.

Procedure

A map of a glaciated area appears in Figure 15-20. There are numbered stations on the map. Geologic data for each station are given in the table in Appendix E. The data were obtained from rock samples taken from road cuts and stream banks or from test holes drilled by geologists.

Plot the elevation of each station on graph paper, using a scale of 1 centimeter = 10 meters. Connect the elevation points to show the shape of the land surface along a given line. This will produce a topographic profile.

Plot the information from the data provided for each station. Next, make a cross section using either different colors or different geologic symbols for the rock and glacial deposits. Be sure to include a key that will identify each color or symbol in your cross section.

Note that there are two kinds of glacial till that can be distinguished by their color.

Discussion

1. Which of the till deposits appears to be older?
2. What evidence shows that a long time elapsed between deposition of the two layers?
3. What could cause this lapse of time between periods of deposition?
4. You will notice from the data that the bedrock has deep grooves at stations 5, 6, 7, and 8. Refer to your cross section. Which till deposit connects with the south-trending grooves?
5. Explain why there are no south-trending grooves at stations 7 and 8.
6. Describe the sequence of glacial activity in this area.

Thought and Discussion

1. How do scientists know that outwash is deposited by water rather than by ice?
2. What evidence indicates that there was more than one glacial advance during the Pleistocene Epoch?
3. How can earth scientists locate the centers from which the ice sheets advanced?
4. Are we living during an ice age, in a period between glacial advances, or at the end of an ice age? What is your evidence?

Unsolved Problems

Because fossils are rare in Precambrian rocks, it is hard to correlate such rocks from different parts of North America. Therefore, there is no detailed history of the Precambrian continent. Many problems relating to the continental origin and its early development are unsolved.

Mountain building has occurred frequently during most of geologic time. However, there is a 500 million year gap between the last known Precambrian mountain building movement and the earliest known Paleozoic mountains in North America. This interval is almost equal to the combined length of the Paleozoic, Mesozoic, and Cenozoic Eras. The lack of fossils and the missing records of sedimentation, mountain building and metamorphism for this period of time are puzzling.

Scientists are trying to discover what changes in climate cause continental glaciers to originate, develop, and grow over large areas. They also need to know what conditions cause glaciers to shrink and finally disappear. For glaciers to form, average annual temperature must be so low that more snow accumulates during winter than melts during summer. One current hypothesis suggests that glaciers in the Northern Hemisphere expand when the Arctic Ocean melts and retreat when the ocean surface is frozen.

Chapter Summary

The North American continent has existed for more than three billion years. It is not known whether it was a feature of the earth's crust when our planet was forming. The ages of igneous rocks from different parts of North America suggest that the continent grew in stages. New crustal material was added by mountain building at the margins of the continent.

Perhaps during Precambrian time, and certainly later, seas alternately advanced and retreated over the continent interior. In these seas animals and plants lived, and sediments were deposited.

The geosynclines on the continental margins were mobile regions for long periods of time. The geosyncline located in eastern North America was subject to folding, faulting, and intrusions by plutonic rocks. It was uplifted during the Paleozoic and early Mesozoic eras. The geosynclines of western North America and the Canadian Arctic Islands were also folded and faulted during the Paleozoic Era. They were uplifted during the Mesozoic and Early Cenozoic eras.

North America assumed its present outline in Late Cenozoic time. Earthquakes, volcanic activity, and movement along active faults—especially along the Pacific Coast—show that our continent is still changing. However, not enough is known about earth processes to precisely predict future developments.

MUSEUM TECHNICIAN

Preparing fossil specimens for exhibition in a museum is usually done by museum technicians. They use various tools to remove rock material from the fossils. Then they brush a preservative, such as resin or shellac, onto the fossils. (The preservative forms an airtight coating to protect the fossil from decaying or disintegrating.) If part of a fossil skeleton is missing, a museum technician may use molding and casting methods to make a substitute piece. Then the technician mounts the skeleton for display. In addition to these tasks, a museum technician may help in maintaining museum files, cataloging and storing specimens, and installing and arranging exhibits.

There is no standard education requirement for work as museum technician. Some education after high school is probably the best background. Several years of on-the-job training is the usual way of gaining the required competence. A college degree is helpful for moving to higher positions.

A museum technician should have a great deal of patience and the ability to do detailed work. The ability to visualize a finished product from a variety of small parts or fragments is also particularly helpful.

The Pleistocene ice sheets were the most recent major events in the development of North America. The glacial advances affected both the development of humans and their migration across the face of the earth. Although most of North America is now free of ice, extensive continental ice sheets still remain on Greenland and Antarctica. It is not known whether glaciers will advance again in the near geologic future. The ice may retreat and not return again until another great Ice Age, or it may advance further before it retreats.

Questions and Problems

A

1. What types of rocks are most likely to be associated with plains and plateaus? with mountain ranges?
2. Why is the trilobite *Olenellus* useful for rock correlation?
3. What geologic principles are used to arrange layers of rock in sequences?
4. In what ways do deposits of till and outwash differ?
5. What features of glaciation are used to indicate the direction in which the ice moved?
6. The Great Lakes and Hudson Bay occupy large depressions. How were these depressions formed?

B

1. Describe the arrangement of mountains and lowlands in North America. Why are they arranged in this way?
2. Igneous and metamorphic rocks commonly are exposed in mountain ranges. How do you explain their presence at the surface of the Canadian Shield?
3. Why is the rock record of the geologic history of a continent less complete for earlier than for later portions of geologic time?
4. Why are some fossils useful for correlating sedimentary rocks but not for explaining the environments in which the rocks formed? Why are other kinds of fossils useful for explaining the environments in which rocks formed, but not for correlation?
5. There have been several explanations for the migration of animals and plants from continent to continent. List some of them and explain what kinds of evidence are needed to support each.
6. What evidence indicates the presence of extensive glaciation in the Northern Hemisphere during the Pleistocene Epoch?
7. Parts of North America are still reacting to the presence of the tremendous weight of Pleistocene ice. What are these reactions and where are they noticeable?

C

1. If the oldest known rocks are sedimentary, does this mean that the first rock-forming process that affected the earth was sedimentation?
2. What does the presence of fossil coral reefs in Arctic regions probably indicate about the ancient climate of those regions? What other possible explanation could there be?
3. How will studies of the ocean floor help to explain the origin of continents?
4. There are indications that climates in the Northern Hemisphere are gradually becoming warmer. What effect could this have on the remaining glaciers and sea level? How might this be related to the Pleistocene Ice Age?

Suggested Readings

Eicher, Don L., and A. Lee McAlester. *History of the Earth.* Englewood Cliffs, New Jersey: Prentice-Hall, 1980.

Stearn, Colin W., and others. *Geological Evolution of North America.* 3rd ed. Somerset, New Jersey: John Wiley & Sons, 1979.

Stokes, W. Lee. *Essentials of Earth History: An Introduction to Historical Geology.* 4th ed. Englewood Cliffs, New Jersey: Prentice-Hall, 1982. Chapters 8–15.

unit V
Plate Tectonics

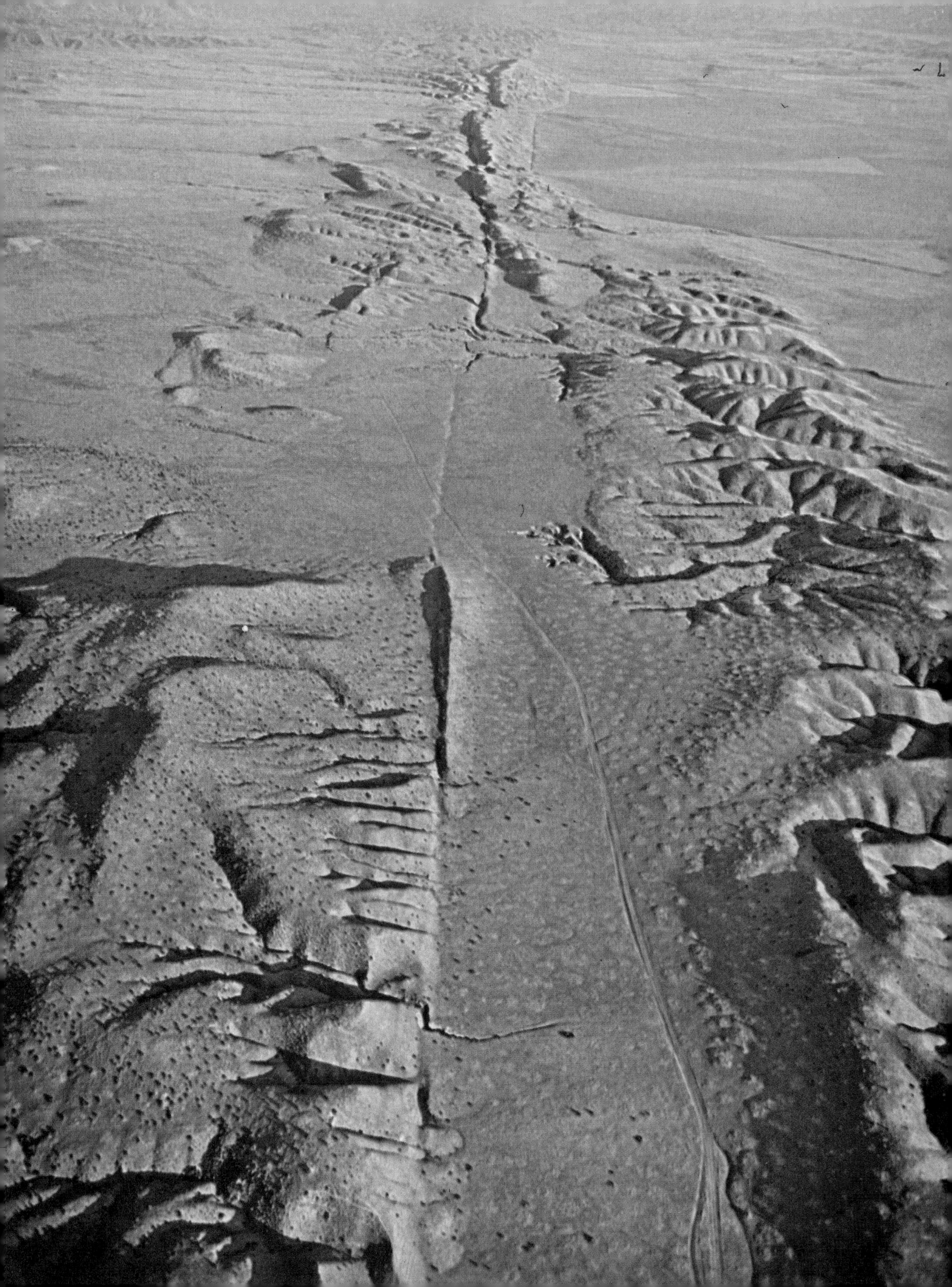

16

Earthquakes and the Earth's Interior

MOST OF GUATEMALA CITY'S *1.5 million people were sleeping soundly at 3* A.M. *on Friday, February 13, 1976. But their dreams quickly turned into a real-life nightmare when they were jolted awake and rolled from their beds as the earth rumbled and shook beneath their houses. As they opened their eyes they were greeted by the crash of collapsing buildings, violent shaking of the ground, and the clanging of church bells as steeples swayed with the trembling earth.*

The main shock lasted little more than half a minute. Yet during those 39 seconds, this huge earthquake and its aftershocks caused great damage over an area of more than 1040 square kilometers. Some 40 towns and villages were destroyed, 76,000 people were injured, 220,000 homes were destroyed, and approximately 23,000 people were dead.

For many years Earth was believed to be a "dead" and "frozen" planet. Spectacular geologic events such as earthquakes and volcanic eruptions were thought to be caused by the small amount of heat left over from our once molten planet.

However, it is now known that earth processes do not happen according to chance or accident. Geologists have learned that most earthquakes and volcanic eruptions occur in definite zones on the earth's surface. For example, California's famous San Andreas Fault Zone is shown on the opposite page. It is one of the best-known examples of the crust's response to a force within the earth. Evidence such as earthquakes and volcanic eruptions shows that the interior of the earth is changing. It suggests that there is a dynamic "machine" inside our planet. It is this "engine" that triggers volcanic explosions and earth tremors. This "machine" is also the driving force behind the rock cycle. In this chapter you will learn more about the natural forces behind earthquakes and what earthquake records can tell about the interior of our constantly changing planet.

AFTER COMPLETING THIS CHAPTER, YOU SHOULD BE ABLE TO:

1. **discuss the causes and occurrences of earthquakes and their effects on humans.**
2. **determine earthquake epicenter locations from the differences between P and S wave arrival times.**
3. **distinguish between the magnitude and intensity of earthquakes and explain the Richter and Modified Mercalli scales.**
4. **explain how information from earthquake records is used to construct a model of the earth's internal structure.**
5. **give several reasons why scientists study the interior of the earth.**
6. **discuss the source of most of the earth's internal heat.**

When the Earth Shakes

16-1 What Causes Earthquakes?

People have tried to explain the cause of earthquakes since earliest times. Some people thought that quakes were sent from heaven to punish the wicked. Others believed that the earth rested on the backs of animals (Figure 16-1). When these creatures moved the earth would tremble, causing earthquakes.

We now have more scientific explanations of the causes of these destructive earth tremors. Earthquakes are evidence that some process is at work below the surface of the earth. Large bodies of magma may be in motion. These and other internal movements can result in a sudden release of energy by breaking the overlying, brittle rocks.

During earthquakes, rocks beneath the surface are clearly bent and broken. In other cases deformation takes place slowly and without recognizable earthquake shocks. For example, in a drill hole in the Great Valley of California, earth movements have been measured for years. The rocks here are moving at a rate of about one meter per century.

Another example of slow earth movements is at a winery in Hollister, California. By chance it was built on a fault associated with the San Andreas Fault Zone. Over the years there have been slow, steady movements without accompanying earthquakes. The winery building, originally a rectangle, has been pulled into a diamond shape (Figure 16-2). The land at this particular location moves about one centimeter a year.

In some places pressure slowly builds up until the rocks break and quickly snap back into position. The faulting releases energy that may have built up in the rocks for hundreds or thousands of years. This sudden burst of energy causes vibrations that travel through the rocks. These vibrations are an earthquake.

Most earthquakes are so small that they can be detected only by sensitive instruments. But when violent earthquakes occur in inhabited areas, they can cause great destruction and human misery.

In large earthquakes most of the stored energy is released in the first movement along the fault surface. However, energy may continue to trickle off in **aftershocks.** These shocks are much less severe than the main shock and may continue for months after the initial earthquake.

Figure 16-1 *The ancient Hindus believed that the earth rested on the head of an elephant that was seated on the back of a tortoise. The movements of the animals caused the earth to tremble.*

16-2 How "Big" Is an Earthquake?

One way that energy travels through matter is in waves. Imagine that you have just dropped a rock in a pond. Watch the waves as they move out from the place where the rock splashed down. The first wave is the highest. As it travels, the height of the wave decreases. The waves that follow the first one tend to be lower. Where did the energy that produced the waves come from? Suppose you drop a smaller rock into the water from the same height. Will the waves be the same size as those that were made by the big rock? Late in the 19th century scientists discovered that some of the energy released by earthquakes takes the form of **seismic** (*SYZ mihk*) **waves.** Seismic waves radiate from the **focus,** or point of origin, of the earthquake. These waves move in all directions through the solid rock, in somewhat the same way that waves radiate from a rock dropped in a pond. Seismic waves are not very

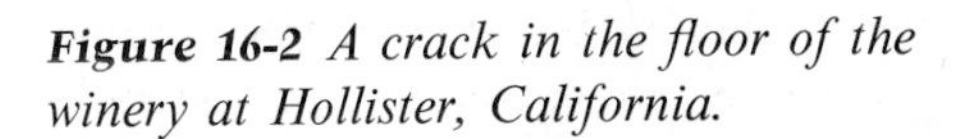

Figure 16-2 *A crack in the floor of the winery at Hollister, California.*

high. Waves that have traveled many kilometers through rock range from only a few thousandths of a millimeter to a few millimeters high. You can see why sensitive instruments are necessary to record them.

There are different types of seismic waves. One type is a **compressional** or primary (**P**) wave. When a compressional wave passes through a substance, individual particles move back and forth as shown in Figure 16-3a. The action is like the expansion and contraction of a spring, except that P waves travel at thousands of kilometers per hour. P waves can travel through solids, liquids, and gases.

Another type of wave is a **shear wave** or secondary (**S**) wave. (See Figure 16-3b.) The slower-moving shear waves cause the individual rock particles to vibrate from side to side at right angles to the direction that the wave is traveling in. Because of this type of motion, S

Figure 16-3 The motions of three kinds of seismic waves.

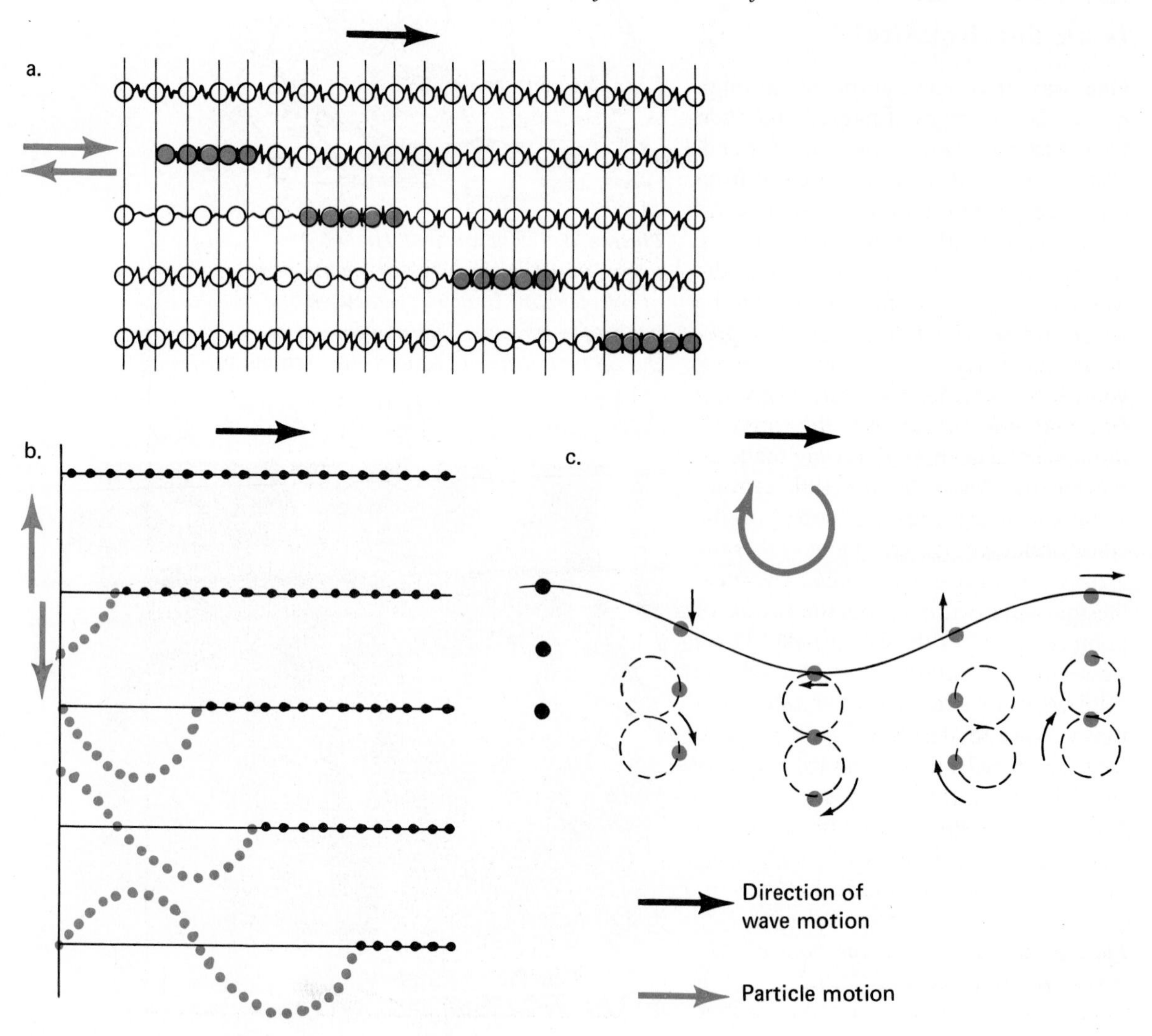

waves cannot pass through liquids. Since molecules of liquids slide past each other so easily, they do not transmit the energy. As you will see later, information derived from P and S waves has revealed much about the composition of the earth's interior.

P and S waves travel *through* the earth. There are two other types of seismic waves that travel at the earth's surface. **Love waves** have no vertical movement and produce a shearing motion in the ground. **Rayleigh waves** pass through surface rocks and produce both horizontal and vertical motions (Figure 16-3c). Both of these relatively slow-moving surface waves greatly disturb the surface rocks and soil. They are responsible for most earthquake destruction.

The size of an earthquake is directly related to the amount of energy released at the focus. Charles F. Richter, an American **seismologist** (*syz MAHL uh jihst*)—a specialist in the study of earthquakes—developed a method for measuring earthquakes. He assigned a number to each earthquake according to its **magnitude,** the amount of energy released at the quake's focus. The magnitude is calculated from seismograph records. A seismograph is a very delicate instrument that detects passing earthquake waves. It transfers the earthquake vibrations to a pen that marks on a moving piece of paper. The **Richter Scale** therefore describes an earthquake independently of its effects on people and civilization.

The Richter Scale uses numbers from 1 up to describe earthquake magnitude. Each number indicates an earthquake ten times stronger than the next lower number. An earthquake with a magnitude of 5 is ten times stronger than an earthquake with magnitude 4. An earthquake with magnitude 7 or higher is considered a major quake.

The Richter Scale is an open-ended scale. That is, there is no fixed upper limit of magnitude. Nobody knows how strong an earthquake might be in the future. The strongest earthquake on record had a magnitude of 9.5 on the revised Richter Scale.

16-3 Measuring Earthquake Damage

At 5:30 in the afternoon of Good Friday, March 27, 1964, many people in Anchorage, Alaska, were busy doing their last minute shopping for Easter Sunday. The scene was peaceful and normal. Suddenly at 5:36 the earth began to shake. One woman waiting in her car for a traffic light to change found herself and the car bouncing sideways across the street. Buildings were rocked so violently that the people inside fell down. Some

Figure 16-4 *This map shows the epicenter and area affected by the 1964 Anchorage, Alaska earthquake. Note the size of the area of moving crust.*

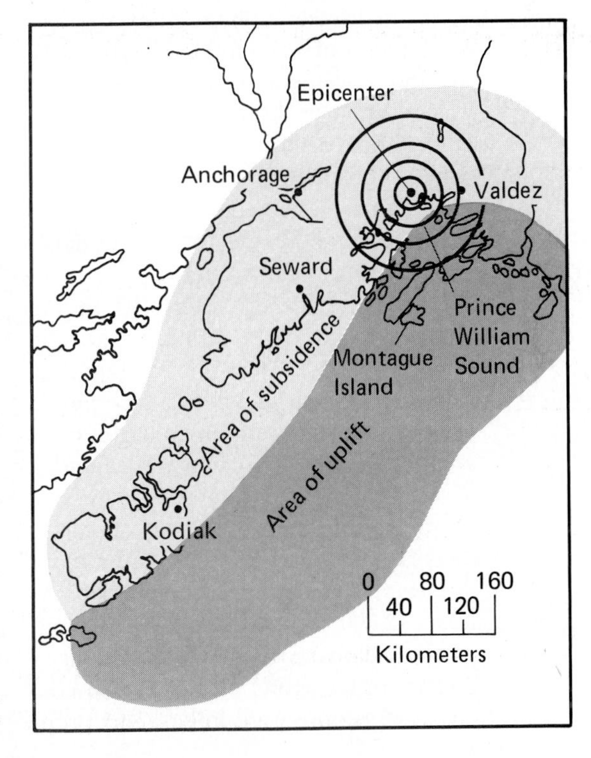

Figure 16-5 *The 1971 San Fernando earthquake did extensive damage to highways.*

buildings were only slightly damaged. Others fell apart. Many might have withstood the shaking, but the ground gave way beneath them.

Seismic waves caused destruction as far as 175 km from the source of the shock in Prince William Sound, about 150 km east of Anchorage (Figure 16-4). Vertical movement of the earth's surface was as great as 12 m on Montague Island. The earthquake set off seismic waves, or tsunamis (*tsoo NAH mees*) in the ocean. When these huge waves smashed the shore, they wiped out docks and buildings near Valdez.

People are always interested in the amount of property damage and the number of lives lost in earthquakes. These figures tell us what has happened to people and property in an earthquake area, but do not give us an accurate idea of what went on inside the earth. For example, the Agadir, Morocco, earthquake of February 29, 1960, had a magnitude of 5.6 on the Richter Scale. At least 12,000 people were killed, and property damage ran to many millions of dollars. In the Good Friday earthquake only 115 people were killed, and property damage was probably not much greater than in Morocco. Yet the total energy released in the magnitude 8.4 Good Friday earthquake was at least 6000 times greater than the Agadir earthquake. The 1971 San Fernando, California, earthquake (magnitude 6.4) was only "moderate," but the damage was great because it took place in a highly populated area (see Figure 16-5).

On July 28, 1976 a killer earthquake struck Tangshan, Peoples Republic of China, killing some 242,000 of the city's 600,000 residents. The quake rated 8.0 on the Richter Scale and left most of Tangshan in ruins.

Earthquake damage is measured in terms of the **intensity** or strength of an earthquake's ground motion in a given place. Estimates of earthquake intensity are usually made by experienced observers and are based on the amount of physical change in a specified location. American seismologists generally use the **Modified Mercalli Scale of Earthquake Intensity** to indicate varying degrees of intensity. This scale uses roman numerals to designate twelve classes of observable effects. An intensity of I is so slight that it will not be felt by most humans. At the other end of the scale is XII, an earthquake intensity that produces nearly total destruction. (See Figure 16-6.)

The Modified Mercalli Scale gives a rough indication of the amount of shaking the earthquake caused at a specific place as observed by people. Many intensity numbers are determined for each earthquake. The maximum intensity for the earthquake region may be given as the intensity of the earthquake. Figure 16-7 shows the distribution of intensities for the 1906 San Francisco and the 1971 San Fernando earthquakes.

Figure 16-6 *The Modified Mercalli Scale of Earthquake Intensity.*

I. Not felt except by a very few under especially favorable circumstances. Birds and animals uneasy. Delicately suspended objects may swing.

II. Felt only by a few persons at rest, especially on upper floors of buildings.

III. Felt noticeably indoors, especially on upper floors of buildings, but many people do not recognize it as an earthquake. Parked cars may rock slightly. Vibrations like the passing of light trucks. Duration of shaking can be estimated.

IV. Felt indoors by many, outdoors by few. If at night, some awakened. Dishes, windows, doors disturbed. Walls creak. Sensation like the passing of heavy trucks. Parked cars rock noticeably.

V. Felt by nearly everyone. Some dishes, windows, etc. broken. A few instances of cracked plaster. Unstable objects overturned. Disturbances of trees, poles and other tall objects sometimes noticed. Pendulum clocks may stop.

VI. Felt by all. Many frightened and run outdoors. Some heavy furniture moved. Books knocked off shelves, pictures off walls. Small church and school bells ring. A few instances of fallen plaster or damaged chimneys. Otherwise damage is slight.

VII. Everybody runs outdoors. Difficult to stand up. Negligible damage in buildings of good design and construction; slight to moderate in well-built ordinary structures; considerable in poorly built or badly designed structures; some chimneys broken. Noticed by persons driving cars.

VIII. Damage slight in specially designed structures; partial collapse in ordinary buildings; great damage to poorly built structures. Panel walls thrown out of frame structures. Chimneys, factory stacks, columns, monuments, and walls fall. Heavy furniture overturned. Small amounts of sand and mud ejected from cracks in the ground. Changes in well water.

IX. Damage considerable in specially designed structures; well-designed frame structures thrown out of plumb; partial collapse of substantial buildings. Buildings shifted off foundations, ground cracked. Serious damage to reservoirs and underground pipes. General panic.

X. Some well-built wooden structures destroyed; most masonry and frame structures destroyed. Ground badly cracked. Rails bent slightly. Considerable landslides from river banks and steep slopes. Shifted sand and mud. Water splashed over banks.

XI. Few masonry structures remain standing. Bridges destroyed. Broad fissures in ground. Underground pipelines out of service. Earth slumps and land slips in soft ground. Rails bent severely.

XII. Damage total. Waves seen on ground surfaces. Lines of sight and level distorted. Objects thrown upward into the air.

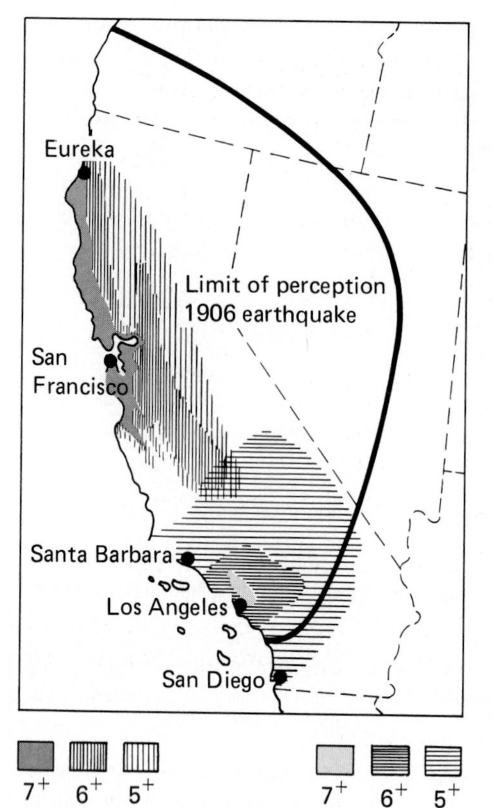

Figure 16-7 *The intensities of two California earthquakes. Which one was the bigger?*

16-4 Locating the Epicenter of an Earthquake

MATERIALS
graph paper, globe, marking crayons, string.

One of the most important tools of a seismologist is a **travel-time graph.** It is used to locate the epicenter of an earthquake, the point on the earth's surface that is directly above the focus of the earthquake.

The P and S waves from an earthquake travel at different velocities from the focus through the earth. As a result, although they both start at precisely the same time and place, they arrive at distant seismograph stations at different times.

When you have completed this investigation you will have made and used a seismic wave travel-time graph.

PROCEDURE

Suppose that a mild earthquake was recorded by a seismograph near Washington, D.C., at 08:00:00 Greenwich Mean Time (the time in Greenwich, England). The shock was registered on seismograph records in different cities at the times shown in Figure 16-8.

Make a graph, showing travel-time on the vertical axis and distance on the horizontal axis, for both the P and S waves.

DISCUSSION

Once you have made the travel-time graph, answer the following questions:

1. How long does it take for a P wave to travel from the focus of an earthquake to a seismograph station 2000 km away?
2. How long does it take for an S wave to travel the same distance?
3. What is the difference in arrival time between P and S waves for an earthquake that is 3000 km away from the station? 5000 km from the station?
4. How is the distance of a seismograph station from the earthquake related to the arrival time of the waves?

Using your travel-time graph and a globe try to locate the epicenter of the earthquake that produced the energy shown on the seismograph records in Figure 16-9. Time marks on each record provide common reference points.

16-5 Investigating Earthquakes

Since beginning your Earthquake Watch, you have been recording the epicenters

City	Distance in Kilometers	P Wave Arrival Time (G.M.T.)	S Wave Arrival Time (G.M.T.)
Buenos Aires, Argentina	8640	8:11:50	8:21:42
Cairo, Egypt	9590	8:12:37	8:23:12
Bogota, Colombia	4840	8:08:05	8:14:25
Chicago, Illinois	988	8:01:54	8:03:32
London, England	6060	8:09:27	8:17:06
Los Angeles, California	3810	8:06:42	8:12:11
Mexico City, Mexico	3120	8:05:48	8:10:32
Houston, Texas	2010	8:04:08	8:07:28
Moscow, U.S.S.R.	8040	8:11:20	8:20:41
New York, New York	339	8:00:38	8:01:18
San Francisco, California	4040	8:07:00	8:12:40
Stockholm, Sweden	6800	8:10:12	8:18:31

Figure 16-8 *Arrival times of earthquake waves, for Investigation 16-4.*

of large and small earthquakes and the depth of earthquake focuses.

In this investigation you will use some of the data used by scientists to outline the great belts of activity in the earth's crust. Your observations will also help in the discussion of forces beneath the surface.

PROCEDURE

Examine the map of the earth on which you have been recording the epicenters

MATERIALS

marking crayons, transparent plastic or tracing paper, graph paper, world map showing epicenter locations

Figure 16-9 *Find the epicenter of the earthquake that produced these three seismograph records.*

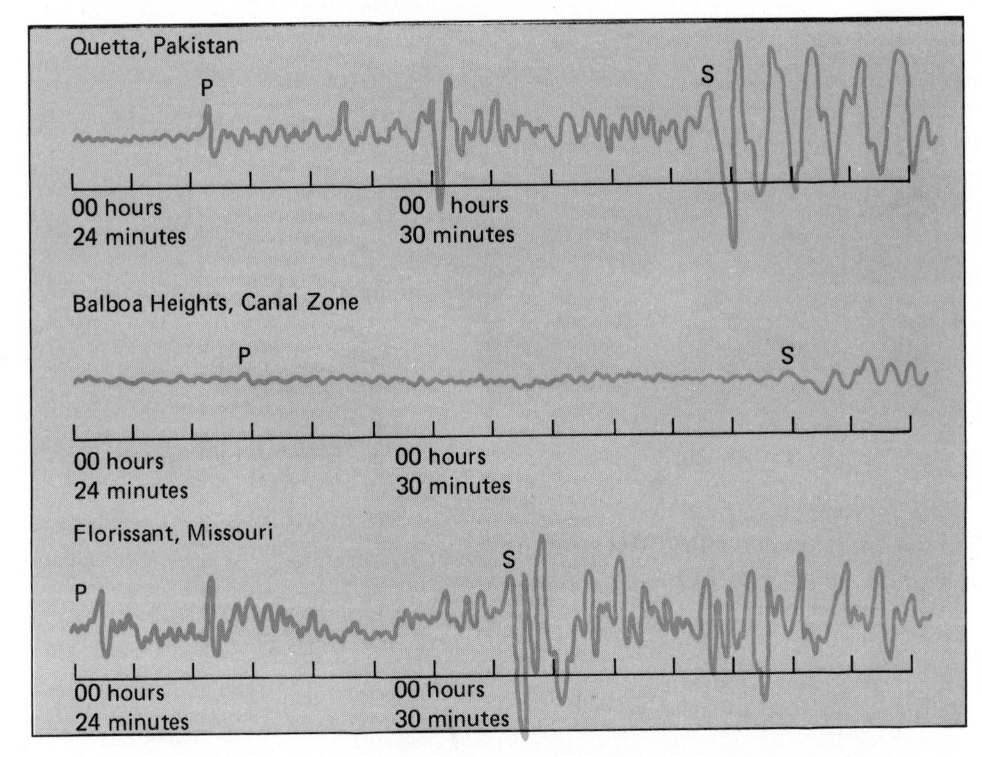

and depths of earthquakes.

1. What is the general pattern of earthquake distribution? Does it resemble any other major patterns of the crust?

2. If so, can you relate these features to earthquakes?

3. Where did the greatest number of earthquakes take place?

You have information on the depth of earthquakes within the earth's crust. Choose an area from the active belt around the edge of the Pacific Ocean where deep, intermediate, and shallow earthquakes have occurred. Place a sheet of transparent plastic over the area you select. Next draw a line at right angles to the coast. Mark the coastline and the location of several shallow, intermediate, and deep-focus earthquakes near this line. Now construct a cross section along the line you have drawn. Using a millimeter rule and graph paper, plot the depth of the earthquake focuses.

From your completed drawing, describe the pattern of earthquake focuses in this area.

4. How do you interpret the pattern?

5. Study Figures 9-4 and 9-10, and describe the distribution of earthquakes with respect to (a) mountains that have been uplifted from geosynclines and (b) volcanoes.

Thought and Discussion

1. What causes earthquakes?

2. Describe the different types of seismic waves.

3. Distinguish between the magnitude and the intensity of an earthquake.

4. Name the two scales that are used to "measure" an earthquake.

5. What is the relation between the focus and the epicenter of an earthquake?

6. Do you think people should live in active earthquake zones?

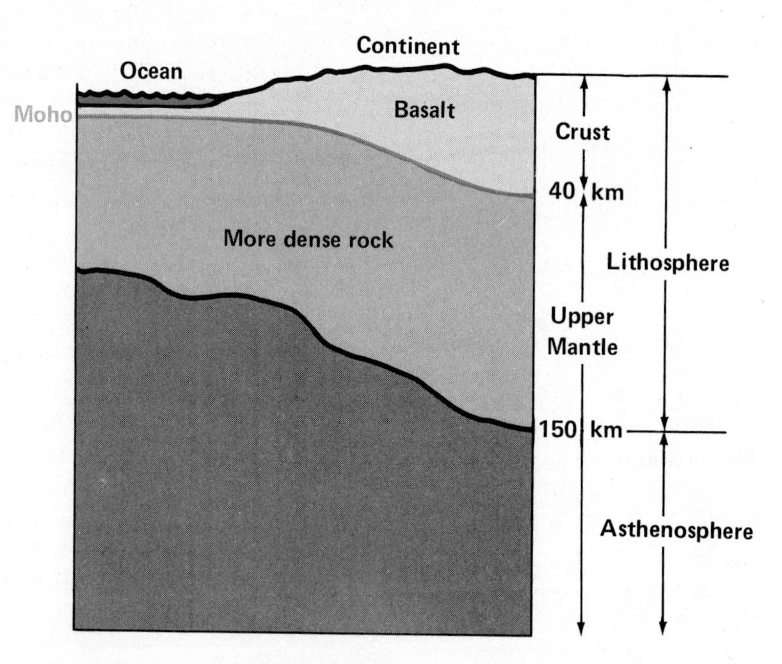

Figure 16-10 *Locate the position of the Moho on this cross section of the outer zones of the earth.*

16-6 Investigating the Inside of a Sphere

MATERIALS
two spheres

The hidden interior of an object can be studied without seeing or sampling it. Your teacher will give you two spheres that seem to be the same. But are they? Experiment with them and see. Make whatever observations or measurements you think are necessary and record any similarities and differences you detect.

16-7 Earthquake Waves and the Earth's Interior

Seismic waves have provided valuable information for studying what geologists are unable to sample: the inside of the earth. By analyzing how seismic waves travel through a substance, you can learn much about the contents of the substance. You can also learn how various parts of it are arranged.

By studying waves that have traveled deep into the earth, scientists have discovered that the earth has a core of very dense material. The inner part of the core is solid. Because S waves do not pass through the outer part of the core, geophysicists think that it is liquid or behaves like a liquid. You may wish to review Figure 1-24.

16-8 The Moho

Seismologists have determined the thickness of the crust by using a method discovered in 1909 by a Yugoslav scientist, Andrija Mohorovičić, (*moh hoh ROH vih chich*). While comparing seismograph records of earthquakes, he noticed that at a certain level in the earth there is an abrupt change in the speed at which earthquake waves travel. This boundary is called the **Mohorovičić discontinuity,** commonly shortened to the **Moho.** The Moho is the boundary between the crust and the upper mantle (see Figure 16-10). Today scientists frequently use the term "lithosphere" to describe the crust and a part of the upper mantle as a unit. The Moho is still an important boundary, however, even though terminology is changing.

The change in the speed of seismic waves is apparently caused by density differences between rocks of the crust and the upper mantle. Seismologists are now using seismographic records made all over the world, on both land and sea, to establish the depth of the Moho. This boundary zone averages about 10 km below sea level in ocean basins and about 30 km below sea level beneath high mountains.

Seismic investigations are carried out not only by monitoring earthquakes but also by setting off an underground explosion on one side of a selected mountain range. The waves radiating outward from the other side of the mountains are measured. Seismic waves travel at different speeds in different kinds of rock. Therefore, by measuring the wave speed, scientists can learn about the nature of the rock and the thickness of the layers. If the explosions and the detectors are moved farther and farther apart, waves that travel deeper in the earth can be detected.

The earth's mantle is denser than the crust. Seismic evidence has shown that the mantle probably consists mostly of rocks rich in olivine, garnet, and pyroxene. There is also evidence that the core is mostly iron with a small amount of nickel.

Thought and Discussion

1. If a seismic wave travels faster in olivine than it does in quartz, what does this tell you about the density of olivine?
2. Why do scientists think the earth's inner core is mostly iron?
3. Where do you think scientists should start to drill if they want to reach the Moho in the shortest distance?

Earth's Inner "Machine"

16-9 Earth's Nuclear-Powered Heat Engine

In the 19th century Lord Kelvin, a famous physicist, tried to calculate the earth's age. He concluded that it was 20 to 40 million years old. His calculations were based on the cooling and crystallization rates of an originally molten Earth. Kelvin assumed that the earth has no internal source of energy. He assumed that any energy or heat it possessed had to be left over from the originally molten Earth. The only exception might be small amounts of heat that were added by the sun.

Kelvin's calculations were correct, but his basic assumption was wrong. He did not know about minerals containing radioactive heat sources. Radioactive isotopes of uranium, thorium, and potassium are present in some kinds of rocks. Much of what we know about radioactivity and the distribution of radioactive heat sources is a result of intensive research on nuclear devices since 1935.

16-10 Taking Earth's Temperature

Field studies have shown how rocks rich in radioactive heat sources are distributed. Figure 16-11 indicates that granitic rocks contain far greater amounts of radioactive isotopes than basaltic rocks. Basaltic rocks in turn have a far higher radioactive content than either mantle

Figure 16-11 *Radioactivity and heat loss of some rocks.*

	Uranium PARTS PER MILLION	*Thorium* PARTS PER MILLION	*Potassium* PARTS PER HUNDRED	*Heat Production* MICROWATTS PER CUBIC METER
A. *Crustal Rocks*				
granitic igneous rocks	4.0	16.0	3.3	2.5
basaltic igneous rocks	0.5	1.5	0.5	0.3
shales	4.0	12.0	2.7	2.1
limestones	2.2	1.7	0.3	0.7
beach sands	3.0	6.0	0.3	1.2
B. *Possible Mantle Rocks*				
dunite	0.005	0.02	0.001	0.004
eclogite	0.04	0.15	0.1	0.04
peridotite	0.022	0.066	0.022	0.01
C. *Rocks Possibly Similar to Core Rocks*				
iron meteorite	0.00011	?	?	0.00006

rocks or meteorites. The several radioactive isotopes in granitic rocks produce nearly five times as much heat as the isotopes in basaltic rocks. And the basalts in turn produce almost 40 times as much heat as mantle rocks or meteorites.

The heat we feel at the surface of the earth, unless we are near an erupting volcano or a hot spring, is chiefly from the sun. However, the earth also produces great amounts of heat in its interior. The latest calculations indicate that 30–40 calories per square centimeter reach the surface of the earth each year from the decay of radioactive elements. This would be enough energy to melt a world-encircling layer of ice about one centimeter thick.

The speed with which heat generated inside the earth flows to the surface and escapes depends on the conductivity of the rock. Rock is such a poor conductor that heat now reaching the surface of the earth has probably taken billions of years to travel from a maximum depth of several hundred kilometers.

Heat flow refers to heat that originates within the earth and that eventually is released at the earth's surface. It takes very sensitive instruments to measure directly the surface heat flow. Temperature is stable in sediments under water more than a few tens of meters deep. Therefore, direct measurements of heat flow in the oceans are now easily made. On the continents a number of precautions must be taken in measuring heat flow in rocks near the surface. Seasonal changes in temperature and the movement of ground water of varying temperatures can affect the results. Thus, subsurface temperatures are best measured well below the zone of surface variability. Heat flow is measured in drill holes, mine shafts, tunnels, and oil wells.

Heat flow is measured in drill holes and mines by instruments called **thermistors** (*thur MIHS turz*). They show that the earth's temperature in any locality, volcanic or not, increases with depth. The rate of temperature increase is called the **geothermal gradient** (Figure 16-12). In non-volcanic areas this

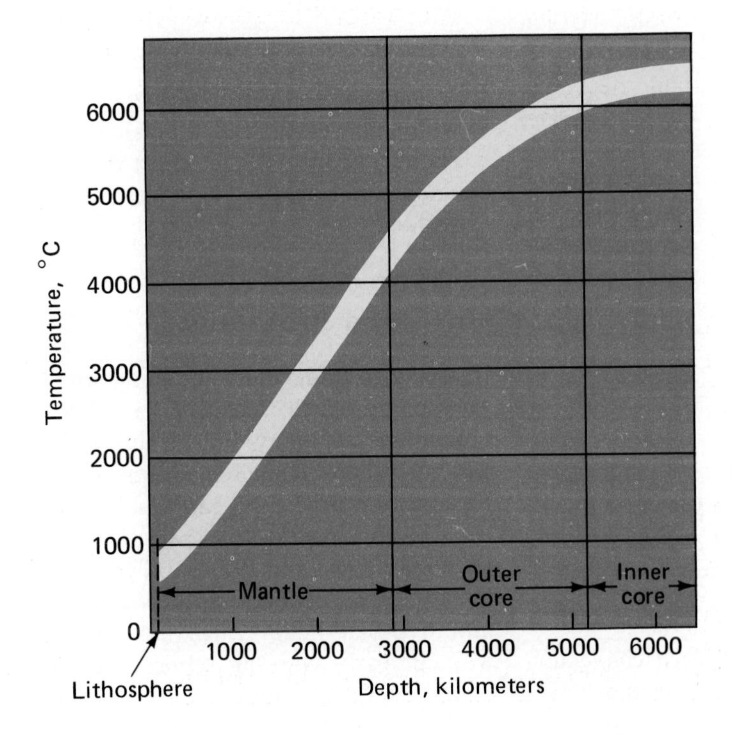

Figure 16-12 *Estimated temperatures in the earth's interior.*

gradient averages 30°C per km of depth or about 1°C per 30 m. However, scientists have been able to drill only a little over nine kilometers into the earth's crust. Any information on heat flow below this depth must be inferred from indirect evidence.

If the geothermal gradient continued at the same rate to greater depths, the temperature at the earth's center would have to be about 200,000°C. Much of the interior of the earth would be molten. But we know from earthquake studies that the only large part of the earth's interior that is molten is the outer core, from about 2900 to 5100 km.

16-11 Where Are the "Hot Spots"?

The amount of heat reaching the surface has been measured at many places within the continents and oceans. An area where all of these measurements have similar values is called a **heat flow province.** The boundaries of these areas are commonly located beneath major geographic landforms. Ocean ridges, for example, are characteristically provinces of high, but variable, heat flow. Ocean basins typically have moderate, relatively uniform heat flow. The trenches are low heat flow provinces. Why is heat flow so variable over the earth's surface? Many scientists believe that giant convection currents operating within the earth may explain this phenomenon.

MATERIALS
beaker, sawdust

ACTION

Fill a beaker with water and sprinkle some sawdust or torn pieces of paper on the surface. Heat the water and observe the movements of the sawdust.

Heating the water caused convection currents. You could observe the movement of these currents by watching the path of the sawdust. Since we know that the interior of the earth is heated by radioactivity, perhaps similar convection currents could be operating within the earth. Remember that even the most rigid substance will flow if stress is applied over long periods of time. Part of the asthenosphere or mantle could be flowing in convective patterns in response to radioactive heating throughout geologic time.

The old nuclei of continents, the Precambrian shields, have low heat flow values. On the other hand, regions of much younger mountain building activity have variable high and low heat flows. The Pacific Border Province and the Cascade Mountains (Figure 16-13) are good examples.

Thought and Discussion

1. How did Lord Kelvin account for Earth's interior heat?
2. What is one major source of the heat inside the earth?
3. What is the geothermal gradient?
4. If convection cells are operating within the earth, and they suddenly reverse direction, what effect do you think this would have on the earth's surface?

Unsolved Problems

In trying to understand the inner processes of the earth, scientists have been able to answer many difficult questions. Yet many problems remain unsolved. Research on predicting earthquakes has been given top priority, especially since the 1964 Alaska and 1971 San Fernando quakes. Scientists hope to provide communities that lie in earthquake zones with an early warning service. People living in those areas could then be

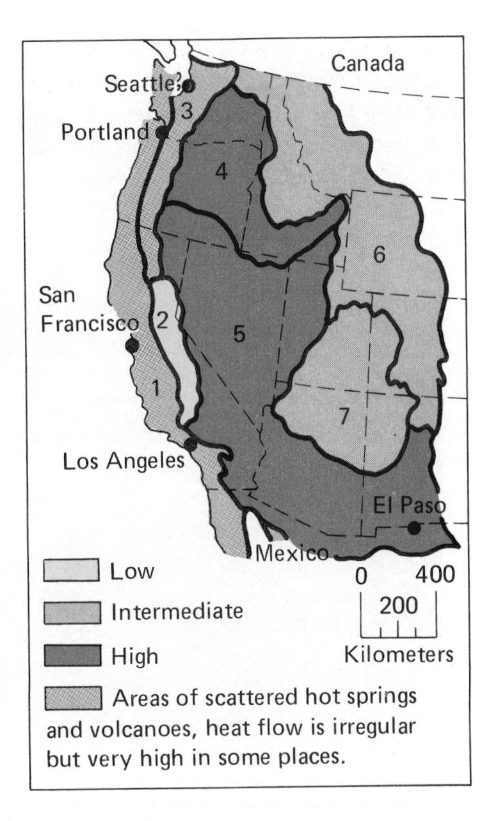

Figure 16-13 *Heat flow provinces in the western United States: (1) Pacific Border Province. (2) Sierra Nevada. (3) Cascade Mountains. (4) Columbia Plateau. (5) Basin and Range Province. (6) Rocky Mountains. (7) Colorado Plateau.*

warned in time to reduce loss of life and damage to property. Although progress is being made, accurate earthquake forecasting is a long way in the future.

Scientists use instruments to measure stresses acting on rocks in certain areas. If the force of these stresses changes suddenly, it could indicate that an earthquake may be about to occur. Some rocks swell or undergo an increase in volume, as many tiny fractures open in the rock before an earthquake occurs. Seismologists have also noted that clusters of small magnitude quakes tend to occur a few hours before a large earthquake hits an area.

Using such clues to forthcoming disturbances, scientists may be able to predict a major earthquake a few hours before it happens. However, the warning signals mentioned may not occur before all earthquakes. Also, even a few hours warning is not enough time to insure the safety of the people living in the quake area. Clearly, more work is needed if scientists are to solve the earthquake prediction problem.

Chapter Summary

In this chapter you have learned that the earth's dynamic interior is responsible for movements that occur on the surface. Deeply buried rocks are under great stress. When they break and snap back into a new position, seismic waves are generated. These waves not only cause earthquakes, they have provided much information about the inside of the earth.

The earth's heat engine—the driving force of the earth's internal processes—is powered basically by radioactive fuel. By studying the geothermal gradient, scientists have been able to estimate the temperature in the earth's interior. The amount of heat that flows from the center of the earth to the surface varies. This may be caused by convection cells in the mantle. The earth can be divided into provinces according to the amount of heat flow. Earthquakes are also an expression of the earth's dynamic interior. By studying arrival times of P and S waves, scientists can learn indirectly about the earth's interior.

SEISMOLOGIST

Seismology is a branch of geophysics—the use of geology combined with physics. Seismologists specialize in the study of earthquakes. They use seismographs and other instruments to record the locations and vibrations of earthquakes. Seismologists work in areas where the earth is particularly active and crustal movements are frequent. The information they supply may indicate fault zones or areas where a building project would be unsafe. Exploration geophysicists are seismologists who explore for oil and minerals. They use explosives to make artificial earthquakes. Seismograph records give information about buried rock layers. Information provided by seismologists is being used to help predict volcanic eruptions in some places. And now seismologists are even studying moonquakes!

Seismologists are employed by many types of organizations. The petroleum industry employs many geophysicists. Some work for geological surveys or other agencies. Research institutions and colleges and universities also employ seismologists.

Seismology, like other areas of scientific research, requires an advanced education. Study is necessary in geology, physics, and math. People interested in seismology but with less education may be seismological technicians. They work with the equipment and may help to interpret the data. Two years' education and/or on-the-job training is usually enough for a job as a seismological technician.

Questions and Problems

A

1. What is a tsunami?
2. What is the major source of heat in the earth's interior?
3. What type of rock contains the highest number of radioactive isotopes?
4. What methods can be used to measure the earth's temperature directly?
5. What explanation is given for the fact that S waves do not pass through the earth's core?
6. Why is it easier to measure heat flow in the ocean basins than on the continents?
7. Name the two types of seismic waves that cause most earthquake damage.

B

1. Name some areas on the earth's surface where you might expect heat flow to be high.
2. What is the difference between compressional and shear waves?
3. What single fact would you have to know to determine the distance to the epicenter of an earthquake with a travel-time graph?
4. How many seismograph stations must record arrival times so that an epicenter location can be determined?
5. If there were no erosion, how many earthquakes exactly like the Anchorage earthquake would it take to raise mountains from sea level to the heights of the modern-day Alps or Himalayas?
6. At the time of the Alaska Good Friday earthquake, there was a meeting of the Seismological Society of America in Seattle. Most of the seismologists did not actually feel the earthquake. A number who were having dinner in the restaurant at the top of the Space Needle did feel it. What would be the Modified Mercalli Intensity rating in Seattle?
7. What is the difference between the Modified Mercalli Scale of Earthquake Intensity and the Richter Scale?

C

1. What evidence can you give to answer the question of whether the earth's interior is liquid or solid?
2. The radius of the earth is 6370 km. The location of the core-mantle boundary is at a depth of 2900 km. What percentage of the earth's volume is occupied by the core? What percentage is occupied by the mantle? (Do not include the volume of the crust.)
3. If you were assigned to evaluate the amount of damage done near the epicenter of an earthquake, which of the two earthquake rating scales would you use? Why?

Suggested Readings

Bolt, Bruce A. *Earthquakes: A Primer.* San Francisco: W. H. Freeman and Co., 1978. (Paperback)

Bolt, Bruce A., ed. *Earthquakes and Volcanoes: Readings from Scientific American.* San Francisco: W. H. Freeman and Co., 1980. (Paperback)

Eiby, G. A. *Earthquakes.* New York: Van Nostrand Reinhold Co., 1980.

Gilfond, Henry. *Disastrous Earthquakes.* New York: Franklin Watts Co., 1981.

Gribben, John. *This Shaking Earth: Earthquakes, Volcanoes, and Their Impact on Our World.* New York: G. P. Putnam's Sons, 1978.

Walker, Bryce, and the editors of Time-Life Books. *Earthquake.* Alexandria, Virginia: Time-Life Books, 1982.

17 Continents Adrift

AT THE BEGINNING *of the 20th century, scientists around the world believed they understood that the earth's structure and evolution were well established. Of course, there remained some nagging questions, such as: Why is there more ocean than land? Why are most of the major land areas north of the equator? Why is the Pacific Ocean so much larger than the Atlantic? How can we explain the earthquake and volcano belts?*

Most geologists believed that the answers would come from continued studies following the idea that the earth was still contracting. According to this theory, the earth had cooled from a former molten mass. During this process, lighter rock materials had moved to the surface, forming the granitic rocks of the continents. Underlying the continents were denser basaltic rocks. As the earth contracted, some parts of the surface collapsed and formed the ocean basins, but other blocks were faulted upward to form the continents. The contractions caused wrinkles, as on the skin of a drying apple, these being the folded and faulted mountain ranges.

In 1912, a young German meteorologist named Alfred Wegener (VAY guh nur), *delivered a lecture that was to shake the very foundations of earth science. He had been struck by the remarkable similarity of the Atlantic coastlines of Africa and South America. He recalled observations he had made a few years before. During an expedition to northern Greenland he observed the ice pack breaking up in the spring thaw. As the slabs split, they slowly drifted apart. They rotated slowly as they moved, but kept the shape along the original crack. To Alfred Wegener, this suggested that in the past the continents had been one, and had drifted apart. For fifty years, some geologists thought Wegener's hypothesis was correct. Others thought that the continents couldn't possibly have moved around. Agreement among geologists came only after the discovery of a strange bit of evidence, evidence concerning the earth's magnetic poles.*

AFTER COMPLETING THIS CHAPTER, YOU SHOULD BE ABLE TO:

1. **describe the old theory of a rigid Earth and its explanation of mountains.**
2. **state several reasons why Alfred Wegener hypothesized that the continents have moved.**
3. **state the major objections to Wegener's hypothesis.**
4. **describe the additions to the hypothesis made by the geologists Holmes and Du Toit.**
5. **describe the kinds of evidence in favor of the hypothesis that have been found since World War II.**

Early Ideas Change

17-1 The Less-than-Rigid Earth

The traditional rigid-earth theory is that the earth was originally molten and is now cooling. In early history it became rigid. The contraction that takes place during cooling creates compressive forces that squeeze up mountains along the weak margins of continents. Isaac Newton first suggested this theory. His calculations showed that an earth-size molten mass would take about 100 million years to cool. The circumference would shrink by several tens of miles during that time. This seemed to account nicely for the mountain ranges of the world, but the distribution and the strange shapes of the continents were still puzzling.

Geologists of the 19th century were satisfied with the idea of granitic continents locked in place. However, paleontologists were having trouble explaining the geographic pattern of some ancient fossils. Especially troublesome were those of some animals that lived in Paleozoic and Early Mesozoic times. To account for the places where the fossils were found, it was necessary to consider that land bridges had once connected continents. Since Early Mesozoic time, the bridges must have sunk beneath the seas. This idea was supported by stories of lost continents, such as Atlantis. As the 20th century dawned, new discoveries began to shake the concept of a rigid, contracting Earth.

The first major break came with the discovery of radioactivity. As you learned in Chapter 13, this allowed scientists to establish the age of rocks with much greater precision than before. The idea of a 100 million-year-old earth disappeared quickly. It was not until the late 1960's, however, that dating techniques became precise enough to give the realistic estimate of 4.5 billion years. Dates from sedimentary rocks indicated that continents had grown through hundreds of millions of years. Then came the most damaging discovery of all to the theory of a cooling, contracting earth. The decay of some radioactive minerals generates heat. A large supply of heat is continuously being added to the earth's crust. If the earth is cooling, it must be cooling more slowly than scientists had formerly thought it was. Therefore, any resulting contraction would not have given enough force to build mountains.

As geologists learned more about the Pleistocene Ice Age, they were able to date it. They also learned why the earth's crust is rising in Scandinavia and central Canada. In these areas, great glaciers melted 11,000 years ago, and the earth is rising at a rate of a centimeter each year. The crust was squashed by the tremendous weight of the ice and is now springing back! If the crust could rise as a weight was removed, then couldn't the crust also sink under the weight of great lava flows?

It began to seem as if the earth was less rigid than had been supposed.

17-2 Early Ideas of Continent Origins

One of the earliest ideas on the origin of continents was published in the mid-1800's. According to this idea, the continents formed on one side of the earth only. So the earth was out of balance. Movements in the crust balanced the earth again by the breaking and pulling apart of the Americas from the Old World. The coastal fit of Africa and South America was even then too obvious to ignore.

The arrangement of the ocean basins and continents still demanded an explanation. In 1879, Sir George H. Darwin suggested that the Pacific Ocean was the

Figure 17-1 *A simplified version of Taylor's map showing amount and direction of crustal movement during the Tertiary Period, according to his theory. The lengths of the arrows indicate the amounts of movement.*

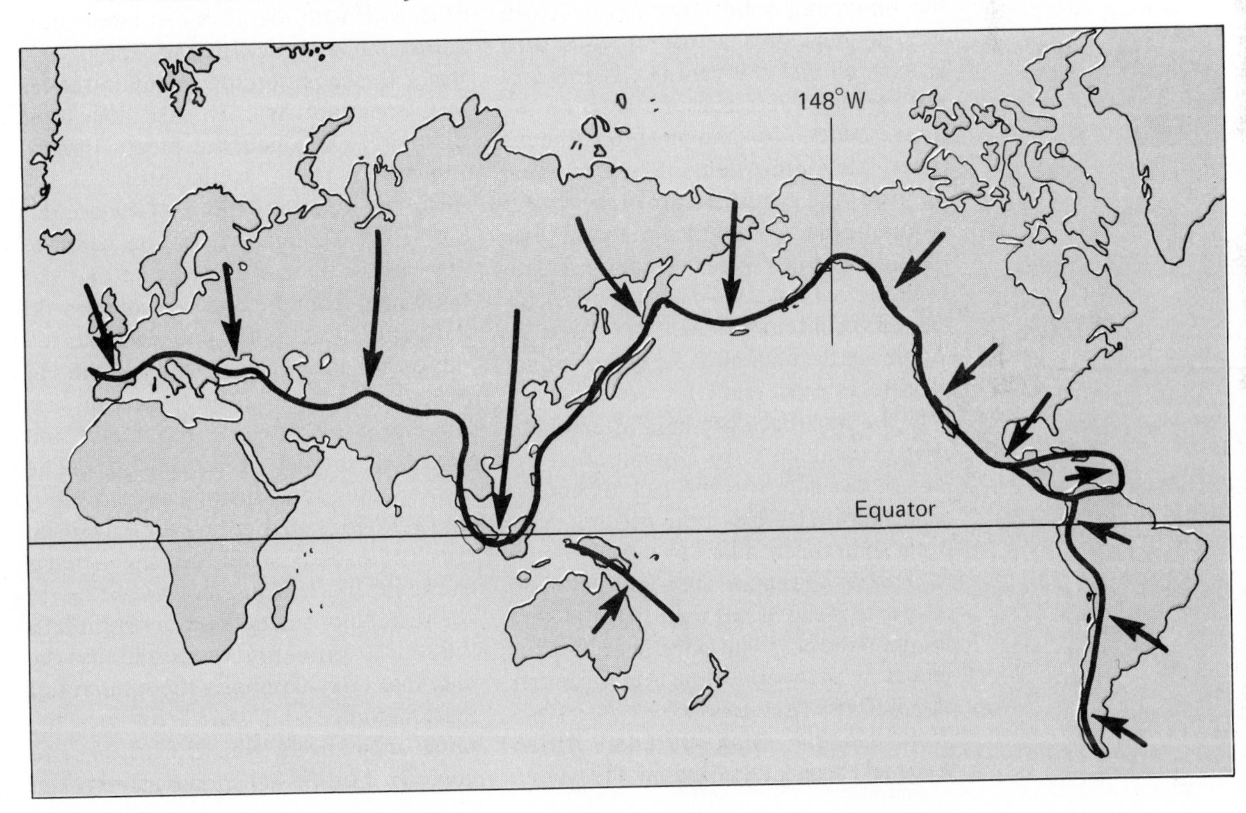

Figure 17-2 *Robeson Channel, between Greenland and the Canadian island Ellesmere Island. The ocean water fills a fault named after Alfred Wegener.*

scar left when the moon was torn away at an early time in the earth's history. A logical result would be the movement of the remaining solid crust as the earth tried to heal such a wound. This idea lasted well into this century. Island arcs and trenches were considered to be the front edges of moving continents. Clearly, the mid-Atlantic Ridge must be where the old continents broke apart, and should be made of granitic material. The Pacific Basin, it was reasoned, must date from the earliest days of the earth. Sediments should have been deposited in that basin throughout most of the earth's history.

As we look back on these attempts to explain continents, it is apparent that the biggest problem was the lack of knowledge about the crust of the ocean basins. But it didn't seem a lack to the geologists of that day because they continued to believe that the basins were places where continental rocks had once been. This is shown by the work of the American geologist F. B. Taylor.

In 1910, the same year that Alfred Wegener began lecturing in Germany, Taylor published the first paper to present a logical concept of what came to be called "continental drift." He was not concerned with the fit of continents bordering the Atlantic. Instead, Taylor was struck by the pattern of mountain ranges in Europe and Asia. At this time, most geologists explained the great Himalayan-Alpine chain as the result of the earth's contraction. This explanation did not satisfy Taylor. Instead, he believed that part of the earth's crust had moved southward from the north to collide with the ancient shield of India. The collision caused the huge, folded mass of the Himalayas, but let the mountains in Malaysia and Indonesia "swing freely" and move southward (Figure 17-1). The movement of the Alps was more complicated, Taylor said, because the European crustal area was small and the African continent blocked the motion.

Crustal movement from northern latitudes was primarily supported by the idea that Greenland was the remnant of a former old land area. This was the same area from which Canada and northern Europe had previously broken

(Figure 17-2). Taylor had less to say about the Southern Hemisphere, but he did note the northeastward shift of Australia and evidence of the westward drift of South America. These points are all of considerable interest. They were keen interpretations, but because there were almost no data to support his ideas, Taylor's 1910 paper created little interest among geologists.

17-3 Wegener Conceives His Hypothesis.

Alfred Wegener was born in Berlin in 1880, the son of a preacher. He was educated in Germany at the universities of Heidelberg, Innsbruck, and Berlin. At the age of 26 he joined a Danish expedition to Greenland to do meteorological research. Upon his return, he became a distinguished member of the University of Marburg, Germany, and wrote a well-known basic textbook on meteorology. He went on a second Greenland expedition and then served in the First World War as a junior officer. During the war he was wounded. Wegener returned to academic life in Hamburg. He also became director of the young but already famous Marine Section of the German Weather Service. Wegener's competence as a scientist grew. In 1924 a specially created position in meteorology and geophysics was established for him at the University of Graz, Austria. It was from that position that he went on his third expedition in 1930, during which he lost his life on the Greenland ice cap.

From his education and experience, Wegener would not seem to be prepared to propose a theory on the origin of continents. On the other hand, his lack of formal training in geology left him without the confining thoughts of the popular theory of a rigid, slowly contracting earth. Also, his university training had come at an important time. The Norwegian meteorologist Wilhelm Bjerknes (*BYURK nehs*) was developing and teaching his new theory on the worldwide circulation of the atmosphere. Bjerknes concerned himself not only with the air, but with the influence of the ocean and land masses as well. He thought of a complex hypothesis and then set forth to collect data that would prove or disprove his ideas. Wegener was thoroughly immersed in these revolutionary concepts on world climate and weather. It was not beyond his experience to consider a broad, unifying hypothesis, in this case, of continent origin.

Wegener's first thoughts were in response to the apparent fit of the Atlantic coasts of Africa and South America. In 1911, he happened to read a report of similar fossils from rocks in Brazil and Africa. He then went on to develop his hypothesis, which he presented in a 1912 lecture in Frankfurt, Germany. He did not know of Taylor's earlier publication. The same year, he published two papers entitled *The Origin of Continents.* Wegener pointed out that he was putting forth an hypothesis, but it was based on arguments that clearly stated the inadequacies in the old ideas of a contracting earth. He put much hope in future precise mapping to prove that the continents were in continuing motion. And, he decided that ancient climates could have resulted only from changes in the positions of continents and the earth's poles.

It would take another 54 years and some remarkable scientific studies to show the great insight of Alfred Wegener into the physics of the solid earth.

Thought and Discussion

1. How did the discovery of radioactivity change our thoughts about the earth's interior?
2. How is it possible to date earth movements?

3. Why do mountain ranges indicate a moving Earth's crust?
4. Why was meteorology good training for Wegener's ideas?

A New Idea of Continental Origin

17-4 Wegener's Continental Drift

Wegener thought that because most mountains are made of *shallow-water* marine sediments (see Chapter 9), the continents between them must be fairly permanent. The continents could not have moved up and down, being alternately dry land and deep ocean basins, as previously thought. This was made clearer by some newly obtained data on the earth's gravity. The rocks of the sea floor were found to be more dense than those of the continents. Certainly the less dense continents couldn't have sunk. But why not horizontally moving continents?

Wegener suggested that beginning in the Mesozoic and continuing to the present time, a supercontinent that he called "Pangaea" (*pan JEE uh*) had split and the fragments had moved apart (Figure 17-3). South America and Africa began to separate in the Cretaceous Period at the same time as North America and Europe split. In the north, however, Canada and northwest Europe maintained contact, with Greenland in the center, until the Ice Age.

As the Americas drifted westward, the great mountain ranges, the Andes and the Rockies, were formed by compression along the front edges of the continents.

In the Indian Ocean, the split began earlier than in the Atlantic, and most of the movement was finished by Tertiary time. The northward movement of India had pushed a large land area upward to form the Himalayas (the reverse of the southward movement imagined by Taylor). Early in the Cenozoic Era, Australia broke off from Antarctica and moved to the north to encounter Indonesia at the end of that period of time (Figure 17-4).

17-5 Wegener's Arguments

There were several geological and biological arguments to help Wegener in his concept. The folded mountains of South Africa apparently are continuations of the range north of Buenos Aires, Argentina (Figure 17-5). Glacial deposits in Africa also matched rocks in Brazil.

The fit of the sides of the North Atlantic was not as obvious. It did seem possible to connect Paleozoic mountains in North America to similar ranges in western Europe. There are great faults that run through Nova Scotia and Scotland. They seem to be continuous. And there was more evidence, from the Ice Age deposits. **Moraines** are piles of glacial till built up at the edges of glaciers. A nice fit of the moraines from the ice sheets of North America and Europe was made by presuming the continents were connected during the Ice Age. You can see for yourself in Figure 17-6.

In these examples, Wegener was trying to point out that, "It is just as if we were to refit the torn pieces of a newspaper by matching their edges and then check whether the lines of print run smoothly across." The "lines of print" were, of course, supposedly continuous geologic features.

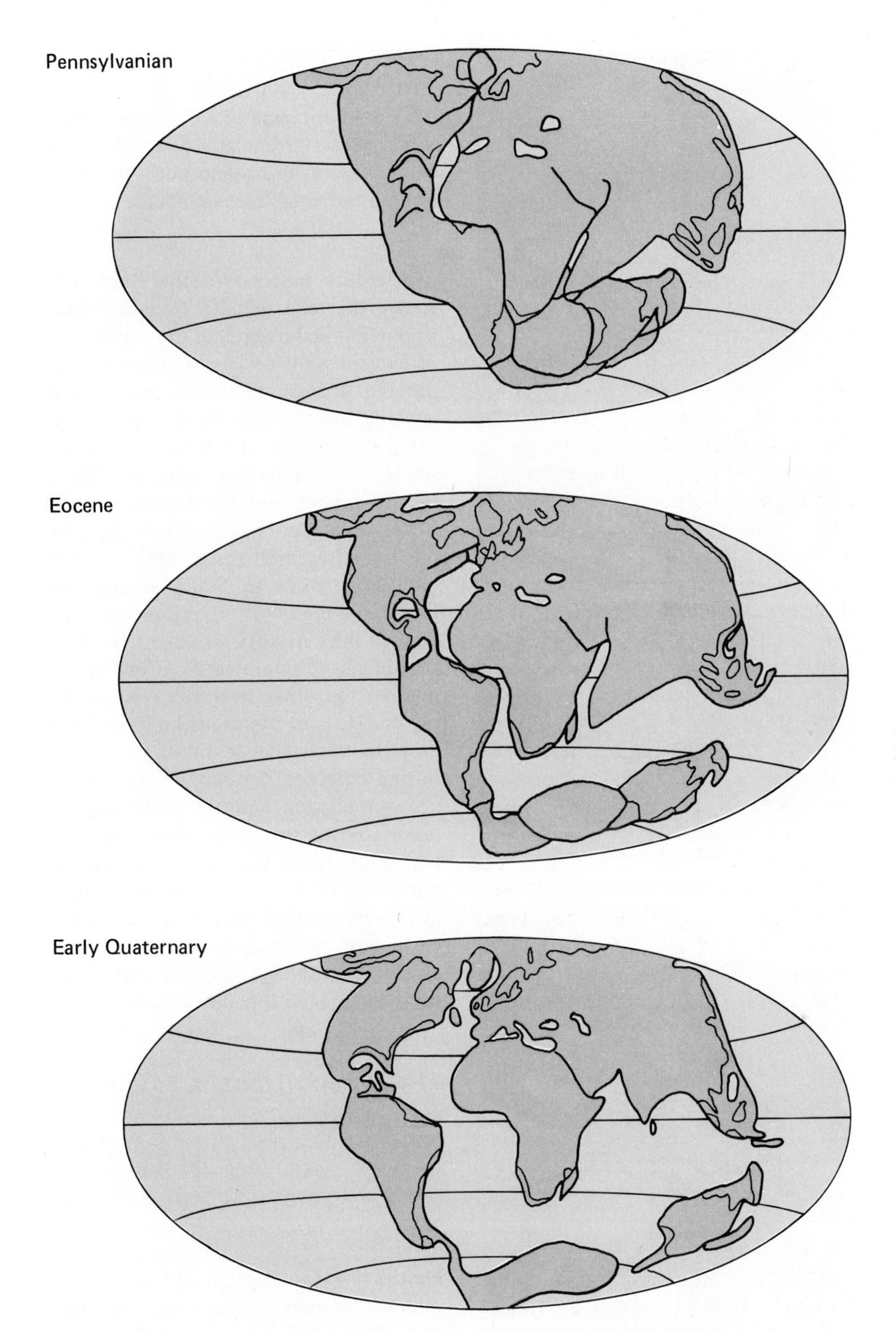

Figure 17-3 *Wegener's idea of the positions of the continents from the Late Paleozoic to the Ice Age.*

MATERIALS
front page of a newspaper

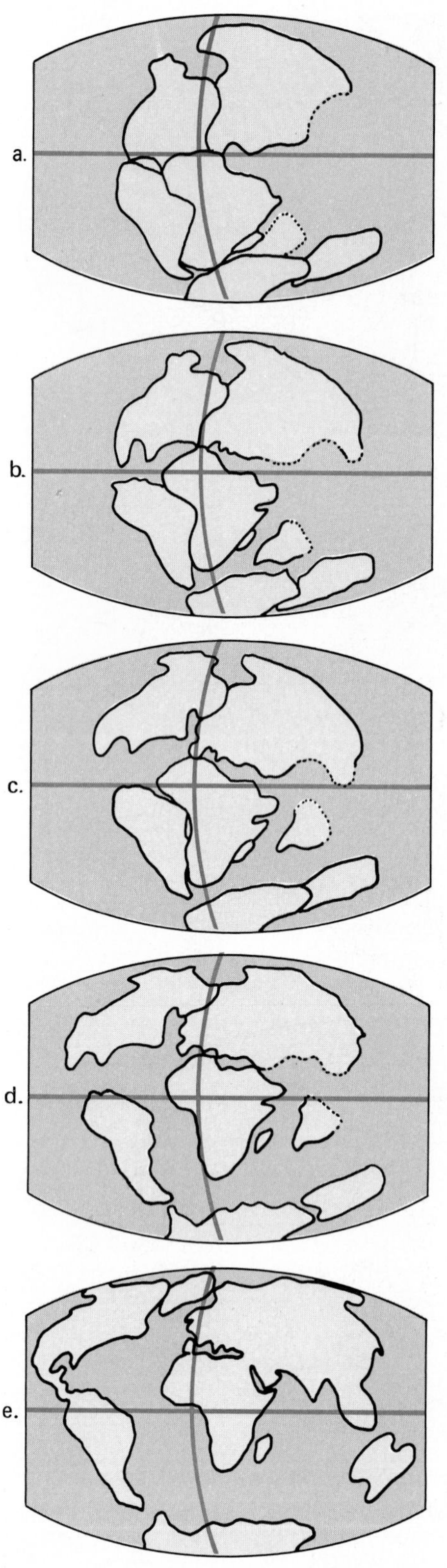

ACTION

Take the front page of a newspaper. Tear it into several irregular pieces. Then ask someone else to try and put it together. Keep a few small pieces hidden.

Scientists before Wegener had presented the idea of land bridges. These bridges helped to explain the similarities of ancient and modern animals on the different continents. If such bridges had actually existed, the fossil record indicated that they would have sunk below sea level in Cretaceous time. To Alfred Wegener, continental drift was a much better explanation. In particular, he mentioned a small fossil reptile known only from rocks in South Africa and Brazil. It wasn't built right for swimming across the Atlantic Ocean. And it couldn't have migrated from one continent to the other over the cold Arctic region. He also mentioned a plant that grew only on the southern continents during Paleozoic times.

Several living animals could also be used. One of the more interesting is a certain earthworm species. It would seem quite difficult for an earthworm to make an ocean crossing. Maps of where this earthworm species lives on all continents seemed to indicate that the continents must have been together in the past.

17-6 Paleoclimates Favor Pangaea.

It is not surprising that Wegener's strongest arguments came from his stud-

Figure 17-4 *Continental drift as suggested by Robert Dietz in 1971:* **a.** *End of Permian Period.* **b.** *End of Triassic.* **c.** *End of Jurassic.* **d.** *End of Cretaceous.* **e.** *Today. How does this compare with Wegener's version in Figure 17-3?*

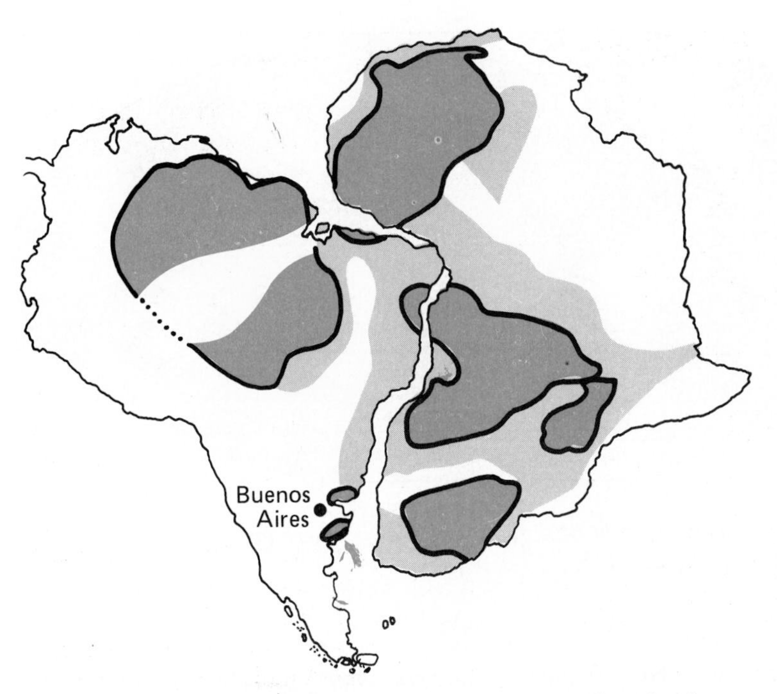

Figure 17-5 *Matching geologic provinces in Africa and South America.*

Figure 17-6 *This map shows how Wegener thought the northern continents were positioned during the Ice Age. He believed that one ice sheet extended from eastern Europe to western Canada. The sea level was lower because of the amount of water frozen into the ice sheet and mountain glaciers elsewhere. Note how the lower sea level changed the coastlines, according to Wegener.*

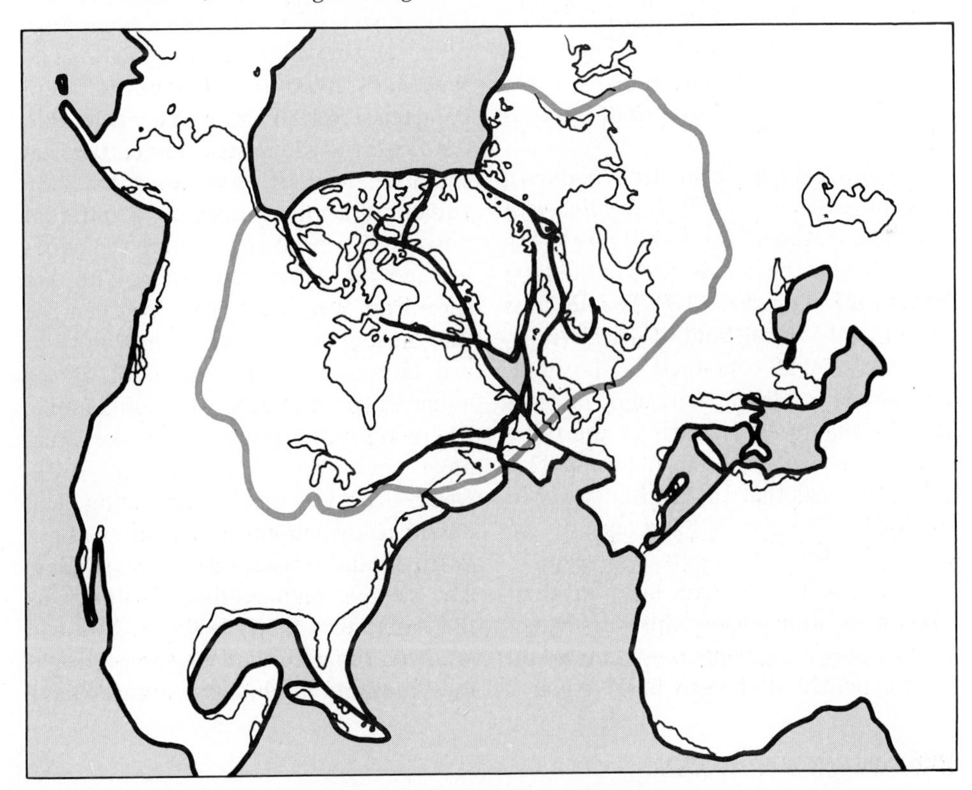

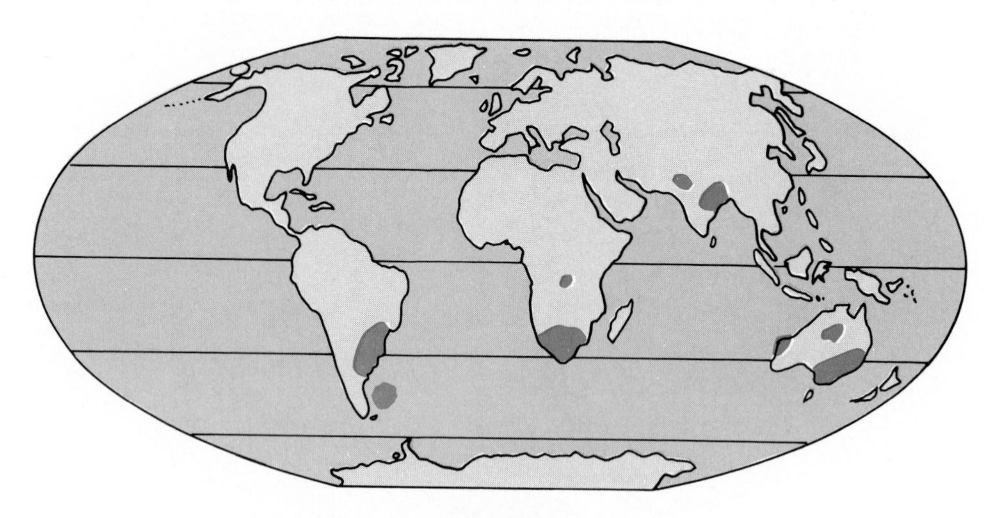

Figure 17-7 *The areas in blue represent places where evidence for Late Paleozoic ice caps is found.*

ies of ancient climates. He recognized that present day climates lie in zones that are roughly parallel to the equator. One would expect similar climate regions in the geologic past. Geologic evidence of ancient climates became available, indicating that glaciers had once grown in areas that are now tropical or arid. The explanation of such a change in climate could come from changes in the positions of the earth's poles. Or, it could come from changes in the positions of the continents.

The best evidence came from beds of a rock called tillite (*TIHL yt*), which is a solidified glacial till (review Figure 7-27). The tillite had been deposited in the Late Paleozoic Era (Figure 17-7). The deposits in South America, South Africa, India, and Australia all contained similar boulders and lay on large, striated pavements (review Figure 7-24). Parts of the tillite were mixed with marine deposits. This told geologists that the ice had reached the ocean. There must have been an ice sheet, not isolated mountain glaciers.

If the continents had been in their present position during this late Paleozoic glacial period, a polar climate would have extended all the way to the equator in one hemisphere. In the other hemisphere, there would have been the usual tropical conditions near the equator. This seemed quite foolish to Wegener. It made more sense to put the continents back together, bringing India southward.

More evidence came from the distribution of coal deposits in the eastern United States, Europe, and China. Suppose you were to mark these locations on Wegener's hypothetical supercontinent Pangaea (Figure 17-8), and consider that the coal was deposited under tropical climates. Then it is obvious that the equator must be moved to a different position. Suppose you assume the equator and the poles were where they are shown on Wegener's map. You can see that most of the coal was formed in tropical latitudes. North and south of the equator, you find desert-type sandstone. There are salt deposits, too, which must have been formed in hot, dry areas where there was much evaporation. The desert sandstone and the salt are about at the latitudes where deserts are today. The ice is at high southern latitudes, as the Antarctic ice cap is today. You can see from the map that Wegener thought the present land areas must have

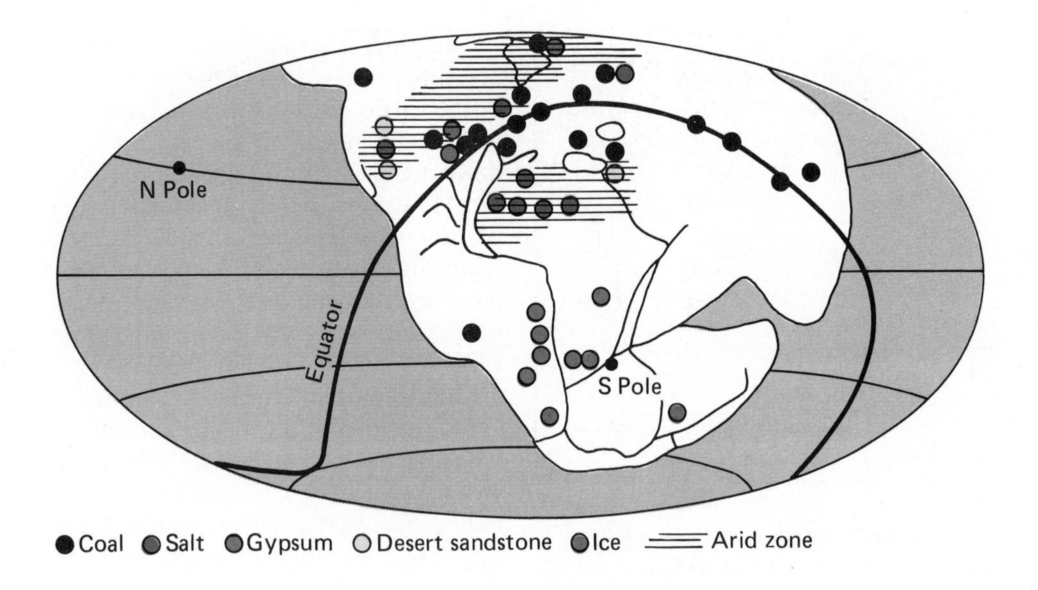

Figure 17-8 *Wegener's reconstruction of Pangaea in the Mississippian and Pennsylvanian Periods. Climate zones are indicated by rocks, fossils, and other evidence.*

changed their positions in relation to the poles and equator.

Thought and Discussion

1. In what places do the edges of the continents appear to fit together?
2. How do folded mountains help the idea of continental drift?
3. What do ancient climates have to do with the idea of Pangaea?

Arguments For and Against

17-7 Critics and Allies

There was little response to the early publications of Wegener. Probably, by chance, they were not read by people who were interested in earth history. It was not until 1924 that *The Origin of Continents and Oceans* was translated into English. Criticism began to grow.

All parts of Wegener's hypothesis were under attack. The jigsaw fit of Atlantic shorelines did not seem appropriate. There was much evidence that uplift, sinking, and mountain building had occured over millions of years. Such activity would certainly distort coastlines and the nice fit proposed could not really be expected.

Both geologists and paleontologists questioned the similarity of rocks, folded mountains, and fossils in the various continents. Even if there were similarities

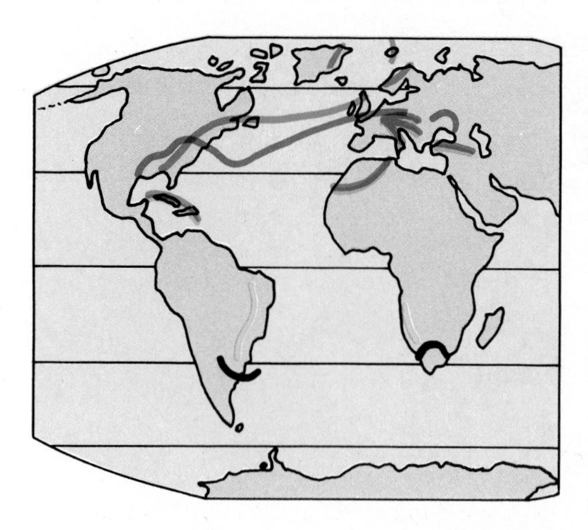

Figure 17-9 *Paleozoic mountain belts on opposite sides of the Atlantic, according to Holmes.*

of ancient animals, this, the critics noted, did not necessarily mean the lands had been joined.

Authorities on glaciation questioned the concept of a huge southern ice sheet. Had there really been such a large continent, most of the area would have been far from the ocean. Moisture-laden winds could not reach the interior of the continent, just as they do not reach central Asia today. The so-called matched glacial deposits must have been formed from small, local glaciers, they said.

The most serious objections were of the mechanism that made the continents drift. Why, it was asked, did Pangaea remain a single continent until the Mesozoic and then break up in just a few million years? How were older mountains formed? How could a continent move so easily over the underlying layers yet crumble at its leading edge to form mountains? Why was it that only the westward movement of the Americas was described when the position of the mid-Atlantic Ridge seemed to imply an eastward movement of Europe and Africa?

The English geologist Harold Jeffreys presented evidence of the great strength of the earth. The crust is strong enough to hold up mountain ranges and hold down ocean deeps, he said. This required great forces. To consider that small forces could move continents, even over a long period of time, was a dangerous assumption, according to Jeffreys. And, polar movement was physically impossible without totally destroying the earth.

Most geologists of that time still favored the idea of a cooling, contracting Earth. They had changed their early views, because of the need to consider radioactive heat. The idea of a fixed, rigid Earth was difficult to remove from their minds. Also, geologists were not really ready to accept such a startling hypothesis from a meteorologist!

Although continental drift was strongly opposed, the idea was attractive to some people. This was especially true of earth scientists who, like Wegener, looked at large-scale structures. Students of geology who had seen evidence of great horizontal pressure in the Alps liked the idea that these mountains were formed as part of greater movements. They saw similarities to the Alps in the ancient mountain belt that was on two sides of the Atlantic Ocean (Figure 17-9). Links were also noted in faults and folds on either side of the South Atlantic.

The greatest advocates of continental drift in the years up to 1950 were Arthur Holmes and Alexander Du Toit (*doo TWAH*). Holmes, who is considered to be the greatest British geologist of this century, had established an absolute time-scale based on the decay rates of radioactive elements. He was well suited, therefore, to study heat flow and large motions in the earth. Du Toit was a South African who was struck by the resemblance of the geology in South America to that in his native land. He became a believer in continental drift.

Holmes suggested that the earth's crust was made of two layers, with granitic rocks on top of a layer of more dense rock. Below those layers was a material that acted as a very stiff fluid, because of its high temperatures. It had become evident to Holmes that volcanic activity was not enough to account for all of the heat generated by radioactive decay. There must be convection currents beneath the crust. If such currents did exist, and if they rose beneath continents, then this could be a mechanism for continental drift (Figure 17-10). Radioactive heat is greater in continents than ocean basins because of the concentration of the radioactive mineral uranium in granitic rocks. It was logical, therefore, to assume that there were currents rising under the continents and moving downward at continental boundaries. Stretching would occur in the continental crust, eventually ripping the continents apart and leaving behind a new ocean basin. This idea answered questions concerning the pushing up of geosynclines, and the great faults in the Red Sea and eastern Africa.

Du Toit approached the problem from a much different point of view. He thought that the best proof of Wegener's hypothesis would be good supporting evidence and some corrections of Wegener's original idea. For example, rather than use the shorelines of the continents as edges of the jigsaw pieces, he fit them at the edges of the continental shelves.

It was Du Toit's idea that if there had been a Pangaea, then it had split during the Late Paleozoic into two large continents, Laurasia (*law RAY zhuh*) to the north and Gondwanaland (*gahnd WAH nuh land*) in the south. Between them lay a great geosyncline into which sediments were deposited for a hundred million years. These sediments formed the Himalayan-Alpine belt of mountains when Africa and India moved northward in Tertiary time.

Du Toit mapped Gondwanaland far more accurately than Wegener had tried to do (Figure 17-11). Then, from many studies of the rocks on South Africa, Australia, and South America, he prepared a list of events during the continental drifting.

Despite the contributions of Du Toit and the logical mechanism suggested by Holmes, the idea of continental drift continued to be considered with great skepticism through the 1940's. No group of scientists came forward to put the hypothesis to a serious test. The years

Figure 17-10 *Holmes' suggestion of how material below the earth's crust made the continents drift.*

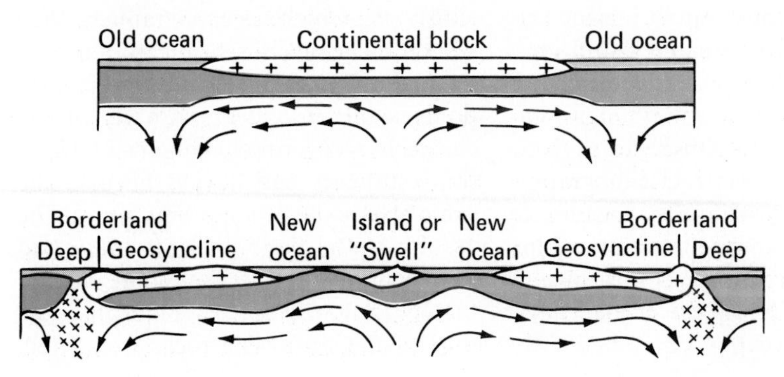

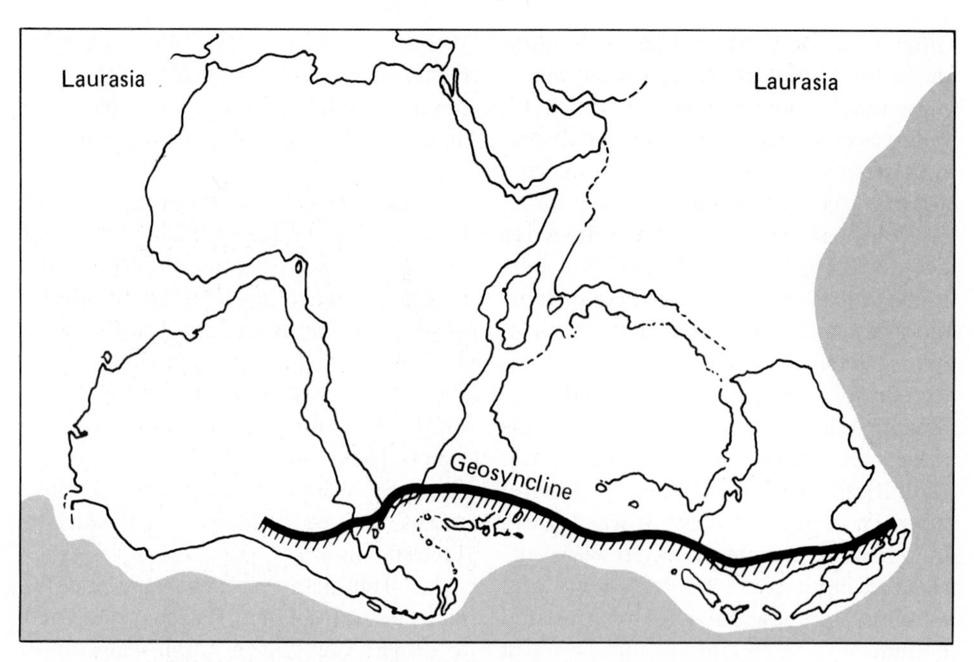

Figure 17-11 *Du Toit's version of Gondwanaland before drifting began in the Mesozoic Era. It shows, in more detail than Wegener could, how the present continents might have fitted together.*

from 1912 to 1950 were to be viewed later as a curious, uncreative phase in the history of earth science.

17-8 New Evidence for Drift

World War II was a turning point in research about the origin of continents and ocean basins. Research laboratories that were established during the war stayed active afterwards. This was especially significant in ocean studies, because three previously quiet marine stations became major research institutions.

Scientists at Scripps Institution of Oceanography (California), Lamont-Doherty Geophysical Observatory (New York), and Woods Hole Oceanographic Institution (Massachusetts) began to conduct cruises throughout all of the oceans. They were provided with surplus Navy ships that had sophisticated navigation and surveying equipment. By 1950, young scientists began to join these institutions. The U.S. Congress supported the rapidly growing research with adequate money and equipment. It created the first U.S. government agency whose only purpose was to support basic research—The Office of Naval Research—to be followed a decade later by the National Science Foundation.

Most of the surplus Navy ships used as research vessels were seagoing tugboats with powerful engines. This was a great change from the small motorized sailing ships on which oceanographers had worked. The scientists quickly took advantage of the power to lower huge dredges into the deepest parts of the oceans to recover rocks (Figure 17-12). If the continents had always been in the same places, the ocean basins must all be as old as the continents. If there had been continental drift then some parts of the ocean floor would be older than the time of the drift. The rocks would tell.

While cruising the oceans searching for locations for dredging, oceanographers used a new instrument, the echo sounder. From the data, they were able to determine the shapes of continental margins and learn of the smooth surface of the deep-sea plains. They also found a great number of seamounts.

Some seamounts were flat-topped and had coral rocks. They must have once been at sea level. The oceanographers therefore reasoned that they must be very old. They were perfect sites to dredge. But, nowhere did dredging recover rocks older in age than the Cretaceous Period! This surprise meant to marine geologists that older rocks must be buried beneath the thick layers of sediment that surely covered the deep-sea plains.

Another aid from war surplus was a great amount of explosives. Starting small artificial earthquakes would permit measurements of the thickness of the sedimentary layers. The resulting information was as puzzling as the mystery of the young rocks. There were not thick layers of sediments on the ocean floor (see Chapter 8). Instead, the layers were so thin that by considering a logical rate of deposition, they had been forming only since Cretaceous time. Could it possibly be that there were no older rocks at all in the ocean basins?

While marine geologists working mainly in the Pacific Basin were thinking about these questions, others were evaluating other sets of echo-sounding data. There appeared to be extensions of the well charted mid-Atlantic Ridge. Scientists were able to show that the Atlantic ridge system was connected to a similar feature in the Indian Ocean. It seemed that the mid-Indian Ridge extended through East Africa and into the Red Sea (Figure 17-13). Scientists found the broad, gentle, elongated rise in the eastern Pacific Ocean Basin (the East Pacific Rise) to be part of a world encircling ocean ridge system. The East Pacific Rise

Figure 17-12 *The rock dredge coming aboard the research vessel* Horizon.

extends into North America, through the Gulf of California! This did not exactly fit all of Wegener's ideas, but it did begin to seem that continents were being split!

One set of measurements gave results that were totally unexpected. A measuring instrument that was sensitive to heat was forced into the deep-sea mud. The temperatures it measured permitted oceanographers to learn the amount of heat flowing from the underlying rocks. They thought the heat flow on the ocean floor would be lower than the flow on the continents, because of the greater radioactivity in continental rocks. This was not the case. Average values of heat flow in the ocean basins were equal to those

Figure 17-13 *The Red Sea and Gulf of Aden, from an orbiting spacecraft. The view shows the fault zone that is a continuation of the mid-Indian Ridge.*

of the continents. And the heat flow at the mid-ocean ridges was greater than in any other part of the ocean basins. It was beginning to seem as if the convection currents suggested by Holmes might really exist. Maybe some currents were rising under the mid-ocean ridges.

The oceanographers, however, were not having all of the fun. A group of geologists at Cambridge University in England were measuring the magnetism of rocks collected from around the world. A rock that has iron minerals becomes magnetized while cooling if it is igneous rock, or during deposition if it is sedimentary. The basaltic lavas of the ocean basins are highly magnetic, as are red sandstones, which contain iron oxide. The magnetic minerals point in the direction of the earth's north magnetic pole at the time of magnetization. So it was possible to determine the pole's position in relation to the rocks throughout geologic time.

The results of the magnetism measurements showed that the north magnetic pole seems to have moved in a long curved path, from Precambrian to Triassic time. The problem was that the path projected from Europe was to the east of the path projected from North America. Study Figure 17-14. The best solution to the problem was to assume that there was no Atlantic Ocean before the Triassic Period. North America and Europe must have been together.

This surprise from British scientists was soon followed by paleomagnetic evidence from Southern Hemisphere rocks that seemed to agree with every suggestion that Du Toit had made. Not only was it easy to imagine a Gondwanaland, but the breakup could be placed during Cretaceous time, exactly as Wegener predicted.

New information on ocean basins, their sediments, rocks, and structure was accumulating at a rate never before experienced by earth scientists. Nearly every bit of data seemed to demand an explanation different from the idea of a rigid Earth. More and more, the idea of continental drift was brought up in scientific meetings. Still, no one put together a theory that everyone could agree with.

By the start of the 1960's, the time was ripe for ingenious, new ideas.

Thought and Discussion

1. How did Du Toit develop his proof of continental drift?
2. Why were scientists surprised at the young age of rocks in ocean basins?
3. What is paleomagnetism?

Unsolved Problems

As you can see, the story presented in this chapter leaves many questions un-

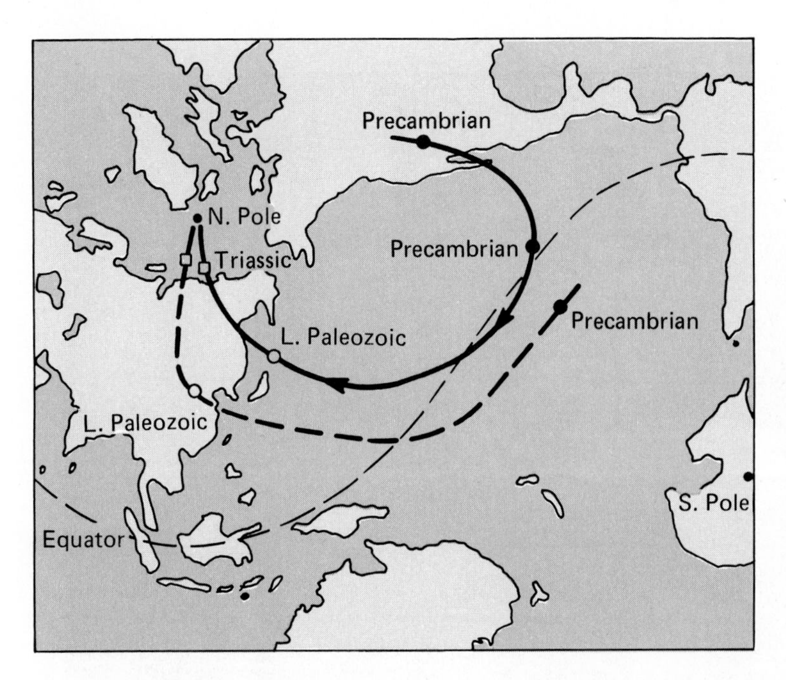

Figure 17-14 *Magnetic minerals show that the magnetic poles seem to have moved since the Precambrian. But the path shown by European rocks (solid line) was different from the path shown by American rocks (dashed line).*

answered. Why do the continents cover only 30% of the earth's surface? How are the mid-ocean ridges formed? Are they related to continental margins? Why are the rocks in the ocean basins so much younger than many rocks of the continents? Where are the sediments that should be on the sea floor if the ocean basins are as old as the continents?

Maybe continents have broken off from Pangaea, or Gondwanaland and Laurasia, and drifted to their present positions. Then there must be answers to questions such as, what is the mechanism that moves the continents? Why did the breakup of Pangaea not begin until the Mesozoic? Why are there deep-sea trenches only in the Pacific Basin? Are the mid-ocean ridges where the splitting took place? If so, where are remnants of the continents?

In the next chapter, you will find out how many of these questions have been answered in the last few years.

Chapter Summary

As the 20th century began, geologists all over the world were content with the idea that the earth had formed as a molten ball about 100 million years ago. It was slowly cooling and contracting. As it contracted, the crust of the earth wrinkled to form mountain ranges. The ocean basins and continents were considered to have been permanent features throughout earth history.

Geologic evidence of similar fossils, rocks, and folded mountains in Africa and South America, and the apparent fit of these continents led Alfred Wegener to propose the idea of drifting continents. He suggested that there had once

SCIENCE LIBRARIAN

Each year the volume of scientific and technical information being published grows larger. So much research is being done that it is impossible for most scientists to keep up with the new information. As a result, science librarians are becoming more important.

An important part of a librarian's job is classifying and cataloging materials. The science librarian must know how to locate information needed to answer a question. Many libraries now use computer systems to store and retrieve information. Once the material is located, the librarian has to interpret and organize it for the library user.

A science librarian must be able to evaluate new information and its importance. In order to help others with their research, the librarian must have a strong background in the subject area. Usually, a bachelor's degree in one of the sciences is required. This is followed by a one-year master's degree program in library science. Science librarians work in college and university libraries and in large public libraries. The number of jobs available in libraries run by science-related industrial and government organizations is increasing rapidly.

Science librarians, as well as other librarians, are helped in their work by library technical assistants. Preparation for these jobs is two years of education after high school or one to three years of on-the-job training. The job outlook for this field is increasing also.

been a single land mass, Pangaea. It began to split during the Mesozoic Era and even today the continents continue to move. Wegener's most convincing evidence was in the matching of ancient rocks indicating similar climates in the geologic past.

Those who did not agree with Wegener were more numerous than those who did agree. Most of the criticism was based on the lack of an understandable mechanism by which continents would slide, or "float" around the world. Wegener's allies were not able to present data that could convince those who believed in a rigid Earth.

World War II brought about a great change in the progress of research. New equipment, many more scientists and laboratories, and enough money to conduct oceanographic expeditions resulted in a rapid increase in our knowledge of ocean basins. We learned that sea floor sediments are thin, and oceanic rocks are no older than one-fifth the age of the earth. A mid-ocean ridge with a high heat flow extends through all ocean basins, and even into Africa and North America. Also, it seems as if the north magnetic pole wandered through a 90° arc during the early part of the earth's history.

None of the data gathered in the 1950's fit the idea of a rigid earth. Instead, all the data suggested that the continents had not been fixed, and that the ocean basins were young. There seemed little doubt. The continents had moved, but no one yet had a logical explanation for the whole pattern of movement.

Questions and Problems

A

1. Why would a cooling, contracting earth have folded mountains?
2. What part of the earth's surface first led Alfred Wegener to think that continents had once fit together?
3. How did earthworms help Wegener develop his hypothesis?
4. What did some people say was wrong with Wegener's idea?
5. Why did Alexander Du Toit favor the idea of continental drift?

B

1. In what way did the discovery of radioactivity influence our ideas of the age of the earth?
2. Why does the idea of convection currents in the earth help the hypothesis of continental drift?
3. How can rocks show where the north magnetic pole was in the past?

C

1. Why would tillite mixed with marine fossils indicate a continental ice sheet rather than isolated valley glaciers?
2. In what ways did the aftermath of World War II aid in developing the hypothesis of continental drift?
3. Describe the differences between the Taylor and Wegener hypotheses.

Suggested Readings

Dietz, Robert S. "Those Shifty Continents." *Sea Frontiers*, July 1971, pp. 204–212.

Tarling, Donald H., and Maureen P. Tarling. *Continental Drift.* New York: Doubleday and Co., 1975. (Paperback)

Wegener, Alfred. *The Origin of Continents and Oceans.* New York: Dover Publications, Inc., 1966. (Paperback) Translated from the 4th revised German Edition of 1929 by John Biram.

Wilson, J. Tuzo. *Continents Adrift.* San Francisco: W. H. Freeman and Co., 1972.

18 Crustal Plates: A Moving Sea Floor

THERE ARE TWO *points of view on the way science progresses. The older view is that if you collect enough data and make enough measurements, you will be able to understand Earth actions. However, what will guide your data collecting? Observing everything at the proper time and in the correct order is impossible. If you try to collect data without a plan, you are more likely to be swamped with too much meaningless information. You may never learn why the facts "behave as they do."*

A modern point of view is that science progresses quite differently. A scientist develops an hypothesis, perhaps an alternative to a popular theory, and then tests the new idea. In this way, data are collected in an orderly, planned manner. At each step of the way, scientists can test data against the hypothesis. The amount of data collecting can be reduced as they become more and more satisfied with their new idea. As the scientist says in considering Newton's law of gravity, "We need no longer record the fall of every apple."

Alfred Wegener's concept of continental drift is an excellent example of the newer idea of the way science progresses.

An example of the older view are some data collected in the late 1950's by two graduate students at Scripps Institution of Oceanography in California. Ronald Mason and Arthur Raff obtained some remarkable data on the magnetism of the sea floor. They found what seemed to be long, thin, parallel areas with their magnetism alternately oriented north and south. In other words, a compass needle would point north in one area and south in the one next to it. The magnetic strips had been cut and moved from side to side in an east-west direction by huge faults. The faults extended thousands of kilometers across the Pacific Basin, yet they seemed to disappear in the continent. No features had ever been observed that could explain such a pattern.

The strange magnetism of the sea floor remained a mystery for another five years. The data were "homeless"—they had no hypothesis in which to fit. But, once the right hypothesis was presented, magnetism of the ocean's crust became the "queen of all earth science data" in the 20th century.

AFTER COMPLETING THIS CHAPTER, YOU SHOULD BE ABLE TO:

1. **state the plate tectonics theory.**
2. **describe recent evidence that led to the final development of the theory.**
3. **describe how the theory explains the geologic history of eastern North America and the formation of Pangaea.**

A New Hypothesis—Sea-Floor Spreading

18-1 A Moving Sea Floor

Geologists studying the ocean basins were very puzzled by the young age of the rocks there. According to any theory of continental and ocean origin, some ocean basins must be as old as the earth. By searching and dredging they finally found rocks formed in the Early Mesozoic Era. They did not find any that were older. So, where were the ancient rocks?

The Pacific Ocean seemed the best place for such a search. It is probably the oldest ocean in any idea of earth history. And, its sea floor is unlike those of all the other oceans. The Pacific Ocean covers nearly half of the earth. The ocean ridge is in the eastern part, not in the center like the mid-Atlantic Ridge. There are thousands of volcanic islands and coral reefs. There are deep trenches and island arcs all around the Pacific Basin. Beneath the surface are thousands of flat-topped volcanoes. Professor Harry Hess of Princeton University (in New Jersey) believed that these strange seamounts were probably very ancient volcanoes. He thought they offered the best chance to find Precambrian rocks.

When research ships continued to return from cruise after cruise with rocks no older than the Mesozoic from seamounts around the Pacific, Hess was surprised. And there was another surprising discovery. Oceanographers couldn't find thick layers of sediments anywhere in the Pacific. The layers were only one-tenth as thick as they should be if sediments had been deposited for billions of years. Could the Pacific basin be no older than Mesozoic?

Most of the facts available seemed to be unrelated. But then Hess was struck by an answer that was almost too simple. If the ocean trenches were over sinking parts of a convection cell, then the ocean ridges must be over the rising parts. Material rising at the ridges would move toward the trenches, carrying the sea floor with it—like a conveyor belt. No wonder the ocean crust was young! It was being remade constantly! Made at the ridges, destroyed in the trenches (Figure 18-1).

Robert Dietz, a marine geologist, had spent many years working on data from the Pacific collected by the Scripps Institution of Oceanography. He came to the same conclusion reached by Hess, and at nearly the same time. And so, in 1961, the term **sea-floor spreading** was proposed by Dietz. It was quickly accepted.

The Hess-Dietz hypothesis answered many of the objections to the early ideas of continental drift. Continents do not cut through the earth's mantle like a ship cuts through the ocean. Instead, the continents ride gently on the tops of conveyor belts. That point had been made by Arthur Holmes some thirty years earlier. But Holmes did not have the data that existed in 1960. Now the hypothesis of sea-floor spreading could be tested with all the new methods and ideas developed in the preceding ten years.

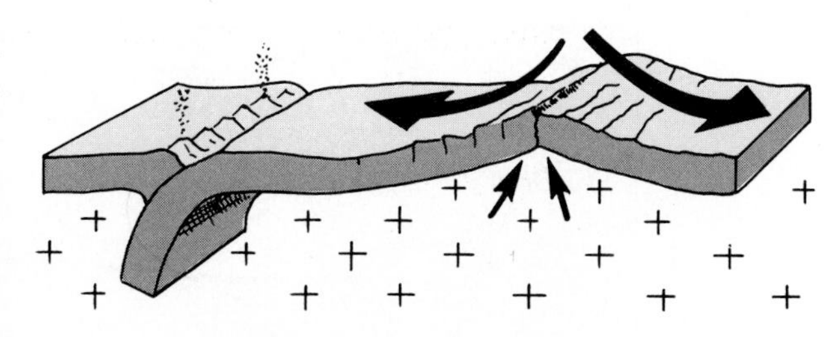

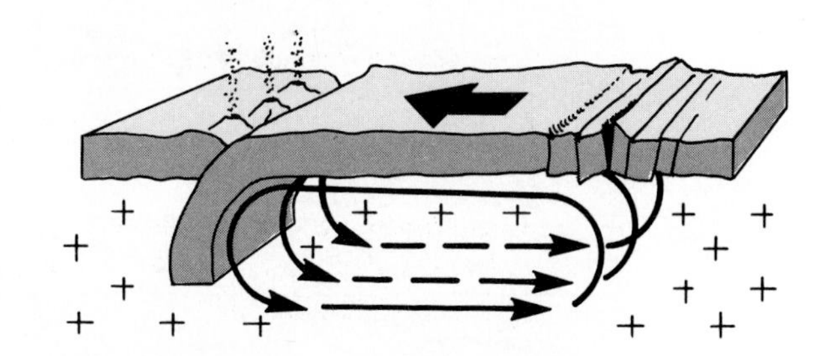

Figure 18-1 *The Hess-Dietz concept. The upper drawing shows how the crust seems to be moving. The lower one shows how a convection cell under the sea floor could be carrying it along.*

18-2 The Great Sea-Floor Faults

At the same time that Hess and Dietz were doing their work, other scientists were studying different features of the ocean floor. Professor H. W. Menard, at Scripps Institution of Oceanography, was investigating the great east-west faults on the Pacific Ocean bottom. These faults are thousands of kilometers long, 100 to 200 km wide, and form cliffs up to 3 km high (Figure 18-2). Horizontal slipping along the faults has been as much as several hundred kilometers.

Menard mapped the fault zones and learned that they extend well out into the central Pacific. The most remarkable is a fault that is 9800 km long!

It was soon learned that few, if any, earthquakes occur along these huge faults. They are, it appears, "fossil faults," or scars left from former movements of the sea floor. They are no longer active.

Other long, straight faults in the ocean's crust were found in other ocean basins. In each case, the faults cut across the mid-ocean ridge. And, wherever this happens, there *are many* earthquakes (Figure 18-3). In the Atlantic Ocean near the equator, the faults cut the mid-ocean ridge into pieces. The largest one has moved the mid-ocean ridge by 860 km. These faults are different from those in the Pacific because they cut an ocean ridge. The great faults in the eastern Pacific Basin do not.

From data collected by scientists at Scripps Institution of Oceanography and Lamont-Doherty Geophysical Institution, Professor Menard presented an idea that surprised all marine geologists. North of Cape Mendocino, California, pieces of a ridge step offshore towards Alaska (Figure 18-4). South of the entrance to the Gulf of California, no ridge had been found. Menard thought, therefore, that the ridge was buried under the continent. The "fossil" faults were all to the west of the ridge, because the eastern

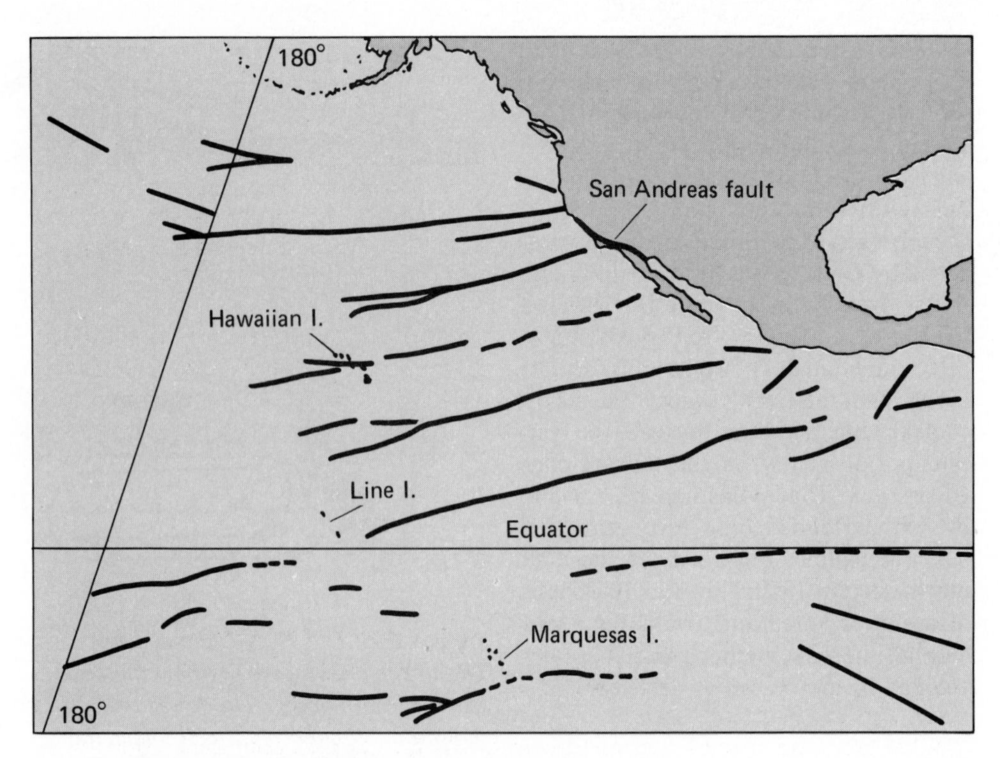

Figure 18-2 *The fault zones in the northeast and central parts of the Pacific Basin.*

Figure 18-3 *Earthquake epicenters in part of the Atlantic Ocean, recorded in the years 1965 to 1975.*

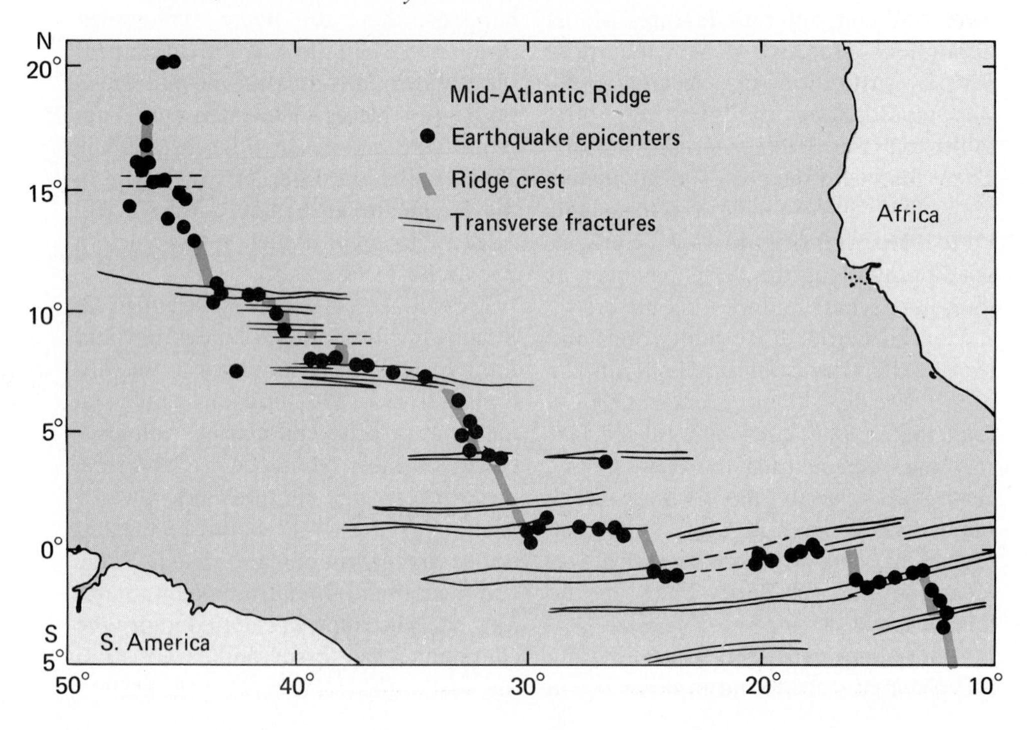

side of the ridge was buried. And, if the ridge had been covered by a drifting continent, then the ridge too must be "fossil."

Across the continent, J. Tuzo Wilson at the University of Toronto, Ontario, was greatly interested in all the new information about the structure of the sea floor. He thought it strange that the "fossil" faults seemed to end abruptly. In 1965 he proposed that the fault zones, ridges, and trenches divide the earth's surface into several large rigid **plates.** The plate boundaries are at one or another of these three structural features.

The sea floor spreads in opposite directions on either side of a ridge and fault zone (Figure 18-5). According to this idea, there would be many earthquakes. That is exactly what happens. Wilson also hypothesized that as the sea floor spread away from the ridge, it would carry with it the traces of former positions of the faults. But, they would now be inactive—the "fossil" faults of the east Pacific Basin.

This interpretation by Wilson neatly solved the problem of the disappearing East Pacific Ridge. California must be split by an active fault zone, the great San Andreas Fault (Figure 18-6). It is cutting the East Pacific Ridge, which comes into the Gulf of California and then extends northward. The ridge goes back into the ocean from Cape Mendocino in northern California.

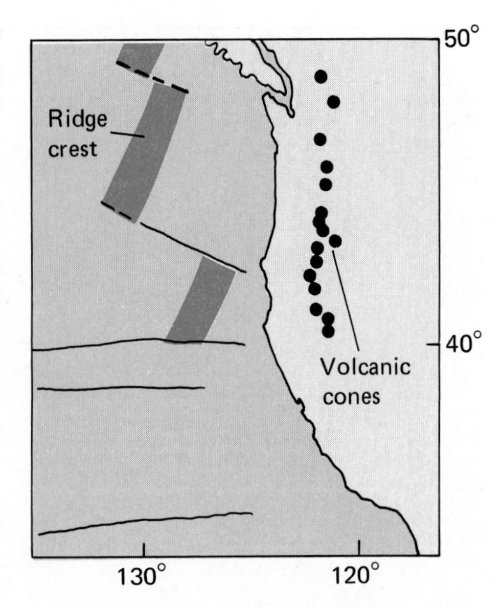

Figure 18-4 *Ridges and faults off the northwest United States and British Columbia.*

By the middle of the 1960's, the ridges, trenches, and fault zones of the sea floor began to make some sense. But although the evidence indicated the ocean's crust had moved, there was no conclusive information that it was still moving.

18-3 Reversals of the Earth's Magnetic Poles

Remember the two graduate students (from the chapter introduction) who found parallel magnetic areas in the Pa-

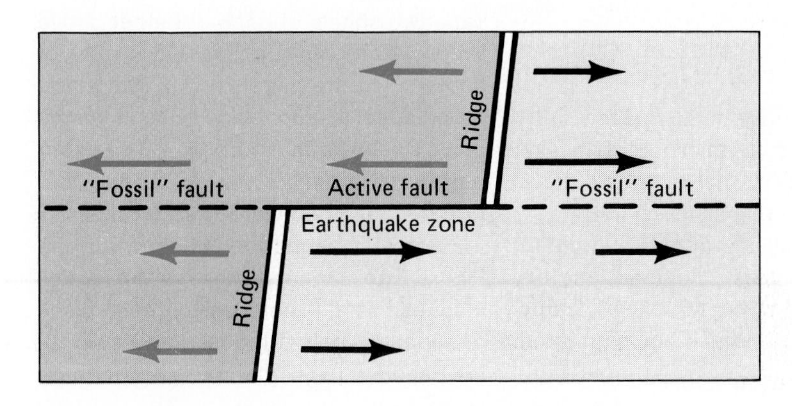

Figure 18-5 *The spreading motion along a fault zone.*

formed when magma rose up through the oceanic ridge. As the magma cooled, the iron-bearing minerals would become magnetized and would line up with the earth's magnetic field. The changes, from north-pointing to south-pointing and back, were caused by periodic reversals of the earth's magnetic poles!

The British scientists based their hypothesis on a discovery that had not been accepted when it was announced sixty years before. Data from lava flows showed that the earth's main magnetic pole had not always been in the north. Throughout geologic time, it seems, there have been many periods millions of years long when south was the important magnetic direction. Magnetic minerals in rocks formed during these periods would point south instead of north. There have been many reversals of the poles during the earth's history, but no one knows why.

Vine and Matthews decided that as the magma welling up from the ridge cooled, it would be magnetized in the direction of the prevailing magnetic pole. As these linear lava masses moved outward from the ridge, a series of parallel blocks would be formed that had alternate north and south magnetized material. Everywhere, the observations agreed with their hypothesis.

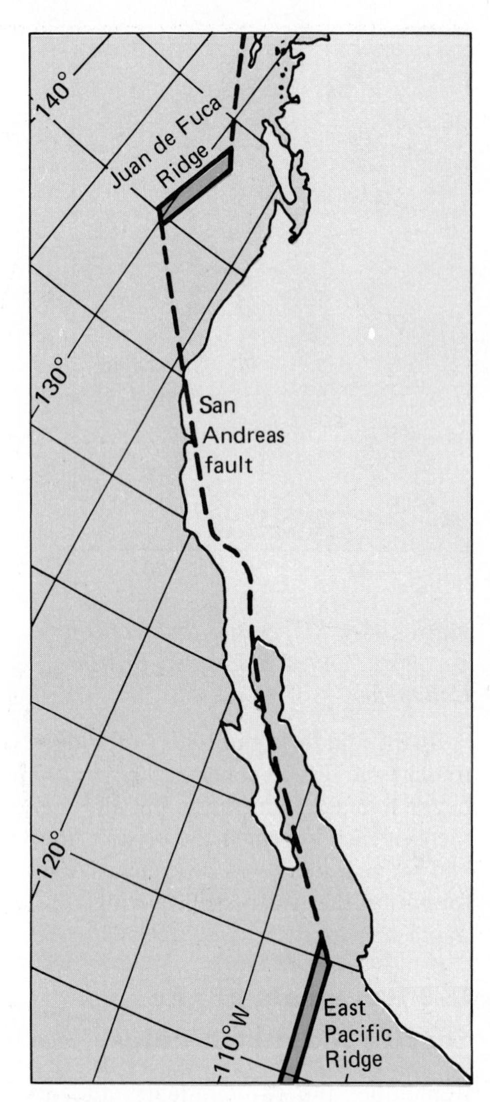

Figure 18-6 *The San Andreas Fault lies directly through California. It appears to be a part of the eastern Pacific ridge and fault zone.*

MATERIALS
plain white paper, tape, small magnet, two compasses (directional)

cific Basin? Scientists were finding similar areas in other ocean basins. It began to seem that the parallel magnetic areas existed in all ocean basins.

Then Fred Vine, a graduate student at Cambridge University, England, and his professor, Drummond Matthews, came up with an hypothesis. They suggested that the parallel magnetic patterns were

ACTION

Tape two sheets of paper together. Move two desks together or make two piles of books. Fold the paper in two where you have taped it, and place it so it hangs down as shown in Figure 18-7. The paper represents a part of the ocean floor before it spreads outward from a ridge.

Place a small magnet as shown in the Figure. The magnet represents the earth's magnetic field. A compass on each piece of paper will detect the magnetic field. (If the needle on each compass is not pointing toward the magnet, turn the magnet around.) Draw a line along each side to represent the edges of the oceanic ridge. Draw arrows in line with the compass needle.

Remove the compasses. Split the sea floor by spreading the paper away from the center for about 2 cm. Turn the magnet around 180°. Place a compass on the new area on each side. Draw arrows in the direction in which the compass needle is now pointing. Repeat the procedure several more times.

The Vine and Matthews hypothesis is an excellent example of fitting data to a previously established hypothesis—the Hess-Dietz idea of sea-floor spreading. If the magnetic reversal hypothesis were correct, there was a way to determine the speed of the sea-floor "conveyor belt." The age of the rocks could be found by using radioactive mineral dating methods. By comparing the ages of the rocks with the distance they had moved away from the ridge, scientists could determine the actual rate of spreading.

But, two questions remained unanswered. Were the magnetic patterns actually parallel with the mid-ocean ridges? Were the patterns of one side formed at the same times as those on the other?

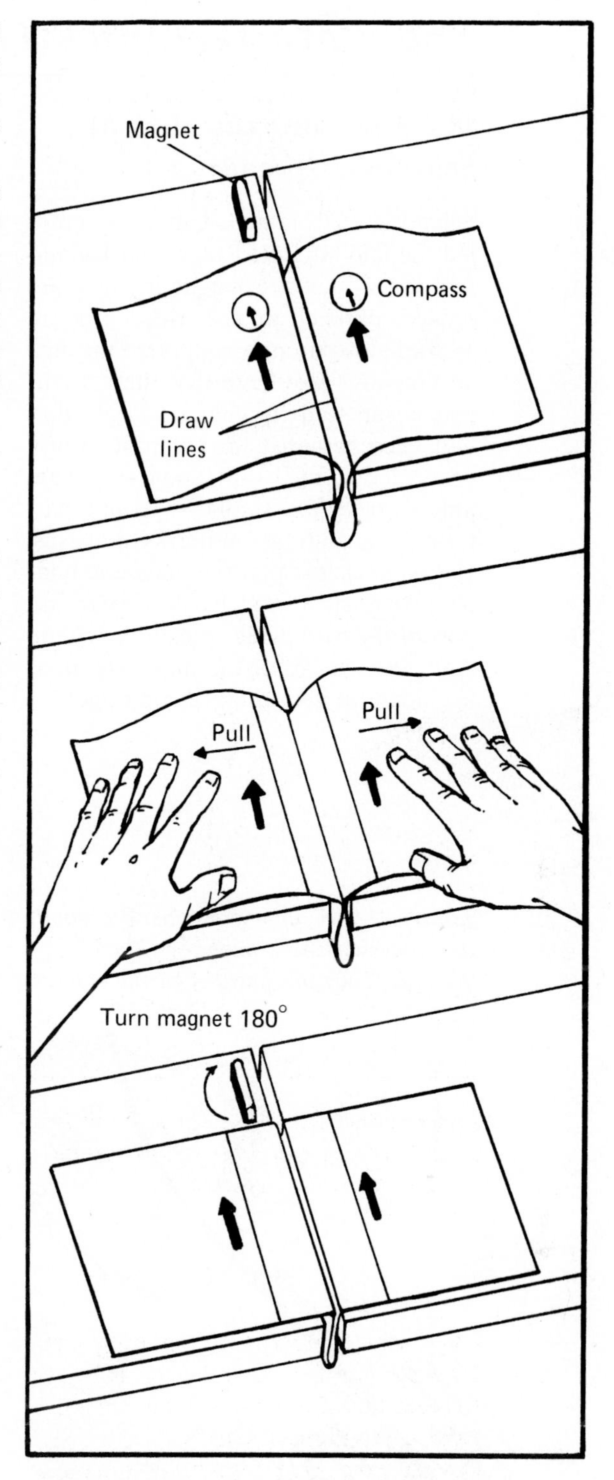

Figure 18-7 *This is a way to model sea-floor spreading.*

Thought and Discussion

1. Why were scientists surprised by the young rocks in the Pacific Ocean?
2. Why does the idea of sea-floor spreading answer the question of why the Pacific has young rocks?
3. What do fault zones tell us about the ocean's crust?
4. What is a magnetic reversal?

18-4 The Confirmation of Sea-Floor Spreading

Scientists at Columbia University studied the mid-Atlantic Ridge near Iceland and found that the magnetic areas are indeed parallel to the ridge (Figure 18-8). This so strongly supported the idea of Vine and Matthews that these scientists began studying magnetic data that had been gathered for the past twenty years all over the world. Soon, they were able to recognize similar magnetic patterns along all ridges in the ocean basins. It was now clear that the magnetic bodies first noted in the Pacific Ocean are part of the worldwide spreading system. And, in every case, the magnetic areas are lined up with the oceanic ridge!

Figure 18-8 *A map of magnetic normal and reversed zones in the sea floor near Iceland. They are parallel to the ridge.*

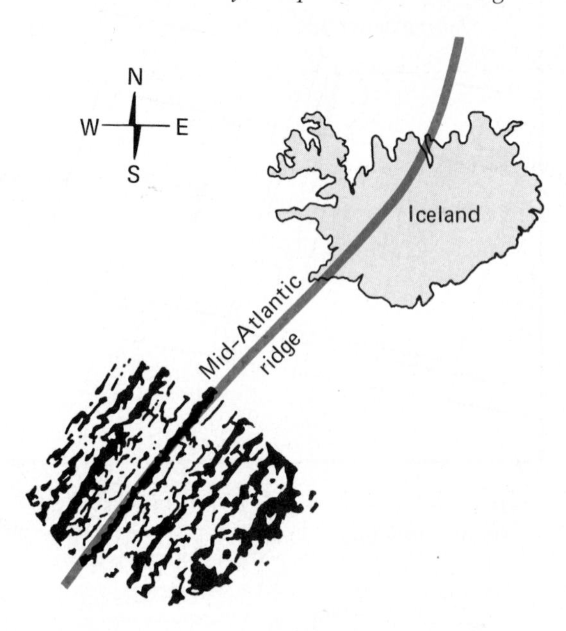

Now the researchers had to determine whether or not the magnetic areas on one side of the ridge matched those on the other side. This was shown clearly in 1966. Data from the Pacific ridge showed the amazingly close match-up. In the same year, Fred Vine at Cambridge presented similar data for ridges in all of the ocean basins. The Vine-Matthews hypothesis was in business.

But, scientific discovery in 1966 was not yet ended. For several years, geologists of the U.S. Geological Survey had been preparing a magnetic reversal time scale using rock dates based on radioactive-decay rates. Now they could figure out the rate at which the sea floor was spreading (Figure 18-9). The rates ranged from about 1 cm per year (in one direction) near Iceland, to 4.5 cm per year in the Pacific. How much has the Pacific Ocean bottom moved during your lifetime?

By carrying those rates back in geologic time, it was clear that the ocean floor could be no older than the Mesozoic. Hess and Dietz had been right, then. The ocean's crust did indeed renew itself about every 150 million years. But this created another question. If the sea floor was renewed, where did the old crust go? Surely, the trenches surrounding the Pacific Basin could be places where the crust was carried down.

18-5 The Basic Crustal Plate Theory

The theory that came out of all the information gathered about the earth's crust went back to the ideas of Wilson. He had suggested that the sea floor spread in segments or plates, with the boundaries of the plates being fault zones, ridges, and trenches. Three young

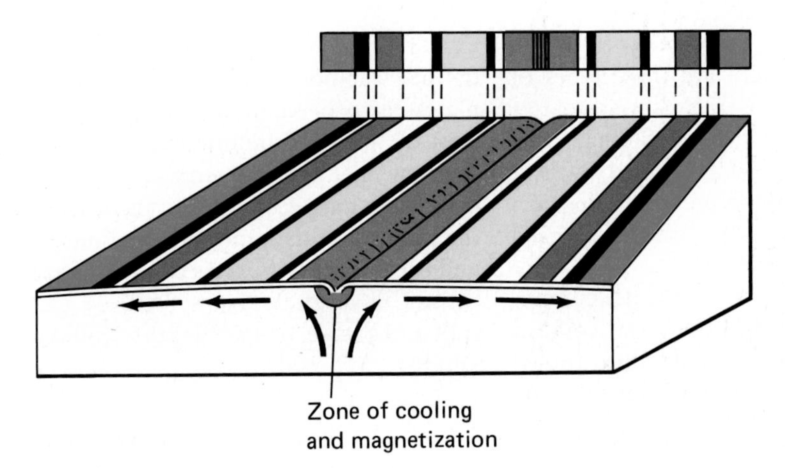

Figure 18-9 *The sea floor spreading from a ridge can be dated if the ages of the rocks are known. In these strips of normal and reversed magnetized crust, the red strips were formed from the present to 750,000 years ago. The blue strips were formed from 3.5 million years ago to 4 million years ago.*

scientists fitted all the facts into the theory. They were Jason Morgan at Princeton, Daniel McKenzie at Cambridge, and Xavier LePichon at the Lamont Observatory. Morgan analyzed the boundaries as they appear on a sphere and saw that the earth's crust could be divided into 20 plates. McKenzie showed that the idea of plates moving around on a spherical earth's surface could be calculated easily. LePichon simplified Morgan and McKenzie's idea into six large crustal plates, named after the continents (Figure 18-10), and many that are smaller. This theory has become known as the **plate tectonics theory.**

According to the plate tectonics theory, the earth's crust is strong and fairly rigid to a depth of about 100 km. Earthquakes that are the result of faulting come from this zone. The rocks at the top of the mantle are so hot that they are close to their melting temperature. They are more "plastic" than the crust. That means they are soft enough to flow without faulting, and slowly carry pieces of the crust with them.

Most volcanic and earthquake activity is confined to narrow zones such as the "Ring of Fire" around the Pacific. These

Figure 18-10 *The earth's surface, divided into the six major crustal plates.*

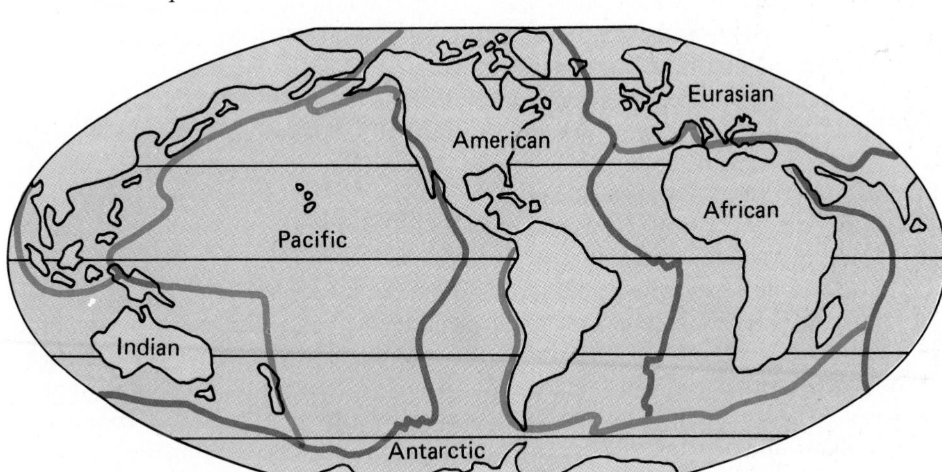

active zones are the boundaries of the crustal plates. There are three types of boundaries:

Ridges from which spreading takes place and new crust is formed by lava rising from the deep mantle.

Trenches where the crust is forced down and is destroyed.

Fault zones where crust is neither created nor destroyed. Rates of motion can be determined where fault zones are active, at ridge crests. Where they are "fossil faults," they indicate earlier plate movement.

The amount of spreading can be estimated quite well from magnetic areas and the magnetic-reversal calendar. However, the actual rate and direction of motion can only be determined by exact measurements, as Wegener suggested so many years ago.

Geologists will be making such measurements during the late 1970's, with lasers and satellites. Even without the exact measurements that will come from the new methods, we know that crustal plates have been in motion during at least the past 150 million years. We can see now that the old term, continental drift, is no longer useful. This is because the plates are made of both continental and oceanic crust. But, the continents are carried on crustal plates. So, they can split and collide, just as Wegener and the Du Toit supposed.

Thought and Discussion

1. What two discoveries confirmed the sea-floor spreading hypothesis?
2. How does the speed of sea-floor spreading compare with the rate of uplift in areas where uplift has been measured?
3. What features of the earth's crust form the eastern and western borders of the plate that North America is on?

Plate Tectonics and Geologic History

18-6 Crustal Plates and the Continents

You have learned now about a series of hypotheses, each of which has been tested and found to make sense out of the complicated nature of the earth. A new pattern for the development of land and ocean basins has come from these bold ideas. The new theory has been applied to many traditional problems. Of these, the most exciting is Wegener's suggestion of continental drift.

The idea that continents can be carried thousands of kilometers by crustal plates can be accepted. The question to be answered, though, is whether or not they ever did fit together.

You know that as the plates move, they spread at mid-ocean ridges and one is forced under another at the trenches. As two plates move from a ridge, the opening is quickly filled by new crust. But, not all ridges are in the centers of oceans. Only in the Atlantic and Indian oceans is this the case. These must have formed where continents had once split apart (Figure 18-11). It would then be natural for the ridge system to be in the center. The Pacific Ocean is not an opening ocean. It must be older, then, than the Atlantic and Indian oceans.

With these ideas in mind, several efforts have been made to reconstruct the continents into Wegener's great continent Pangaea. The fit has been made by using a continental borderline that is about halfway down the continental slope. There the continents probably have not been changed much by erosion or deposition. British and American scientists used computers in making this fit. Some questions still need to be answered, but the fit seems good, as you can see in Figure 18-12.

Pangaea was surrounded by one great ocean, which is the ancestral Pacific. Pangaea began to break up at the end of the Paleozoic Era. By the end of the Triassic Period the Atlantic and Indian ocean basins began to form (Figure 18-13a). A major split from east to west created Laurasia and Gondwanaland. Soon afterward, India broke from Antarctica and began to drift rapidly to the north.

Thirty-five million years later, the east coast of the present United States and Canada ran almost east and west at a latitude of about 25° north. Coral reefs could grow along the edge of the continental shelf. In the Southern Hemisphere, a split began between Africa and South America.

Figure 18-11 *Stages in the splitting of a continent. A new ocean is formed, with its mid-ocean ridge.*

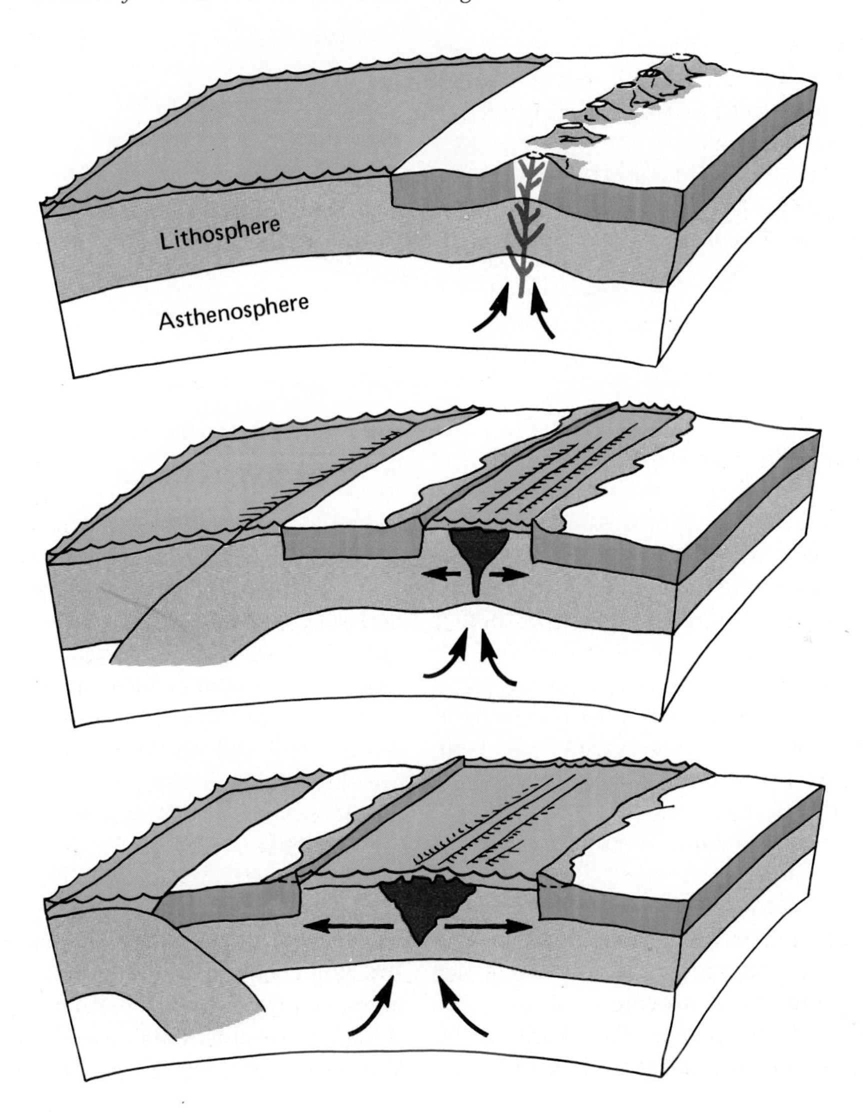

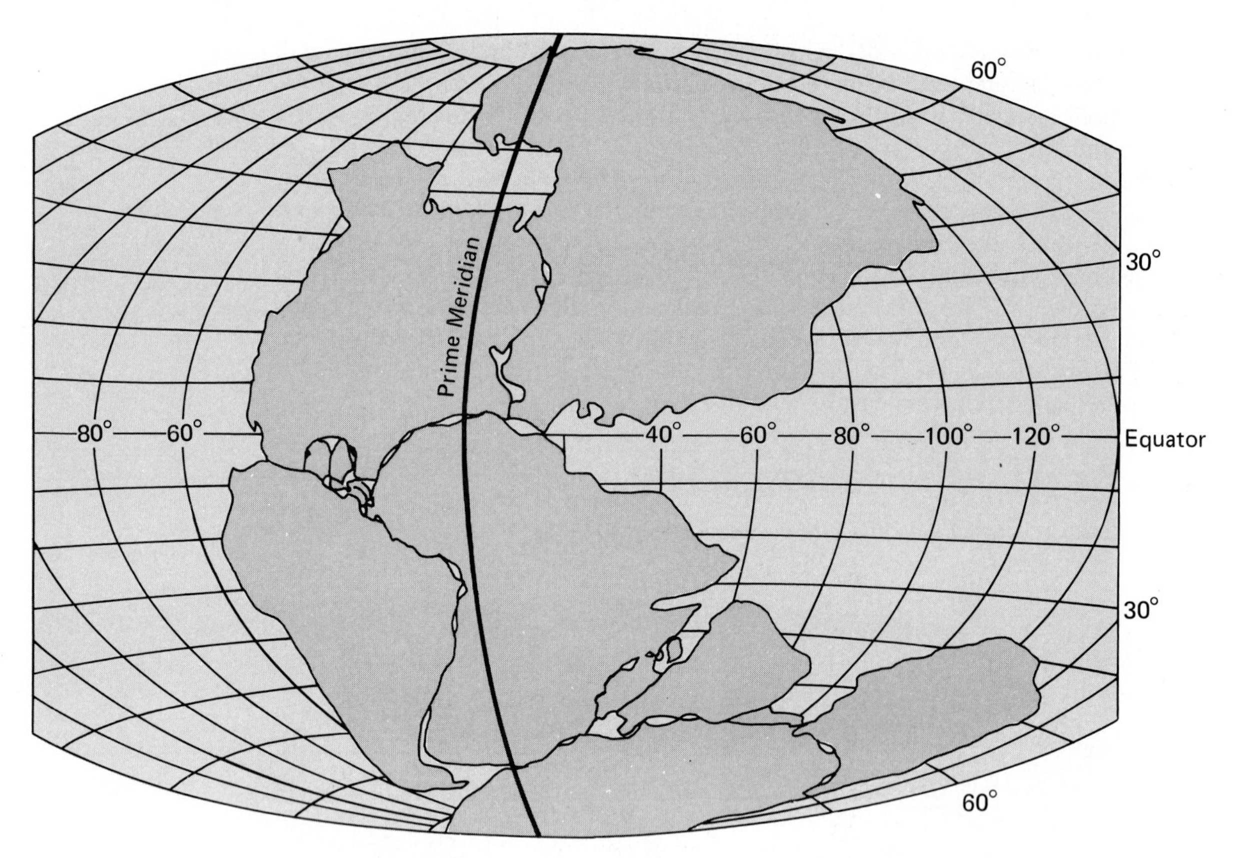

Figure 18-12 *Wegener's Pangaea, as restored by modern scientists.*

By the end of the Mesozoic Era, some 65 million years ago, the separation of Africa and South America was complete (Figure 18-13c). North America still remained attached to Eurasia with Greenland caught in the middle. Australia and Antarctica were still connected to each other.

During the Cenozoic Era, the plates have moved to their present positions. In its last move, the Indian Plate completed its northward journey by bumping into Asia. This collision formed the Himalaya Mountains. Also during the last 65 million years, Australia split from Antarctica.

Many details of the reconstruction of Pangaea remain to be explained. Yet basically, the amazing foresight of Alfred Wegener in fitting the continents together has been proven correct.

18-7 Plate Boundaries and Geosynclines

When you studied the rocks of a geosyncline during the James Hall field trip (Chapter 9), you learned that the deposits were thicker toward the east than they were on the west. Now that you know about crustal plates, let's revisit geosynclines.

The shelf and continental rise off the eastern United States today run along the coast for 2000 km. The rise alone forms a sedimentary apron 250 km wide (review Figures 8-5 and 6). The rise is made up of sediments whose greatest thickness is about 10 km. The sediments are formed from turbidity currents that have moved down submarine canyons and poured out onto the rise. Mixed with those layers are deposits of clay, and the

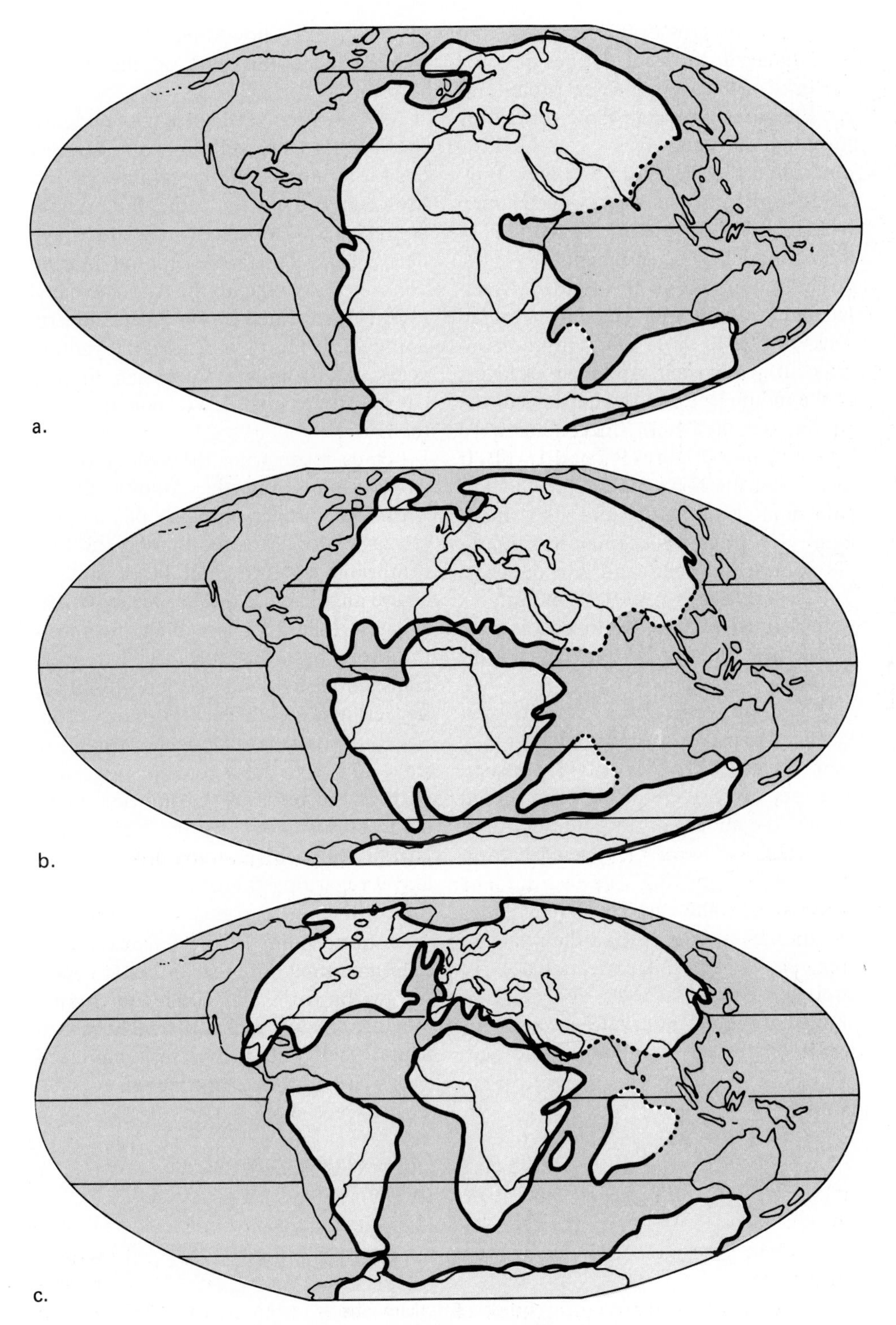

Figure 18-13 *The moving continents, according to recent data.*

remains of microscopic organisms that have fallen down from the ocean.

Rocks of the Appalachian Mountains have features similar to those formed by turbidity currents. There can be little doubt that the Piedmont and New England-Acadian Provinces (review the map in Figure 11-18) contain the features of old continental-rise sediments.

The mountains in the Folded Appalachian Province are formed from folded rocks similar to those found in the modern continental shelf. And the rock layers of the mountains and the deposits of the continental shelf both thicken eastward before ending (Figures 9-7 and 18-14). It seems that the rocks making up the Appalachian Mountains and the Piedmont-New England-Acadian regions are old continental shelf and rise deposits. The eastern North American mountains were formed during the Paleozoic Era. Were they caused by moving crustal plates?

If you study the folds of the Appalachian Mountains you can tell that they were formed by a horizontal pressure. The rock layers were pushed and folded against the middle of the continent. But, the Atlantic Ocean is *opening;* North America is moving westward. The mountains of the eastern United States are in the middle, rather than on the front edge, of the American Plate. To explain the Appalachians, you must assume that the Atlantic was once a *closing* ocean. Plates were colliding! The ages of the rocks show that the collision happened at the end of the Paleozoic Era.

As you have seen, geologic evidence shows that the splitting of Pangaea began at about the beginning of the Mesozoic Era. At that time, the original Appalachian Mountains would have been inside Pangaea. Yet, the marine deposits that make up the Appalachians' rocks were formed in the Paleozoic Era. Some rocks are as much as 500 million years old, from the Cambrian Period. Clearly, there must have been an ocean then.

Let us reconstruct the history of the Appalachians and the Atlantic Ocean with our understanding of moving crustal plates. At some time in the Precambrian, a continental block split to create an ancient Atlantic Ocean. By the process of sea-floor spreading, the ancestral ocean basin was formed. There were continental shelf and rise deposits on the bordering plates. The earth's crust probably bulged upward along the mid-ocean ridge and sank between the ridge and each of the parts of the splitting continent. This sinking, by the way, solves part of the problem noted in Chapters 8 and 9 regarding the constant sinking of the continental margins.

At some point in the Paleozoic Era, the spreading of the Atlantic basin stopped and the old continents started moving toward each other. The continental shelf and rise were compressed

Figure 18-14 *A cross section of the Folded Appalachian rocks in Tennessee, from northwest to southeast.*

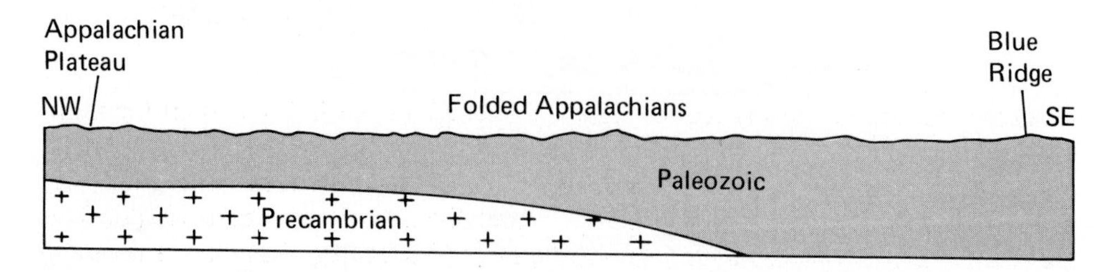

and the formation of the Appalachians began. The final scene was completed when the two continental plates collided as Pangaea came into existence. This took place between 350 million and 200 million years ago. The final act in this sequence began with the splitting of Pangaea. Today the Atlantic continues to open at a rate of about 3 cm per year, and new wedges of geosynclinal sediments are forming along the edges of the Atlantic Ocean.

Thought and Discussion

1. Why would the modern continents not fit at their coastlines if you could push them together to form Pangaea?
2. North America has moved westward since the Triassic Period. In what other way has it moved?
3. How do the present Appalachian Mountains suggest the way that Pangaea formed?

Unsolved Problems

As with all theories, there are many details that remain to be fit properly into the overall scheme. In the theory of moving crustal plates, there are many questions that are still unanswered.

In any reconstruction of the fit of continents, the Gulf of Mexico and the Caribbean Sea make problems. No matter how Yucatán, Honduras, and the other land masses are moved about, the gap between North and South America is never quite filled. Scientists think that Panama and Costa Rica are younger than Pangaea, formed as new lands along the Central American Trench.

Another mystery involves the Bahamas. This large mass of coral and limestone creates an overlap when the continents are refitted. The solution considered most logical is that the Bahama Banks were formed long after the split between Africa and North America took place. The base of the Banks is probably volcanic. The Banks could have formed in a way similar to the present-day growth of the volcanic island, Iceland.

Then, there are several islands whose fit into Pangaea is unsolved, such as the Canary Islands and Madagascar.

And what of Antarctica? There are no trenches around that continent. It does not seem to be moving, nor has it moved for many, many years. It is locked in by surrounding plates and can move only when they move. All the other plates move independently of each other.

These and other unsolved problems are under active study. In 1978, the spreading boundary of three plates in the south Atlantic Ocean was studied. Major expeditions are under way in the Indian and Pacific oceans, and in 1983, detailed studies began on Antarctica's boundary.

Despite the unsolved problems, oceanographers continue to be excited by the new theory because it does seem to bring harmony to geology.

Chapter Summary

Rocks in the ocean basins are younger than the Paleozoic Era. On top of these rocks of the ocean's crust are sediments that are but $\frac{1}{10}$th as thick as you might expect if the ocean basins date from the origin of the earth. To account for these unexpected facts, two American scientists suggested that the sea floor was spreading, from the ocean ridges. The rate of spreading that they suggested

INSTRUMENTATION TECHNICIAN

Scientific research has been greatly changed in recent years by the development of complicated instruments to measure, record, and analyze data. Technicians with special training work with the scientists and engineers who develop and use the equipment. They form an important link between the scientists and the people who actually construct the equipment. The technician's job is to turn ideas into results.

Instrumentation technology is important to many scientific fields. In recent years, oceanographic and space exploration have come to depend on new-instrument development. Meteorology, atomic energy research, and oil refining are other examples of fields employing large numbers of instrumentation technicians.

Training for work as an instrumentation technician can be gotten on the job. However, most employers prefer to hire people with some special education. Two years of training after high school is recommended for most jobs.

would account for the young age of the sea floor.

Great fault zones on the sea floor indicate that the ocean's crust is composed of different plates that move next to each other. The speed of movement was answered when British geologists were able to show that reversals of the earth's magnetic poles could be directly related to long, narrow magnetic strips on the sea floor.

By a number of careful studies, some Americans showed that the magnetic areas in the ocean's crust were parallel to the ocean ridges, and that the areas on one side matched up with those on the other. This permitted a magnetic reversal calendar to be constructed and sea-floor magnetic patterns to be dated back in time for 80 million years.

Finally, three young scientists recognized the plate boundaries as fault zones, trenches, and ocean ridges. They also studied the different spreading rates. They put the facts together into an idea of crustal plates in motion, all moving from a world-wide oceanic ridge.

Their theory shows that Alfred Wegener's half-century-old idea of a single land area, Pangaea, does indeed seem to be correct. It also answers many questions about the sources of earthquakes and volcanoes. It gives us ideas about the origins of ocean basins, continents, mountain ranges, and deep ocean trenches.

It is the most important new earth science theory since the theory of uniformitarianism.

Questions and Problems

A

1. Why did Harry Hess think seamounts were good places to find Precambrian rocks?
2. Why were the great fault zones in the Pacific Basin called "fossil faults"?
3. What is happening at the mid-Atlantic Ridge?
4. What is happening at the trenches around the edge of the Pacific Basin?
5. How does the plate tectonics theory explain the sinking of the geosyncline that eventually became the Appalachian Mountains?

B

1. What was the simple answer that Hess and Dietz devised to account for the young sea floor?
2. Where does the reversal of magnetic poles fit in the Vine-Matthews hypothesis?
3. The ridge systems of the Atlantic and Indian oceans are in different positions inside their oceans from the ridge system of the Pacific Ocean. What does this tell you about what is happening in the basins, and their relative ages?

C

1. The idea of continental drift was conceived by Wegener, a young nongeologist. The solution to parallel sea-floor magnetism was found by a young geology student. Three other young scientists were involved in the final details of the plate tectonics theory. Is there a message here, about how science progresses?
2. How does the plate tectonics theory solve the problem of the formation of ancient mountain ranges?
3. Estimate where on the earth's surface your home area will be ten million years from now.

Suggested Readings

Ballard, Robert D. "Window on Earth's Interior." *National Geographic Magazine* (August 1976), pp. 228–249.

Colbert, Edwin H., et al. *Our Continent.* Washington, D.C.: The National Geographic Society, 1976.

Heirtzler, James R. "Where the Earth Turns Inside Out." *National Geographic Magazine* (May 1975), pp. 586–603.

Sullivan, Walter. *Continents In Motion; the New Earth Debate.* New York: McGraw-Hill Book Co., 1974.

unit VI
Exploring the Universe

19 The Moon—The First Step Into Space

THE MOON *has a strikingly different appearance from that of planet Earth. Its stark, nearly colorless landscape is covered with craters. These craters range in size from hundreds of kilometers in diameter to only a few centimeters across. The rugged parts of the moon are marked with deep valleys and high peaks. Other parts of the moon's surface are relatively smooth. These dark plains make up the features of "the man" people see in the moon. The moon has no oceans, no river valleys, no water, no clouds, no air—and therefore no sound, no life.*

The rate of change on the surface of the moon is almost at a standstill compared to here on Earth. On Earth the clouds change noticeably in a few minutes or hours. Hardwood trees change noticeably in months or years. The rivers, oceans, and even mountains have all changed through the centuries, but the moon looks the same as always. It has probably not changed noticeably since the Apollo teams left their footprints on its surface. Now, with scientific instruments on the moon's surface, we may be able to detect changes that were too small to detect from Earth.

We see the same side of the moon and the same features that Cleopatra saw when she watched the night sky from the banks of the Nile. The moon's face is a real link with history. Queen Elizabeth I, Napoleon, and George Washington all saw the same moon you see. Over the centuries people have looked at the moon with awe. Some people worshiped it as a goddess or god. Countless lovers have courted under its reflected light. It continues to inspire artists and poets. And now people have even traveled to it, landed on it, studied it, and brought pieces of it back to Earth.

AFTER COMPLETING THIS CHAPTER, YOU SHOULD BE ABLE TO:

1. **describe the general topography of the moon.**
2. **compare and contrast the landscape of the moon with that of the earth.**
3. **describe lunar soil and the major types of surface rock present on the moon.**
4. **use the law of superposition to put overlapping moon features in a time sequence.**
5. **compare several hypotheses of the moon's origin, and outline its history.**
6. **describe ways of measuring the distance to the moon.**
7. **explain how the interaction of the sun and moon causes tides and affects other conditions on Earth.**
8. **explain how the relative positions of the earth, moon, and sun cause the different phases of the moon.**
9. **tell the approximate time by checking the phase of the moon and its position in the sky.**
10. **explain how the relative positions of the earth, moon, and sun cause lunar and solar eclipses.**

From the Earth to the Moon

19-1 Features of the Moon

You have been familiar with the moon for as long as you can remember. Even if you don't know much about the moon, you have noticed that it has dark areas and light areas. You probably know that it is covered with circular pits called **craters.**

If you look at Figure 19-1a, and if you look closely at the moon at night, you will notice that some of the dark areas look like very large flat craters. The dark regions are officially called **maria** (*MAH ree uh*), which is the Latin word meaning "seas." The singular of maria is **mare** (*MAH ray*)—one mare, two maria. The ancient observers thought they were oceans. Maria have an approximately circular shape. This shape shows very well in the Sea of Serenity and the Sea of Crises. The circular shape also shows, but less well, in the Sea of Rains and the Sea of Tranquillity. The larger maria form the "man in the moon."

The light areas of the moon are regions with much rougher landscapes than the maria. You can see in Figure

19-1a that they have mountains and a great number of craters mixed together. These regions are called the lunar **highlands.**

Figure 19-1b shows the farside of the moon, which we never see. As the moon moves around the earth, the nearside always faces us. The moon both rotates on its axis and revolves around the earth in 27⅓ days. How would you describe the difference in appearance between the nearside and the farside?

The landscape of a mare is shown in Figure 19-2. It appears to be a flat plain, covered to the horizon with sandy soil and rocks. It also has numerous craters punched into it.

In contrast, the highlands are very mountainous. (See Figure 19-3.) Moon mountains look different from Earth mountains in two ways. First, moon mountains have smoothed and rounded profiles, which are different from the rough and angular shapes of many Earth mountains. Second, there seem to be few large outcrops of bare rock. It appears that the moon mountains are covered to their tops with soil.

Everywhere in the mountains, as on the maria, there are craters of various sizes. Look at the crater in Figure 19-4a. Compare it with the photograph of Meteor Crater in Arizona (Figure 11-17). Note the raised rim and the smoothly curved interior of Meteor Crater. The crater Copernicus, shown in Figure 19-4b, is named after the famous astronomer. It is 92 km wide and is one of the most obvious features of the moon. It looks as if its inner walls had slumped downward, suggesting that it is a collapsed volcanic crater. What do these types of craters tell you about the history of the moon?

Here and there on the moon's landscape are canyon-shaped valleys called **rilles** (*RIHLZ*). They are thought to be lava tunnels that collapsed, or breaks in the surface caused by faulting. Hadley Rille (Figure 19-3) is as much as 4000 m deep.

Figure 19-1 a. *The nearside of the moon is the side you see from the earth.* **b.** *The farside. How does it differ from the nearside?*

***Figure 19-2** The Sea of Tranquillity, photographed by the first astronaut team to reach the moon, in 1969. The clean break between the two rocks in the foreground indicates that they may have been a single large rock that split. Beyond the rocks is a small crater.*

How high are moon mountains? Since there is no water on the moon, there is no sea level. Therefore, scientists decided upon an imaginary sphere that is about halfway between high and low points on the moon's surface. It is a sphere of 1737 km radius. Using that as "moon level," the Apennine (*AP uh nyn*) Mountains are about 2.5 km high, and some mountains on the farside are as much as 5 km high. The maria whose depths have been measured are 2.5–4.0 km deep.

19-2 A Closer Look at the Moon

Lunar maria are flat, with no outcrops of bedrock, and are covered with soil. The soil consists of small pieces of rock and rounded blobs of natural glass. The glass must have formed from rock material that heated to the melting point and then cooled very fast. The melting could have happened when meteoroids struck the moon. A **meteoroid** is a chunk of mat-

***Figure 19-3** A lunar highlands landscape. The largest mountain, Hadley Delta, rises about 3500 m above the plain in the foreground. The canyon on the right is the Hadley Rille. This area was visited by astronauts in 1971.*

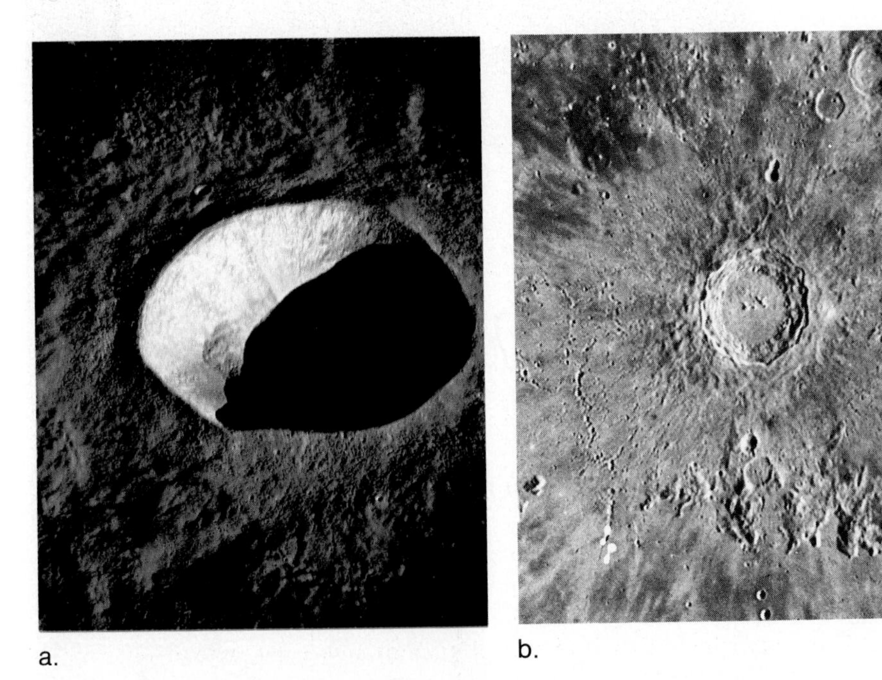

Figure 19-4 *Craters on the moon.* **a.** *How does this simple crater compare with Meteor Crater in Arizona (Figure 11-17)?* **b.** *The crater Copernicus. Note the central peak and the rays.*

ter traveling through space. It is reasonable to assume that the meteoroids were traveling so fast and hit the moon's surface so hard that the rock or soil where they hit was melted. Droplets of the molten rock flipped up and cooled immediately into glass blobs. The meteoroid impacts probably broke up whatever rock did not get heated enough to melt. The impacts of meteoroids that were too small to cause much rock melting might also break moon rocks into soil fragments.

Scientists who have studied moon rocks and soil have suggested that the maria have crusts of basalt. The bedrock has been ground up into soil and broken rock by meteoroid impacts. Samples that the astronauts brought back from the Sea of Tranquillity show that the larger rocks there and a large number of the soil particles are basalt.

Look at Figure 19-5. In the middle of the photograph you can see an outcrop

Figure 19-5 *Lava layers in the wall of Hadley Rille.*

a.

b.

Figure 19-6 *Rock samples from the moon.* **a.** *Basalt.* **b.** *Anorthosite.* **c.** *Breccia.*

c.

of rock layers. They look like layers of hardened lava here on Earth. From the edge of the rille to the horizon is a plain that extends out into the Sea of Rains. Figure 19-6a is a photograph of a sample of basalt from the Hadley area.

A second common rock type on the moon is a plutonic igneous rock called anorthosite (*an AWR thuh syt*). (See Figure 19-6b.) Remember, a plutonic igneous rock cooled below the surface from molten material.

The last astronauts to visit the moon landed in a valley between the Sea of Serenity and the Sea of Crises. The bedrock there is a series of basalt layers 1–2 km thick. The valley may be a down-dropped fault block. (Review Figure 9-9.) The mountain on one side of the valley appears to be a mass of anorthosite-like rock. The mountain on the other side is formed of another kind of rock, described in the next paragraph.

All of the travelers to the moon's surface have discovered a third major rock type. It is called breccia (*BREHCH ee uh*), from an Italian word referring to a rock similar to conglomerate, but whose pebbles are angular (with sharp edges and points) instead of rounded. Conglomerate is shown in Figure 12-2a. Breccia (Figure 19-6c) occurs in the maria as well as the highlands. It seems to be solidified moon soil because it has rock fragments of various sizes cemented together by natural glass.

Pieces of breccia were found in the Sea of Tranquillity, and in the plain next to the Apennine Mountains. It also comes from the Apennine Mountains themselves. Near the Descartes (*day KAHRT*) Crater is a series of layers of breccia at least 200 m thick.

The minerals that have been found on the moon are also found on Earth. Three common moon minerals are feldspar, olivine, and pyroxene, which are familiar Earth minerals. The moon is different from the earth in that no quartz was collected by the astronauts. (You may wish to review the minerals in Chapter 6, especially Figures 6-8 and 6-9.) Figure 19-7 is a microscopic view of a moon rock with three distinct minerals in it. The minerals are feldspar, pyroxene, and an oxide of iron and titanium called ilmenite. Compare this microscopic view of the moon rock with the similar view of the granite in Figure 6-3c.

For years people have jokingly said that the moon is made of "green cheese." During the Apollo program, several scientists studied sound velocities in lunar rocks. (This is the speed at which sound travels through the samples.) The scientists discovered that the sound velocities in moon rocks were considerably different from all 24 Earth rocks used in comparison. They did find some Earth materials, however, that had similar sound velocities to those found in lunar samples. Based on experiments with seven kinds of cheese, the scientists determined that the sound velocities of lunar rocks were very similar to cheese! The average sound velocity of two lunar rocks examined was 1.54 km/sec and the sound velocity for Wisconsin Muenster cheese was 1.57 km/sec.

19-3 Investigating Lunar Landscapes

MATERIALS

sheet of transparent acetate

Few places on Earth have features like the moon. Does this mean that the geologic principles used to study the earth's history do not work on the moon? Does the law of superposition hold true?

After completing this investigation, you will be able to use the law of superposition to establish the sequence of surface features as they were formed on the moon.

Figure 19-7 *A section of an igneous rock from the moon, seen through a microscope.*

Figure 19-8 *A lunar landscape in the north central part of the moon. The large crater is Archimedes.*

PROCEDURE

Place a piece of transparent acetate over the large photograph (Figure 19-8) showing part of the north central region of the moon, including the crater Archimedes (*ahr kih MEE deez*). Study the photograph carefully.

DISCUSSION

1. What seems to be the last, and therefore the youngest, feature formed on the surface?
2. What is the oldest feature?
3. Starting with the most recent event, try to list the sequence of events that took place to produce the landscape as you see it.
4. List features in order of formation. Look for features such as large craters, small craters, maria, and mountains.

19-4 The Origin of the Moon

We know the least about the moon's beginning. There are several possible explanations for its formation. One theory is that moon and Earth formed in the same environment and at the same time. According to this theory, clouds of gases pulled together under the forces of gravitation and condensed into two bodies. One was the earth; the other was the moon.

According to a second theory, during their early history the moon was pulled out of the earth. The moon left a hole that is now the basin of the Pacific Ocean. Perhaps there was a star that passed nearby and provided the gravitational force needed to break off part of the earth.

A third attempt to explain the moon's origin deals with a still molten earth

(Figure 19-9). At that time gravity concentrated the most dense elements at the core. this caused the earth to spin faster on its axis. (You can compare this effect to spinning ice skaters who spin faster when they draw in their arms.) Then, as the earth's speed increased, its molten material began to flatten to a disk, until a mass of material, the moon, was thrown off.

A fourth theory, which is rarely discussed, is that the moon was originally an independent planet around the sun, traveling in an orbit close to the earth's orbit. According to this theory, at one time the independent "moon" came extremely close to the earth, within about 48,000 km. At that distance, the earth's gravity was strong enough to change the moon's orbit. The moon moved farther away, but remained with the earth.

A general history of the moon can be suggested, on the basis of what is now known about it. There are radioactive minerals on the moon that can be used to date rocks. You might say there were three stages in moon history.

When the moon first formed, it was very hot, but hotter on the outside than inside. The anorthosite formed at this stage, from about 4.6 billion to 4 billion years ago. The outer crust cooled and solidified and the deep interior became hotter.

During the second stage, the mare basalt was formed and some of the mountains were raised. The moon was hit by meteoroids of varying sizes. This bombardment may have caused some of the maria and large craters. The meteoroid that crunched through the moon's crust to form the Sea of Rains was the size of Rhode Island. When the crust was broken, lava flowed out to become basalt. Mountains such as the Apennines were uplifted by faulting around the edge of the crater. Ages of the mare basalts show that these changes took place from about 3.7 to 3.1 billion years ago.

It is likely that most craters were caused by bombardment by material from space rather than from volcanic activity in the moon's crust. Volcanoes on Earth tend to rise up in patterns related to the patterns of the crustal plates. Lunar craters seem to be dotted randomly over the entire surface. A bombardment by meteoroids would produce a random pattern like the crater pattern on the moon.

The third stage has lasted up to today. For the last 3 billion years, apparently, no extra-large objects from space have struck the moon. But a rain of medium-size to tiny meteoroids has changed the landscape. They have formed the soil and small craters that are everywhere, and the layers and pieces of breccia. Movements in the crust have formed rilles and fault blocks, a process that probably started in the second stage.

Is there still geological activity on the moon? About 20 years ago, a Russian

Figure 19-9 *According to one theory, the fast spinning, still molten earth threw off material to form the moon.*

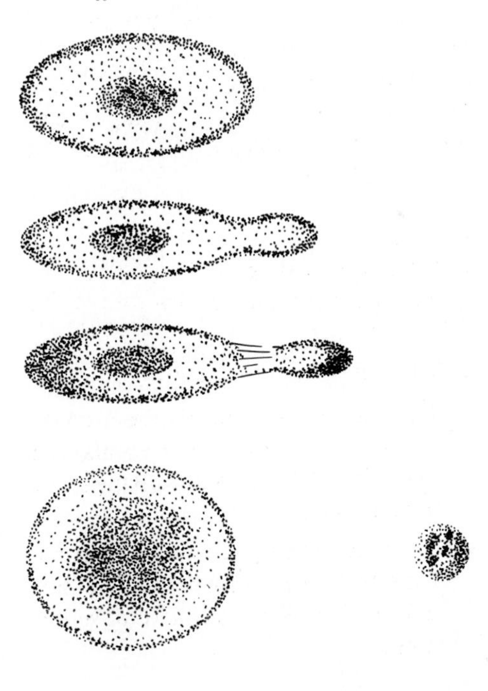

astronomer, N. A. Kozyrev, reported measuring volcanic gases from the crater Aristarchus. In 1966, astronomers found that the crater had a higher temperature than the surrounding surface. And in 1971, the Hadley astronauts flying over the crater detected a gas called radon that may have come from a volcano. So there may still be some volcanic activity.

Instruments left on the moon have registered a number of moonquakes. They are less powerful and occur less often than earthquakes. There may be 300 to 400 moonquakes per year. Are there moving plates in the crust of the moon like those on Earth? We may not find out for certain for many years, if ever.

Thought and Discussion

1. How are many of the mountains on the moon different from those found here on Earth?
2. Much of the moon's rock contains lots of glassy material. What does that fact suggest about the rate of cooling?
3. What do you think the chances are for finding sedimentary rock on the moon? Explain your answer.

The Earth-Moon-Sun System

19-5 The Distance to the Moon

The distance from the earth to the moon changes every day, even from minute to minute, because both the earth and the moon travel in oval orbits. Since the moon's orbit is not circular, but oval-shaped, the moon is closer to the earth at some times and farther away at other times. At the nearest approach to the earth, the moon is 360,000 km away. At its farthest point, the moon is 404,800 km away.

Accurate distance measurements to the moon were made as early as 1590 by Tycho Brahe (*TEE koh BRAH eh*). He sent two people to two different observation points on the earth. He had already determined the distance between these two points. From their positions the two people observed the moon at the same time. The observers determined the angles between their lines of sight to the moon and the horizontal plane. Study Figure 19-10. If two angles and one side of a triangle are known, a person can calculate the lengths of the other sides. The moon is at the tip of this triangle. The distance between the earth and the moon is the length of either the second or the third side of the triangle. Thus Brahe was able to arrive at a good estimate of the distance to the moon.

There is another way to measure the distance. Scientists have used laser beams and radar signals that bounce off the moon and return to the earth. Radar and lasers involve radiation that travels at the speed of light. At one point in the moon's oval orbit, the round-trip signal takes 2.5 seconds. If the speed of light is 300,000 km/sec, what is the total round-trip distance at that point? What is the distance one way?

19-6 The Moon's Influence on Earth

Even though the moon is smaller than the earth, its gravitational pull reaches us, some 386,400 km away. The sun also influences the earth. The sun's gravita-

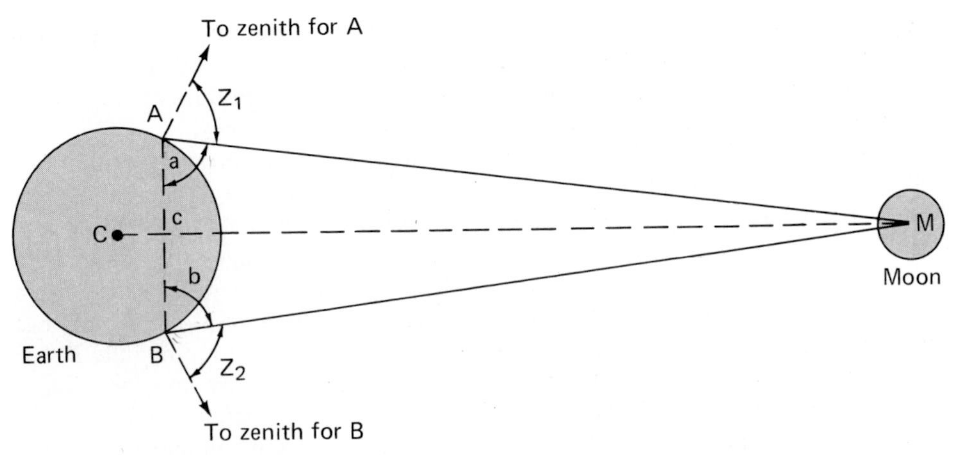

Figure 19-10 *How to measure the distance to the moon. People at points A and B measure the angles between their zeniths and the moon. That permits them to calculate angles a and b. They know the distance from A to B. Since they know two angles (a and b) and one side (AB) of the triangle AMB, they can calculate its height. To get the distance from the center of the earth to the center of the moon, you must add the distance cC. That is the distance from line AB to the center of the earth.*

tional pull on the earth is relatively weak in spite of the sun's huge size because it is much farther away. The sun is 400 times farther away from the earth than the moon is. The moon, because it is much closer to the earth, pulls 2.4 times harder than the sun does. Together, the gravitational pulls of the sun and moon create a stretching effect, raising and lowering the water, air and land.

It is easier to see the stretching effect on water than on solid land or invisible air. The moon makes the water in the oceans bulge upward on the side of the earth that is facing the moon at that time. The highest part of the bulge in the ocean water is **high tide.** High tides, such as those seen entering the narrow Bay of Fundy between New Brunswick and Nova Scotia, sometimes rise as much as 16 m in a very short time. (See Figure 19-11.) In a narrow area with steep sides, such as the Bay of Fundy, the water has nowhere to go but up.

Figure 19-11 *The Bay of Fundy at high tide and low tide.*

There is always a second high tide on the opposite side of the earth. The side of the earth that is away from the moon experiences less of the moon's gravitational pull. This allows the water to bulge upward on this side also. Water would spin off the surface of the turning earth if it were not for the force of gravity.

Since water is bulging up on the side of the earth facing the moon and also on the side away from the moon, there must be less water somewhere else. The earth's ocean layer is thinnest at right angles to the moon. (See Figure 19-12.) **Low tides** occur at these times when the ocean reaches its lowest level. Just as there are two high tides each day, there are two low tides in between the high tides.

It is important to remember that the tides at each point on the earth are always changing. A point on Earth turns away from the moon and toward it again each day as our planet turns on its axis. The sun's gravity also affects the tides. When the moon, the sun, and the earth are lined up, the highest high tides occur. This is because the moon's gravity and the sun's gravity have a combined pull on the earth. Sometimes, the earth, moon, and sun are aligned at right angles with respect to one another. Then the gravity of the sun and the moon seem to cancel each other's pulls. At those times, the high tides are the lowest that high tides can be.

Although you probably cannot notice a difference, the moon's gravity affects the earth's land and atmosphere, too. The atmosphere bulges upward, making the air thinner, or less dense, in those areas. Even mountains and buildings become higher. The Empire State Building in New York City and the ground underneath it both stretch so that the top of the building is 10 cm higher when the moon is overhead.

The following example may help you understand how tides work the way they do.

Imagine two people on a seesaw, as you see in Figure 19-13. This seesaw spins around on a pivot instead of moving up and down. One person, a fat one, is called Earth and the other, a skinny one, is called Moon. Obviously the pivot point of the seesaw is closer to the fat person. Now imagine what happens when the seesaw spins around the pivot point. The faster the seesaw spins, the more quickly the people tend to slide off it.

Now assume that they are both glued to the seesaw, so that they don't slide off as they revolve. The force of gravity

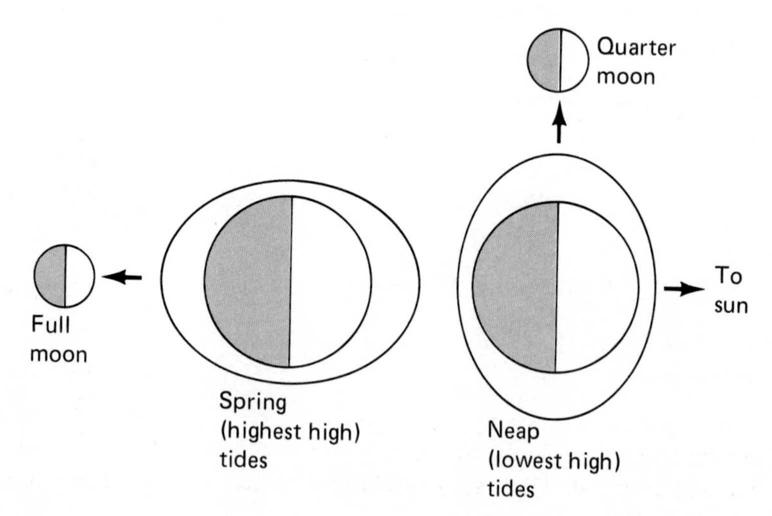

Figure 19-12 *When the earth, moon, and sun are in line, the high tides are the highest. When the moon and sun are at right angles, they weaken each other's gravitational pulls, and the lowest high tides occur.*

Figure 19-13 *What would happen to the fat boy if the thin boy got off the seesaw?*

holds the moon and the earth together as they revolve around their pivot point.

The side of the earth facing the moon is pulled more than the farside simply because it is closer to the moon's center of gravity. The solid land cannot bulge very much, but the water on the nearside is free to bulge toward the moon. A person on this side of the earth would see a rise in water called high tide.

On the side of the earth away from the moon there is less gravitational pull toward the moon and so the water is more free to bulge outward. Also, remember that the earth and moon are revolving around a point, like the seesaw in Figure 19-13. So, there is a tendency of water and earth to fly off into space, like the fat person would if he weren't glued on. The solid land under the ocean on the farside cannot respond very much to the tendency to fly off, but the water can, because it can flow freely. That is why there is a high tide on the farside of the earth as well as on the nearside.

19-7 Phases of the Moon

ACTION

You can show how sunlight changes the appearance of the moon. All you need is a light-colored ball, such as a tennis ball, and a strong source of light, such as light from a window or a lamp. (See Figure 19-14.) You are the earth. The ball represents the moon and the light represents the sun. Stand with your back to the light and hold the ball straight in front of you, at arm's length. Turn to your left until the "moon" has made one complete circle around the "earth." Describe what happens.

Figure 19-14 *How to model the phases of the moon.*

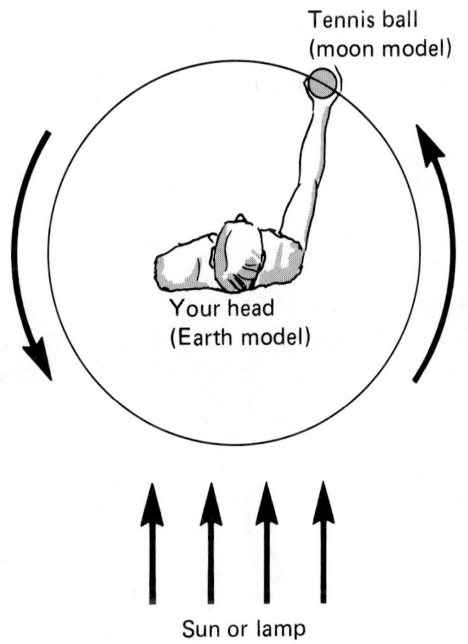

MATERIALS
tennis ball (or other light-colored ball), lamp (or other light source)

Since earliest times, people have been fascinated by changes in the moon's appearance. They noticed that the moon seems to gradually grow smaller and then grow larger again. It seems to change from a round disk to a narrow crescent. These apparent different shapes are called the **phases** of the moon.

The moon does not actually change shape. It is the pattern of reflected light that changes. The moon does not give off light of its own. It receives light from the sun, just as the earth and other planets do. The moon's barren surface reflects much of the light into space and some of that light reaches the earth. One half of the moon is always lighted by the sun and one half is always dark, just as the earth is. But the same half of the moon is not lighted all of the time because the moon is traveling in an orbit around the earth while the earth travels around the sun.

In order to understand the phases of the moon you should follow the changing positions of the earth and moon, relative to the sun. At one time of the month, the moon is in position A in Figure 19-15. The lighted side is the side facing away from the earth. The side facing Earth is dark. This phase is called **new moon** because the new cycle of phases is about to start.

A few days later (position B), most of the lighted side is still turned away from Earth, but a small portion of it is visible. That portion is called a **crescent.** Between seven and eight days after new moon, the moon is in position C. People on Earth see one-half of the half of the moon that is illuminated. This phase is called **first quarter.**

When the moon is at position D, the side that always faces Earth is fully lighted. This is **full moon.** Of course, to see the fully lighted side of the moon, you must be between the sun and the moon. The full moon rises, therefore, as the sun sets, since they are in opposite parts of the sky.

Between seven and eight days after full moon, the moon is at position E, and

Figure 19-15 *A diagram of the phases of the moon. At which point is the hidden side of the moon the same as the dark side?*

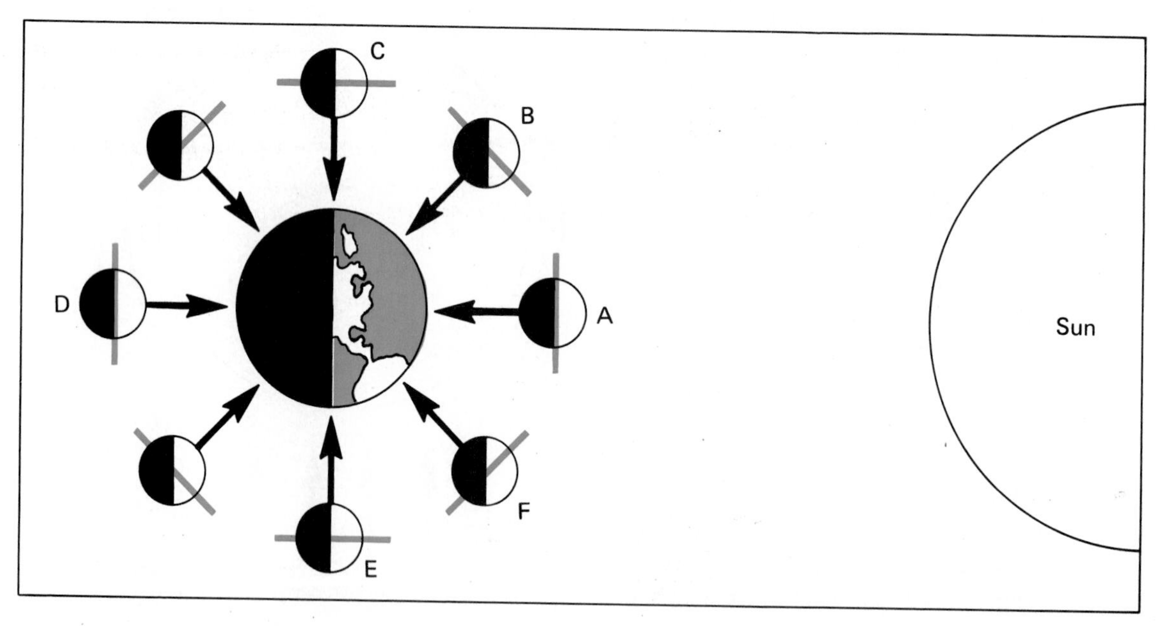

Figure 19-16 *The phases of the moon as seen from the earth during a phase month.*

the side facing Earth is half light and half dark. This phase is called **last quarter.** The moon in last quarter rises at midnight. Each night after that the moon passes closer and closer to the sun in the sky, finally rising just before the sun does. The moon appears as a crescent (F), since almost all of the lighted side is turned away.

What does the moon look like when it is halfway between positions C and D or D and E? Look at Figure 19-16 and choose the views of the moon that best show how it would appear. Then match the other views to positions A through F in Figure 19-15.

One month is based essentially on the phase period of the moon, which is $29\frac{1}{2}$ days instead of $27\frac{1}{3}$ days. (See Figure 19-17.) A month is defined as the time for the moon to go from full moon to full moon. This duration could be called a "moonth" rather than a month. If there were no moon, our concept of a month would have no meaning.

19-8 Investigating How to Tell Time by the Moon

MATERIALS
index card

For thousands of years people have used the sun and the moon as clocks. If you think about the predictable motions of the sun and moon in the sky, you can see how easily people can tell time by them.

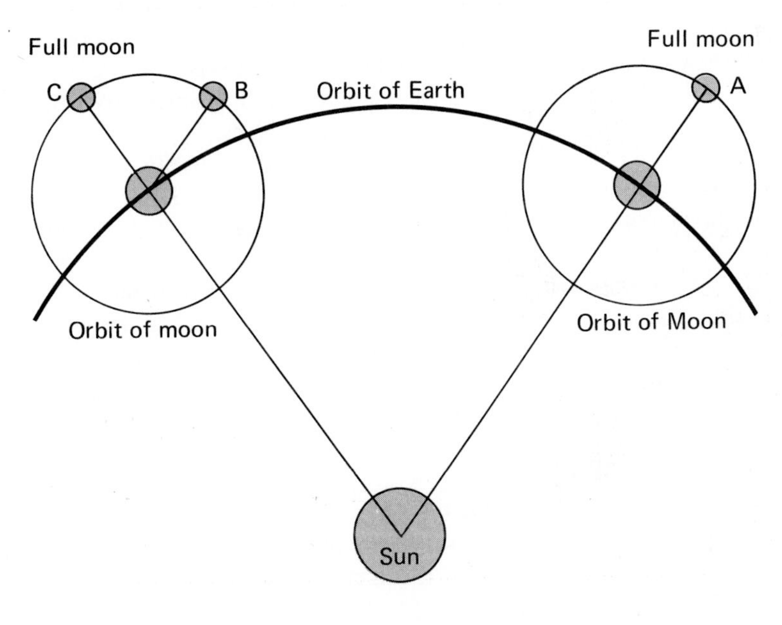

Figure 19-17 *One rotation of the moon occurs in 27.3 days (A to B). One phase period, from full moon to full moon, occurs in 29.5 days (A to C).*

People can estimate the hour during the daytime by checking the altitude of the sun in the sky. People can also estimate the hour during nighttime by checking the height of the moon in the sky. Many people do not try to tell time by the moon because it rises and sets 50 minutes later as each day passes. They do not recognize that the rising and setting of the moon is predictable on a monthly basis.

When you have completed this investigation, you will be able to tell time by the moon, by noting its phase and its position in the sky.

Imagine the earth to be a 24-hour clock. The hour-marks stay in the same place while you ride the hour hand of the clock, making one full turn in 24 hours. The earth's motions in space are actually much the same. When the earth's passengers ride to the point nearest the sun, it is noon. When they are turned opposite from the sun, it is midnight. Because the earth is turning like a merry-go-round, everything else—the sun, moon, planets, and stars—appears to be moving in the opposite direction.

Meanwhile, the moon is making a slower journey around the earth. The moon is actually moving in the opposite direction from what you see in the sky! It seems to be moving westward along with the sun and stars. In reality, it is moving eastward. You can check this by noticing the moon's position in the sky relative to a reference object such as a tree or house. The next night, stand in the same position at the same hour. Has the moon moved east or west?

Figure 19-18 *The moon timetable for Investigation 19-8.*

Phase	Rise	Highest Point	Set
new	6 A.M.	noon	6 P.M.
first crescent	9 A.M.	3 P.M.	?
first quarter	?	6 P.M.	midnight
first gibbous	3 P.M.	?	3 A.M.
full	?	?	6 A.M.
last gibbous	9 P.M.	?	?
third quarter	?	?	noon
last crescent	?	?	?

Procedure

Now copy the Moon Timetable shown in Figure 19-18. Leave blank spaces wherever there are question marks. You will also need an index card to help you interpret the rising and setting of the moon in Figure 19-19.

Start with the new moon. The side of the moon that faces the earth is completely dark. Ignore the other moon positions. You will use them later. Now place the index card over one half of the earth so that the half of the earth that is away from the moon is covered. The half of the earth that faces the moon is uncovered. Starting at the 6 A.M. position, a person on the earth can begin to see the new moon. As the earth turns, the moon appears to rise. As the earth turns counterclockwise toward the noon position, both the new moon and the sun are high in the sky. Six hours later, a person in the same position on the merry-go-round earth sees that the sun is setting. The new moon is also setting. Therefore, the new moon rises at 6 A.M., is at its highest point at 12 noon, and sets at 6 P.M. (Of course, rising and setting times vary with time zones and seasons.)

Turn the diagram so that the first crescent, the next moon to the left, is at the top. Move the index card so that the half of the earth that is facing the crescent moon is uncovered. Now check the rising and setting times for the first crescent and record them in the Moon Timetable. Then check the other phases of the moon in the same way and complete the timetable.

Discussion

1. When the moon is full and just rising, what time is it?

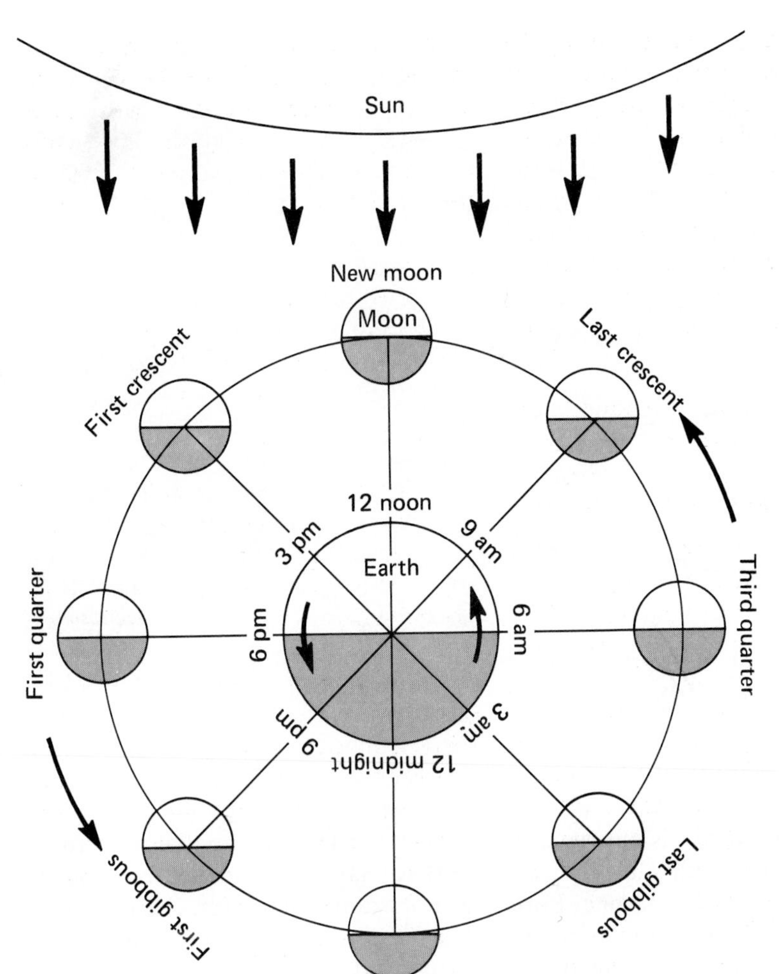

Figure 19-19 *Use this diagram to tell time by the moon.*

2. When the moon is at third quarter, and is at its greatest height, what time is it?
3. Suppose the moon is a crescent and is in the southwestern part of the sky. What is the time of day?
4. Suppose the full moon is midway between rising and setting (high in the south). What is the time of day?
5. If the first quarter moon is 45 degrees below the western horizon, what is the time of day?
6. The time of day is 3:00 P.M. Where would you find the 3rd quarter moon?

19-9 Eclipses of the Sun and Moon

During an **eclipse** (*ih KLIHPS*), light from the sun or moon is temporarily blocked due to the alignment of the earth, moon, and sun. To a person who does not understand what is happening, an eclipse of the sun can be quite frightening. It can become so dark that the birds go to sleep even though it is the middle of a clear day.

An eclipse of the sun is called a **solar eclipse.** An eclipse of the moon is called a **lunar eclipse.** For a lunar eclipse (Figure 19-20a), the phase of the moon

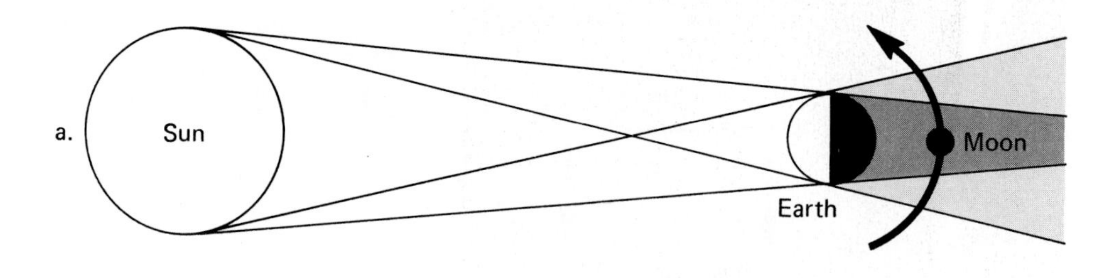

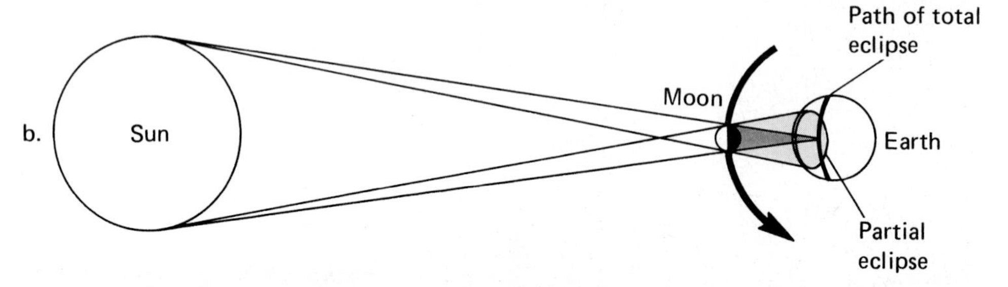

Figure 19-20 *The positions of earth, moon, and sun in* **a.** *a lunar eclipse and* **b.** *a solar eclipse.*

must be full. The moon is opposite the sun as seen from Earth and all three bodies must be in a straight line. For a solar eclipse (Figure 19-20b) the moon's phase must be new. The moon is between the earth and the sun with all three bodies in the same alignment.

You may be surprised to know that the moon and the sun appear to be the same size against the sky. You could cover up the moon with an aspirin tablet held at arm's length. The actual sizes of the moon and sun are much different. However, the moon and sun appear to be the same size, because the sun is so much farther away. So it is possible for the moon to cover the sun. This causes an eclipse of the sun. Since the earth is so much larger than the moon, the earth has a greater chance of stopping sunlight from reaching the moon and causing a lunar eclipse.

Both kinds of eclipses can be partial or total, depending on the alignment of the earth, moon, and sun, and the distance of each from one another. A **total eclipse** happens when the sun or moon is in maximum darkness. For the moon, this means it is completely in the earth's shadow. It does not receive any light directly from the sun, but it does receive a little light reflected from the earth. During a lunar eclipse, the moon turns a strange reddish color. (See Figure 19-21.)

During a total solar eclipse, the disk of the sun is blocked by the moon, but solar flares extend into space. (See Figure 19-22.) If you are able to observe a solar eclipse, NEVER LOOK DIRECTLY AT THE SUN! ITS RADIATION CAN DAMAGE YOUR EYES. A **partial eclipse** happens when only part of the sun or moon is covered. A total eclipse can only be seen in a small part of the earth where the alignment of the earth, moon, and sun is perfect for an eclipse. Usually, the angles at which the earth and moon pass by each other do not cause eclipses. A partial eclipse happens once in a while. A total eclipse is rare.

Figure 19-21 *The moon in almost total eclipse.*

The earth-moon distance is quite variable during the month because of the moon's oval orbit. The apparent size of the moon changes during this motion. Suppose that all conditions were met to produce a solar eclipse, but that the moon was the farthest it could get from Earth. How would its smaller apparent size affect the appearance of the eclipse? (See Figure 19-23.) Try to imagine what eclipses would look like if you were on the moon. Try to describe them. Could there then be such a thing as an eclipse of the earth?

Figure 19-22 *The sun in total eclipse.*

Thought and Discussion

1. Explain how it is possible to have two high tides at the same time—one on the side facing the moon and one on the other side.
2. How is it possible for the sun and the moon to have the same apparent size?

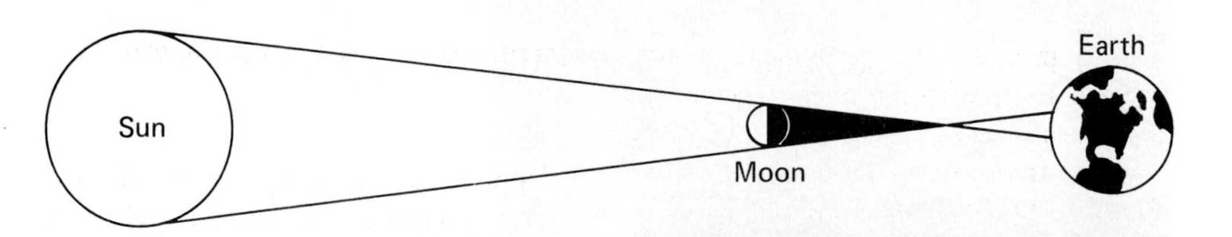

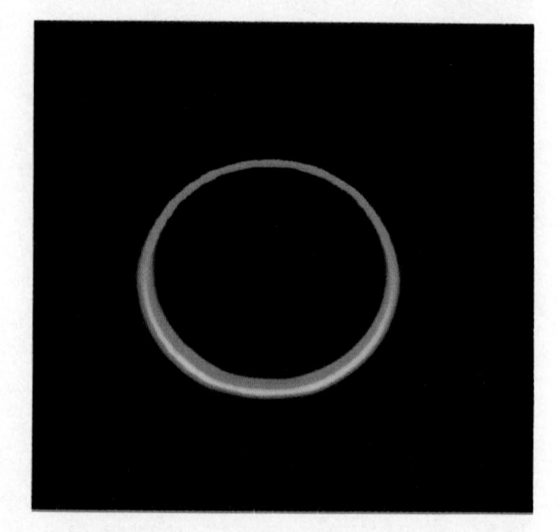

Figure 19-23 *When the moon is at its farthest from the earth, it does not completely cover the sun during an eclipse.*

3. Explain why it takes longer to complete a moon phase cycle than it does for one actual revolution of the moon around the earth.
4. In reality, in which direction does the moon revolve around the earth, eastward or westward? In which direction does it appear to move? Explain.

Unsolved Problems

People are still searching for answers to questions about the moon's origin, the evolution of its landscape, the composition of material in its interior, and its place as a future home for people.

In a single lifetime, people have reached out to other worlds. But many of the answers to the mysteries that we had hoped to solve by going to the moon remain locked within the confines of its environment. Our imaginations are stirred and we are still somewhat amazed at the success of the Apollo missions to the moon. They were carefully planned and executed, with precise engineering, excellent technology, and sound decisions. This was indeed a great step forward in human knowledge.

Chapter Summary

The moon has two kinds of landscapes. There are relatively smooth dark maria and the lighter colored mountainous highlands. The whole moon is dotted with craters of all sizes. Most of the craters were caused by meteoroids, but a few may be volcanic. There are also winding, canyon-like rilles.

The bedrock of the maria is basalt, with some of the special lunar breccia.

The highlands are geologically more complicated, with basalt, breccia, and anorthosite rocks and evidence of faulting. Common moon minerals are those found in igneous rocks on Earth.

The moon seems to be covered with soil that consists of small broken rock fragments and droplets of natural glass. Its origin is thought to be from showers of meteoroids over billions of years.

The origin of the moon itself is not known, but several theories have been presented to describe it. It was formed about 4.6 billion years ago. It is believed to have been very hot at first. Its original crust was bombarded by large meteoroids over 3 billion years ago.

Accurate distances to the moon can be calculated by measuring angles from two known locations or by using radar or lasers. These methods involve measuring the time it takes for a signal sent to the moon to be reflected back to Earth.

Along with the earth's rotation, the gravity of the moon (and to a lesser extent that of the sun) is responsible for tides on Earth. There are tides on land and in the air, as well as in the ocean. The tides change as the relative positions of the sun, the moon, and points on the earth change.

We can see the sunlit half of the moon from various angles. These are the phases of the moon. The phases change as the moon revolves around the earth.

The time it takes the moon to revolve around the earth ($27\frac{1}{3}$ days) is not equal to its phase period ($29\frac{1}{2}$ days) because of the earth's movement around the sun during a month ($\frac{1}{12}$ of a year).

You can tell what time it is if you know three things about the moon. You must know its phase, its height above the horizon, and whether it is rising or setting.

Solar eclipses occur when the moon blots out the sun wholly or partially. Lunar eclipses occur when the moon passes through the earth's shadow.

Questions and Problems

A

1. Suppose an Apollo team found a series of rocks on the moon in which the oldest were on top rather than on the bottom. What would you assume?
2. Why do geologists think most of the craters on the moon are meteoroid craters instead of volcanic craters?
3. Is there still activity in the moon's crust or beneath it? How can you tell?
4. How old are the oldest moon rocks and how do they compare in age with those found on Earth?
5. Explain the relative positions of the earth, the sun, and the moon necessary to produce the highest tides.
6. When the moon is opposite the sun in the sky, what is the moon's phase?
7. Why does the moon change its phase as the month progresses? Explain.
8. What time of day does the new moon set?
9. What is the shape of the moon's orbit around the earth?

B

1. Explain why many moon rocks are composed of angular fragments, rather than rounded pieces.
2. What was the origin of the moon soil? How does it differ from Earth soil?
3. Discuss the characteristics of the rocks brought back to Earth by the Apollo teams.
4. If you found a moon rock composed of crystals tightly grown together, where would you assume they were formed?
5. What part does the sun play in producing the lowest possible high tides?

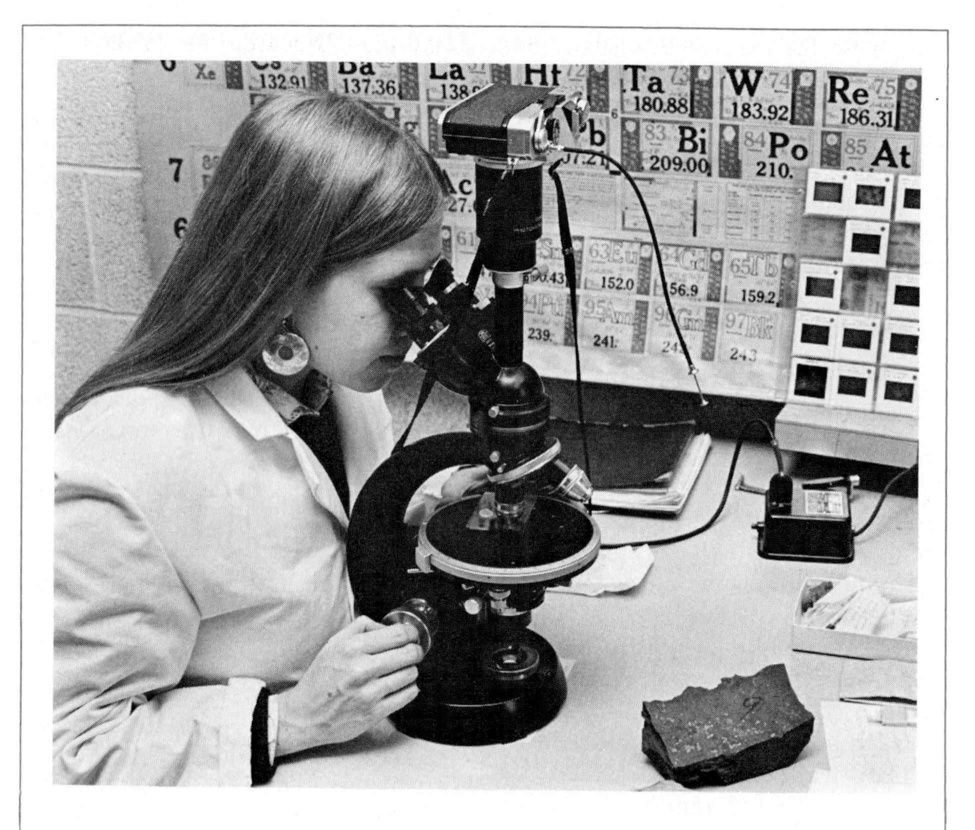

SCIENTIFIC PHOTOGRAPHER

People studying career possibilities frequently find that they have more than one area of interest, and so they look for careers that combine two or more fields. Scientific photography is an example of this type of career. Scientific photographers must plan photographs carefully to meet specific objectives. They may photograph fossils and rock outcrops to illustrate geologic processes. Some scientific photographers specialize in photographing microscopic specimens. Medical research is another area that employs photographers. Photographers may also work to develop photographic techniques for use in space research. There is a good deal of variety in the work of scientific photographers.

Many photographers are employed by or own their own commercial studios. The government, private industry, and hospitals are other employers of scientific photographers. Preparation for this type of work includes scientific knowledge as well as photographic skill. A bachelor's degree in one of the sciences and formal or apprenticeship training in photography are the usual requirements. Scientific photography is growing in importance, so the job outlook is excellent.

Personal qualifications for a career in scientific photography include artistic ability as well as the ability to obtain scientific knowledge. Good eyesight and excellent color vision are particularly helpful.

6. Suppose the moon was midway between rising and setting and the time of day was 6 P.M. What is the phase of the moon?

7. What kind of eclipse could occur when the moon is opposite the sun as seen from the earth?

8. Describe the appearance of an eclipse of the sun when the moon is at its farthest distance from the earth.

C

1. What is a possible explanation for the fact that astronauts didn't find metamorphic rocks on the moon?

2. Why is it so difficult to apply the concept of superposition when you are trying to interpret the moon's features from the earth?

3. What would you consider to be the most reasonable theory of the moon's origin? Explain your choice.

4. Describe some sports events as they might be held on the moon.

5. If the moon's orbit about the earth were more oval than it is, how would you describe the difference between the moon's nearest and farthest distances from the earth?

6. Discuss in some detail why people would have wanted to study the moon on the spot, instead of through telescopes and other earth instruments.

Suggested Readings

Alter, Dinsmore. *Pictorial Guide to the Moon.* 3rd ed. New York: Thomas Y. Crowell Company, 1979. (Paperback)

Cadogan, Peter H. *The Moon: Our Sister Planet.* New York: Cambridge University Press, 1981. (Also available in paperback)

Gamow, George, and Harry C. Stubbs. *The Moon.* Rev. ed. New York: Harper & Row, 1971.

Kopal, Zdeněk. *A New Photographic Atlas of the Moon.* New York: Taplinger, 1971.

Levinson, A. A., and Ross S. Taylor. *Moon Rocks and Minerals.* Elmsford, New York: Pergamon, 1976.

Lewis, Richard S. *The Voyages of Apollo: The Exploration of the Moon.* New York: New York Times Book Company, 1974.

Moore, Patrick. *New Guide to the Moon.* New York: W. W. Norton Co., 1977.

Mutch, Thomas A. *Geology of the Moon: A Stratigraphic View.* Rev. ed. Princeton, New Jersey: Princeton University Press, 1973.

20 The Solar System

BEFORE THERE WERE TELESCOPES, *people noticed seven objects in the sky that moved among the stars. They were the sun, moon, and five planets. These seven objects may account for the special regard often shown for the number seven. They almost certainly account for the seven days in a week. We now know that there are nine planets and a lot of other objects in the solar system. This system includes all of the celestial bodies that are controlled and dominated by the force of the sun's gravity.*

The solar system covers a lot of space—as a matter of fact, it is mostly space. If all of the objects in our solar system were ground into dust and scattered randomly throughout the same volume of space, there would still be a far better vacuum than any found on Earth. The dust from the millions of objects in the solar system would be unnoticed by a visitor passing through the area where the sun and planets had been.

AFTER COMPLETING THIS CHAPTER, YOU SHOULD BE ABLE TO:

1. **use the astronomical unit and compare relative sizes and distances in the solar system.**
2. **describe how the asteroids may have formed and why they were looked for at the end of the 18th century.**
3. **describe the motion of the planets around the sun and explain the apparent backward motion of some.**
4. **describe how Kepler's discoveries explain the true motions of the planets.**
5. **show why some planets have phases like the moon's phases and others do not.**
6. **describe the general properties of the planets from Mercury out to Jupiter and predict the properties of the others.**
7. **explain how the solar system may have formed.**

Dimensions and Motions in the Solar System

20-1 Making a Model of the Solar System

Suppose you could travel far into space and look back at the solar system. You would see the planets traveling in their orbits, at different distances from the sun. All of the planets travel in the same direction. Would some planets be too small to be seen from space? A model may help you appreciate the sizes and relative distances in the solar system.

When you have completed this investigation, you will be able to compare relative sizes and distances in the solar system. You will also be able to use the astronomical unit to show distances on a scale model.

PROCEDURE

A soccer field is 100 m long. Let the length of a soccer field represent the radius of the solar system from the sun to the orbit of the planet Pluto. Pluto is the farthest out. You will use Figure 20-1 to calculate where the planets and the asteroids fit on this scale model and how large they would be. (**Asteroids** are thought to be the remains of a broken planet between Mars and Jupiter.) Notice that all distances are compared to the Earth-sun distance, which is one unit. This unit is called an **astronomical unit** (1 AU). It equals 150 million km. The diameters of the sun and planets are expressed in earth units. The diameter of Earth is 1 earth unit. One earth unit equals 12,800 km.

Imagine that the sun is at one goal on the soccer field and the planet Pluto is at the other one. The earth would be 2.5 m from the sun's goal. Where would all the other planets be? Calculate their distances from the sun in meters.

Example: Calculate the distance between Mercury and the sun on the soccer field.

We know that the Earth-sun distance (1 AU) is equal to 2.5 m on the soccer field. Since Mercury's distance from the sun is 0.4 AU, its soccer field distance equals four-tenths of 2.5 m.

Therefore, $0.4 \times 2.5 \text{ m} = 1 \text{ m}$ on the soccer field.

Object	*Distance from Sun (Earth = 1)*	*Diameter (Earth = 1)*
Sun	—	110.0
Mercury	0.4	0.4
Venus	0.7	1.0
Earth	1.0	1.0
Mars	1.5	0.5
Asteroids	2.8	?
Jupiter	5.2	11.2
Saturn	9.5	9.5
Uranus	19.2	3.7
Neptune	30.1	3.5
Pluto	39.5	1.0 ?

Figure 20-1 *Solar system distances and diameters.*

DISCUSSION

1. If 2.5 m is equal to the distance from the earth to the sun (1 AU) and the earth's diameter is roughly $\frac{1}{10,000}$ of an AU, how big would the earth be on your scale model?
2. Calculate the diameters of the other planets on your scale model.
3. Find an object that is the same size as Pluto on your scale model. If you have a soccer field, put the object where it belongs on the field. Stand where the earth would be. Can Pluto be seen from the earth?

In this scale model the moon would be 2 mm from the earth. You would need nearly 7000 soccer fields in a row, end to end, to represent the distance to the nearest star beyond the sun.

20-2 Bode's Trick

In 1772, Johann Bode (*BOH duh*), a German mathematician, used a mathematical trick for finding the distance of the planets from the sun. He began with the series of numbers 0, 3, 6, 12, 24, and so forth. He added 4 to each number and divided the total by 10 to get the approximate relative distances of the planets from the sun. Try his method and compare your answers to the first column in Figure 20-1. Start with 0, which will give you Mercury's relative distance. For Venus, start with 3; for Earth, start with 6, and so forth. Notice that Bode's trick does not work for Neptune and Pluto.

Bode's trick came before the asteroids were discovered. Because of Bode's trick, some scientists thought that there must have been an undiscovered planet in the gap between the orbits of Mars and Jupiter. For over 20 years following Bode's discovery there was an all-out search for the "lost planet." In 1801, Giuseppe Piazzi (*PYAHT zee*), an Italian astronomer, discovered a tiny planet. He named it Ceres (*SEER eez*). It was the first of a group of thousands of tiny planets found between the orbits of Mars and Jupiter. These tiny planets are the asteroids. Ceres is the largest asteroid, but its diameter is only 768 km. The combined mass of all known asteroids is less than $\frac{1}{100}$ of the mass of the earth. Maybe these few remaining chunks are all that is left of a much larger body.

Professor Michael Ovenden of the University of British Columbia thinks that he may have found the "footprint" of the lost planet. He thinks that this planet disappeared in a powerful explosion about 16 million years ago. According to Professor Ovenden's hypothesis, moons orbiting around a planet, or planets orbiting around a star, tend to get as far away from each other as possible. Ovenden's calculations on the basis of his hypothesis correctly predicted the positions of the moons of the planets Uranus and Jupiter.

When he tried the calculations on the planets, the mathematical predictions were wrong. He finally found out that his hypothesis worked if he assumed that there was a large planet between Mars and Jupiter until about 16 million years ago. The other planets are still in a period of adjusting themselves to get as far away from each other as possible after the explosive disappearance of the now missing planet.

What happened to this missing planet? Why did it explode? Where did all of its mass go? We can only account for about $\frac{1}{900}$ of the estimated original amount. Did the material fall into the sun, or into

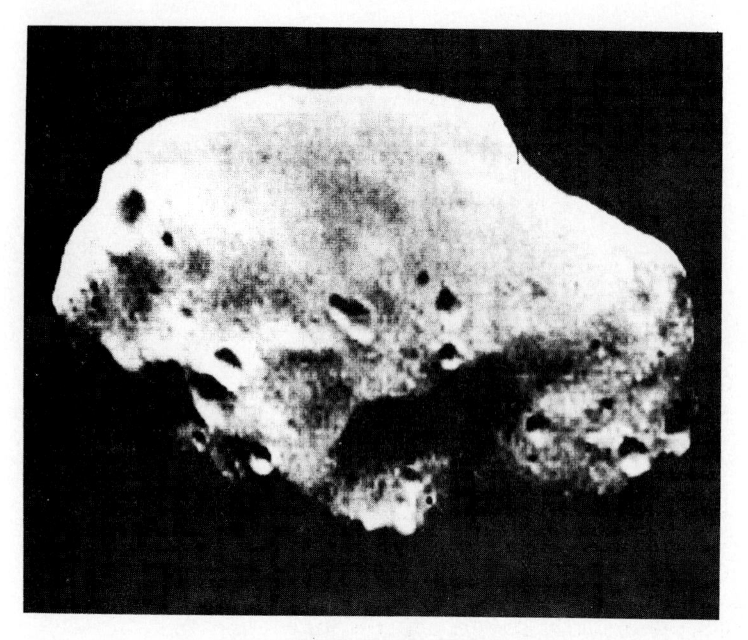

Figure 20-2 *Phobos, one of the two moons of Mars, is believed to be a captured asteroid.*

Figure 20-3 *A grid for plotting the movements of Mars. The star is Delta Capricorniae, in the eastern part of the constellation Capricornus (see Chapter 21).*

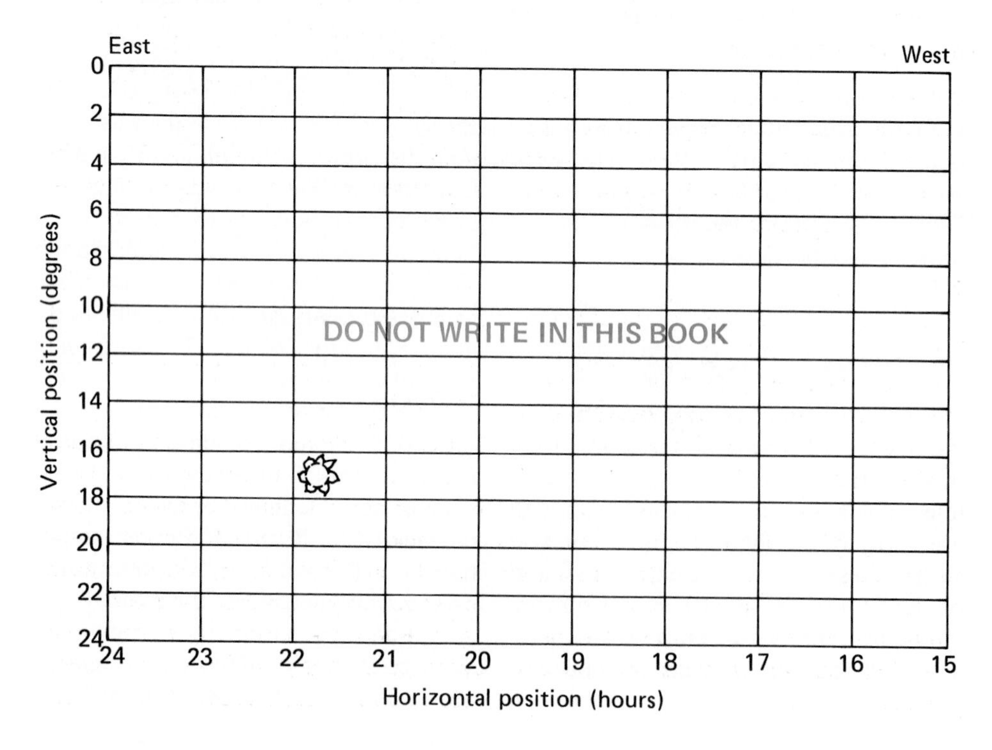

Jupiter, or was it thrown clear of the sun's gravity? Perhaps we can account for all of this mass as the material that makes up comets, meteoroids, and even some natural satellites. Maybe that is where some of Jupiter's satellites came from. Dr. Ovenden suggests that we might call this "lost" planet Krypton, after an exploding planet in a comic strip, the home of Superman. (See Figure 20-2.)

20-3 Investigating Planetary Motion

On a clear night, a planet looks like a bright star in the sky. As a single night goes by, a planet appears to move westward, along with the apparent movement of the stars. But after a few nights, you will find that it has drifted eastward to a new position among the stars. The name "planet" comes from the Greek word *planetes* meaning "wanderer." You would expect planets to keep drifting eastward, but at regular intervals they make westward loops.

When you have completed this investigation, you will be able to describe the orbit of Mars as seen from Earth, and explain why the orbit seems to loop.

PROCEDURE

Make a chart like the one in Figure 20-3. The chart represents an area in the sky. You are facing south. West is on the right and east is on the left. The vertical positions are measured in degrees, like latitude on the earth. The horizontal positions are expressed in hours. They are like longitude on the earth. From the coordinates given in Figure 20-4, plot the position for Mars on each date. Connect the points with a smooth line. Describe what happens to Mars as the months progress.

Month	*Horizontal Position*	*Vertical Position*
January	15.5 hr	18.3° south
February	16.9 hr	22.2° south
March	18.1 hr	23.5° south
April	19.5 hr	22.8° south
May	20.5 hr	20.8° south
June	21.5 hr	19.0° south
July	21.8 hr	19.6° south
August	21.4 hr	22.6° south
September	21.1 hr	22.3° south
October	21.5 hr	18.1° south
November	22.5 hr	11.4° south
December	23.4 hr	3.5° south

Figure 20-4 *Positions of Mars throughout a year.*

MATERIALS
graph paper

To understand the reason for the loop, make a drawing like the one in Figure 20-5. Draw a line from Earth position 1 through Mars position 1 to the "Apparent orbit" line. The line represents your line of sight from Earth to Mars. Do the same for the other Earth and Mars positions, up to 8.

DISCUSSION

1. Does Mars actually change direction?
2. For the year of the positions in Figure 20-4, what time of year is Mars' apparent motion westward?
3. Considering that the earth is also moving in its orbit, explain why the orbit of Mars seems to loop.

20-4 Kepler's Laws

ACTION

Put two thumbtacks in a piece of cardboard, as shown in Figure 20-6. Tie a piece of string together and loop it over the tacks. The string should be longer than twice the distance between tacks. Insert your pencil in the string and pull it tight. Keep the string tight while you swing your pencil around on the paper.

MATERIALS
two thumbtacks, cardboard, string, pencil

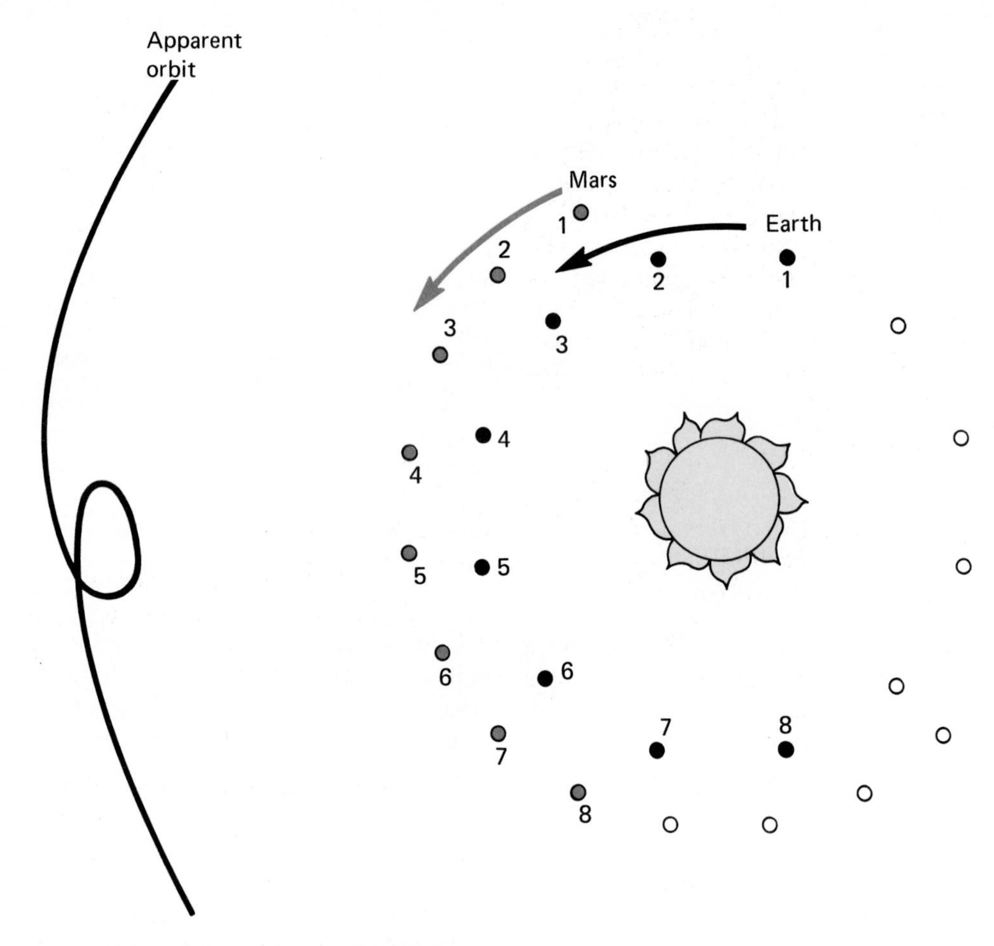

Figure 20-5 *With a drawing like this one, you can explain the apparent loop in the motion of Mars.*

The figure you have drawn is an **ellipse** (*ih LIHPS*). Unlike a circle, an ellipse has two centers. For the orbits of the planets, the sun is always located at one of the two centers.

Early astronomers believed that orbits were circular. But using circles led to minor errors in predicting planetary positions. Johannes Kepler (1571–1630) refused to overlook the errors. From his research, he discovered some important laws of planetary motion. You should know about two. Kepler's first law is this: *Planets move about the sun on paths that are ellipses.* Kepler's second law tells how planets move on these paths. *Each planet moves in such a way that a line joining the planet to the sun would pass over equal areas of space in equal amounts of time.*

MATERIALS

large sheet of wrapping paper, string, crayon or piece of chalk, ruler, three small balls, small electric lamp (or other small light source)

Notice in Figure 20-7 that an imaginary line joining the earth to the sun sweeps out equal areas in equal times. The time for traveling through part a of the ellipse equals the time for part b. The planet speeds up as it approaches the sun and slows down when it is farther away. From time period b to period a, the sun's gravitational force is speeding up the earth. What happens from period a to period b?

20-5 The Motions and Phases of Three Planets

In 1610, Galileo, using his newly constructed telescope, discovered surprising changes in the appearance of a planet.

Figure 20-6 *This is the way to draw an ellipse, like the orbits of the planets.*

Galileo reported his findings in a coded sentence. It read, "The mother of love imitates the forms of Cynthia." Perhaps you will be able to decode this statement after you have done this investigation.

When you have completed this investigation, you will be able to describe the line-up of planets needed to produce planetary phasing. You will also be able to explain why a sun-centered system is necessary to produce a planet's phase as seen from Earth.

PROCEDURE

Place a large sheet of wrapping paper on a table or desk. Put a large dot at the center of the paper. Now take a piece of string about 40 cm long and tie one end of it around a crayon or piece of chalk. You will use the string and the marker (crayon or chalk) to construct circular orbits around the center point. (Remember that orbits of real planets are elliptical, not circular.)

Find the point on the string 10 cm from the marker and hold it on the dot on the paper. Construct an orbit 10 cm from the center. Call it orbit X. Next, construct an orbit 20 cm from the center. Call this the earth's orbit. Then construct a third orbit 30 cm from the center. Call it orbit Y. Place a small ball on each of these orbits. These represent planets. Place a small electric lamp on the center dot. This lamp represents the sun.

Lean close to the wrapping paper so that your eye is next to the orbit of the

Figure 20-7 *Kepler's second law. Area a is equal to area b.*

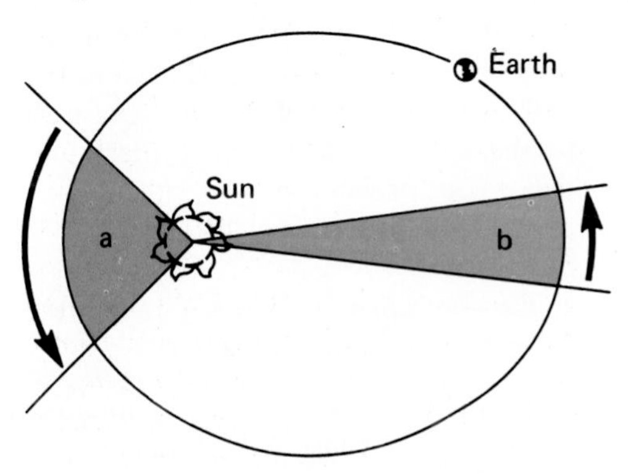

Figure 20-8 *The phases of planet "X."*

earth. Look at the other planets. Try to line up planet X (on the inner orbit) between the earth and the sun. Move the planet around in its orbit as you make your observations from Earth's orbit. Does it have phases? Try to make both planets X and Y look like the photographs in Figure 20-8.

Suppose the earth were the center of the solar system instead of the sun. Would the planets have phases in the same way? Place the lamp (the sun) on the earth's orbit and put the earth in the middle where the sun was. Have another look at the system. Move your eye next to the earth (now in the center) and make your observations.

DISCUSSION

1. Can both planets X and Y match the phases in the photographs?
2. Do the phases of X look anything like the moon's phases? Explain.
3. Do the phases of Y also resemble the moon's phases? Explain.
4. If there were a planet Z in an orbit closer to the sun than planet X and Earth, would it have phases like X or Y?
5. Have you discovered any evidence in support of the sun being the center of the solar system? Explain.
6. Can you decipher Galileo's coded sentence better now? What does it say?
7. Which planets would you now assign to X, Y, and Z?

Thought and Discussion

1. What was the distance in kilometers from the sun to the planet Krypton?
2. Explain why planets seem to loop and change direction as seen from Earth.

20-6 The Planets

Until recently, people could only look at the planets through land-based telescopes. People had to make their observations through the layer of atmosphere. The atmosphere acts like a blurred window, making all images fuzzy.

We have learned more about our planetary neighborhood in the past ten years than in the preceding ten thousand years. Close-up views of the planets, once imagined only in science fiction stories, have become a reality. The NASA missions Mariner, Pioneer, Viking, and Voyager, as well as Soviet space probes, have given us a close-up view of the solar system out to Saturn and beyond. These close-up views provide exciting new data for solar-system research. They also raise at least as many questions as they answer about the precise conditions on the planets.

Although each planet is unique, the planets can be divided into two main groups: the rocky planets and the giant, gaseous planets. The rocky planets are Mercury, Venus, Earth, and Mars. They are comparatively small in size and mass. Pluto might also fit in this category, but it is so far away that little is really known about it.

The giant planets Jupiter, Saturn, Uranus, and Neptune are thought to be huge balls of hydrogen and a few other elements and molecules.

20-7 Mercury

Mercury is the nearest planet to the sun and the smallest. It is named for the winged messenger god in Roman mythology. The surface of Mercury is covered by craters, as shown in Figure 20-9. It resembles the surface of the moon.

Figure 20-9 *A photo of Mercury, taken by the spacecraft Mariner 10 in 1974.*

Also, like the moon, one side of Mercury has more mare-like features than the other side does.

Mercury is 4800 km in diameter and has a density 5.5 times that of water, making it by far the densest object for its size in the solar system. This high density suggests that Mercury may have a core made of iron and nickel, similar to the earth's core.

Mercury rotates on its axis once every 58 Earth days and has an orbital period of 88 days. One day on Mercury occupies nearly two-thirds of Mercury's year. The force of gravity on Mercury is only $\frac{2}{5}$ that of Earth. How much would you weigh on Mercury?

Because of the high daytime temperatures, Mercury has a very thin atmosphere—one-millionth of the density of the earth's atmosphere. Some of the gases that have been found include hydrogen, helium, argon, and neon.

A thermometer on the side of Mercury that faces the sun would show a temperature of 800°C, hot enough to melt lead or tin. On the side away from the sun, the thermometer would read −200°C. All in all, Mercury would not be a very comfortable planet to live on.

Figure 20-10 *A photo of Venus, taken by Mariner 10 while it was on its way to Mercury.*

20-8 Venus

The planet Venus is named for the Roman goddess of love. Sometimes it is called the "evening star" or "morning star" when it shines in the sky near sunset or sunrise. It can appear to be brighter than any star in the sky. Venus always appears near the sun because it travels in an orbit smaller than the earth's.

Mysterious Venus is the least revealing planet. A dense cloud layer completely covers its surface (Figure 20-10). Venus shines brightly because sunlight reflects off this cloud layer. Even when Venus is photographed from spacecraft traveling near it, it appears to be a featureless white ball. But photographs taken with ultraviolet light show a stormy cloud structure. These clouds sometimes move at wind speeds of 325 km per hour. That is three times the speed of hurricane winds.

More information about this cloud-hidden planet has come from space probes. The atmosphere of Venus is mostly carbon dioxide. Two Soviet probes that landed on Venus measured the atmospheric pressure. It is about 60 to 140 times the atmospheric pressure on Earth at sea level. That would be similar to the pressure of sea water on Earth at 1200 meters below sea level. The surface of Venus has mountains and jagged rocks. (See Figure 20-11.)

The Soviet spacecraft Venera-8 found the surface temperature of Venus to be about 470°C. The high surface tempera-

Figure 20-11 *The rocky surface of Venus, photographed by the spacecraft Venera 9. The stripes are breaks in the TV signal, when it was sending other information.*

ture on Venus is due to the "greenhouse effect" caused by its thick, insulating atmosphere. Venera-8 also detected radiations from the radioactive elements uranium, thorium, and potassium. These elements are found on Venus in the same proportions in which they are found in granite here on Earth.

Venus is very similar to Earth in size and mass, and its diameter is only 640 km less than Earth's. It is also Earth's nearest planetary neighbor. At its nearest approach to Earth it is only 42 million km away. Venus rotates on its axis once every 243 Earth days in the opposite direction to that of Earth. Imagine the sun rising in the west, rather than in the east! Venus revolves around the sun once every 225 Earth days.

20-9 Mars

Mars was named for the Roman god of war. Mythology about Mars is still with us. On Halloween Eve in 1938, some people actually believed that Martians had landed when they heard the Orson Welles radio presentation of H. G. Wells' *War of the Worlds.*

When Mars is viewed through a telescope, its most noticeable features are its red color, unusual markings, and polar ice caps (see Figure 20-12). Photographs taken closer to the surface of Mars during the Viking missions now show that it looks much like the moon (see Figure 20-13). The craters on the surface of Mars show that it has been bombarded and scarred by meteors. The surface also has signs of volcanic activity. Nineteen large volcanoes have been found. The

Figure 20-12 *The planet Mars, photographed by the Viking 1 orbiter.*

Figure 20-13 *The surface of Mars, photographed by the Viking 1 lander.*

largest of these, Mount Olympus, is believed to be the largest volcano in the solar system. It is three times as high as Mt. Everest, and its base would cover the whole state of Missouri (see Figure 20-14). It is the largest known volcano. There are also canyon lands thousands of kilometers long, with splits and fractures that resemble dried creek beds. They may have been carved by ancient rivers. The largest canyon is called "The Great Rift." It is about 3600 km long

Figure 20-14 *Mount Olympus on Mars.*

and stretches one fifth of the way around the Martian equator. It may be evidence of massive faulting.

Mars is about half the size of the earth and has a clear, thin atmosphere that is mostly carbon dioxide. It has north and south polar ice caps that are probably frozen water, and not frozen carbon dioxide as astronomers formerly believed. Turbulent winds blow the fine Martian dust at speeds of nearly 400 km/hr. The thin Martian air would have to blow much faster than winds on the earth to move loose rock from the surface. These high velocity winds pick up pieces of rock, which chip away at other rocks and grind them to sand.

Will we really find life on Mars? Mars has no ozone layer like the earth's to filter out high-energy ultraviolet radiation. It has almost no oxygen. The prospect does not look good. But on Earth we have found organisms that do not need water and can live in extreme temperatures. On Mars liquid water could be available from permafrost or from minerals that attract water. If Martian organisms exist, they would have evolved differently from those here on Earth. So far, the possibility of life on Mars has not been ruled out, although the chances of finding life seem to be very small.

20-10 Jupiter

Figure 20-15 Jupiter and its four largest moons, not shown to scale.

The planet Jupiter's name was taken from that of the major Roman god. When Voyager 1 passed near Jupiter in 1979 it took many photographs of Jupiter and its satellites. Some of the photographs were assembled into the collage shown in Figure 20-15. Jupiter is a ball of whirling gases and liquids with no solid outer surface. Its outermost atmosphere consists of a 1000-km thick layer of hydrogen, helium, and clouds of ammonia, hydrogen sulfide, and water ice. The rest of the planet can be pictured as a boiling mass of liquid hydrogen and helium with a small core of rock. Its stripes are rising clouds of hydrogen gas drawn into rings by Jupiter's rapid rotation. This enormous planet completes one rotation in only nine hours and fifty minutes. Its atmosphere is more violent than any hurricane imaginable. The winds of Jupiter average 480 km per hour. Sometimes bolts of lightning extend all the way to its nearest moons. The puzzling red spot is now thought to be the eye of a gigantic hurricane.

Jupiter, a ruler of 16 moons, has a violent radiation belt of ionized particles, which nearly surround the planet. It has a tremendous magnetic field that extends outward some 10,000 km. Its internal temperature may be as much as 35,000°C, which is more than the temperature at the surface of the sun! Jupiter's atmosphere has a tremendous greenhouse effect. The average temperature of the outer atmosphere is about −150°C, but 200 km farther down into

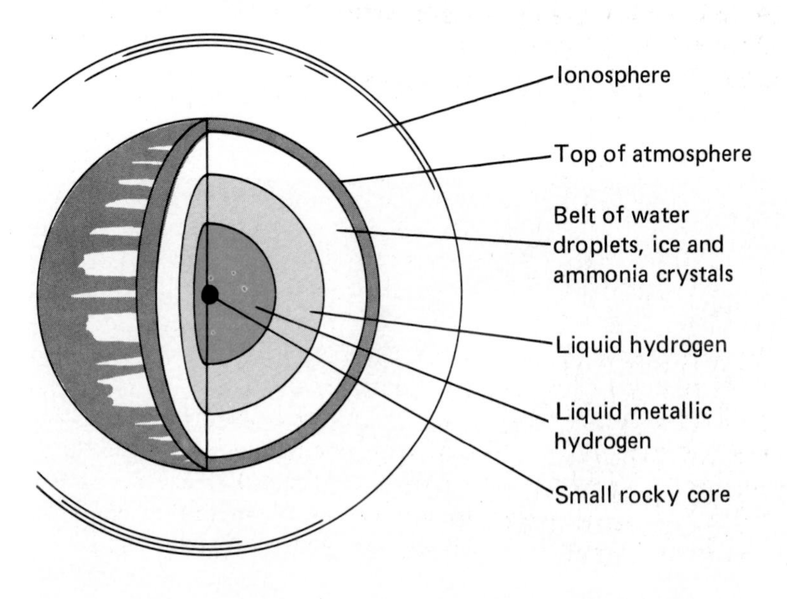

Figure 20-16 *A cross section of Jupiter.*

the atmosphere the temperature rises to 260°C. Farther down than that, the temperature soars to 800°C.

Inside the atmosphere there is probably a slushy substance that surrounds the core. This is a very dense, almost metallic-like hydrogen (denser than lead). The core is probably dense rock material. See Figure 20-16.

Jupiter has probably changed very little since it was formed, for even the largest asteroid in the solar system could slam into Jupiter and have no significant effect. External forces simply do not change Jupiter. It is almost like a star whose nuclear furnace never lighted. Some scientists think that if Jupiter were any more massive than it is, it would really be a star rather than a planet.

In 1980, Pioneer 10, which passed by Jupiter in 1973, passed into space beyond the solar system to wander there for eternity. Pioneer 10 carries a plaque that tells the story of its origin from Earth. Perhaps in a million years it will be found by another civilization.

20-11 Saturn

Saturn, named for the Roman god of agriculture, is shown in Figure 20-17. It is the most spectacular body ever photographed in space. Saturn's ring system contains an estimated thousands of rings. Some of the rings cannot be seen at all when viewed edgewise. Others seem to be intertwined, as though they were braided together. Saturn's rings are thought to be the remains of crushed moons that may have come too close to Saturn and were destroyed by its strong gravitational force. Findings from Voyager I and II confirmed that the rings are made of chips of rock and water ice.

Astronomers think that Saturn is composed of lower density materials than the earth—it is only 70% as dense as water. If we could find a pond large enough, Saturn would float. It is made of light gases, mostly hydrogen, with some helium, methane, and ammonia. Various hydrogen-rich ices may be the main ingredient for the whole planet.

Saturn's atmosphere consists of a series of cloud levels composed of hydrogen-rich ice particles. These are very similar to the clouds of Jupiter, though much less dense. Inside, as in Jupiter, is probably metallic hydrogen.

Saturn rotates on its axis in only 10 hours. Because of its low density and rapid rotation, its shape is highly distorted. Its polar diameter is 12,000 km shorter than its equatorial diameter, giving it a squashed appearance.

Figure 20-17 *A montage of images of Saturn and four of its moons, not shown to scale.*

	Period of Revolution Around Sun	Period of Rotation on Axis	Number of Moons	Mass (Earth = 1)
Mercury	88 days	59 days	0	0.05
Venus	225 days	243 days	0	0.82
Earth	365 days	1 day	1	1.00
Mars	687 days	1.03 days	2	0.10
Jupiter	12 years	9.9 hours	16	318.30
Saturn	29½ years	10.2 hours	15	95.30
Uranus	84 years	10.8 hours	5	14.60
Neptune	164 years	15.8 hours	2	17.30
Pluto	247 years	6.4 hours	1	0.90

Figure 20-18 *Solar system information, to add to that in Figure 20-1.*

20-12 Uranus, Neptune, and Pluto

The remaining two large planets, Uranus (Greek sky-god) and Neptune (Roman sea-god), are twins. Through a good quality telescope they both appear as tiny green disks against the black sky. But, like all planets, they can be seen to change position against the background of stars. Based on what you know about all the other planets we have discussed and considering the known properties of Uranus and Neptune given in Figure 20-18, what do you think their conditions and temperatures would be like?

The tiny planet Pluto was discovered in 1930 at the Lowell Observatory in Flagstaff, Arizona. Very little is known about Pluto, since it appears as a mere pinpoint of light. In 1980, an elongated smear was observed on a photographic image of Pluto. From this evidence it was determined that even the most distant planet has a satellite, since named Charon.

20-13 Minor Members of the Sun's Family

You may never witness the spectacular passage of a large meteoroid through the air and its landing on Earth, but you have probably seen meteors. **Meteors** are streaks of light on the sky that many people call "shooting stars." (See Figure 20-19.) The meteors are light from burning pieces of rock from outer space. They are called meteoroids before they zip into our atmosphere. They create so much heat by friction that they usually burn themselves up completely long before they hit the ground. If a meteoroid is large enough to reach the earth it is called a **meteorite** after it lands (Figures 20-20 and 20-21).

Several times each year, the earth intercepts a large mass of meteoroids. They are also in orbit around the sun and appear in swarms. When the earth and these belts of particles collide, there is a meteor shower. It is not uncommon for observers to see as many as 50 meteors per hour during a meteor shower.

Figure 20-19 *A meteor flashing across the atmosphere.*

Figure 20-20 The Willamette meteorite, the fourth largest known meteorite in the world. It weighs 13.6 metric tons.

ACTION

Check Figure 20-22 for the date of the next meteor shower. Go out and count the number of meteors you can see.

Like strange ghosts, comets appear as fuzzy figures against the sky. **Comets** are floating piles of gravel, dust, and gas. Their orbits around the sun are extremely flattened ellipses. Some travel far out, near the edge of the solar system, and return periodically to the sun.

Planets can be seen almost any night, but comets are rarely seen. They appear suddenly when they get close enough so that they reflect light from the sun.

Halley's comet, the most famous comet of all time, is shown in Figure 20-23. Its orbital period is 75 years. It will appear in our skies again in 1985. The tails of comets sometimes extend for 100,000,000 km. The tails seem to burst into fluorescence as the comet returns to the blazing sun. At Halley's comet's closest approach to the earth, its tail stretched two-thirds of the way across the horizon.

Figure 20-21 *How to tell meteors from meteorites and meteoroids.*

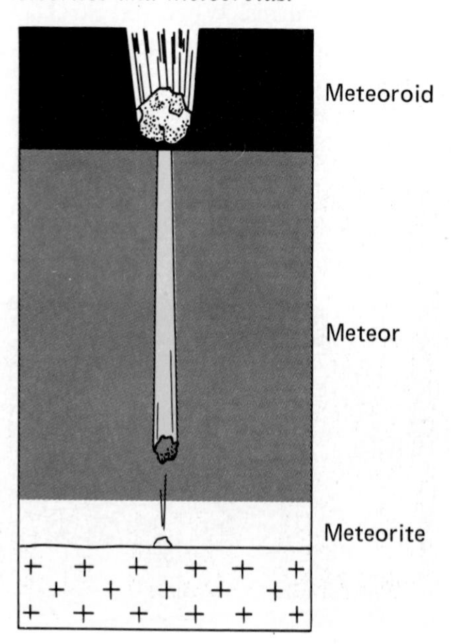

Thought and Discussion

1. What are the features on Mercury that suggest it had a history similar to that of our moon?
2. As strange as it may seem, the Venusian day is longer than its year. Explain.
3. What are some of the Martian surface features that suggest it may have had a geological history similar to that of the earth?
4. Why is it difficult to say whether Jupiter really has a surface or not?
5. How would you explain the history of the rings of Saturn?
6. Explain what causes a "shooting star."

Name	Date	Description
Quadrantids	Jan. 3	slow, long paths
Lyrids	April 20	fast streaks
Aquarids	May 1–11	fast, long paths
Aquarids	July 28–30	slow, long paths
Capricornids	July 25–Aug. 4	slow, brilliant, long paths
Perseids	Aug. 9–13	fast
Orionids	Oct. 19–23	fast
Leonids	Nov. 14–17	fast
Geminids	Dec. 11–13	fast, white, short paths

Figure 20-22 *Meteor showers.*

Unsolved Problems

We are discovering more and more about the nearer planets, but much more is yet to be found out. And you know that people would like to find out more about Neptune, Uranus, and Pluto.

Astronomers have discovered a great deal about the solar system and the forces that hold it together and govern its motions. But they can only guess at how it all started. According to one theory, the planets came into being as a result of a near collision between our sun and a passing star. As the star passed the sun, it pulled away a piece of the sun. The hot gas formed whirlpools, which condensed into planets and other smaller bodies.

Another hypothesis states that our solar system formed from a cloud of dust and gas that whirled in space. Finally it condensed to form the sun, the planets, and the smaller members of the solar system. As the sun became more and more dense, pressure built up at its core. Its energy helped condense the planets, especially the closer ones, and it swept away much of the surrounding gases. The result was the planets.

We will never have all the answers, but much more will be discovered during your lifetime. Perhaps you may read about (or even help to form) new hypotheses about the solar system.

Figure 20-23 *Halley's comet, photographed in 1910. Halley's comet returns every 76 years.*

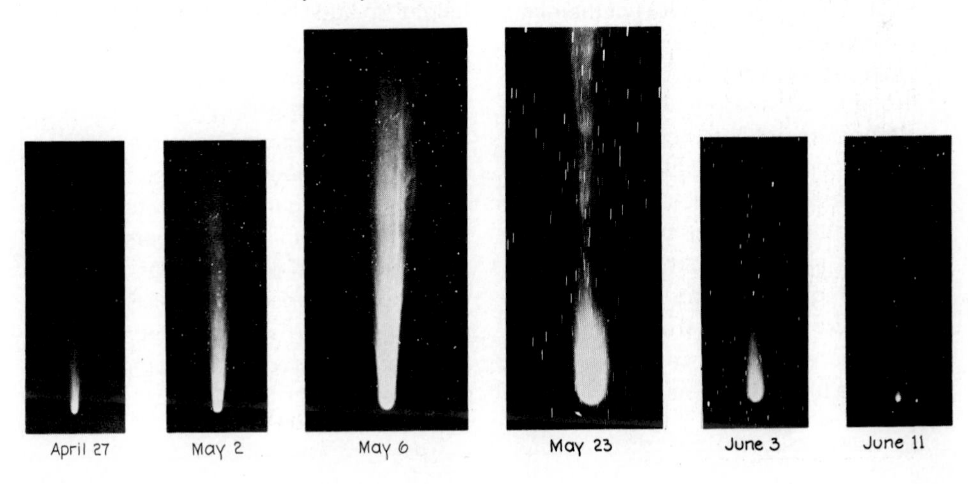

SCIENCE TEACHER

Most people today agree that a general knowledge of science is important for every citizen. Science-related problems are constantly in the news, and are frequently issues in elections. This makes the job of science teachers increasingly important. High school science teachers spend most of their time teaching science. They must also, as other teachers do, assist in administrative and supervisory work: checking attendance, supervising study halls, chaperoning school functions. In small schools a science teacher may also teach other courses, such as mathematics. In larger schools a science teacher often has the opportunity to specialize in one particular science, such as earth science or biology. A science teacher may be called upon to assist in extracurricular activities such as science clubs and science fairs.

A science teacher must be well prepared in a field of specialization. This generally involves a bachelor's degree in one of the sciences. Many science teachers find a more advanced degree helpful. In addition, most states require education and practice teaching courses. Some areas require licensing examinations as well.

Besides being competent in a particular field, a good science teacher should be imaginative and creative. Keeping up with new research and development is of particular importance to a science teacher. Enthusiasm for working with students and a high degree of patience are indispensable assets.

Chapter Summary

The earth and other planets are members of a solar system, an orderly arrangement of bodies of greatly varying size. The volume of space they occupy is mostly empty and it is isolated in space. All planets in the system move in elliptical orbits about the sun at various speeds.

Planets that lie closer to the sun display a cycle of phases like the moon. This concept supports the idea that the sun is the center of the solar system.

Through recent space technology and space missions, we have a much better understanding of the nature of many of the members of the solar system. We can measure their masses, distances from the sun, their compositions, and we can calculate and observe their motions. This information along with the data from NASA missions provides us with what we know about the planets. Mars and Mercury are crater-riddled, Venus is covered by dense clouds of carbon dioxide, and Jupiter and Saturn are mostly hydrogen.

Questions and Problems

A

1. Compare the sizes of Jupiter and Earth.
2. How many astronomical units are there between the sun and Saturn?
3. What is the atmosphere around Mercury like?
4. Why do we have so much difficulty looking at the surface of Venus?
5. Discuss the possibility of finding life on Mars, giving both sides.
6. What do scientists now think the huge red spot on Jupiter is?

B

1. Describe one of the theories for the origin of the solar system.
2. How is it possible to suggest that a large planet once orbited between Mars and Jupiter?
3. How large would the sun appear on your soccer field scale model?
4. Explain why Mars apparently changes direction in the sky as seen from Earth.
5. How can the planets show phases as they move around the sun?

C

1. On a model of the solar system the size of a soccer field where would you find the moon?
2. Explain why planets move faster at some times and slower at other times, when orbiting the sun.
3. Do planets that lie farther from the sun than Earth show phases? Explain.
4. Which planet gets closest to the earth during its revolution about the sun?
5. Since there is no liquid water on Mars, how could the apparent river beds have been formed? Explain.
6. What would the conditions on Jupiter be like if it were any more massive?

Suggested Readings

Menzel, Donald H. *A Field Guide to the Stars and Planets.* Boston: Houghton Mifflin Company (Peterson Field Guide Series), 1975. (Paperback)

Miller, Ron, and William K. Hartmann. *The Grand Tour: A Traveler's Guide to the Solar System.* New York: Workman Publishing, 1981. (Paperback)

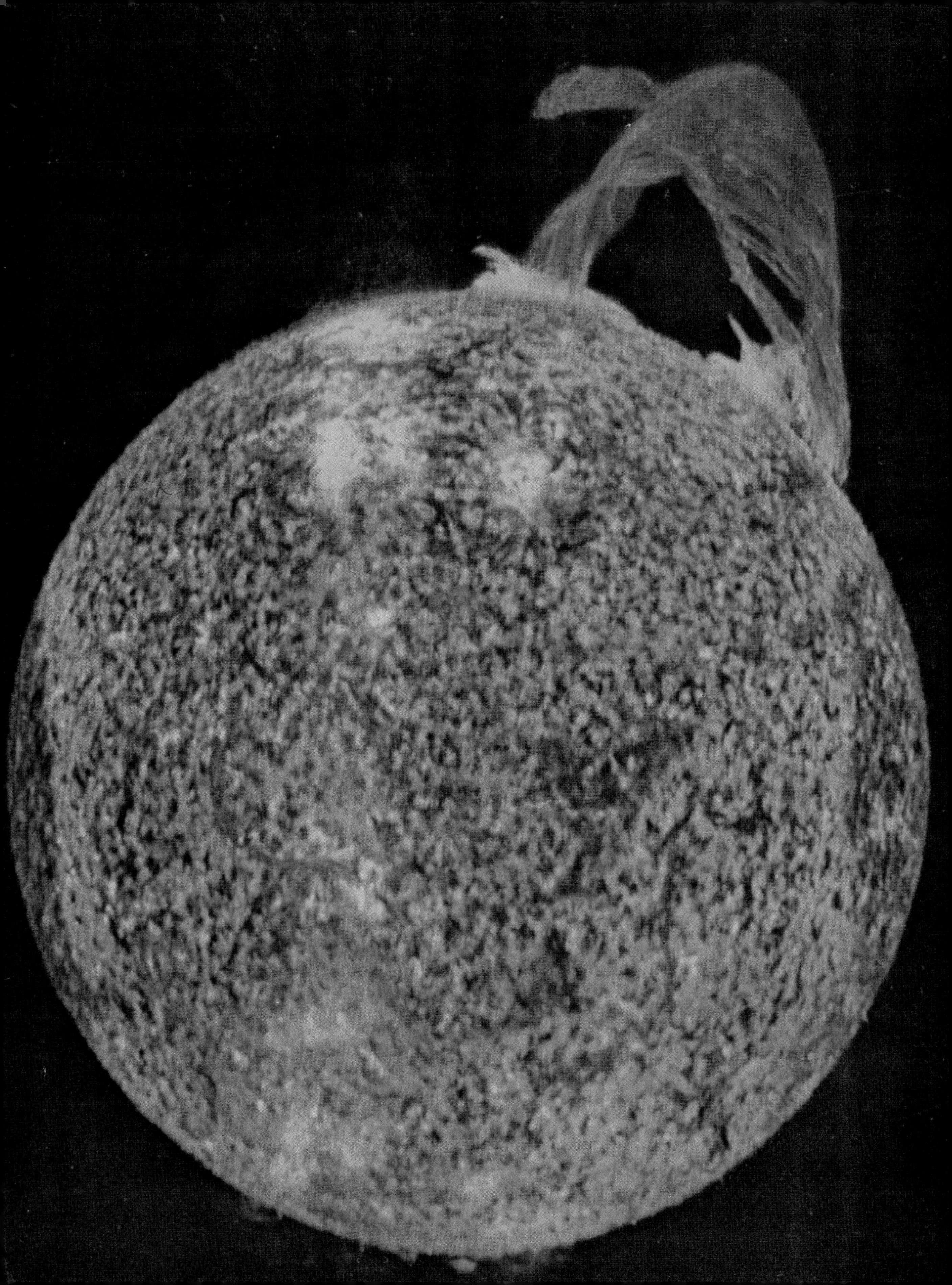

Stars

WHAT ARE STARS LIKE? *Except for the sun, the stars are so far away that even the greatest telescopes can show them only as points of light. Do they all have the same size, brightness, color, and temperature? Are they all at the same distance from Earth?*

Suppose you looked at a small portion of the sky where no stars were clearly visible to the eye alone. Would that mean there were no stars there? If you looked at that portion of sky through binoculars and then through telescopes of increasing power, you would see more and more stars. A long-exposure photograph taken with a powerful telescope would show thousands of stars where your eye could not see even one.

Since the stars are so far away, it might seem almost impossible to learn much about them. Yet scientists have found out what stars are made of, how large and massive they are, how bright and how hot they are, and that they are moving through space. People have learned all these things by studying only the feeble starlight that reaches us from such great distances.

AFTER COMPLETING THIS CHAPTER, YOU SHOULD BE ABLE TO:

1. **describe some important characteristics of the sun.**
2. **estimate the number of stars visible in the sky at different times.**
3. **find an object on a star chart and in the sky by using coordinates and star time.**
4. **explain and give examples of methods used in measuring distances to stars.**
5. **explain how stellar brightness (magnitude) is expressed, in both visual and absolute terms.**
6. **compare the absolute magnitudes of stars in order to find their luminosities.**
7. **compare and contrast the quantity and quality of starlight and summarize what can be determined by studying star spectra.**

The Sun, a Nearby Star

21-1 Our Energy Source—the Sun

People have wondered for many years about the nature of the sun and the source of its tremendous energy. Anaxagoras (*an ak SAG uh ruhs*), a Greek who lived around 450 B.C., thought the sun was a burning hot stone 10,000 km across and 10,000 km away. Of course, if the sun were a hot stone, it would cool off in time and stop giving off radiant energy. Until the mid-20th century, scientists could not account for the tremendous amount of energy given off by the sun. Even if it were a huge lump of burning coal, for example, it would last only about 6000 years.

The source of the sun's energy is a process known as **fusion.** In this process hydrogen atoms combine to form helium atoms. The sun contains plenty of hydrogen fuel. In its intensely hot central region, or **core,** the fusion reaction takes place, giving off tremendous amounts of energy.

The sun, with a mass 332,000 times that of the earth, is, in effect, a huge nuclear power plant. Scientists believe that it has enough fuel to keep "shining" for another ten billion years.

When you are feeling the heat of the sun on a hot summer day, you may have difficulty imagining that you are receiving only a tiny part of the solar energy. Almost all of the sun's energy spreads out into empty space. The light that reaches the earth is only one two-billionth part of what is produced by the sun. If we could learn to use this light energy, there would no longer be an energy shortage here on Earth. Just one week's supply of the solar energy that reaches the earth is equal to the energy of all past and present fossil fuel and nuclear fuel resources.

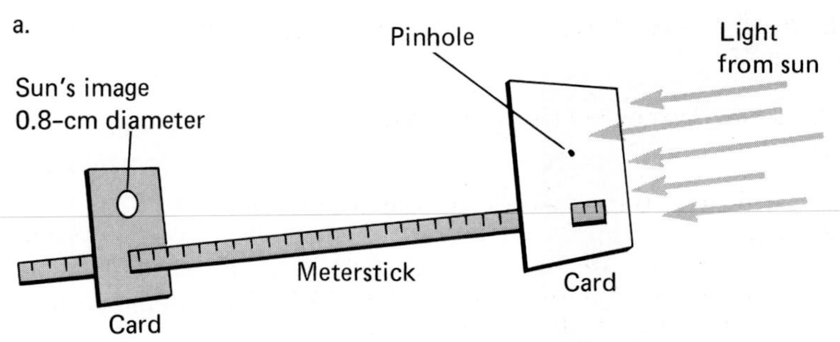

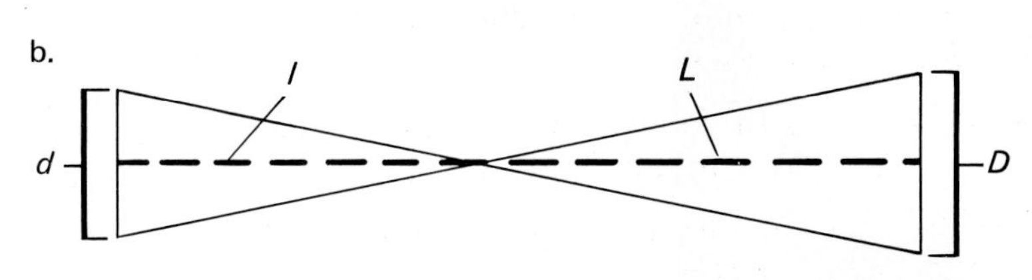

***Figure 21-1* a.** *Meterstick-and-card apparatus for measuring the diameter of the sun.* **b.** *Similar triangles formed by the diameter of the sun, D, the diameter of the image, d, the earth-sun distance, L, and the distance from pinhole to image, l.*

21-2 Measuring the Diameter of the Sun

MATERIALS
straight pin, meterstick, small index card, large index card, ruler, compass (drawing)

It is possible to measure the diameter of the sun from the earth. You can do it yourself using the simple materials shown in Figure 21-1a. As shown in Figure 21-1b, you can use measurements from a small triangle to calculate a corresponding measurement on a similar large triangle that is too large to measure directly.

When you have completed this investigation, you will be able to use the concept of similar triangles to measure the diameter of the sun.

PROCEDURE

Put a good-sized pinhole in the larger card, as shown in Figure 21-1a. Draw a circle with a diameter of 0.8 cm on the other card. Place the two cards on a meterstick so that the hole in one card and the circle on the other are about the same distance above the meterstick. Point the meterstick toward the sun, so that the sun shines through the pinhole onto the drawn circle. **Be careful never to look directly at the sun. It can very quickly damage your eyes.** Move the cards back and forth, keeping them perpendicular to the meterstick, until you can exactly fill the small circle with the sun's image. Note the exact distance in centimeters between the cards.

To find the sun's diameter, draw triangles like those illustrated in Figure 21-1b. In this drawing, the diameter of the sun is D. The diameter of the projected image of the sun (0.8 cm) is d. The distance to the sun (150 million km) is L, and the distance between the cards on the meterstick is l. Since you are using similar triangles, you can set up the proportion $D/L = d/l$. Solve for D, the diameter of the sun.

21-3 Measuring the Size of a Sunspot

MATERIALS
2 pieces of white cardboard, ruler, small telescope

Sunspots appear as dark blotches on the sun. (See Figure 21-2.) They represent areas of cooler temperatures, but they are regions of violent activity. You

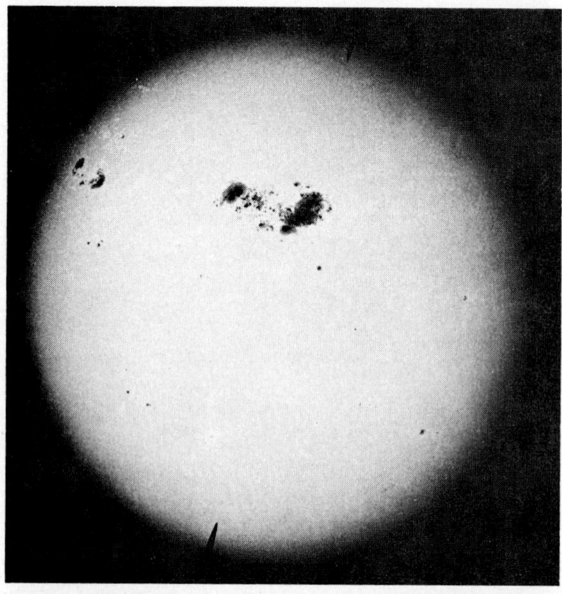

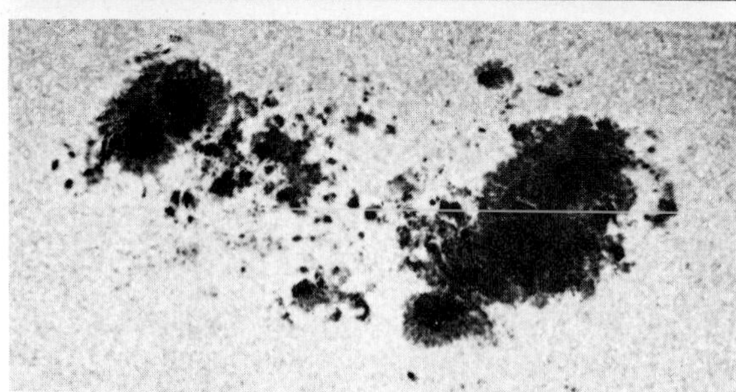

Figure 21-2 *These photographs, made on April 7, 1947, show one of the largest sunspot groups ever observed.* (*The bottom photo is an enlargement of the sunspot group in the top photo.*) *At this scale, the earth would be a dot not much larger than the spot visible on the lower right part of the sun.*

can almost always find at least one on the sun, but they are more common during the peak of an 11.3-year cycle.

Recent studies of the sunspot cycle have shown that it is related to the positions and periods of the planets Mercury, Venus, Earth, and Jupiter. The gravitational forces from the planets produce tides on the sun. The other planets are too small or too far away to have any significant effect on the sun. The "tidal" planets have their greatest effect when they are in a line, so their forces all act together. Their least effect is when they are scattered all around the sun.

The dates of sunspot cycle peaks can be predicted, within an error of about one year, from planet positions. Thus, the tidal effect of the planets appears to be an important influence in the production of sunspots.

Occasionally, extra large bursts of radio microwaves, ultraviolet light, and protons from hydrogen atoms erupt from the sunspot areas. When the waves and particles reach the earth's atmosphere two days later, they interfere with radio, television, radar, and telegraph communications. They cause "fadeouts" and various kinds of static.

Even with a small telescope, you can indirectly measure the approximate size of sunspots.

When you have completed this investigation, you will be able to measure the size of a sunspot.

PROCEDURE

Project the image of the sun on a piece of white cardboard as shown in Figure 21-3. **(Caution: Keep away from the eyepiece—it will burn you.)** With a ruler, measure the diameter of the sun's projected image (in cm) and the approximate diameter of the projected image of a sunspot (in cm).

You can set up a proportion to solve for the size of the sunspot. Let D be the real diameter of the sun (the value you calculated in the previous investigation). Let d be the diameter of the sunspot. Let P be the diameter of the projected image of the sun. Finally, let p be the diameter of the projected image of the sunspot. Then $d/D = p/P$. Solve for d, the approximate diameter of the sunspot. You might want to compare the approximate surface area of the sunspot with the area of the entire earth.

Figure 21-3 *The sun's image can be projected onto a piece of white cardboard with the use of a small telescope.*

Thought and Discussion

1. Suppose the projected image of the sun in your sun diameter investigation had been twice the size that you found it to be. What would the diameter of the sun be?
2. Why is the sun like a nuclear power plant? How does it differ from a nuclear power plant?
3. What do you think will happen to the earth when the hydrogen fuel of the sun is used up?

The Stars Around Us

21-4 Investigating the Number of Stars

MATERIALS
paper towel tube or rolled up cardboard

There are so many stars in the sky that it may be beyond our imagination to even guess the number. For most of us, large numbers have very little meaning. Suppose you began to count now at the rate of one number per second. It would take you about thirty-three years of continuous counting to reach a billion. Yet it would take two hundred times longer than that to count just the stars in our own galaxy, the Milky Way. (See Figure 21-4.)

How many stars are there in the whole universe? The number of stars may be more than the number of drops of water in all the oceans or all the grains of sand on all the beaches in the world. How many stars can you see from your own backyard on a clear night? Go ahead—make a guess: a hundred, a thousand, a million?

Figure 21-4 *You could not count all the stars in the Milky Way in a lifetime, if you tried to count them one by one.*

When you have completed this investigation, you will be able to estimate the number of stars that you can see with the unaided eye at one time and list some factors that may affect your result.

PROCEDURE

Make a table like the one in Figure 21-5 listing the eight points of the compass: N, NE, E, SE, S, SW, W, NW. For each compass direction in your table, there should be three altitudes: 22½° (one-fourth of the way up), 45°(halfway up), and 67½° (three-fourths of the way up). You will have one extra box for the **zenith** (90°), that is, directly overhead.

Pick a good clear night. Get away from the glare of the city lights if you can. With a paper towel tube or a rolled-up piece of cardboard, look at each of the twenty-five portions of the sky indicated in your table. Record the number of stars you can count in each portion.

DISCUSSION

What was your total number for all twenty-five portions? Multiply that number by the constant that your

Figure 21-5 *Make a table like this for recording your star count.*

	N	NE	E	SE	S	SW	W	NW
67½°								
45°			DO NOT WRITE IN THIS BOOK					
22½°								

90°		Zenith

teacher will give you. The product of this multiplication is a rough estimate of the total number of stars that you can see with the unaided eye. You may want to do the same investigation later the same night or perhaps six months from now. Think of some reasons why your answers might vary from viewing to viewing.

21-5 The Constellations

With so many stars in the sky, locating any given star might seem like an impossible task. One way to deal with a large number of objects is to find patterns or pictures in the sky. The groups of stars that form such patterns are called **constellations.** These patterns come from peoples' imaginations. Experts say there are as many as 88 recognizable constellations. Most constellations are named for characters from Greek and Roman mythology. The stories about these characters have been passed on from generation to generation.

Find the constellations on the star charts of Figure 21-6 (a–d). The constellations can serve as landmarks to help you find other stars in the sky. Remember, we only see where stars and constellations appear to be. We can't tell how far away they are simply by looking at them. You may not see much connection between the constellations and their names. We live in a much different society from that of the people who invented these names and stories thousands of years ago. Over the past 2000–3000 years some of the stars have actually changed their positions with reference to one another.

You see different constellations and stars at different times of night. The progression of the constellations throughout each night is the result of the rotation of the earth. Since the earth rotates from west to east, the stars seem to move from east to west. If the earth turns 360° in a 24-hour day, how many degrees will it turn in one hour? How far will the stars seem to move in an hour?

You will not see the same stars at midnight in October that you see at midnight in April. The reason is that as the earth rotates, it also revolves around the sun.

Look at Figure 21-7. Imagine you are standing at the spot marked with an X when the earth is in position 1. If you look south, you will be looking in the direction indicated by the arrow from X. If the season is fall, you should be looking at the constellations Capricornus and Sagittarius. When the earth has reached position 2, and you look south at midnight, you obviously see a different part of the sky. You should see Pisces. As the earth continues its revolution, when you get to position 3, you should see Leo and Gemini.

21-6 The Celestial Sphere

To more accurately determine where stars seem to be on the sky, you will need to understand some of the concepts involving the celestial sphere.

Suppose the earth were like a huge balloon and you could blow it up until positions on the earth became positions on the sky. The earth's North and South poles would become the sky's north and south **celestial poles** and the earth's equator would become the sky's **celestial equator.** The sky would become the **celestial sphere** and stars could be located on it. (Star charts, such as those of Figure 21-6, are maps of the celestial sphere.) Figure 21-8 shows the earth in relation to the celestial sphere. The **ecliptic** (*ih KLIHP tihk*) is the apparent path of the sun on the sky. The celestial equator is inclined 23.5° from the ecliptic because the earth is tilted 23.5° with respect to the plane of the solar system. The ecliptic is also the path along which many familiar constellations appear to move throughout the year.

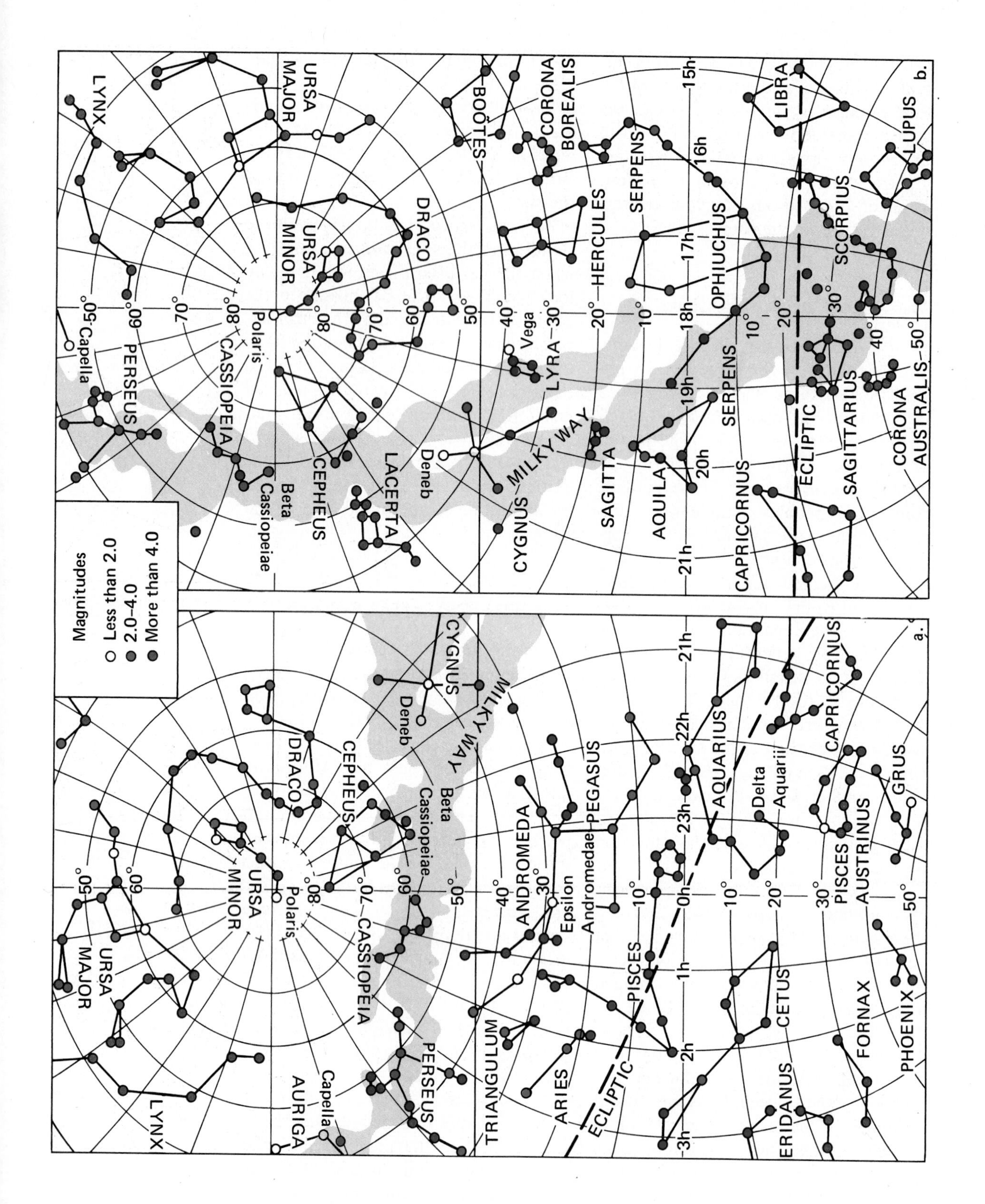

Figure 21-6 *The top of each chart is at the left.*

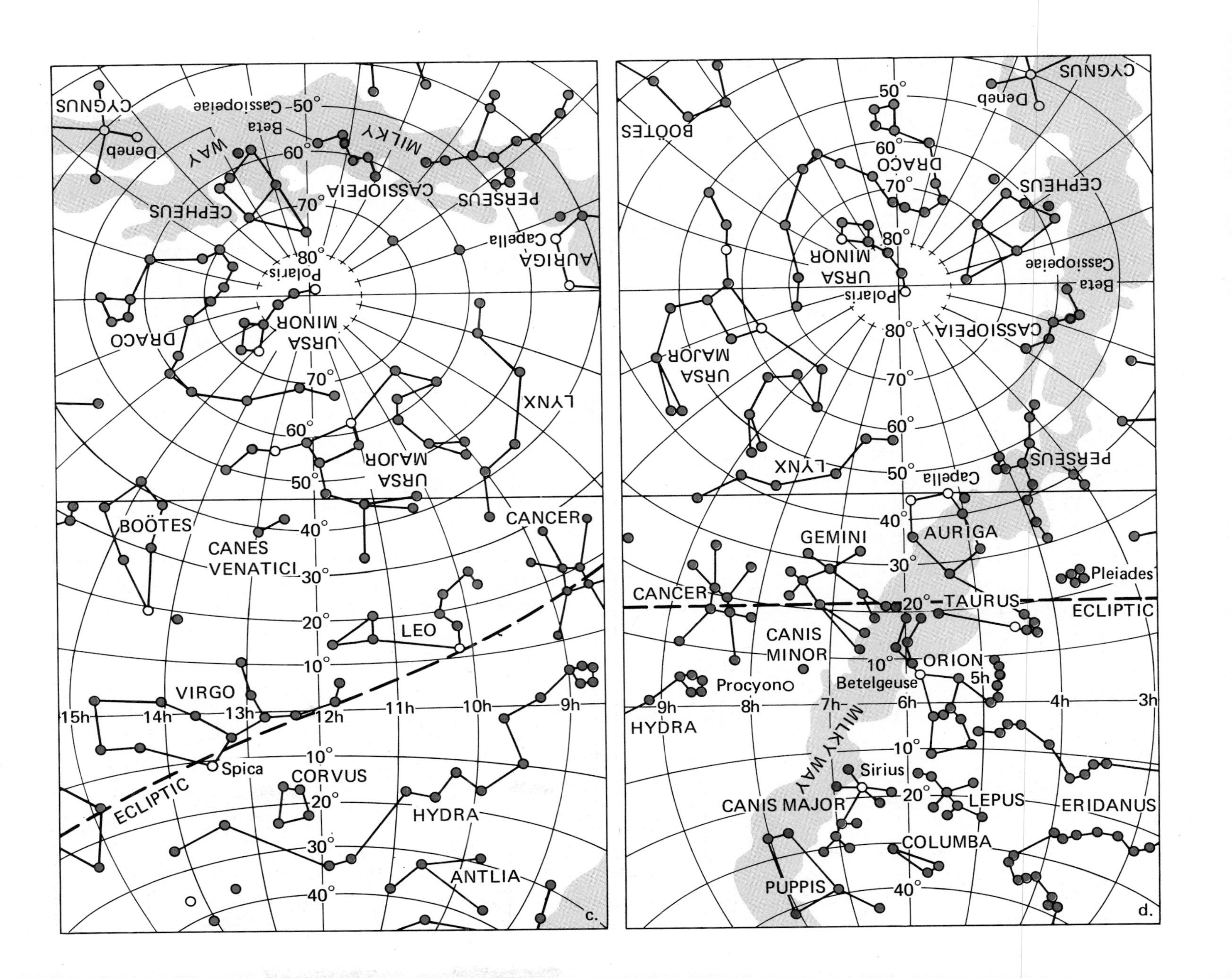
CYGNUS
Deneb
Beta Cassiopeiae
MILKY WAY
CASSIOPEIA
PERSEUS
CEPHEUS
Capella
AURIGA
Polaris
URSA MINOR
DRACO
LYNX
URSA MAJOR
CANCER
BOÖTES
CANES VENATICI
LEO
VIRGO
Spica
CORVUS
ECLIPTIC
HYDRA
ANTLIA
c.
BOÖTES
CYGNUS
Deneb
DRACO
CEPHEUS
URSA MINOR
Beta Cassiopeiae
Polaris
CASSIOPEIA
URSA MAJOR
LYNX
PERSEUS
Capella
GEMINI
AURIGA
Pleiades
CANCER
TAURUS
ECLIPTIC
CANIS MINOR
ORION
Procyon
Betelgeuse
HYDRA
MILKY WAY
Sirius
CANIS MAJOR
LEPUS
ERIDANUS
COLUMBA
PUPPIS
d.

On the earth, north-south distances are measured in terms of latitude and east-west distances in terms of longitude. Measurements on the sky are similar, but different terms are used. **Declination** (*dehk luh NAY shuhn*) is a vertical angle, like latitude, and is measured in degrees above and below the celestial equator. (See Figure 21-9.) **Right ascension** is a horizontal angle and is measured in hours, minutes, and seconds around the celestial equator.

Right ascension is measured eastward from the place on the sky where the sun crosses the celestial equator on March 21 (the vernal equinox). Note in Figure 21-8

Figure 21-7 *Constellations you can see at various times of the year, when you look at the southern sky.*

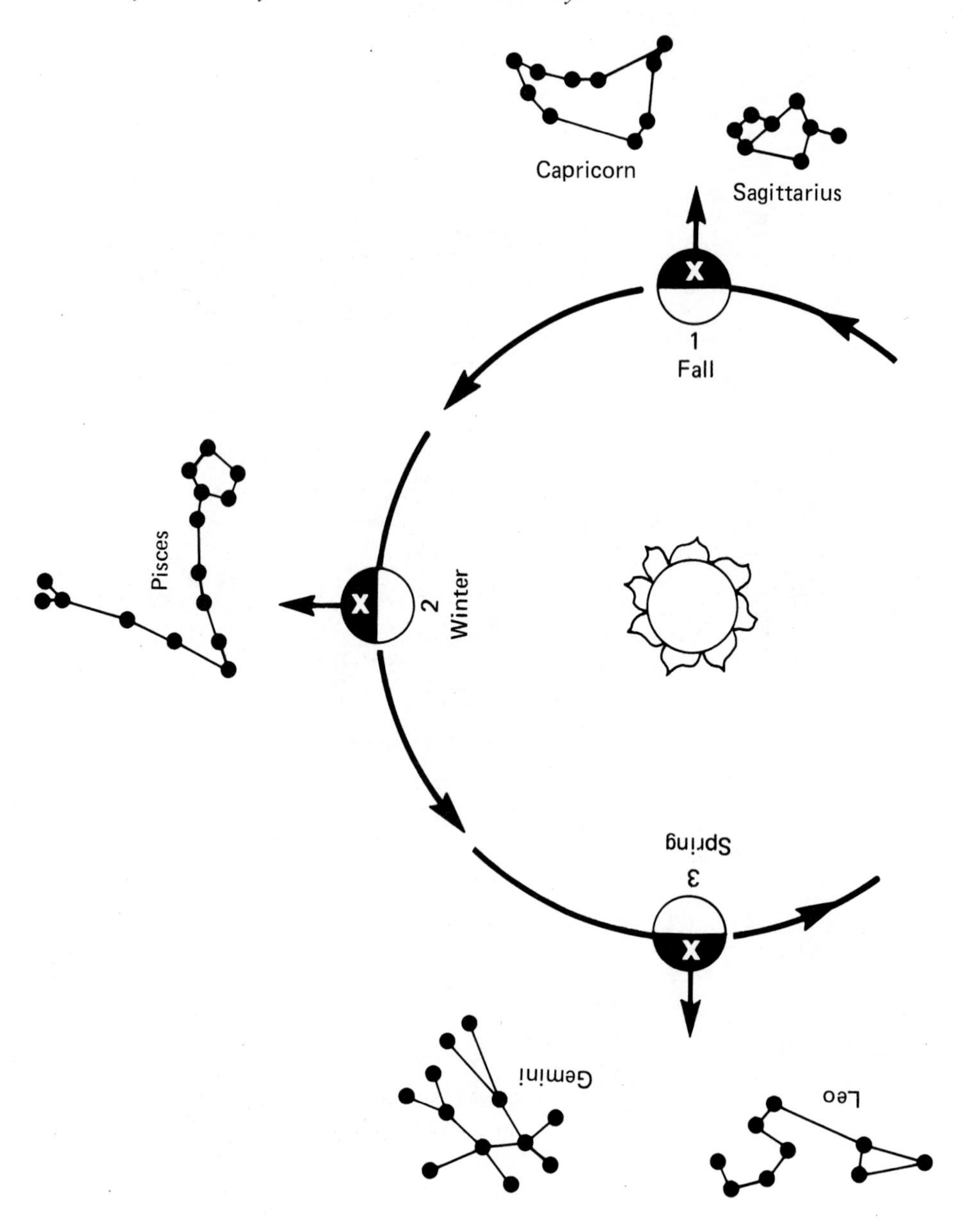

the point at which the ecliptic crosses the celestial equator (from south to north). This is the point from which right ascension is measured and it is called zero hours, 0 h. (Similarly, the Prime Meridian of Earth longitude at Greenwich, England is called zero degrees.)

ACTION

Can you find the bright star Procyon (*PROH see ahn*) on one of your star charts? Its location is 7 h 30 min right ascension and +4° declination.

Figure 21-8 *The celestial sphere.* **a.** *The part of the celestial sphere that you see from the Northern Hemisphere middle latitudes.* **b.** *The plane of the ecliptic is the plane of the earth's orbit around the sun. The plane of the celestial equator is different because of the tilt of the earth's axis.*

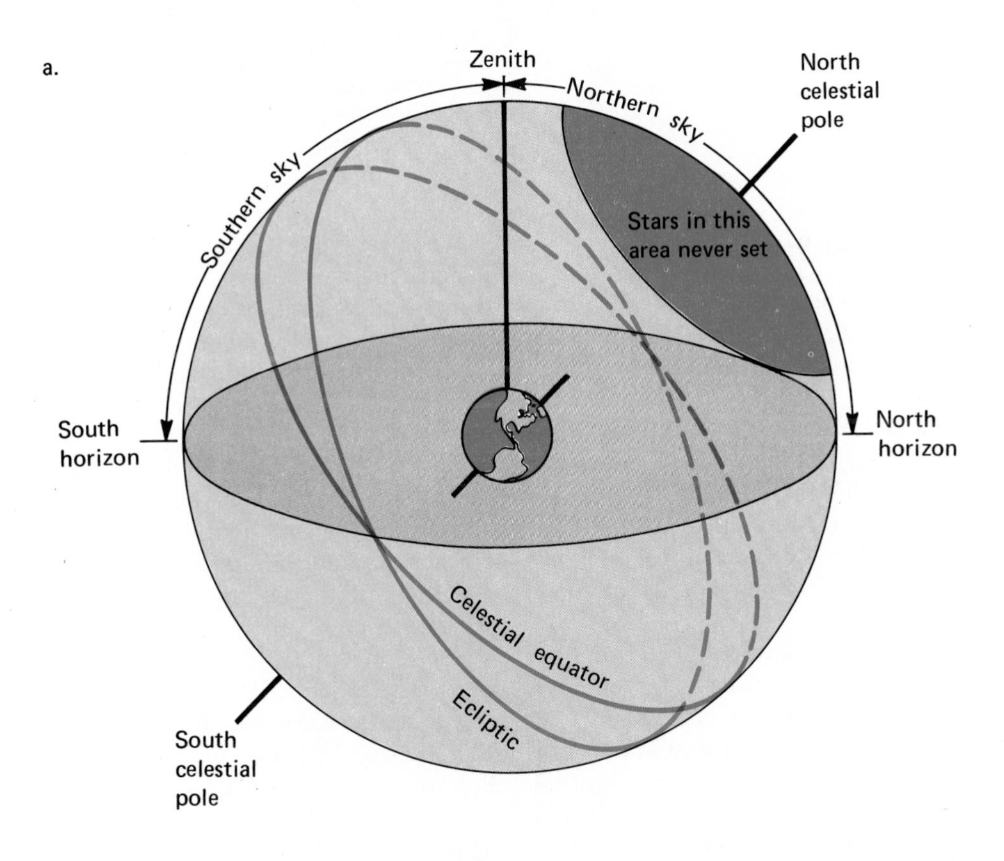

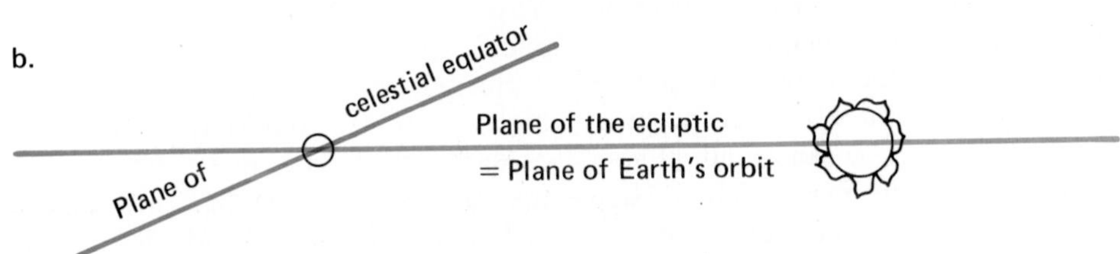

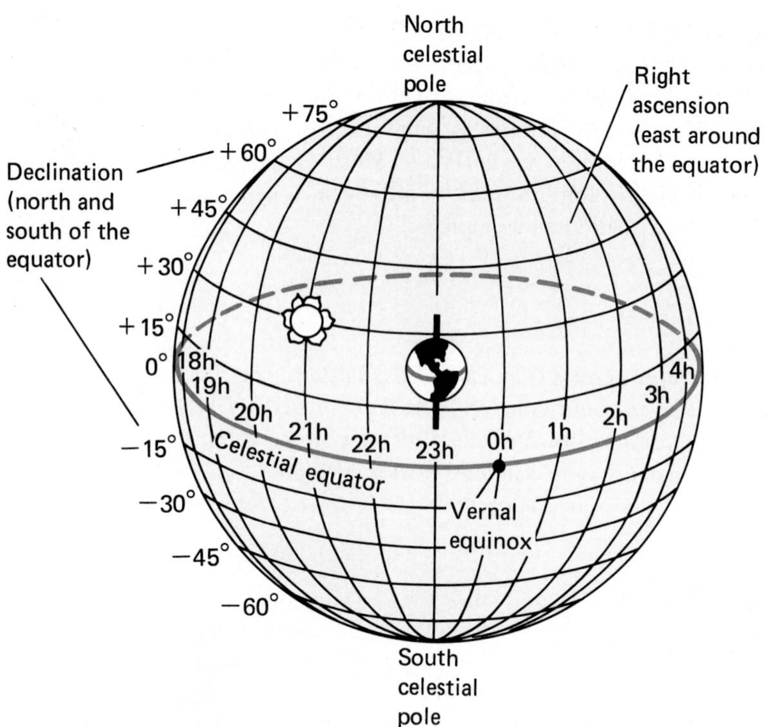

Figure 21-9 *The celestial sphere makes a globe for mapping the sky like the globe used for mapping the earth.*

21-7 Local Star Time

In order to determine which portion of your star charts is visible at your location, you first have to find the Local Star Time.

To approximate the Local Star Time for a corresponding clock time, you can use the following method: Multiply the number of months since March 21 by two. To this add the number of hours past noon. If the total is 24 or more, subtract 24.

Suppose the date is May 21 and the clock time for your observation is 9:00 P.M. What is your Local Star Time? The number of months past March 21 is two. Two times two gives four and the number of hours past noon for 9:00 P.M. is 9. Nine plus four gives 13 hours. Your approximate Local Star Time is 13 hours. You can try it for another date and time of night. For your purpose, using star charts, you can round off your calculations to the nearest month and hour, as in the preceding example.

Whatever your Local Star Time is, imagine that this hour on the sky extends from north to south. It appears to separate the sky into halves. There are roughly six hours of stars and constellations on either side of the calculated hour. When you use 13 hours as your Local Star Time, the 13th hour is the north-south meridian that passes directly above you. The 19th hour is on the eastern horizon and the 7th hour is on the western horizon. You can see 12 hours of sky at one time.

The 13th hour is close to the middle of chart c in Figure 21-6. You should be able to see the constellations Leo (12 h to 10 h), Cancer (9 h to 8 h), and the star Procyon. Cancer and Procyon are on chart d. You should also be able to see Virgo (13 h to 15 h), Libra (15 h to 16 h) and Scorpius (16 h to 18 h—chart b).

Here's how to use the charts, once you have determined your local star time and chosen a chart. Note that the names on the top part of the chart are upside down. Turn the book upside down. This upper part of the chart shows you what you can see toward the north. See if you can find Cassiopeia and Ursa Major (the Big Dipper), with the north star (Polaris) in between. Polaris is a medium-bright star that appears to be in a mostly empty space. If you are stargazing in a city or suburb, you may not be able to see Ursa Minor.

Turn the book rightside up again, and the lower part of the chart shows you what you should see toward the south. Remember, you may have to look at other charts to find all the stars to the east and west of your north-south meridian.

21-8 Measuring Stellar Distances

No one can tell where the stars really are simply by looking at them. We see only their apparent place in the sky. In a way, they all appear to be the same distance from us. Some simply appear brighter than others. Where they really are is not where they appear to be. By looking, we can only tell their direction.

The simplest way to measure the distance to the sun involves the difference in direction of the sun from two different places on the earth. This method involves something known as **parallax,** the apparent displacement of an object as seen from two different points that are not on a straight line with the object.

You can do a simple experiment to demonstrate parallax. First, hold your pencil (pointing upward) in front of you, at eye level and at arm's length. Then close your left eye and look at the pencil with your right eye. Next, without moving the pencil, close your right eye and look at the pencil with your left eye. What seems to happen to the pencil? Repeat the procedure, but first bring your pencil closer, to half the original distance. How does the difference between the apparent positions of the pencil change? How does this difference relate, in general, to the distance between your eyes and the pencil?

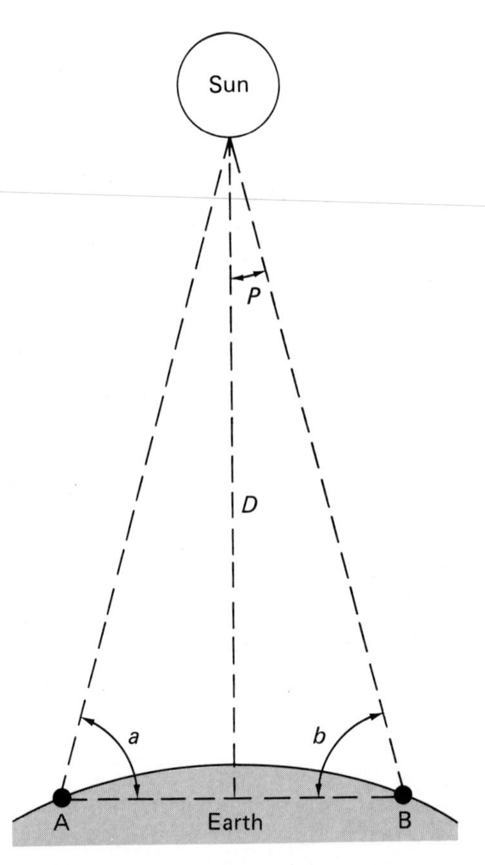

Figure 21-10 *Using parallax, the distance* (D) *to the sun can be calculated, from the angle of each sighting* (*at A and B*) *and the distance between A and B. Angle* p *is the "parallax angle."*

The use of the parallax method to measure the distance between the earth and the sun is illustrated in Figure 21-10. Naturally, the drawing is not to scale, but the method should be clear. Since the distance AB and the angles a and b can be measured, the distance (D) between the earth and the sun can easily be calculated. The angle (p) formed at the sun by each of the lines of sight is called the **parallax angle.** How does the parallax angle change as an object moves closer? farther away?

The sun is relatively close to the earth. It is only about 150 million km (1 AU) away. The next nearest star is approximately 274,332 AU away. Thus the astronomical unit is too small a unit to be practical for measuring distances to stars other than the sun.

The distance unit used by most astronomers is derived from the corresponding parallax angle. In order to get useful parallax angles, the measurements are made from two places on the earth's orbit, six months apart. This unit is obtained by dividing the parallax angle (in seconds of arc) into one. (Remember that one second of arc is only $\frac{1}{3600}$ part of a degree and there are 360 degrees in a full circle.) The unit is called the **parsec.** The word "parsec" is derived from the term "parallax-second." Can you find the distance in parsecs between Earth and the star Vega? The parallax angle is 0.123 sec.

One parsec is equal to 3.26 light-years. A **light-year** is the distance that light can travel in one year. Since light travels at 300,000 km/sec, a light-year is almost 10 trillion km.

The parallax method is good for stars that are relatively near. But the parallax angle for very distant stars is so small that it is impossible to measure. Even the angle to the nearest star, Proxima Centauri, is only 0.75 sec. Fortunately, there are other methods that can be used to measure distances to stars. One such method is discussed in Section 21-9.

Thought and Discussion

1. Why do we see different groups of stars during different times of the year?
2. How does right ascension compare to longitude here on Earth?
3. How far away is a star for which the parallax angle is 0.115 sec?

The Quantity and Quality of Starlight

21-9 The Brightness of Stars

When you look up at the sky you see that some stars are brighter than others. A star's **magnitude** is a measure of its brightness. The ancient Arabs had five magnitudes for the stars they could see. The brightest were assigned first magnitude, the second brightest, second magnitude, and so forth. The fifth magnitudes were the dimmest stars that they could see. After the development of the telescope, dimmer objects could be detected and were given the numbers 6, 7, 8 and so on.

The modern astronomer's magnitude scale is also based on the idea of assigning the brightest stars the smallest numbers. The larger the number, the dimmer the star. The difference in brightness, or brightness ratio, between any two magnitudes is about 2.5. So a star of magnitude 3 is about 2.5 times brighter than a star of magnitude 4. Remember that all these magnitudes are based on how bright a star appears when it is seen from Earth. They are called **apparent magnitudes.** Do you think all stars are really as bright as they look?

MATERIALS
four identical flashlights with new batteries

ACTION

Get permission for yourself and four friends to use a school athletic field some evening. Take four flashlights, preferably all of the same kind and with new batteries. Stand at one end and ask a friend to stand at the other end, or about 100 m away. Have the others stand at midfield and at 25 and 75 m. At a signal from you, have each person turn on his or her flashlight, directing the light beam at you. What observation can you make about the brightness of the four flashlights?

The apparent brightness of a star depends on two things: its luminosity and how far away it is. **Luminosity** (*loo muh NAHS uh tee*) is the total amount of energy a star is sending out. It is independent of how far away the star is.

What would all stars look like if they were all at the same distance from us? Then their differences in brightness would be due to their actual luminosities and not their distances from us. Astronomers have chosen a standard distance in order to compare brightness in all stars. This distance is 10 parsecs, or 32.6 light-years. The brightness that any star would have when it is viewed from the standard distance is called its **absolute magnitude.**

In order to find the absolute magnitude of a star, you need to know the star's distance and its apparent magnitude. At the standard distance (10 parsecs) a star's apparent magnitude and its absolute magnitude are the same. If a star is closer than 10 parsecs, it looks brighter than it would if it were moved back to 10 parsecs. So its *absolute* magnitude is *less* than its *apparent* magnitude.

If a star is farther away than 10 parsecs, it looks dimmer than it would if it were moved up to 10 parsecs. So its *absolute* magnitude is *greater* than its *apparent* magnitude.

When you know the absolute magnitudes of stars, you can compare their luminosities. Remember, the luminosity of a star is the total amount of energy it is sending out. If you compare absolute magnitudes, as if all stars were 10 parsecs away, you are also comparing luminosities. To be very bright, a star must be giving off a large amount of energy. A dimmer star at the standard distance must be giving off less energy. In other words, the greater the luminosity, the greater the absolute magnitude. The lesser the luminosity, the lesser the absolute magnitude.

You know that the sun is much closer than any other star. Imagine how much dimmer it would appear if it were moved to the standard distance of 10 parsecs. The sun's absolute magnitude is only 5.0. If it were at standard distance, it would be barely visible on a very clear night.

Figure 21-11 *Use this chart to relate distance and luminosity to apparent magnitude.*

Name	Apparent Magnitude	Luminosity	Distance in Light-Years
Sun	−26.70	1.00	0.00002
Sirius A	−1.43	23.00	8.70
Canopus	−0.72	1500.00	100.00
Procyon A	+0.38	7.30	11.30
Betelgeuse	+0.41	17,000.00	500.00
Deneb	+1.26	60,000.00	1600.00
Beta Cassiopeiae	+2.26	8.20	45.00
Delta Aquarii	+3.28	24.00	84.00
Epsilon Andromedae	+4.37	1.30	105.00
Barnard's Star	+9.54	0.00045	6.00

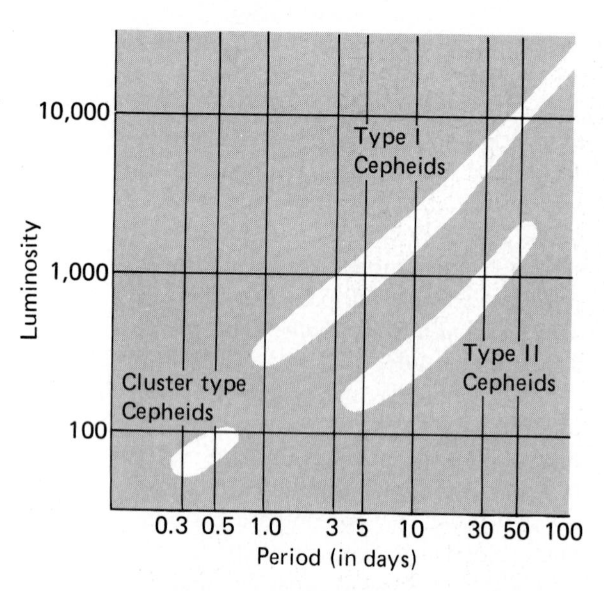

Figure 21-12 *There are several different kinds of Cepheids. Each kind has a different relationship between period and luminosity.*

Astronomers compare star luminosities with that of the sun. They call the luminosity of the sun 1.0, to make it convenient. If another star at standard distance has an absolute magnitude of 5.0, its luminosity is also 1.0. If a star has an absolute magnitude of 3.0, it is two magnitudes brighter than the sun. (Remember, the brightness ratio between any two magnitudes is about 2.5.) To calculate the luminosity, you square the brightness ratio, since the difference in magnitudes is 2:

$$(2.5)^2 = 6.25$$

You can round it off to 6.3 for the luminosity.

If a star has an absolute magnitude of 0.0, the difference between its absolute magnitude and that of the sun is 5.0 magnitudes. Its luminosity is $(2.5)^5$, or about 100. Suppose a star had an absolute magnitude of 10. What would its luminosity be?

Study Figure 21-11 and compare the luminosities and apparent magnitudes of several stars. Can you see how much difference the distance makes?

Some stars have a changing luminosity. These stars do not appear equally bright at all times. Some even follow a definite cycle of change. They pulsate on a regular schedule. These stars are called **Cepheid** (*SEE fee ihd*) **variables.** They are named after a star in the constellation Cepheus. The time it takes a Cepheid variable to change from its greatest brightness to its least brightness and return to its greatest brightness again is called its **period.**

Henrietta Leavitt, an American astronomer, discovered a relationship between the period and the brightness of Cepheid variables. As a result, once the luminosities of a few Cepheid variables had been determined, a graph like the one in Figure 21-12 could be made. With such a graph, you can find the luminosity of any Cepheid if you know its period.

By measuring the period of a Cepheid variable, you can find its distance from Earth. Once you have found the luminosity of the star, from the graph, you can calculate its absolute magnitude. Then, by comparing its absolute magnitude to its apparent brightness, you can calculate its distance from Earth. In this way, you can determine the distance of stars too far away for use of the parallax method.

21-10 The Spectra of Stars

Anyone who has looked carefully at the stars has noticed that they are not all the same color. The star Betelgeuse (*BAY tuhl jooz*) is red, Capella is yellow, and Spica is blue-white. These color differences are due to differences in temperature. Stars that are cooler look red; hotter ones appear blue. This does not mean, however, that stars shine with only red or blue light. They just emit more red than blue or more blue than red. By analyzing the light from a star,

Figure 21-13 *When white light is passed through a prism, a spectrum of colors is produced.*

scientists can learn a great deal about its composition and temperature, among other things. How are scientists able to learn such things from starlight?

When white light is passed through a prism, it appears as a continuous band of colors ranging from violet to red (Figure 21-13). This **visible spectrum** is also formed when light is bent as it passes through raindrops. You see it as a rainbow. This suggests that white light is a mixture of all the colors of the rainbow. There are other ways to show that this is so. When white light passes through a piece of red cellophane it emerges as red light. The red cellophane acts as a filter, and cuts off all colors but red. If white light is passed through a blue filter, it becomes blue, and so on.

Scientists have found that each of the colors of which white light is made has a characteristic wavelength. You may re-

Figure 21-14 *Various types of radiation can be arranged in order of increasing (or decreasing) wavelength and energy. Such a diagram illustrates the electromagnetic spectrum.*

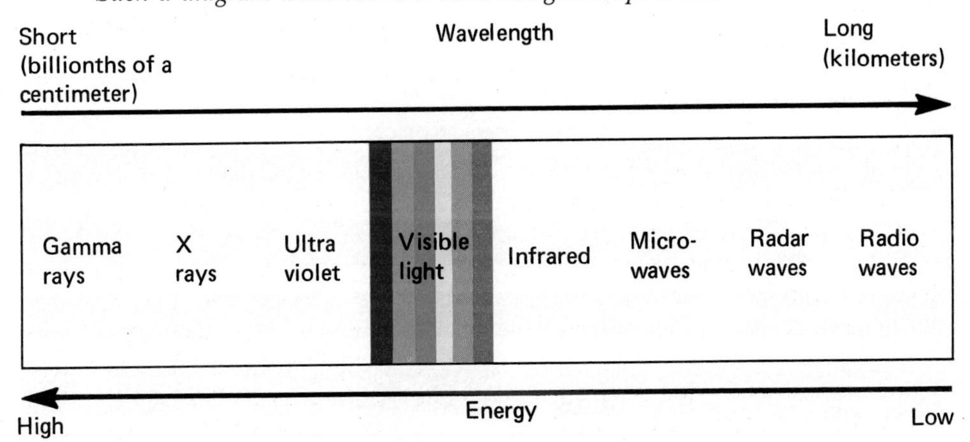

member that wavelength is also a property of water waves. Review what you learned about water waves in Section 5-4. Although light waves are much too small to see, scientists are able to measure their wavelengths with special instruments such as **spectroscopes.** (If your school has spectroscopes, you will use one in the next investigation.) The wavelengths in the visible spectrum range from about 40 millionths of a centimeter (violet) to about 80 millionths of a centimeter (red).

All of these forms of radiation can pass through empty space (while water waves can hardly be imagined to exist without the water). Since electric and magnetic properties of this radiation apparently enable it to travel through space, it is known as **electromagnetic radiation.** The various types of electromagnetic radiation, arranged in order of increasing (or decreasing) wavelength, form an **electromagnetic spectrum.** See Figure 21-14. Notice that the wavelength increases from left to right, but the energy of the radiation increases from right to left. That is, the smaller the wavelength, the greater the energy of the radiation. Thus, as you read earlier, blue stars are hotter than red stars.

One more topic must be considered before you can apply what has been discussed so far to the analysis of starlight. Suppose you supplied an element with enough energy (in the form of heat or electricity, for example) to cause it to give off light. You would find that it does not produce a continuous solid spectrum, with all of the colors. Instead, when you passed the light given off by the element through a narrow slit and then through a prism, you would get a series of colored lines. This is called a **bright-line spectrum;** four are illustrated in Figure 21-15. No two elements give the same line spectrum. Thus by measuring the wavelength of light in the bright-line spectrum of an element you can tell which element is producing the light.

You can produce a **dark-line** spectrum instead of a bright-line spectrum. Suppose you passed white light through an element that is a gas, like oxygen. The element would *absorb* those same wavelengths of light that appear in its bright-line spectrum. So, if the light that has passed through the element is then passed through a slit and prism, dark lines will appear in the otherwise continuous spectrum. These dark lines correspond to the wavelengths that are missing because they have been absorbed by the element.

Dark-line spectra are of interest to scientists because stars produce such spectra. ("Spectra" is the plural form of *spectrum.*) Elements (and compounds) in the star's atmosphere absorb certain wavelengths of light. As a result, the

Figure 21-15 *Each element gives a characteristic line spectrum, which can be used to identify that element. Each line represents a specific wavelength.*

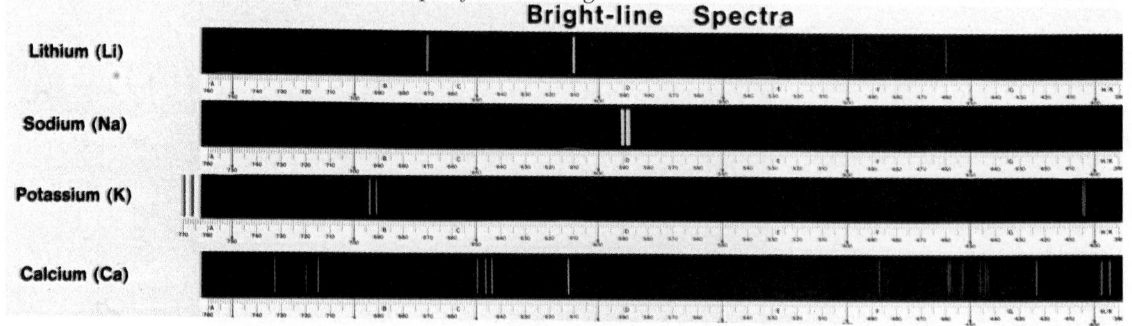

Figure 21-16 *Stars produce dark-line spectra. The letters identify the type of star that produces each spectrum.*

starlight produces a dark-line spectrum. Cooler stars contain more elements and compounds and so have more dark lines in their spectra. Several spectra produced by stars can be seen in Figure 21-16.

21-11 Investigating the Fingerprints of Stars

When you have completed this investigation, you will be able to use a spectroscope to determine the composition of certain compounds and to use a spectrogram to determine the temperature and color of a star.

PROCEDURE

Bright-line spectra are produced by elements. These spectra can be used as fingerprints to identify the elements. You will try to be a detective and identify four elements from their line spectra. A number of metal elements produce spectra when certain of their compounds are heated in an ordinary burner flame. Only the metal portion of each compound gives off light, so the metals can be identified from the resulting spectra.

You will heat four compounds and observe the spectra they produce, using a spectroscope. The compounds are: sodium chloride, strontium nitrate, lithium chloride, and copper chloride. Each of these compounds produces a different spectrum when heated in a flame. Sodium gives a bright yellow line (actually, two lines very close together). Strontium gives two brilliant orange lines. Lithium gives a strong red line, among others. Copper produces a series of bright, intensive lines, extending from green to orange. (The lithium spectrum is shown in Figure 21-15.)

MATERIALS
spectroscope, sodium chloride, strontium nitrate, lithium chloride, copper chloride, alcohol or bunsen burner

Use a spectroscope to observe the light from each compound that is held in the flame. You will not be told which compound is which. You will try to identify each one from the line spectrum you observe.

DISCUSSION

Check with your teacher to see whether you have identified the compounds correctly. What could account for any

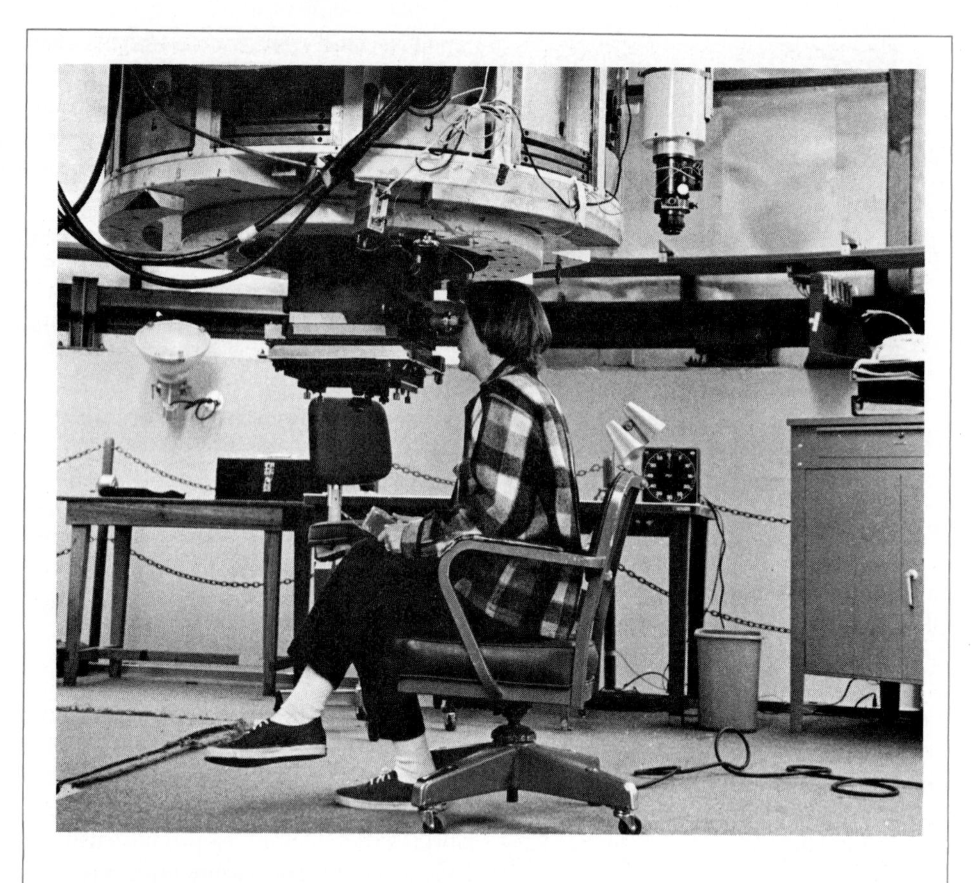

ASTRONOMER

Astronomy is a branch of physics that holds a strong fascination for many people. Current research in astronomy covers a variety of areas: the origin and characteristics of the solar system, the birth and evolution of stars, galaxies and their motions, the origin of the universe, and many more. Astronomy is different from other sciences in that astronomers cannot experiment with the objects they study. Practically all of our contact with the universe comes indirectly, through light and other forms of radiation. Moon rocks and meteorites are the only objects from outside the earth an astronomer can actually touch.

The work of astronomers has two aspects: theory and observation. Most astronomers do both types of work, but some are involved primarily in one or the other. Theoretical work in astronomy is developing mathematical equations to compute models. Mathematical models of stars and the ways in which they evolve are examples of this type of work. Observational work is observing stars or other objects, and analyzing and explaining the results.

Astronomy is a difficult field of study. Astronomers typically have a bachelor's degree in physics or mathematics, and hold a Ph.D. in astronomy. Opportunities for individuals with only a bachelor's degree are extremely limited. There are also some jobs available at the technician's level for people with two years' training.

wrong or uncertain results you might have gotten?

PROCEDURE

Continuous spectra are produced when white light is passed through a prism or through the grating of a spectroscope. (Gratings disperse, or "spread out," light just as prisms do. They are used instead of prisms in many spectroscopes.)

Use your spectroscope to look at the light produced by a lamp. Note the nature of the spectrum that you see. Next, place a colored filter, or piece of cellophane, between the lamp and the spectroscope and again note the spectrum that you see. Repeat the procedure for other colored filters that may be available.

DISCUSSION

1. What kind of spectrum did you see when the unfiltered white light passed into your spectroscope?
2. What effect did the colored filters have on the spectra that you observed? How do you explain the effect of the filters?

PROCEDURE

In Figure 21-16 you see the spectra produced by six different types of stars. Each spectrum (or **spectrogram**) is identified by one of the following letters: A, B, F, G, K, or M. These letters identify different types of stars. Examine the spectrograms and see if you can answer the following Discussion Questions.

DISCUSSION

1. Which of the star types contain the greatest number of different elements and compounds? How did you arrive at your answer?
2. More elements and compounds are found in the cooler stars. Which of the spectrograms represent the hotter stars? Which represent the cooler ones?
3. The hottest stars are blue-white and the coolest stars are red. Which spectrograms represent blue-white stars? red stars?
4. Some astronomers use the following sentence to code a sequence of stellar spectra: "Be A Fine Guy/Girl, Kiss Me." Can you decode the sentence? How does the coded sequence run in terms of temperature (hot to cold or cold to hot) and color (blue-white to red or red to blue-white)?

MATERIALS

spectroscope, lamp, colored filters or cellophane of various colors

Thought and Discussion

1. If star A is magnitude 3.0 and star B is magnitude 5.0, which star is brighter and by how much?
2. Explain the difference between apparent magnitude and absolute magnitude.
3. How is it possible to study stars' luminosities?
4. What is the relationship between color and temperature in stars?
5. What is the relationship between the temperatures of stars and their spectra?

Unsolved Problems

Most scientists believe that fusion is the source of the sun's energy. However, observations in recent years have failed to detect one of the by-products of the fusion reaction in the expected amount. This by-product is the **neutrino** (an uncharged particle smaller than the electron), which is difficult to detect anyway. There are a number of possible explanations for the failure to find the expected number of neutrinos, but the actual explanation has not yet been found. It is possible that the explanation could change some of our basic ideas concerning the sun and other stars.

Although the earth receives a lot of solar energy, the energy is not concentrated in any one place but is distributed all over the earth's surface. This makes collection of solar energy and its conversion into usable energy difficult and thus expensive. Moreover, the amount of solar energy received at any given place varies with the weather and time of day. This makes it necessary to provide some means for storing the energy. Nevertheless, various methods for utilizing solar energy are being investigated throughout the world. In several countries, solar energy is now commonly used to heat water for household use. As solar energy technology becomes further developed, more and more of the world's energy needs will probably be met with energy from the sun.

Chapter Summary

The sun is an ordinary star, one in a hundred billion. It sends its energy outward into space, and only a small portion of that energy is received by Earth.

Our universe is filled with billions of stars of all types, sending out their energy in all directions. From Earth we see but a tiny fraction of the number of stars. The stars appear to be in every conceivable direction in the sky. Some are brighter than others. Some seem to fit into imaginable groups called constellations. Certain of these star groups are seen during particular times of the night and year. If you know some constellations, Local Star Time, and the coordinates of right ascension and declination, you can find stars on star charts and on the sky.

There are many methods for measuring the distances to stars, but the most widely used method involves parallax. By dividing the parallax angle (in seconds of arc) into one, you can obtain the star's distance in parsecs. Stellar distances are quite variable.

Star brightness depends on both distance and luminosity. Some stars appear bright because they are close, while some are bright because they are putting out huge amounts of energy. Astronomers have designed the magnitude scale for star brightness. The larger the number, the dimmer the star. Stars have two basic kinds of magnitudes. Apparent magnitude is the brightness of the stars as they appear from Earth. Absolute magnitude is the brightness of stars when all are compared at an imagined standard distance (10 parsecs). A knowledge of absolute magnitude enables the astronomer to compute star luminosity.

Some stars produce their energy in a variable way. One type of short-period variable, called a Cepheid variable, is used to measure the distances to stars in deep space.

Our knowledge of the general characteristics of stars can be attributed to our understanding of the stars' spectra. This defines nearly all the basic properties of stars, including their colors and temperatures.

Questions and Problems

A

1. What factors influence how many stars you can see each night?
2. Why are constellations important to learn?

3. Which has the larger parallax angle, a nearby star or a more distant one?
4. What is the standard distance, expressed in parsecs? light-years?
5. What would be the magnitude of the sun at standard distance? Would it be visible to the unaided eye at this distance?
6. Which is brighter, a star of magnitude 6.0 or one of magnitude 5.0?
7. Why do astronomers want to visualize stars at a standard distance?

B

1. Why do we see different constellations at different times of night?
2. What is the Local Star Time on Halloween night at 12 midnight?
3. How much brighter or dimmer than magnitude 3.0 is magnitude 6.0?
4. When Columbus first landed in America (1492), a distant star was just giving off light that finally reached the earth in 1976. How far away is that star, in light-years? parsecs?
5. What do the spectra from stars tell about their composition?

C

1. If the projected image of the sun is 25 cm in diameter and the projected image of a sunspot is 0.5 cm, how large is the sunspot?
2. What is a star's distance in parsecs if it has a parallax angle of 0.275 sec?
3. Why is there an effective limit to measuring the distance to stars using the parallax method?
4. How many times brighter is a star of magnitude 1.0 than one having a magnitude of 5.0?
5. What is the difference between luminosity and absolute magnitude?
6. How can astronomers use Cepheid variables to measure the distances to deep space?

Suggested Readings

Claiborne, Robert. *The Summer Stargazer: Astronomy for Absolute Beginners.* New York: Penguin, 1981. (Paperback)

Fisher, David E. *The Creation of Atoms and Stars.* New York: Holt, Rinehart & Winston, 1979.

Frazier, Kendrick. *Our Turbulent Sun.* Englewood Cliffs, New Jersey: Prentice-Hall, 1982.

Gallant, Roy A. *The Constellations: How They Came to Be.* New York: Scholastic Book Services, 1979.

Gribbin, John. *The Death of the Sun.* New York: Delacorte, 1980. (Also available in paperback)

Kerrod, Robin. *Stars and Planets.* New York: Arco, 1980.

Mitton, Simon. *Daytime Star: The Story of Our Sun.* New York: Charles Scribner's Sons, 1981.

Moore, Patrick. *The New Guide to the Stars.* New York: W. W. Norton Co., 1976.

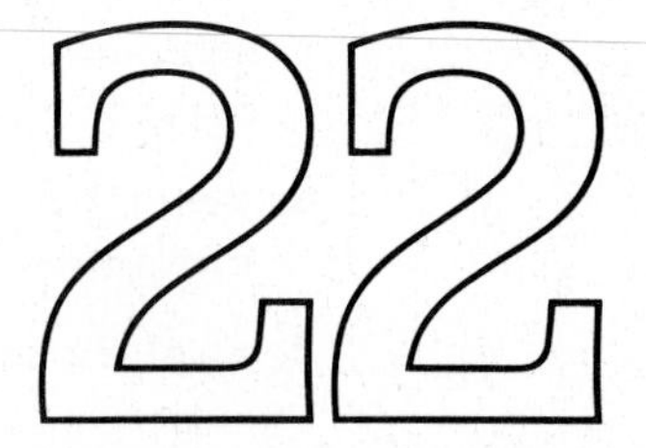

Galaxies and the Universe

IF YOU HAD THE FREEDOM *to go beyond the moon, the sun, and the nearby stars, you would come to an endless sea of night. Scattered through this darkness lie clusters of stars, such as the ones in the photograph on the opposite page. The clusters, or galaxies, come in several shapes and spin slowly.*

Between galaxies is empty space. In some places, there may be a light-year between two molecules of matter! Between the galaxies are thousands and sometimes millions of light-years. Space seems endless. Even when we look through some of the world's largest telescopes, there seems to be no end. There are billions of galaxies, and in each galaxy are billions of stars. Is it any wonder that many scientists believe other civilizations exist elsewhere in deep space?

The universe contains everything that exists—space, galaxies, stars, planets—including the earth and all its plants and creatures (see Figure 22-1).

As far as we can tell, even the most distant galaxies are made of elements just like those found on earth. We can't prove *the elements are the same. On the other hand, we have no reason to believe they are different. We assume that the laws of physical science are the same "out there" as they are here on Earth. There are some strange goings-on in space that challenge our imaginations, but most of what we observe follows known laws of science.*

AFTER COMPLETING THIS CHAPTER, YOU SHOULD BE ABLE TO:

1. **describe in general terms the extent and components of the universe.**
2. **explain the universality principle—that the same matter exists everywhere.**
3. **describe the nature of the Milky Way galaxy and compare it with other galaxies.**
4. **explain the known properties of pulsars and quasars.**
5. **explain how motions of deep space objects are measured.**
6. **support with known data the theory of the expanding universe.**
7. **describe the life cycle of stars.**

Galaxies

22-1 The Milky Way

If you look at the sky on a clear night, you will see a broad band of faint light extending across the sky like the band in Figure 22-2. This band of light is the Milky Way. The Milky Way is a **galaxy,** a collection of about 100 billion stars. It is shaped like a disk about 100,000 light-years in diameter and several thousand light-years thick. Astronomers have known for 200 years that the sun is part of the Milky Way. The sun is about 30,000 light-years from its center; you can see its position in Figure 22-3.

In the middle of the galaxy, there is a large group of globe-shaped clusters of stars. Each cluster may contain as many as 50,000 stars. You see most of the globular clusters when you look toward the center of the Milky Way, toward the constellation Hercules. A globular cluster found in Hercules is shown in Figure 22-4. It looks like a swarm of thousands of bees traveling in space together.

The Milky Way spins like a huge Ferris wheel. However, its inner parts are rotating faster than its outer parts. Stars that lie closer to the center are passing us, and we pass those farther away. The sun and nearby stars are really moving—at about 240 km/sec! But because the galaxy is so large, it still takes the sun about 200 million years to make one journey around the center. Just think of it! The last time the earth was in the position it is in now, dinosaurs walked about and Pangaea was beginning to split up! Since the birth of the sun, believed to be about five billion years ago, our solar system has made some 23 trips around the galaxy.

MATERIALS
photographs of nine different galaxies

▣ 22-2 Investigating Other Galaxies

On a clear night you may be able to see a fuzzy patch of light in the constellation Andromeda with the unaided eye. It is the Andromeda galaxy, the giant spiral galaxy shown in Figure 22-5. It is a slightly larger version of our Milky Way.

Most galaxies are so far away that they show only on photographic film exposed to their light for a long time. The faintest galaxies that you can see are many billions of light-years away. Generally, the fainter and smaller the galaxy appears, the farther away it is. Distances to galaxies are measured by using Cepheid variable stars. Remember that the period of a Cepheid variable star tells us how bright it really is. Comparing its actual brightness with its apparent brightness provides us with a key for finding its distance.

Galaxies come in many different shapes and sizes. There are some that seem to have no pattern at all. They look like shapeless fuzzy patches, like the one in Figure 22-6. Others are oval-shaped (Figure 22-7), or shaped like thin lenses, such as the spiral galaxy you have already seen in Figure 22-5.

More than half of all known galaxies appear to be spirals. There are two kinds of spirals. Normal spirals have rounded centers and sometimes several spiral arms as shown in Figure 22-8. Barred spirals have elongated centers and only two spiral arms (Figure 22-9).

When you have completed this investigation, you will be able to classify galaxies according to visible characteristics and use these classifications to estimate the relative ages of galaxies.

PROCEDURE

You will be given photographs of nine galaxies. Try to place the photographs in an order that relates to characteristics you observe.

DISCUSSION

1. Assume that the order you set up was related to the ages of the galaxies. Which galaxies seem oldest? Which seem youngest?

Figure 22-1 *The earth is only a tiny part of the universe.*

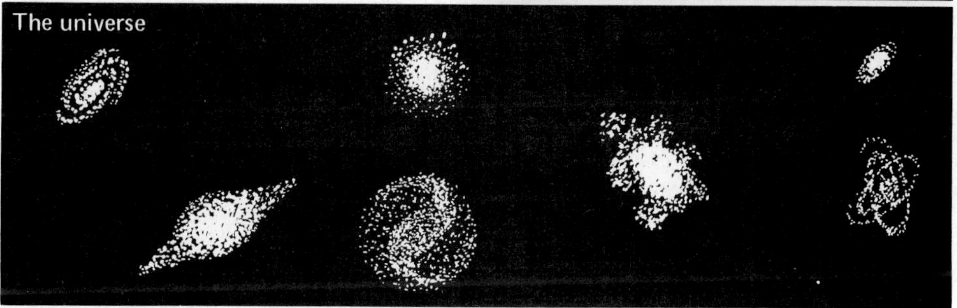

Figure 22-2 *The Milky Way.*

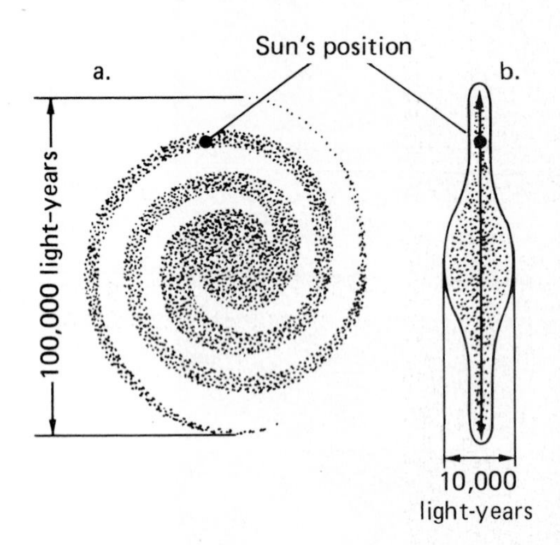

Figure 22-3 *The spiral Milky Way galaxy:* **a.** *as it might appear from outer space and* **b.** *a cross section.*

Figure 22-4 *A globular star cluster in the constellation Hercules.*

Figure 22-5 *Andromeda is the nearest spiral galaxy.*

Figure 22-6 *The Large Magellanic Cloud is an irregular galaxy.*

Figure 22-7 *The companion galaxy to Andromeda is elliptical.*

Figure 22-8 *The nebula in the constellation Canes Venatici is a normal spiral.*

Figure 22-9 *A barred spiral galaxy in Eridanus.*

2. Describe another system of classifying the galaxies. What is good about your system?

22-3 The Motions of Stars and Galaxies

Until this century, astronomers had no way of knowing how galaxies moved—or even if they moved at all. In Chapter 21 you learned about spectral lines. Astronomers use pictures of star spectra, called spectrograms, to find the velocity and direction of the stars. Because stars make up galaxies, we can use star spectrograms to study the movements of their home galaxies, too.

Imagine you are riding along a road in a car. Objects far away from the road seem to move by slowly. Things next to the road zip by you. In a similar way, stars closer to Earth seem to move through the sky faster than stars farther away from Earth. Barnard's star moves very quickly—for a star—and has been called a runaway star (see Figure 22-10). Even with its fast motion, 65,000 years will pass before this nearby star moves all the way across the sky. Some stars are so far away that astronomers using the best modern instruments cannot find any movement at all.

Side-to-side movement is easy to think about. But how can the motion of a star toward you or away from you be measured? Have you heard of the **Doppler effect?** It works this way. Think of a train approaching, with its horn blasting. The sound waves seem to have a higher pitch as the train comes toward you. After the train passes, the pitch of the horn gets lower. A scientist named Doppler first noted this. How does it work? The speed of the approaching train is added to the speed of the horn's sound waves. The number of waves hitting your ear each second increases. Thus, the approaching train's whistle seems higher in pitch. A train moving away from you decreases the number of waves hitting your ear each second. So the sound pitch is lower.

Light waves from stars operate in a similar way. A spectrogram of a star's light may show that the dark absorption lines are all shifted toward the higher frequency (violet) end of the spectrum. Then, astronomers believe the star is approaching. This happens because the velocity of the approaching star "pushes" more wavelengths of the starlight per second toward the spectrograph. If the star's spectrogram shows lines shifted toward the low frequency (red), astronomers believe the star is moving away. As you might expect, the farther the dark lines are shifted from their usual positions, the faster the star is moving. The shift in spectra from most galaxies is toward the red. They are rushing away from us. There are some exceptions. For example, the galaxy in Andromeda is coming toward us at 224 km/sec. Or is the Milky Way going toward the Andromeda galaxy? By studying the movement of many other galaxies we can find out how the Milky Way is moving in relation to these galaxies. Astronomers have found that the faster galaxies move away from us, the farther away they are. You will learn how this relationship came to be in the next section.

Thought and Discussion

1. What tool do astronomers use to figure out the speed and direction of galaxies?
2. Describe the motion of most galaxies in relation to the Milky Way.
3. Why can't we be sure if the Milky Way is doing most of the moving in space, or if the other galaxies are doing most of the moving?

Figure 22-10 *These photographs show how Barnard's Star moved in 22 years.*

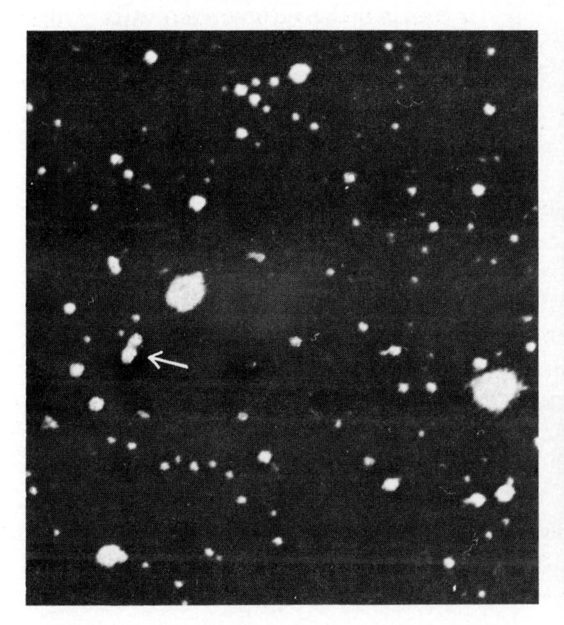

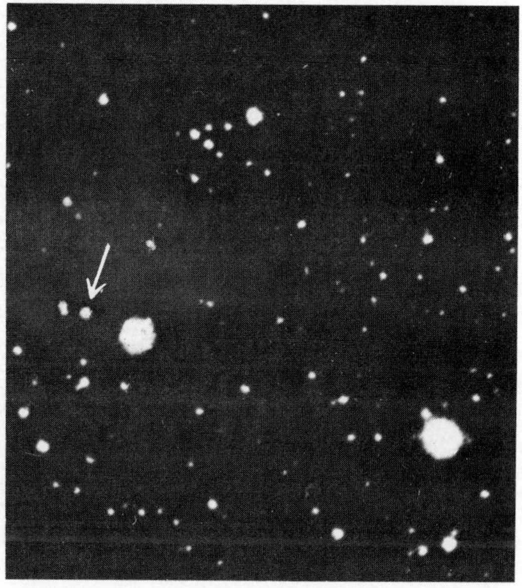

22-4 Determining the Size of the Universe

Galaxies seem to belong in groups containing many members. Our Milky Way is in a group of about 21 galaxies. These galaxies are no more than about three million light-years from each other. Recently, a dwarf galaxy called Snickers was discovered by radio telescopes. Dr. S. Christian Simonson III of the University of Maryland called it Snickers because it was "like the Milky Way, only peanuts." Snickers, thought to be our nearest neighboring galaxy, is still 55,000 light-years away.

For the last 50 years astronomers have been picking up radio signals from space (see Figure 22-11). But they found it difficult to locate the source of these signals. Because the first radio sources that were located looked very much like ordinary stars, they were called quasistellar objects or **quasars** (*KWAY zahrz*). Astronomers have located at least 200 of these strange objects. Quasars are strong sources of radio waves. The red shifts of quasars' spectra are very large; they are therefore very far away. One recently discovered quasar is thought to be moving at $4/5$ the speed of light. This one must be at least 8 billion light-years away! Most strangely, each quasar appears to have the mass of an entire galaxy—but is only about the size of a solar system! Quasars cannot be clearly photographed, because they are too far away. In fact, they may be the most distant objects we know!

Figure 22-11 *The Harvard University radio telescope in Massachusetts.*

In 1929 Edwin Hubble discovered a relationship between the shift in spectral lines and the speed of galaxies. He found that the greater the spectral shift toward the red (longer wavelengths), the faster the galaxy was moving away from us. Astronomers believe the faster a galaxy is moving, the farther away it is. Thus, once the galaxy's velocity is determined, its distance can be calculated. Figure 22-12 shows spectrograms of five galaxies. It also shows the spectral shifts of calcium toward the red. The two black lines in each of the horizontal streaks represent forms of the element calcium.

Hubble's study showed that for every million light-years a galaxy was away from us, its velocity increased 30 km per second. Study the distances in light-years and the red shift velocities for the galaxies in Figure 22-12.

Astronomers learned many years ago that the masses and luminosities of most stars are related. The more mass a star has, the greater is its luminosity. This idea was first investigated by the

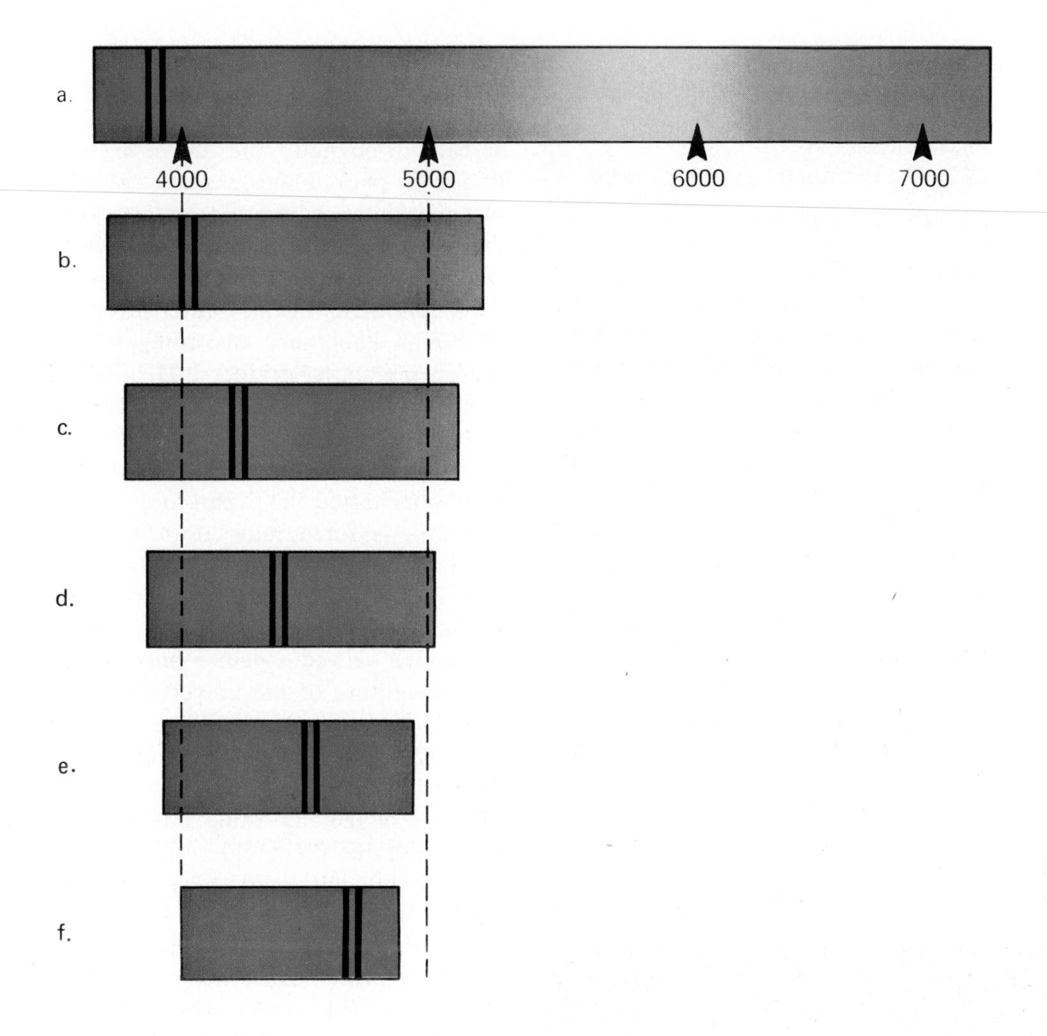

Figure 22-12 *The red shift, shown by spectra from five galaxies.* **a.** *The light (emission) spectrum from the sun. The numbers refer to wavelengths in units called angstroms. One angstrom equals one hundred-millionth of a centimeter. The black lines are absorption lines of the element calcium.* **b.** *The spectrum of a galaxy in the constellation Virgo. The calcium absorption lines are to the right of where they are in the solar spectrum because the wavelength has changed about 60 angstroms as the light has traveled from the galaxy to the earth. This galaxy is 43 million light-years away; its speed is 1200 km/sec.* **c.** *A galaxy in Ursa Major; distance 560 million light-years, speed 15,000 km/sec.* **d.** *A galaxy in Corona Borealis; distance 728 million light-years, speed 21,500 km/sec.* **e.** *A galaxy in Boötes; distance 1 billion 290 million light-years, speed 39,000 km/sec.* **f.** *A galaxy in Hydra; distance 1 billion 960 million light-years, speed 61,000 km/sec. The red shift here is about 800 angstroms.*

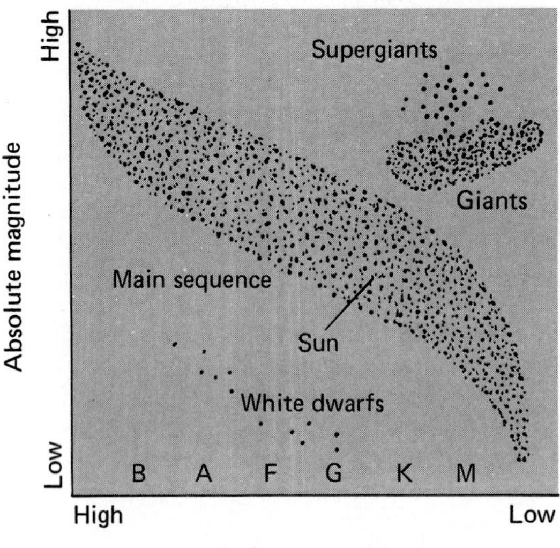

Figure 22-13 *The Hertzsprung-Russell diagram. How are absolute magnitude and the surface temperature of a star related?*

Danish astronomer Ejnar Hertzsprung (*HURTS spruhng*) and the American astronomer H. N. Russell. They plotted on a graph the absolute magnitudes of some stars against their spectral classes (see Figure 22-13). The letters refer to the classes illustrated in Figure 21-16. They found that hotter stars were indeed more luminous. Remember, the spectrum tells us the star's color, hence its temperature. Once the star's temperature is known, an astronomer can find the star's absolute magnitude using the H-R for (Hertzsprung-Russell) diagram.

MATERIALS

pieces of cotton, balloon, rubber cement

22-5 Investigating the Expanding Universe

Because the spectra from most deep space objects show red shifts, astronomers believe the universe is expanding. Wherever we look, galaxies are rushing away from us. The farther away they are, the faster they are moving.

Why is the universe expanding? According to the widely accepted "Big Bang" theory, all of the matter now spread throughout the universe was originally packed into a small volume. No one knows where the matter came from. Temperatures and pressures were too high to imagine. The mass of every cubic centimeter of this matter was over 90 billion kilograms. According to the theory, the matter exploded. Debris was flung violently in all directions. This is where we are today: the universe still seems to be expanding. If the Big Bang explosion started the expansion, in time gravitational forces may slow it down. If gravitational forces eventually stop the expansion, the universe may collapse again. All the galaxies and stars would be tightly packed together once more.

Is the future of the universe to be a series of accordion-like collapses and expansions? If so, how old is the universe? About 10 billion years have passed since the time of the Big Bang. But the universe might be much older, if it has expanded and contracted many times!

From this evidence, do you think we are in the center of the universe? We aren't! You can make a model of what is happening. The model shows why we can't be in the center.

When you have completed this investigation, you will be able to make a model of the universe to show that galaxies are expanding away from each other and to explain why more distant galaxies appear to move away faster than closer ones.

Procedure

Take little pieces of cotton of various sizes. Shape some of the pieces into forms representing the types of known galaxies—spiral, oval, and irregular.

Blow up a balloon part way. With rubber cement, fasten the pieces of cotton (galaxies) to the balloon at random

Figure 22-14 You can make a model of the expanding universe.

places as shown in Figure 22-14. Continue to blow up the balloon. What happens to each galaxy (cotton) as the universe (balloon) gets bigger? Pick a typical spiral galaxy on your model. Call it the Milky Way. Imagine that you are in this galaxy looking toward all the others.

DISCUSSION

1. Do all the galaxies move away from the Milky Way at the same speed?
2. Which galaxies appear to move away faster? Which appear to move away the slowest?

22-6 Investigating the Size of the Universe

Suppose you were a navigator from deep space in search of the earth. From the billions of galaxies you would have to choose just the right one, the Milky Way. Of the 100 billion stars in the Milky Way you have to find the sun. So, finding the sun would be more difficult than finding a toothpick in a haystack. Once the sun was located, it would be an easy matter to find Earth. But what a job to get started! Imagine the entire universe was the size of the earth. How large do you think the earth would appear in that scale? It would be so small we couldn't see it with a microscope!

How far is the most distant object in light-years? How close is the nearest object?

When you have completed this investigation, you will be able to compare the distances from earth of certain objects in space.

PROCEDURE

Draw a red line across one end of a piece of adding machine tape 10 m long. Next to the line write the word "Earth." Mark each consecutive meter of tape with a red line. Each meter of tape represents a distance of one billion light-years. Label each billion light-years until you reach the far end of the tape, which should be 10 billion light-years.

MATERIALS
10-m length of adding machine tape, red pencil or crayon, meterstick

Write on the tape the names of stars and galaxies from Figure 22-15 at the proper distances from Earth. Start with "Galaxy X" and work your way down the table. Don't try to figure ahead! After you have plotted the distance of all the objects, answer the following questions:

DISCUSSION

1. How many light-years does each millimeter on this scale represent?
2. Why did you have difficulty fitting all the objects on your tape?
3. What would you do to the scale to plot all the objects?

22-7 The Life Cycle of Stars

Is there any limit to the future? We believe the sun will provide energy to the earth's surface for another five to ten billion years. After that, the sun may change greatly, if today's astronomers are correct. Let's see how they reached their conclusions about stars.

Astronomers think that a star begins when a huge mass of dust and gas (mostly hydrogen) condenses in space. A mass like this, shown in Figure 22-16, is called a **nebula** (*NEHB yuh luh*). The word *nebula* comes from Latin and means "mist, cloud, or vapor." As gravitational forces pull the nebula together, it becomes compressed into a smaller and smaller volume. As the nebula is compressed, its temperature rises. When the temperature becomes very high, hydrogen in the mass changes into helium. And what a change it is! When hydrogen atoms fuse together to form helium atoms, huge amounts of energy are produced. Some of the energy is given off as light: a star is born! The young star proceeds to use up the fuel (hydrogen) in its core. Gradually other, heavier elements evolve from the hydrogen and helium. This is the longest single stage in the life of a star. Some stars use up their fuel more rapidly than others, but about 8 billion years is a useful estimate.

After most of the hydrogen in the star's core has been used up, the star's gravity begins to squeeze it. As the core contracts, it gets hotter. Unused hydrogen is forced outward. The shell of the star expands as this hydrogen continues fusing into helium. Then the star is thought to expand to a huge size. As the star expands, it cools. These huge, cooling stars appear red in color and are called **red giants.** The core of the red giant continues collapsing. The core becomes a battlefield between gravity and nuclear fusion reactions. Gravity squeezes the helium, which fuses to become carbon. The carbon fuses to become oxygen. Finally, if the mass of the core is large enough, iron is produced. That is the end of the line. Small stars become incredibly dense, white-hot spheres about the size of a planet. This stage is called the **white dwarf** stage.

For most stars the game is over at this stage. The white dwarf cools to become a dense, black cinder. However, a white dwarf that develops from a larger star continues to collapse because of gravity. Pressure becomes so high that electrons are squeezed into atomic nuclei to produce neutrons. The result of this squeezing is a **neutron star** (see Figure 22-17). A neutron star is thought to be about 15 km across. Neutron stars are,

Figure 22-15 *Astronomical objects and their estimated distances in light-years.*

Name or Number	*Distance (Light-Years)*
Galaxy X (position not known)	9.70×10^9
3 C 9 (quasar)	7.00×10^9
Blue Stellar Object No. 1	5.50×10^9
3 C 295 (quasar)	4.50×10^9
3 C 273 (quasar)	1.50×10^9
Elliptical galaxy in Draco	2.66×10^8
Elliptical galaxy in Sculptor	1.85×10^8
Clusters of galaxies in Pegasus	1.21×10^8
Snickers galaxy	5.50×10^7
NGC 2903 (barred spiral galaxy)	1.90×10^7
Andromeda galaxy	1.80×10^6
Large Magellanic Cloud (irregular galaxy)	1.60×10^5
Great Nebula in Orion	1.80×10^3
Deneb (star)	1.60×10^3
Pleiades star cluster	4.90×10^2
Mira (first observed variable star)	103
Sirius (brightest star)	8.7

therefore, very dense. Their density is about the same as would be the density of all the world's automobiles packed into a thimble!

A few years ago at the Jodrell Bank radio observatory in Manchester, England, a regular blinking signal was received from what many thought was another civilization. Today we know that these outbursts of energy are really produced by **pulsars** (*PUHL sahrs*). Some astronomers think pulsars are rapidly spinning neutron stars. Their pulsing effect may be caused by hot spots on their surfaces. They may remind you of the flashes of light from a lighthouse beacon.

The outer portions of a neutron star may finally explode with unimaginable force, producing a **supernova.** A supernova is the brightest known object in space. Material from the explosion may get together and form a new star. The core of the supernova—the neutron star—may then become a pulsar. The Crab Nebula (see Figure 22-18) is thought to be the remains of an old supernova that exploded over 900 years ago. At its center is . . . a pulsar!

Perhaps the strangest celestial object is a **black hole.** It looks like a small round region of black surrounded by scattered stars. Some astronomers believe black holes evolve from large neutron stars and represent the final stage in the star life cycle. Remember, a neutron star is incredibly dense and is only about 15 km across.

But gravity sometimes wins out and the star's mass collapses even more—inwardly upon itself. Then, suddenly, the star disappears from sight! Why? The gravitational pull of the mass becomes so great that even the star's light cannot escape! If you could drop a thimbleful of the dense material in a black hole on the ground, it would fall all the way to the center of the earth!

The collapsing neutron star becomes extremely tiny. Its core temperature soars to billions of degrees. At this point astronomers can only guess what may happen because they can't see anything. Gravity in a black hole is so strong that physical laws do not apply. Every spoonful of the black hole mass is equal to millions of metric tons! The star at this stage has become a pinpoint source of gravity. Has the star really vanished? We simply don't know. It has reached what some astronomers call the end of a star's evolution. What was once a shining beacon has become a mysterious black hole. Is this really the end? We don't know for sure.

Figure 22-16 *The Great Nebula in Orion.*

Thought and Discussion

1. How do quasars help us understand the size of the universe?
2. When the spectra from most galaxies are examined, what do they show?
3. What stage of a star's life cycle is represented by a nebula?

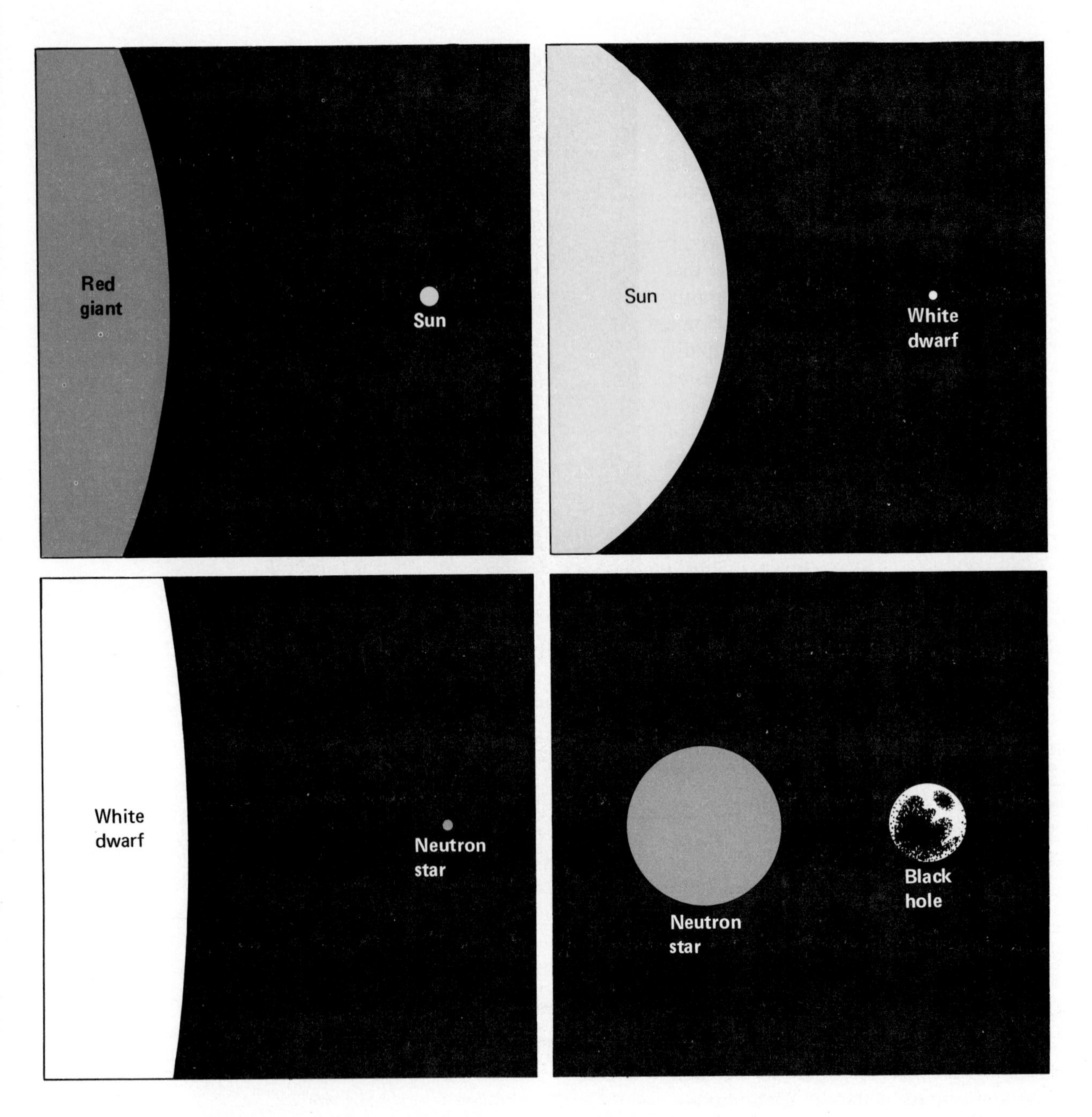

Figure 22-17 *The different sizes of stars compared to the size of a black hole.*

Unsolved Problems

After finishing this chapter, you probably think astronomers know a great deal about objects in space. It's true that many observations have been made of stars, galaxies, and other things far away from Earth. But how sure can we be that the *conclusions* of the astronomers based on these observations are good ones?

For example, consider distances to stars. We can be comfortable with measurements to nearby stars based on parallax observations. But what about finding distances to the very distant stars? We use spectral shifts to find the velocity of these distant objects. We then use the velocity to figure out their distance. The Doppler effect explains the change in pitch of train whistles. How can we be sure that it also explains the shift of absorption lines in a star's spectrogram? Applying the Doppler effect to spectrograms seems like a safe thing to do, but can we be positive that light waves behave so similarly to sound waves? After all, our picture of the size and state of the universe is based on this one principle!

Our ideas about the life cycle of stars seem to make sense to us. Together, these ideas make up the story of the universe. Yet many questions remain. For example, our physical laws may apply on Earth, but can we be sure that they will explain the goings-on in the extremely powerful gravitational fields of neutron stars and black holes?

We believe that the speed of light is both constant and also the greatest speed possible in the universe. But, remember, astronomers think that black holes have such high gravitational forces that light cannot even escape from them! We can only imagine what goes on in such an unearthly environment!

The British astronomer Sir James Jeans, in his book *The Universe Around Us,* described how people should take today's ideas about the universe. His ideas are a fitting end for this chapter . . . and this book.

Figure 22-18 *The Crab nebula in Taurus is left over from the supernova of 1054* A.D.

As inhabitants of a civilised earth, we are living at the very beginning of time. We have come into being in the fresh glory of the dawn, and a day of almost unthinkable length stretches before us with unimaginable opportunities for accomplishment. Our descendants of far-off ages, looking down this long vista of time from the other end, will see our present age as the misty morning of the world's history; they will see our contemporaries of to-day as dim heroic figures who fought their way through jungles of ignorance, error and superstition to discover truth, to learn how to harness the forces of nature, and to make a world worthy for mankind to live in. We are still too much engulfed in the greyness of the morning mists to be able to imagine, even vaguely, how this world of ours will appear to those who will come after us and see it in the full light of day. But by what light we have, we seem to discern that the main message of astronomy is one of hope to the race and of responsibility to the individual—of responsibility because we are drawing plans and laying foundations for a longer future than we can well imagine.

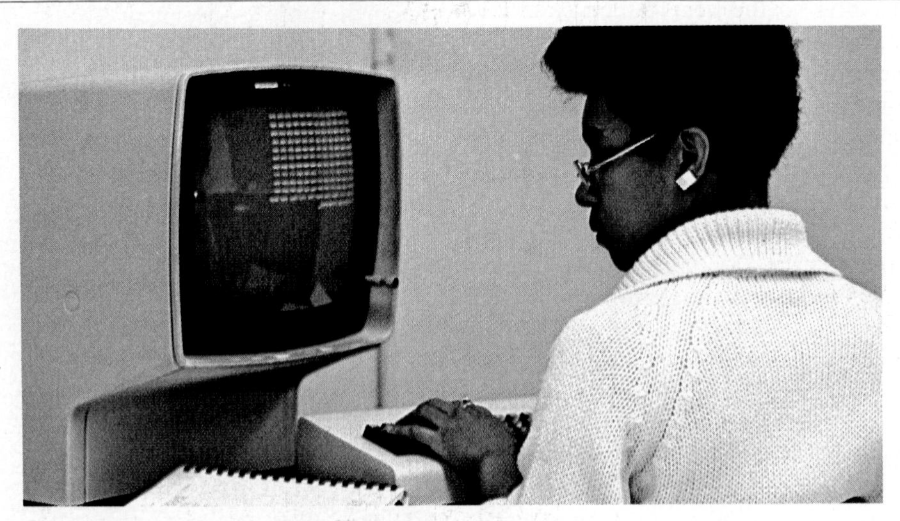

SCIENTIFIC PROGRAMMER

The work of a scientific programmer involves converting scientific, engineering, and technical problems to computer programming language. The scientific programmer prepares computer flow charts and block diagrams using advanced mathematics such as differential equations. Once a program has been written, the programmer observes or operates the computer during test runs to correct programming errors. The programmer may select and supervise the installation of special computer equipment needed for a particular purpose. The scientific programmer also reviews the results of computer runs with other personnel.

A scientific programmer usually has a college degree with a major in science, engineering, or mathematics. Two to four years of additional on-the-job training are required in order for a scientific programmer to become familiar with all the available equipment and its applications. Scientific programmers are employed in many fields, and may work for government agencies, or in private industry. Aside from proficiency in mathematics and science, the most important qualification for this type of work is the ability to represent ideas using symbols.

Chapter Summary

Stars group together to form galaxies. Galaxies have shapes and rotate slowly. Galaxies seem to occur in small groups. Between groups of galaxies are vast distances of empty space. Most galaxies appear to be rushing away from us at great speeds. The farther away the galaxies are, the faster they seem to be moving. Quasars are the most distant objects in space we have yet detected. Astronomers have investigated the speed and direction of galaxies by observing and measuring the shift of absorption lines in spectrograms of starlight. The brighter a star is, the hotter it is, also. The color of a star is used to find the temperature of the star. By using the H-R diagram, astronomers can find the star's absolute magnitude, once the star's temperature is known. By comparing the absolute magnitude with the apparent magnitude of the star, astronomers can figure out the distance to the star.

Many astronomers believe the universe began as a huge explosion of material about 10 billion years ago. The explosion scattered the original material throughout space. The force was so great that the material is still rushing away!

Astronomers believe stars may condense from nebulas, burn brightly for perhaps 8 billion years, expand to become red giants, collapse to become either white dwarfs or neutron stars, and then either explode as supernovas or shrink from the known universe to become black holes. A pulsar is the remains of the core of a star that exploded as a supernova.

Questions and Problems

A

1. What is a galaxy?
2. What are the differences between a pulsar and a quasar? How do both differ from regular stars?
3. What is the most common shape for galaxies?
4. What is the difference in distance between a near galaxy, such as the large Magellanic Cloud, and a very far galaxy, such as Galaxy X?
5. What is the longest single stage in the life of a star?
6. Describe the changes stars may have.

B

1. Describe the space between galaxies.
2. Describe the size and position of the earth in relation to the whole universe.
3. What is the relationship between the distance of a galaxy from Earth and the speed with which it travels away from Earth?
4. What is the Doppler effect?
5. How can we tell whether the Andromeda galaxy is traveling toward us or away from us?
6. What is the heaviest material in the universe? Why is it the heaviest?

C

1. How do astronomers compare the chemical composition of the earth with the chemical composition of the rest of the universe?
2. How do astronomers measure the distances to other galaxies?
3. What evidence do we have that galaxies rotate?
4. Do you think that elliptical galaxies might be in a more advanced state of evolution than spiral galaxies? Why?
5. Why do scientists believe that if you were at almost any point in the universe, you would feel as if you were at the center of it?
6. What evidence do we have to support the theory of the expanding universe?

Suggested Readings

Couper, Heather, and Terence Murtagh. *Heavens Above! A Beginner's Guide to the Universe.* New York: Watts, 1981.

Jastrow, Robert. *Red Giants and White Dwarfs: Man's Descent from the Stars.* New York: Warner Books, 1980. (Paperback)

Murdin, Paul, and David Allen. *Catalog of the Universe.* New York: Crown, 1979.

Shipman, Harry L. *Black Holes, Quasars, and the Universe.* 2nd ed. Boston: Houghton Mifflin Co., 1980. (Paperback)

MEN AND WOMEN IN EARTH SCIENCE

I. *Who Was Who in Earth Science*

All of the men and women shown here have made important contributions to the development of earth science. Many of these contributions are described in greater detail in the text.

A

B

A Louis Agassiz (1807–1873) was a Swiss-American scientist who concluded that glaciers had once covered much of Europe and North America. His teaching stressed the importance of direct observation and influenced the methods of all researchers in earth science.

B Aristotle (384–322 B.C.), Greek philosopher and naturalist, wrote the earliest known book on what we today call earth science. He was the first person to use careful observation as a scientific method.

C

D

C Florence Bascom (1862–1945) was the first woman in the United States to receive a Ph.D in geology, and the first woman appointed as geologist with the U.S. Geological Survey. The geology program she created at Bryn Mawr College became internationally known for its high level of teaching and research.

D Wilhelm Bjerknes (1862–1951) was a Norwegian physicist who defined the system of "air masses" and "fronts" that is the basis for modern weather forecasting.

E Nicolaus Copernicus (1473–1543) was a Polish astronomer who devised a mathematical system to show that planetary positions could be calculated by assuming that the planets revolve around the sun.

F William Morris Davis (1850–1934), an American geologist, was a pioneer in the science of geomorphology, the study of landform development.

E

F

G Tilly Edinger (1897–1967), German-American paleontologist, was for many years the resident paleontologist at the Museum of Comparative Zoology, Harvard University. She was recognized as a leading authority on the comparative anatomy of fossil vertebrates.

H James Hall (1811–1898) was the leading invertebrate paleontologist of his day. He is remembered now primarily for the geosynclinal theory of mountain building. This concept is considered an essential part of modern geology.

G

H

I Caroline Herschel (1750–1848), German-English astronomer, was the first important woman astronomer. She was particularly interested in comets, and discovered eight of them.

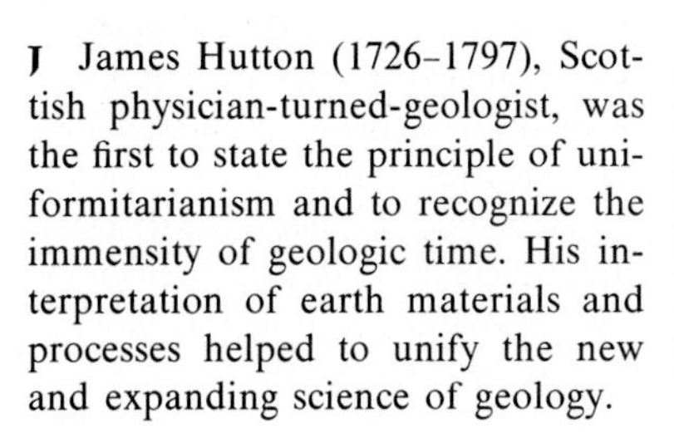

J James Hutton (1726–1797), Scottish physician-turned-geologist, was the first to state the principle of uniformitarianism and to recognize the immensity of geologic time. His interpretation of earth materials and processes helped to unify the new and expanding science of geology.

I

J

K Johannes Kepler (1571–1630) was a German astronomer who stated mathematically the "laws" that describe planetary motions. His concept of the solar system has been followed by astronomers ever since, without major changes.

L Henrietta Swan Leavitt (1868–1921) was an American astronomer who found the relationship between the period and brightness of Cepheid variable stars, and from this determined their distances. This was the first method of determining really vast stellar distances.

M Sir Charles Lyell (1797–1875), a Scottish geologist, collected data that supported the principle of uniformitarianism stated earlier by Hutton. Through his book *The Principles of Geology,* Lyell popularized and expanded Hutton's views.

N Maria Mitchell (1818–1889) was an astronomer and professor at Vassar College. She charted many stars, studied sunspots and solar eclipses, and discovered a comet, which was named after her. She was the first woman elected to the American Academy of Arts & Sciences in Boston.

O Sir Isaac Newton (1642–1727) was a brilliant English scientist and mathematician who established the basic laws of modern physics. His book *Mathematical Principles of Natural Philosophy* dealt with the motions of objects in space and introduced his three laws of motion, as well as the law of gravitation.

P John Wesley Powell (1834–1902) was a geologist who explored and studied the Rocky Mountains, the Great Plains, and the Grand Canyon. His work with William Morris Davis and others helped to establish the basic principles of structural geology. Powell was one of the founders of government science in the United States, and was the second director of the U.S. Geological Survey.

Q William Smith (1769–1839) was an English surveyor who became interested in the rocks exposed along the banks of canals. He showed that the fossils in rocks could be used to correlate rock layers many kilometers apart.

R Alfred L. Wegener (1880–1930) was a German meteorologist and Greenland specialist who proposed the theory of "continental drift."

Q

R

II. Who Is Who in Earth Science

Today there are many people doing research in the earth sciences. Each of the scientists shown here is trying to answer unsolved problems in oceanography, geology, astronomy, or meteorology. Perhaps someday your name will appear in a similar group.

A June Bacon-Bercey, international aviation specialist for the National Weather Service, was the first woman to receive the Seal of Approval for TV-Radio Weathercasting from the American Meteorological Society.

B Celso Barrientos, research meteorologist with the National Weather Service, researches hurricane dynamics, ocean circulations, winds, and waves.

A

B

C

D

C Jocelyn Bell-Burnell, an astronomer, discovered pulsars when she was a graduate student at Cambridge University.

D Randolph Wilson Bromery is a professor of geophysics and Chancellor at the University of Massachusetts. His research involves studies of surveying methods for geologic maps.

E

F

E Elizabeth T. Bunce is a geophysicist who studies marine seismology and the way sounds travel under water at Woods Hole Oceanographic Institution, where she is a senior scientist.

F Helen L. Cannon is a geochemist with the U.S. Geological Survey. She was formerly chairperson of the subcommittee on geochemical environments in health and disease of the National Research Council. She is a recipient of a Meritorious Award and a Distinguished Service Award from the Department of the Interior, and was nominated as Federal Woman of the Year in 1970.

G

H

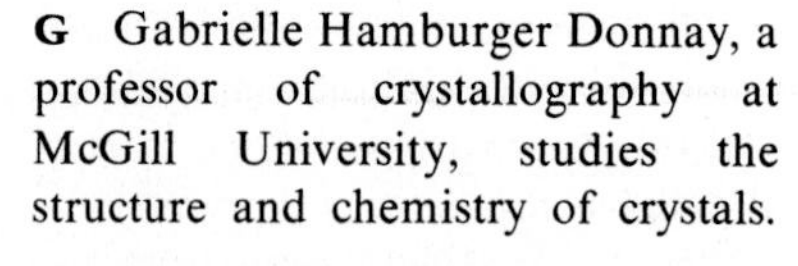

G Gabrielle Hamburger Donnay, a professor of crystallography at McGill University, studies the structure and chemistry of crystals.

H James R. Heirtzler, a geophysicist, is senior scientist in the Department of Geology and Geophysics at Woods Hole Oceanographic Institution. He was the U.S. chief scientist on Project FAMOUS, a joint U.S.-French undersea program, from 1971–1975.

I Zofia Kielan-Jaworwska, director of the Paleozoological Institute of the Polish Academy of Sciences, has led eight joint Polish-Mongolian expeditions into the Gobi Desert, where she made significant finds of fossil mammals and dinosaurs.

J Konrad Krauskopf, professor of geochemistry at Stanford University, has done research on the chemical composition of rocks, and on the trace elements present in seawater. He is past president of the Geological Society of America, the American Geological Institute, and the Geochemical Society.

I

J

K Michael Ovenden studies the dynamics of the solar system as an astronomy professor at the University of British Columbia.

L Katherine V. W. Palmer is director of the Paleontological Research Institution, where she does research on fossil sea snails. She has received awards from the Paleontological Society and the Western Society of Malacologists (a group of scientists who study shellfish).

K

L

M Joseph L. Reid is a professor of physical oceanography at Scripps Institution of Oceanography. He has shown that waters that form off Greenland and the coast of Antarctica flow through the deepest parts of all oceans.

N Elizabeth Roemer is an astronomer at the University of Arizona. She has received numerous awards for her work on comets and planets, and has been chairperson of the National Research Council and a member of the astronomy panel of the National Science Foundation.

M

N

O

P

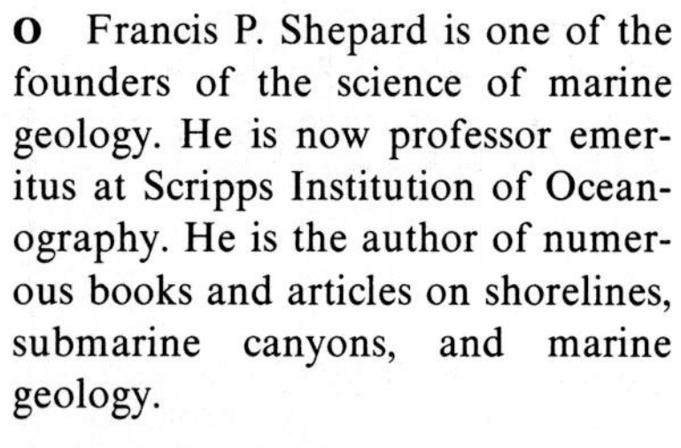

O Francis P. Shepard is one of the founders of the science of marine geology. He is now professor emeritus at Scripps Institution of Oceanography. He is the author of numerous books and articles on shorelines, submarine canyons, and marine geology.

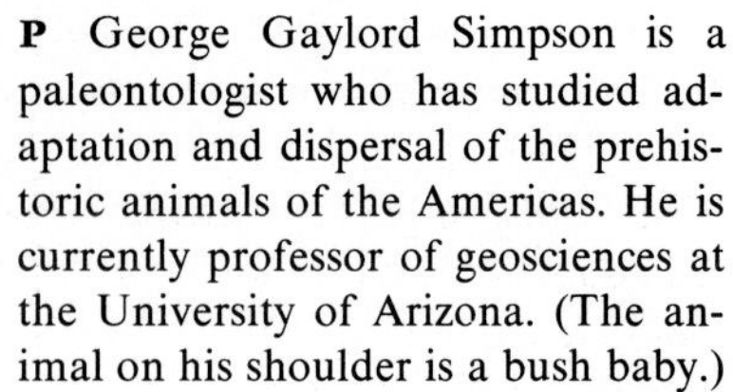

P George Gaylord Simpson is a paleontologist who has studied adaptation and dispersal of the prehistoric animals of the Americas. He is currently professor of geosciences at the University of Arizona. (The animal on his shoulder is a bush baby.)

Q

R

Q Alonzo Smith is a meteorological services coordinator for the National Weather Service. His work involves developing computerized aircraft flight plans.

R J. Tuzo Wilson, presently Director-General of the Ontario Science Center, has developed theories of continental structure and plate tectonics. He has received many awards and honorary degrees.

OTHERS OF NOTE . . .

Try and find out more about these people.

Martine, Baroness de Beausoleil (1602–1640), mineralogist, first French mining engineer

Marland Pratt Billings, geologist

Edwin H. Colbert, vertebrate paleontologist

Maria Cunitz (c. 1610–1664), Silesian astronomer

Katharine Fowler-Billings, geologist

Winifred Goldring, paleontologist

Irving Langmuir, specialist in weather modification

Dame Kathleen Lonsdale, crystallographer

Lou Williams Page, geologist, well-known author

Harrison Schmitt, first geologist on the moon

Marie Tharp, geologist

Florence Van Straten, specialist in atmospheric physics

APPENDIX A MATHEMATICAL INFORMATION

A-1 Powers of Ten

In earth science it is often necessary to use very large and very small numbers. The area of the earth's surface is 361,000,000 square kilometers. A convenient shorthand for writing numbers like this one is to use powers of ten. For example:

Number				*Equivalent Power of 10*
1000	=			1×10^3
100	=			1×10^2
10	=			1×10^1
1	=			1×10^0
0.1	=	$\frac{1}{10^1}$	=	1×10^{-1}
0.01	=	$\frac{1}{10^2}$	=	1×10^{-2}
0.001	=	$\frac{1}{10^3}$	=	1×10^{-3}
0.0001	=	$\frac{1}{10^4}$	=	1×10^{-4}

Thus, 361,000,000 is the same as 3.61 times 100,000,000. Since this is 3.61 times $10 \times 10 \times 10 \times 10 \times 10 \times 10 \times 10 \times 10$ or 3.61 multiplied by 10 eight times, we call this 3.61×10^8.

coefficient → 3.61×10^8 ← *exponent* (8), ↖ *base* (10)

The **exponent** tells how many times to multiply by 10, which is called the **base.** To change a number from the usual long form to the standard form, move the decimal point to the left until you have a number between one and ten. The number of places that you moved the decimal point is the exponent, or power of ten. The **coefficient** is the number between one and ten used with the power of ten. In the example, the decimal point was moved eight places to the left, so the exponent is 8. The base is 10 since we are using the decimal number system. The coefficient is 3.61.

If the original long number is less than one, it can be written as a number between one and ten divided by ten to some power.

$$0.008 = 8 \times \frac{1}{1000} =$$

$$8 \times \frac{1}{10^3} = 8 \times 10^{-3}$$

That is, if you have to move the decimal point to the *right* to get a number between 1 and 10, the exponent has a negative sign.

A-2 Metric and Other Units of Measure

The Metric System

Prefixes

PREFIX	MEANING
kilo-	1000 or 10^3
centi-	0.01 or 10^{-2}
milli-	0.001 or 10^{-3}

Units of Length

1 kilometer (km) = 1000 meters (m) = 10^3 m
1 centimeter (cm) = 0.01 m = 10^{-2} m
1 millimeter (mm) = 0.001 m = 10^{-3} m
1 angstrom (Å) = 0.0000000001 m = 10^{-10} m

Units of Area

1 square meter (m^2) = 10,000 square centimeters (cm^2) = 100 cm × 100 cm

Units of Volume

1 cubic meter (m^3) = 1,000,000 cubic centimeters (cm^3) = 100 cm × 100 cm × 100 cm
1 liter (l) = 1000 milliliters
1 milliliter (ml) = 1 cm^3

Units of Mass

1 metric ton = 1000 kilograms
1 kilogram (kg) = 1000 grams
1 gram (g) = approximately the mass of 1 cm^3 of water
1 milligram (mg) = 0.001 g

Units of Time

1 hour (hr) = 60 minutes
1 minute (min) = 60 seconds (sec)

Other Frequently Used Units of Measure

Distance

1 Astronomical Unit (AU) = 149.6 × 10^6 km (mean distance from earth to sun)
1 light-year = 9.46 × 10^{12} km

Angle Measurement

1 degree (1°) = 1/360 of a circle = 60 minutes
1 minute (1′) = 60 seconds (60″)

Pressure

1 millibar (mb) = 1000 dynes/cm^2
Average atmospheric pressure at sea level = 1013.25 mb

	Unit	*Symbol*	*Equals*	*About the same as*
Length	millimeter	mm	$\frac{1}{1000}$ m	the thickness of a dime
	centimeter	cm	$\frac{1}{100}$ m	the thickness of a slice of bread
	meter	m		the height of a teacher's desk
	kilometer	km	1000 m	the distance you can walk in about 12 minutes
Volume	milliliter	ml	$\frac{1}{1000}$ l	the volume of two nickels
	liter	l		the volume of four medium-size glasses of water
Mass	gram	g		the mass of two raisins
	kilogram	kg	1000 g	the mass of a 500-page book
Force	newton	N		the weight of an apple at the earth's surface (= the force needed to lift an apple)

Adapted from *Physical Science Investigations*, Revised Edition, Houghton Mifflin Co., 1979

A-3 Relative and Percentage Error

If you divide the difference between your answer to a problem and the correct answer by the correct answer, you obtain the **relative error.** If you multiply the relative error by 100, you obtain the **percentage error.** For example, in Investigation 1-2 you calculate the circumference of the earth. If you consider the actual value of the earth's circumference to be 40,000 km and your measurement was 38,000 km, the difference between them is 2000 km. Dividing this difference (2000 km) by the actual value (40,000 km) gives a relative error of 0.05 $\left(\frac{2000}{40{,}000} = \frac{1}{20} = 0.05\right)$. Multiplying this relative error by 100 gives a percentage error of 5% ($0.05 \times 100 = 5\%$). A student who obtained a measured value of 42,000 km would have the same percentage error as your value of 38,000 km. Can you see why?

It is often desirable to calculate the percentage error to see if your answer is reasonable in relation to the possible sources of error of your instruments and measurements. If all your measuring instruments are reasonably accurate and your answer has a percentage error of 40%, you should review your work and look for errors. If, however, your measuring instruments are crude and your percentage error is only 8% or so, then it is likely that your work is as accurate as your instruments will permit. As a general rule, the greater your percentage error, the more carefully you should recheck your work for mistakes.

APPENDIX B WEATHER DATA

B-1 *Recording Weather Watch Data (Investigation P-7)*

Your teacher will provide you with the necessary wall chart on which to record your data for this investigation. Gather data carefully to avoid errors that might affect your analysis of the data later. Record data legibly on the chart.

The specific information to be gathered as part of the investigation includes:

1. **Date.** Record the day of the month at the top of each column.
2. **Time.** Record the exact time your observation is made. Use the 24-hour clock notation. In this notation, 10:15 A.M. is 1015 and 9:02 P.M. is 2102. The hours after noon (P.M.) are numbered from 1300 to 2400.
3. **Air Temperature.** A thermometer is commonly used to measure air temperature. Air temperature can also be measured by a thermograph, a thermometer that automatically records air temperature on a continuous graph attached to a rotating drum. Temperatures should be measured outdoors in a shaded shelter about 1.5 m above the ground. The thermometer bulb should be kept dry and the air should be free to circulate through the shelter. All temperatures should be recorded in degrees Celsius.
4. **Atmospheric Pressure.** Air pressure can be measured with an aneroid or mercurial barometer or by a barograph. A barograph, like a thermograph, records air pressure on a sheet of paper attached to a revolving drum. Unlike the thermometer or

Figure 1 *The Beaufort Wind Scale.*

Beaufort Number	*Name*	*Effects of Wind at Various Speeds*	*Wind Speed*	
			km/hr	*knots*
0	Calm	Smoke rises vertically	Under 1	Under 1
1	Light air	Wind direction shown by smoke drift	1–5	1–3
2	Light breeze	Wind felt on face; leaves rustle; ordinary vane moved by wind	6–11	4–6
3	Gentle breeze	Leaves and twigs in constant motion; wind extends light flag	12–19	7–10
4	Moderate breeze	Dust and loose paper; small branches are moved	20–29	11–16
5	Fresh breeze	Small trees in leaf begin to sway	30–39	17–21
6	Strong breeze	Large branches in motion	40–50	22–27
7	Moderate gale	Whole trees in motion	51–61	28–33
8	Fresh gale	Twigs broken off trees; progress generally impeded	62–71	34–40
9	Strong gale	Slight structural damage occurs	72–87	41–47
10	Whole gale	Trees uprooted; considerable structural damage	88–101	48–55
11–17	Hurricane	Very rarely experienced; widespread damage	102–218	56–118

thermograph, the barometer or barograph can be placed indoors. Air pressure should be measured and recorded in millibars.

5. **Wind.** Both wind direction and speed are needed in this observation. Record wind direction and speed along with sky condition for each day on your wall chart. **Wind direction** can be measured with a wind vane or some means you can devise. It is recorded on the chart by a line representing the compass direction *from* which the wind is blowing, with north at the top of the chart.

   Wind speed can be measured with an anemometer or a wind speed meter. If neither is available, you can observe local conditions and estimate the velocity of the wind from Figure 1. The "flags" indicating the wind speed as shown on the daily weather maps should be drawn on the end of your wind direction line—the end from which the wind is blowing.

   An explanation of symbols and entries on weather maps is published by the National Oceanic and Atmospheric Administration. You can also refer to *Weather Maps: How They are Made and Used* by Miles F. Harris and John O. Ellis (ESCP Reference Series, RS-10) or *Weather Map Study* by Fred W. Decker (Paperback).

6. **Sky Condition.** This observation includes the amount of the sky covered by clouds and the type of cloud. For ease of recording, determine if the sky is clear, partly cloudy, or cloudy. Use the symbols in Figure 2 for your three categories. The type of cloud should be noted as billowy or sheet-like. Sketch the type of cloud on the chart as simply as possible.

Sky Condition	Cloud Cover	Symbol
Cloudy	More than eight-tenths of sky covered.	●
Partly cloudy	Two-tenths to eight-tenths of sky covered.	◑
Clear	Less than two-tenths of sky covered.	○

Figure 2 *Symbols for sky condition.*

7. **Weather.** Observe and enter on the chart the state of the weather at the time of observation, such as rain, snow, thunderstorms, clear, cloudy, fog, gaze, smog, and so forth.

8. **Precipitation.** Moisture that has fallen to the earth's surface in the form of rain, hail, drizzle, sleet, or snow is considered to be precipitation. Dew or fog is not. Precipitation is recorded as the quantity of water deposited in the gauge since the last reading. If the gauge has snow in it, melt the snow to get a liquid reading.

B-2 Finding the Dew-point Temperature and the Relative Humidity

The dew point can also be obtained by using a **psychrometer.** This instrument measures indirectly the amount of latent heat that would be required to produce saturation. The sling psychrometer consists of two thermometers. One (the **dry-bulb**) is an ordinary glass thermometer. The other (the **wet-bulb**) has its bulb covered with a piece of muslin. The muslin must be soaked with pure (distilled) water just before the measurement is made. Then the psychrometer is whirled until the wet-bulb temperature stops falling.

The difference between the dry and the wet-bulb temperatures is called the **wet-bulb depression.** It is a measure of the amount of energy needed to evaporate enough water to produce saturation at the wet-bulb temperature. If you know the dry-bulb temperature and the wet-bulb depression you can find the dew point and the relative humidity. The examples in Figure 4 will help you learn how to use Figure 3.

Figure 3 *Dew-point temperature chart.*

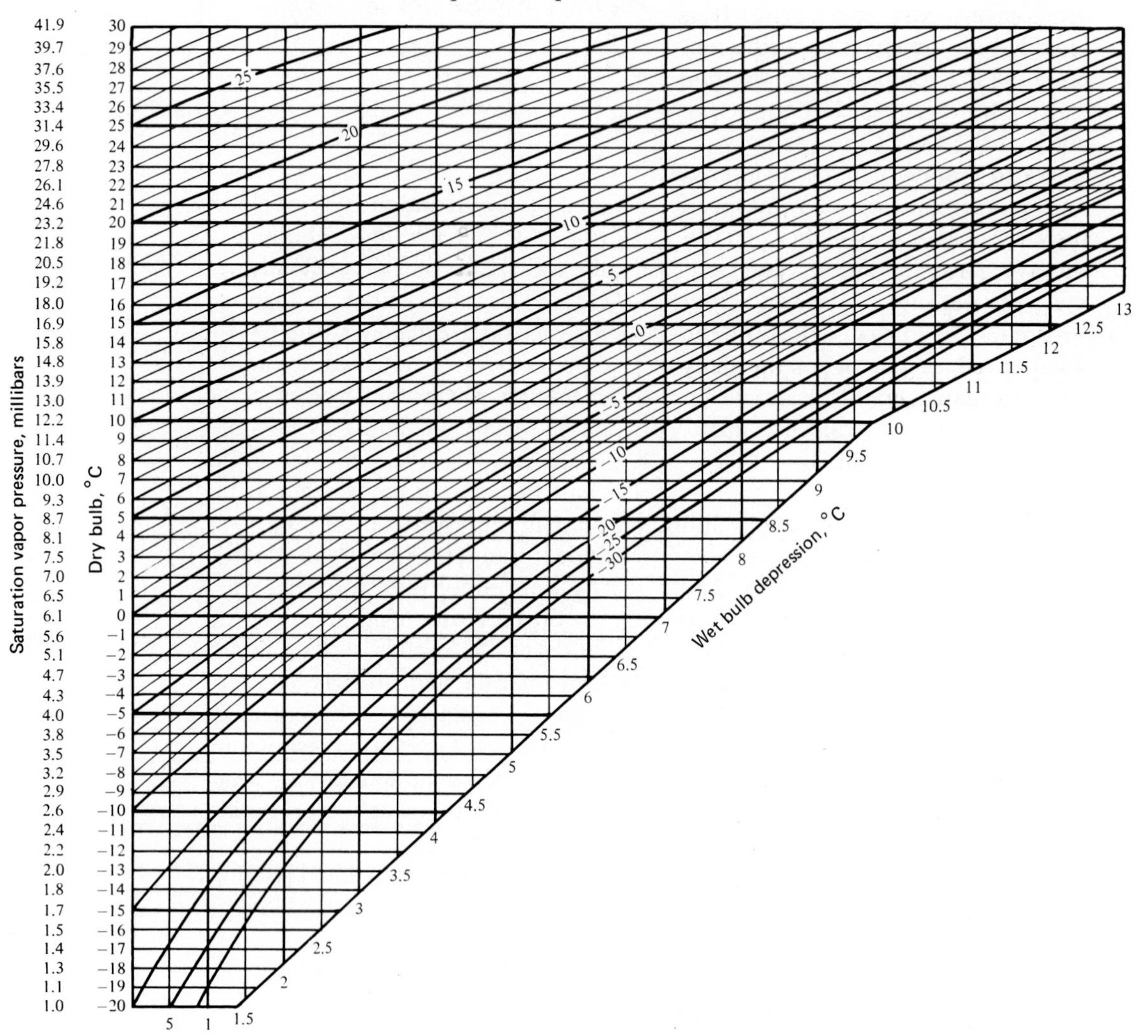

1. To find the dew-point temperature (See Example A): Find the dry-bulb temperature along the left side of the chart (12 °C). Follow horizontal line to the vertical line for the wet-bulb depression (difference between dry-bulb and wet-bulb temperatures, or 3.5 °C). Read the dew-point temperature from sloping line at this intersection (5 °C).
2. To find the relative humidity (See Example B): Read the value of the saturation vapor pressure for the dry-bulb temperature at left side of chart. (13.9 mb is saturation vapor pressure for air at 12 °C.) Read the value of saturation vapor pressure for dew-point temperature also at the left side of chart. (8.7 mb is saturation vapor pressure for air at 5 °C.) Divide the second value (8.7) by the first (13.9) and multiply by 100. Answer: 63%.

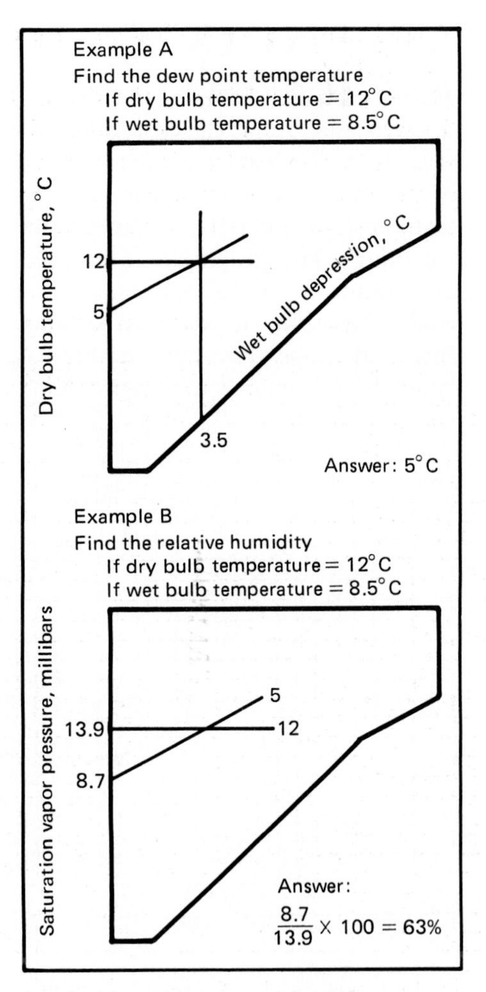

Figure 4 *How to use Figure 3.*

APPENDIX C HOW TO MAKE A GRAPH

A graph shows the relationship between two things. This graph shows how the temperature of a heated substance is related to the time it has been cooling. *Time* and *temperature* are variables. A variable is something that can change.

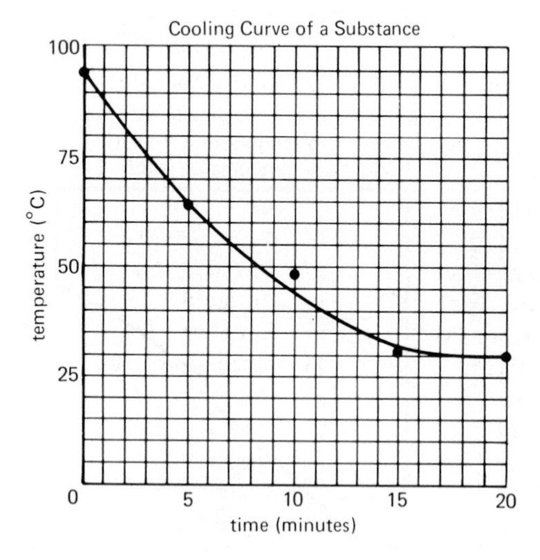

The line at the left side of the graph is called the vertical axis. The line across the bottom of the graph is called the horizontal axis. In most graphs, one variable depends on the other variable. In this graph, the temperature depends on the time. As the time changes, the temperature changes. Therefore, the temperature is the "dependent" variable. The dependent variable goes on the vertical axis.

Follow these steps when you make a graph:

1. Find out which variable goes on which axis. On each axis, put the name of the variable and the unit in which it is measured.
2. Choose scales that let you use as much of the graph as possible. Choose a scale that makes it easy to locate points. For example, never let 5 squares equal 7 units. A scale having 5 squares equal to 10 units is better. It is not necessary to label every line on the graph.
3. Mark each point on the graph in pencil. The points on the cooling curve must have come from this table of observations.

Time (minutes)	Temperature (°C)
0	95
5	65
10	49
15	32
20	30

   When you are sure that the points are in the right places, mark them in ink. Put a circle around each point. The circled points are the only part of the graph that must be in ink.
4. Decide whether your points fit on a straight line or a curve. If they fit on a straight line, use a transparent ruler to draw the line very lightly. If you do not like the line you have drawn, erase it and try again. Because your points are in ink, you will not lose them. With every measurement there is a chance of error. So the line will probably not go exactly through all the points. If the points do not fit on the straight line, draw a smooth curve.
5. Give your graph a title.

Adapted from *The Chemical World, Activities and Explorations,* Houghton Mifflin Co., 1977

APPENDIX D KNOWN ELEMENTS AND THEIR SYMBOLS

(Listed alphabetically with the atomic number preceding them)

89 Actinium (Ac)
13 Aluminum (Al)
95 Americium (Am)
51 Antimony (Sb)
18 Argon (Ar)
33 Arsenic (As)
85 Astatine (At)
56 Barium (Ba)
97 Berkelium (Bk)
4 Beryllium (Be)
83 Bismuth (Bi)
5 Boron (B)
35 Bromine (Br)
48 Cadmium (Cd)
20 Calcium (Ca)
98 Californium (Cf)
6 Carbon (C)
58 Cerium (Ce)
55 Cesium (Cs)
17 Chlorine (Cl)
24 Chromium (Cr)
27 Cobalt (Co)
29 Copper (Cu)
96 Curium (Cm)
66 Dysprosium (Dy)
99 Einsteinium (Es)
68 Erbium (Er)
63 Europium (Eu)
100 Fermium (Fm)
9 Fluorine (F)
87 Francium (Fr)
64 Gadolinium (Gd)
31 Gallium (Ga)
32 Germanium (Ge)
79 Gold (Au)
72 Hafnium (Hf)
2 Helium (He)
67 Holmium (Ho)
1 Hydrogen (H)
49 Indium (In)
53 Iodine (I)
77 Iridium (Ir)
26 Iron (Fe)
36 Krypton (Kr)
57 Lanthanum (La)
103 Lawrencium (Lw)
82 Lead (Pb)
3 Lithium (Li)
71 Lutetium (Lu)
12 Magnesium (Mg)
25 Manganese (Mn)
101 Mendelevium (Md)
80 Mercury (Hg)
42 Molybdenum (Mo)
60 Neodymium (Nd)
10 Neon (Ne)
93 Neptunium (Np)
28 Nickel (Ni)
41 Niobium (Nb)
7 Nitrogen (N)
102 Nobelium (No)
76 Osmium (Os)
8 Oxygen (O)
46 Palladium (Pd)
15 Phosphorus (P)
78 Platinum (Pt)
94 Plutonium (Pu)
84 Polonium (Po)
19 Potassium (K)
59 Praseodymium (Pr)
61 Promethium (Pm)
91 Protactinium (Pa)
88 Radium (Ra)
86 Radon (Rn)
75 Rhenium (Re)
45 Rhodium (Rh)
37 Rubidium (Rb)
44 Ruthenium (Ru)
62 Samarium (Sm)
21 Scandium (Sc)
34 Selenium (Se)
14 Silicon (Si)
47 Silver (Ag)
11 Sodium (Na)
38 Strontium (Sr)
16 Sulfur (S)
73 Tantalum (Ta)
43 Technetium (Tc)
52 Tellurium (Te)
65 Terbium (Tb)
81 Thallium (Tl)
90 Thorium (Th)
69 Thulium (Tm)
50 Tin (Sn)
22 Titanium (Ti)
74 Tungsten (W)
92 Uranium (U)
23 Vanadium (V)
54 Xenon (Xe)
70 Ytterbium (Yb)
39 Yttrium (Y)
30 Zinc (Zn)
40 Zirconium (Zr)

APPENDIX E DATA FOR INVESTIGATION 15-9

Station	Surface Elevation	Rock Description From Surface Downward			
		Layer I	Layer II	Layer III	Layer IV
1	248 m	Weathered Gray Outwash (2 m)*	Unweathered Gray Outwash (2 m)	Weathered Red Outwash (7 m)	Bedrock
2	246 m	Weathered Gray Outwash (2 m)	Unweathered Gray Outwash (2 m)	Weathered Red Outwash (7 m)	Bedrock
3	250 m	Weathered Gray Outwash (2 m)	Unweathered Gray Outwash (3 m)	Weathered Red Till (4 m)	Bedrock
4	270 m	Weathered Gray Till (3 m)	Unweathered Gray Till (15 m)	Weathered Red Till (7 m)	Bedrock
5	272 m	Weathered Gray Till (2 m)	Unweathered Gray Till (15 m)	Weathered Red Till (7 m)	Bedrock S Grooves
6	276 m	Weathered Gray Till (3 m)	Unweathered Gray Till (20 m)	Weathered Red Till (7 m)	Bedrock S Grooves
7	264 m	Weathered Gray Till (2 m)	Unweathered Gray Till (5 m)	Bedrock SW Grooves	
8	264 m	Weathered Gray Till (1 m)	Unweathered Gray Till (3 m)	Bedrock SW Grooves	

* thickness of layer

APPENDIX F DATA FOR INVESTIGATIONS P-8 AND 16-5

The following seismic data represent approximately one-fifth of the earthquakes that occurred during November, 1975. They range in magnitude from 4.0 to 6.1 on the Richter scale. The data were obtained from the U.S. Geological Survey in Alexandria, Virginia.

Date	*Latitude*	*Longitude*	*Depth (km)*
1	13.8N	144.7E	113
1	11.9N	125.7E	41
1	18.4S	177.8W	424
1	26.9N	56.2E	33
1	17.2S	172.6W	33
1	4.1N	103.5W	33
2	49.4N	154.0E	178
2	38.4N	70.4E	20
2	16.2N	145.4E	405
2	19.7S	178.5W	603
2	6.6N	72.9W	156
2	68.4N	18.2W	33
2	68.2N	19.4W	33
2	4.7S	102.1E	67
2	6.3S	155.1E	71
2	6.4S	154.9E	87
3	19.8N	109.3W	33
3	19.9N	109.6W	33
3	41.4S	85.9W	33
3	7.3S	128.5E	170
3	43.8N	74.6W	3
4	59.3S	18.0W	33
4	43.7N	138.2E	261
4	54.3N	167.5E	24
4	60.0N	160.3E	33
4	33.9N	136.4E	385
5	6.2N	76.9W	44
5	0.6N	90.5W	33
5	7.1N	34.1W	33
5	38.2N	15.8E	135
5	1.2N	126.1E	33
5	7.2N	94.3E	30
5	14.3S	13.3W	33
5	16.9N	92.8W	12
5	38.2S	93.7W	33
6	51.8N	176.2E	61
6	8.1S	114.2E	170
6	35.8N	53.0E	3
6	43.3N	147.3E	33
6	7.2N	94.3E	62
6	19.4N	155.3W	25
6	0.3S	80.7W	54

Date	Latitude	Longitude	Depth (km)
7	3.9S	76.9W	113
7	50.6N	175.4E	22
7	21.6S	169.4E	33
7	12.2S	167.2E	269
8	20.9S	179.0W	574
8	6.6N	126.8E	97
8	10.4S	118.5E	22
8	44.5N	129.4W	33
9	31.3S	76.5E	33
9	21.0S	68.4W	77
9	55.6N	162.4E	43
9	17.9N	146.5E	82
10	41.3S	88.8W	33
10	51.3N	179.1W	57
10	4.3S	105.6W	33
10	6.5S	124.7E	606
11	46.6N	145.4E	355
11	41.8N	143.7E	69
11	51.6N	176.0E	63
11	38.7N	22.5E	33
11	0.4N	67.1E	33
12	56.9N	7.2E	33
12	21.7S	170.3E	134
12	9.0S	108.6W	33
12	3.9N	125.8E	83
12	40.3N	125.2W	33
12	36.3N	28.1E	68
12	34.5N	137.4E	44
12	58.5N	153.4W	85
12	71.7N	2.4W	33
12	71.6N	1.2W	33
12	71.6N	2.4W	33
13	31.7N	50.9E	17
13	62.9S	158.0W	33
13	33.5N	22.9E	33
13	24.3N	121.6E	66
13	10.8N	57.3E	33
13	51.7N	173.3W	37
13	51.6N	173.3W	27
14	40.6N	124.3W	23
14	12.8S	166.5E	30
14	10.3N	103.5W	33
14	38.6N	40.6E	41
15	39.4N	121.5W	5
15	26.3S	27.3E	33
15	34.3N	116.3W	6
15	53.8S	141.5E	33
15	32.1N	54.4E	43
16	53.6S	140.9E	33
16	44.7N	9.5E	19
16	36.4N	71.0E	128
16	39.0N	74.9E	50
17	29.6S	179.0W	33
17	13.1N	125.9E	26
17	34.3N	23.2E	33
17	13.5N	125.6E	89

Date	Latitude	Longitude	Depth (km)
18	40.3N	29.8W	33
18	0.1N	66.9E	33
18	2.7S	125.7E	33
18	36.9N	116.0W	5
18	10.9S	165.7E	50
19	81.9N	4.8W	26
19	31.6S	178.1W	26
19	54.3N	161.3E	62
21	7.5N	77.5W	36
21	12.6S	168.5E	628
22	8.0N	126.4E	74
22	57.2N	149.4W	33
22	40.5S	176.8E	40
22	20.3S	68.1W	33
22	43.4N	126.7W	33
23	32.1S	72.3W	20
24	18.9S	172.6W	33
24	51.5N	130.5W	33
24	3.7S	11.9W	33
25	36.3N	140.1E	67
26	13.6N	56.6E	33
26	56.6S	25.3W	46
26	25.6N	142.5E	33
26	7.8N	38.8W	33
27	42.2N	90.5E	33
27	43.7N	28.6W	33
27	12.8N	87.3W	121
27	18.2S	174.4W	109
27	21.7N	142.9E	317
28	49.0S	127.2E	33
28	39.4N	39.4E	21
28	5.2S	139.7E	33
28	17.7S	174.7W	78
28	15.2S	70.6W	201
29	31.8S	178.4W	49
29	57.7S	25.3W	51
29	49.6N	126.3W	33
29	19.3N	155.0W	5
29	37.5N	141.4E	79
29	28.9S	177.4W	71
30	5.0S	145.1E	47
30	52.2N	176.2W	99
30	49.3N	123.5W	32
30	38.9N	142.5E	32
30	27.1N	100.3E	18

Glossary

The number in parentheses at the end of each definition refers to the page on which the word is defined.

absolute magnitude The brightness of a star when it is viewed from the standard distance of 10 parsecs or 32.6 light-years. (487)

accretion Growth by addition of new material around a central core. An hypothesis states that the North American continent developed in this way from two or more small nuclei. (358)

acidic Soil low in certain substances, especially calcium and magnesium. (178)

adaptation The process by which the descendents of organisms become better suited to survive in their environments over a series of generations. (333)

aftershock A less severe quake that follows an earthquake. (373)

amber A fossilized resin (pitch) that flowed from certain ancient cone-bearing trees. (325)

anticline An upward fold of sedimentary rocks. The layers slope down on both sides. (255)

anticyclone Winds that spiral in a clockwise motion (in the Northern Hemisphere) outward from a high pressure center. (74)

apparent magnitude The brightness of a star when it is viewed from the earth. (486)

aquifer Earth material that stores large amounts of water and through which water flows easily. (107)

artesian system A porous rock layer that is between two non-porous layers. (108)

asteroids Small celestial bodies with diameters from 150 to several hundred kilometers and orbits lying mostly between those of Mars and Jupiter. (454)

asthenosphere The zone of the earth that geologists believe flows like thick tar. (148)

astronomical unit (AU) A unit of length (the earth-sun distance) used to measure astronomical distance. It equals about 150 million km. (454)

atmosphere The gaseous layer that surrounds the earth. (39)

atom A particle of matter consisting of protons, electrons, and neutrons. Each element is made up only of atoms of that kind of element. (9)

bedding plane The surface that is formed between layers of sediment when conditions of deposition change. (278)

bedrock Solid rock underneath soil or exposed at the earth's surface. (103)

black hole A small, extremely dense region in the universe that does not radiate light. It is considered to be the final stage in the life cycle of a star. (509)

bright-line spectrum A series of colored lines gotten by passing the light given off by an element through a narrow slit and then through a prism. (490)

Canadian Shield The shield area of the North American continent. A region of Precambrian rocks that lies mostly within Canada. (348)

capillary water Water stored in the form of tiny droplets that are held together by the molecular attraction between the water and soil particles. (106)

carbonate A compound containing a unit consisting of one carbon atom surrounded by three oxygen atoms. (154)

carbon cycle The cycle of carbon atoms from the atmosphere into living organisms and then back again. (322)

carbonization The formation of a fossil through the gradual decay of organic material, leaving only a thin film. (329)

cast (fossil) A mold that is filled with minerals deposited by ground water. (330)

celestial equator A circle on the celestial sphere in the same plane as the earth's equator. (479)

celestial poles Two opposite points where the extensions of the earth's poles intersect the celestial sphere. (479)

celestial sphere An imaginary sphere with the planets and stars appearing to be located on its surface and with the earth at its center. (479)

Cepheid variable A star with changing

luminosity that pulsates in a regular way. (488)

change of phase A change in a substance that occurs as a result of such processes as evaporation and condensation. Changes are between solid, liquid, and gas phases. (58)

cleavage The tendency of some minerals to break along smooth, flat, parallel surfaces. (154)

climate The history of weather over a period of time. (86)

cold front The interface at which cold air replaces warm air. (72)

comet A floating pile of gravel, dust, and gases that orbits the sun in a flattened ellipse. It is visible only when relatively close to the sun, and then appears as a head with a long vapor tail. (468)

compressional wave A seismic wave whose action is like the expansion and contraction of a spring, which travels at thousands of kilometers per hour. It can travel through solids, liquids, or gases. It is also called a P-wave. (374)

condensation The process by which a substance changes from a gas to a liquid. (48)

condensation nuclei The particles on which water vapor condenses to form clouds or fog. (49)

conduction The process by which higher-energy molecules transfer energy to lower-energy molecules. (53)

constellation A pattern of stars that looks like or is named after some object, usually a character from mythology. (479)

contact metamorphism Baking, or changes that occur when molten rock comes in contact with solid rock. (242)

continental climate The climate of the interior of a continent. It has relatively large temperature changes from winter to summer. (92)

continental rise An apron of sediments that have moved down and come to rest in deep water at the base of most continental slopes. (199)

continental shelf The edge of a continent that gently slopes into the ocean. (198)

continental slope The region beyond the continental shelf where the sea floor dips more steeply. (199)

contrail A white trail of ice-crystal clouds made by high-flying jet planes. (50)

convection The actual movement of heated substances. (54)

convection current A stream of moving heated substance. (55)

core Central region, as the core of the earth. (149)

Coriolis effect The deflection of air moving over the earth, due to the earth's rotation. (73)

correlate To match up. (312)

covered shield A part of a continental shield that is covered by a relatively thin layer of flat-lying sedimentary rocks. (346)

creep The slow downward shifting of earth material, due to gravity. (183)

crescent (moon) The appearance of the moon during the first quarter, when only a small portion, with convex and concave sides ending in points, is visible. (442)

crest (wave) The top of a wave. (127)

cross-bed A thin rock layer that lies at an angle to the larger layer that contains it. (281)

crust The upper part of the lithosphere. (39)

cumulonimbus A huge, wedge-shaped, cumulus cloud, a "thunderhead." (78)

cumulus cloud A separate puff or towering mass of clouds made up mainly of water droplets. (50)

cyclone A low-pressure area in which the inward-flowing air moves upward as it gets near the center. (74)

dark-line spectrum A spectrum in which dark lines represent light absorbed by elements in a star's outer regions. (490)

decay product Material that is formed from decay of a radioactive element. (306)

declination A vertical, or north-south angle on the celestial sphere, measured in degrees above or below the celestial equator. (482)

decomposer A small organism that lives off dead plants and animals and releases carbon dioxide into the atmosphere. (322)

decomposition (rock) The breaking up of rock due to chemical weathering. (170)

deformed (rocks) Rocks that are bent, broken, squeezed or stretched. (223)

degree 1/360 of a circle. (17)

delta The sediment deposit formed at the mouth of a river. (197)

dendritic drainage The type of drainage (tree-like) that develops in an area where the bedrock has uniform resistance to erosion. (263)

density The mass of a substance per unit of volume. (5)

desert soil Soil that develops in dry regions; it is rich in minerals and low in nitrogen and humus. (180)

dew point The temperature at which the air becomes saturated with water vapor. (60)

diatom A microscopic plantlike single-celled organism with a silica shell. (206)

direction The geographic relationship of a point to any other point on a map, such as north, south, east, or west. (30)

discharge The amount of water that flows past a stream gauging station in a second. (113)

disintegration (rock) The type of weathering by which rocks break up due to physical processes. (170)

Doppler effect An apparent change in the frequency of light or sound waves when the observer and the source are in motion relative to each other. As the source approaches, the frequency increases; as the source moves away, the frequency decreases. (502)

drainage pattern A pattern that forms due to the way water drains across the land in a stream or river system. (263)

drift Rock debris that is deposited by a melting glacier. (362)

eclipse The temporary partial or complete blocking, to the observer, of either the sun or moon by the other. (445)

ecliptic The apparent path of the sun on the sky. (479)

electromagnetic radiation Radiated energy waves with electric and magnetic properties. (490)

electromagnetic spectrum The various types of electromagnetic radiation arranged in order of increasing wavelength. (490)

element Matter made up of atoms that are all essentially the same. (121)

epicenter The point on the earth's surface that is directly above the focus of an earthquake. (12)

epoch A time unit that is a subdivision of a geologic period. (312)

era The largest time unit in geologic history. (312)

erosion The movement of rock and soil particles from one place to another. (183)

eutrophication The last stage in the aging process of a lake, during which there is an overabundance of algae that reduces the oxygen level. (114)

evaporation The process by which a substance changes from a liquid to a gas. (57)

evapotranspiration The combination of water evaporation from the earth's surface and transpiration from plants. (109)

face (crystal) A smooth surface of a solid mineral crystal. (149)

fault A crack in the crust of the earth along which rocks have moved. (223)

fault block A block of crust that tilts or moves up or down. (223)

first quarter The phase of the moon occurring during the first seven to eight days after the new moon. The moon appears to grow from a crescent to a half-moon. (442)

flash flood A sudden, violent flood after a heavy rainstorm. (114)

focus The point of origin of an earthquake. (12)

fog A cloud that rests on the earth's surface. (52)

folded mountain Part of a mountain belt that consists largely of sedimentary rock that has been squeezed into tight folds during the mountain-building process. (346)

foraminifera Microscopic single-celled organisms that form carbonate skeletons. (206)

force of gravity The force by which objects attract each other. (6)

forest soil Soil that forms in regions of high rainfall and is acidic, being low in calcium and magnesium. (177)

fossil The remains or traces of once-living organisms found in sedimentary rock. (324)

fossil correlation Matching rock layers

by using fossils that are characteristic of particular layers. (312)

fossil fuel Fuel formed from the remains of prehistoric plants and animals. (322)

fossiliferous Containing fossils. (277)

frequency (waves) Wavelengths per second. (503)

front (weather) A boundary between air masses. (75)

frost Water vapor that changes directly to ice as a result of freezing before condensation can occur. (60)

full moon The phase of the moon when the earth is between the sun and the moon. The lighted side faces the earth and the moon appears as a complete disk. (442)

fusion The combining of two or more atomic nuclei to form a single, more massive nucleus, with a release of energy. (474)

galaxy A large collection of gases, dust, and billions of stars, that has a more or less definite overall shape. (498)

geosyncline A large depression in the earth's crust where marine deposition occurs. (217)

geothermal gradient The rate of temperature increase with increasing depth toward the center of the earth. (383)

glacier A huge, slowly flowing mass of ice. (102)

grassland soil Thick, fertile soil, rich in humus from the decay of grass plants. (178)

ground water Water below the surface of the water table. (107)

guide fossil A fossil that is characteristic of a certain geologic time. It is also called an index fossil. (315)

Hadley cell A circulation system due to convection, in which air rises at one place and sinks at another. (84)

hail Balls or irregular lumps of ice that fall from cumulus clouds. (63)

half-life The amount of time it takes for half of the radioactive atoms in a sample to decay. (305)

halide A compound that contains chlorine or fluorine together with sodium, potassium, or calcium. (154)

hardness The resistance of a mineral to scratching. (154)

heat flow Release of heat from within the earth by conduction or radiation at the earth's surface. (383)

heat flow province An area within a continent or ocean where the amount of heat reaching the surface has been mapped. (384)

heat sink A place that has the capacity to remove a large amount of heat from the atmosphere and hold it. (56)

heat source A place that loses heat to the atmosphere. (56)

highland (lunar) A light area of the moon that has mountains, plateaus, and craters. (431)

high tide The tide when the water in the ocean reaches its highest level along a particular shoreline. (439)

horizon (soil) A layer of soil. (175)

humus Decayed organic matter. (176)

hurricane A tropical cyclone in the Atlantic Ocean region. (79)

hydrosphere The surface water on the earth. (39)

hypothesis A possible but not a positive answer to a question. (6)

ice sheet A glacier so large that it covers all or a large part of a continent. (186)

igneous rock Rock that solidified from a molten state. (159)

immature soil Poorly developed soil with only a few horizons. (175)

impact crater A crater formed by a meteoroid. (267)

index fossil A fossil that is characteristic of a certain geologic time. It is also called a guide fossil. (315)

inertia The tendency of an object to remain at rest, or in uniform motion in a straight line, unless a force acts upon it. (21)

inorganic compounds Chemical compounds that are not made up of organic material. (321)

insolation The incoming solar radiation. (86)

intensity (earthquake) The strength of an earthquake's motion in a given place based on the amount of destruction that occurs. (376)

interface Any place where two or more earth-spheres meet. (146)

International Date Line The meridian that is opposite the Prime Meridian. To the east of this line, the calendar date is one day earlier than it is to the west. (27)

Intertropical Convergence Zone The zone where the trade winds of the Northern and Southern hemispheres meet. (85)

intrusive igneous rock Plutonic rock. It cooled below the surface of the earth. (239)

invertebrate An animal without a backbone. (335)

ion An atom or a group of atoms that has an electrical charge because it has gained or lost one or more electrons. (121)

isobar An isoline connecting points of equal atmospheric pressure on a map. (72)

isoline A line on a map that connects points where a particular condition, such as temperature, pressure, or sound, is the same. (71)

isotherm An isoline connecting places where temperatures are equal. (71)

isotope One of two or more kinds of atoms of an element that differ only in mass. (151)

jet stream The strong current within the upper level westerly winds. (80)

kinetic energy The energy that molecules or other objects possess due to their motion. (49)

last quarter The phase of the moon occurring during the first seven to eight days after the full moon. The moon appears as a crescent. (443)

latent heat The heat energy that is taken up by a substance during melting and evaporation. (59)

latitude Locations or distances on the earth north or south of the equator. (26)

leaching The process by which ground water carries soluble material when it moves through soil or rocks. (173)

light-year The distance that light can travel in one year. One light-year is equal to about 0.3 parsec or almost 10 trillion km. (486)

lithosphere The solid portion of the earth's crust. (39)

loam A soil that is a sandy and clayey mixture. (106)

longitude Locations or distances on the earth east or west of the Prime Meridian. (26)

Love wave A seismic wave that has no vertical movement and produces a shearing motion in the ground. (375)

low tide The tide when the water in the ocean reaches its lowest level along a particular shoreline. (440)

luminosity The total amount of energy a star is sending out. (487)

lunar eclipse An eclipse of the moon caused by the earth being lined up directly between the sun and the moon. (445)

luster The shiny appearance of a mineral that reflects light. (156)

magma Molten rock. (237)

magnitude

- **earthquake** The amount of energy released at an earthquake's focus. (375)
- **star** A measure of a star's brightness. (486)

mantle A thick shell of the earth below the asthenosphere, that behaves like a solid. (149)

mare (pl. maria) A large dark flat area on the moon. (430)

marine climate A climate in a seacoast region, that has relatively small temperature changes from winter to summer. (92)

mass The amount of matter a substance contains. (5)

measured time scale A scale that tells how long ago events took place. (302)

meridian An imaginary line on the earth that marks longitude and that extends from one pole to the other. (26)

metamorphic rock Sedimentary or igneous rock in which the minerals or texture or both have been changed by high temperature and pressure without melting. (159)

meteor The luminous tail or streak caused by the burning of a meteoroid that has entered the earth's atmosphere. (467)

meteorite A meteoroid that survives passage through the earth's atmosphere and falls to the earth's surface. (467)

meteoroid Any piece of rock or mineral matter traveling through space. (432)

microfossil A fossil that is so small it must be studied with a microscope. (323)

mineral A naturally occurring inorganic material with a definite chemical composition and molecular structure. (149)

minute 1/60 of a degree, in measurement. (17)

mobile belt An area of the earth's crust where large amounts of force and heat energy result in a mountain system. (236)

model An illustration (on paper or three-dimensional) that helps a person understand an idea. (8)

Modified Mercalli Scale of Earthquake Intensity A scale that uses roman numerals to indicate varying degrees of earthquake intensity from slight to nearly total destruction. (376)

Moho A commonly used short name for "Mohorovičić discontinuity." (381)

Mohorovičić discontinuity The zone in the lithosphere where there is an abrupt change in the speed at which earthquake waves travel. (381)

mold (fossil) Space left in a rock when the remains of an organism have dissolved. (330)

moraine A pile of glacial till built up at the edge of a glacier. (394)

mountain soil Soil that is usually rocky and thin, formed on slopes where mature soil cannot develop. (177)

natural selection The process by which the individuals in a population that are best suited to their environment are most likely to survive and pass their life-saving characteristics on to their offspring. (333)

nebula A huge mass of dust and gas in space. (508)

neutrino An uncharged particle in an atom, smaller than an electron. (493)

neutron star The last stage in the life of some stars, in which the star consists mostly of neutrons and is very dense. (508)

new moon The phase of the moon when the moon is between the sun and the earth. The lighted side is away from the earth and the moon appears invisible. (442)

ore mineral A mineral that has enough of a particular element in it to be worth mining. (154)

organic compound A compound containing carbon atoms that join with one another and with atoms of other elements, especially hydrogen and oxygen. (321)

organic matter The remains of dead organisms broken down by bacteria, molds, and funguses. (173)

outwash Clay, sand, and gravel that have been deposited by streams flowing from melting glaciers. (363)

oxide A compound made up of oxygen and one or more other elements. (154)

P (wave) A compressional earthquake wave. (374)

paleontologist An earth scientist who studies fossils. (323)

parallax The apparent movement of an object caused by a change in the position of the observer that gives a new line of sight. (485)

parallax angle The angle formed at an object whose distance is to be measured, from two lines of sight. (485)

parent material Original radioactive material that is decaying. (306)

parent rock The original bedrock from which a soil forms. (177)

parsec An astronomical distance unit obtained by dividing the parallax angle of an astronomical object into one. (486)

partial eclipse An eclipse during which the sun or moon is partly blocked. (446)

peat A mixture of plant fragments formed from partly decayed vegetation, that is slowly built up in swamps. (335)

period

- **of a Cepheid variable star** The time it takes a Cepheid variable star to change from its greatest brightness to its least brightness and return to its greatest brightness again. (488)
- **unit of geologic time** A smaller unit of time than an era; the next unit in length below an era. (312)
- **of waves** The time it takes for the crests of two waves to pass a given point. (127)

permafrost The deeper layers of tundra soil that remain permanently frozen. (180)

permeability The rate at which water can pass through a porous material. (104)

petrify To turn to stone. (329)

phase (lunar) One of the apparently different shapes of the moon that recurs in cycles. (442)

plain A large area that is flat or gently inclined. (257)

plate (geology) A section of the earth's crust that moves on top of a convection cell that flows through the upper mantle. (413)

plateau A large elevated tableland. (258)

plate tectonics theory The idea that there are six large crustal plates, and many smaller ones, that move around on the surface of the earth in a way that can be calculated. (417)

pluton A mass of magma that cools and solidifies beneath the earth's surface to form a large body of rock. (237)

plutonic rock Igneous rock that solidifies below the earth's surface. (159)

pore A hole in loose soil or rock. (103)

porosity The percentage of space between soil or rock particles. (104)

porous A term used to describe rocks with many pore spaces. (103)

prairie soil Fertile soil in a region of high rainfall. It has a deep topsoil rich in humus. (180)

precipitation The process by which water vapor molecules condense to form drops or ice crystals that are heavy enough to fall to the earth's surface. (48)

pressure The force with which air or other molecules strike a surface. (59)

Prime Meridian The zero meridian on the earth, which passes through Greenwich, England. (26)

principle of superposition The idea that the oldest bed in a sequence of rock layers is the one on the bottom. (312)

principle of uniformitarianism The idea that the same processes that affect the earth today also affected the earth in the past. (312)

prognostic map A weather map that predicts the weather, based on past weather data. (82)

pulsar A rapidly spinning neutron star that gives out bursts of energy. (509)

quasar A star-like object that gives off radio waves and visible light. It has great speed, energy, and is a great distance from the earth. (504)

radial drainage The drainage pattern, typical of volcanic mountains, in which streams flow outward in all directions from a central area. (264)

radiation The waves of energy that travel through space. (53)

radioactive decay The process by which a radioactive element breaks down to form another, more stable element. (304)

radioactivity The giving off of energy and charged particles by certain atoms. (304)

Rayleigh wave A seismic wave that passes through surface rock and produces both horizontal and vertical motions. (375)

red giant A huge, red, cooling star. (508)

regional metamorphism Metamorphism that occurs over a large area when the heat from a pluton slowly flows outward and changes rocks over a long distance. (243)

relative humidity The ratio of the actual amount of moisture in the air to the maximum amount it could hold at the same temperature. (60)

relative time scale The listing of events in history in a sequence. (302)

replacement (fossilization) The process by which hard parts of dead organisms are replaced by mineral matter. (329)

Richter Scale A scale that describes the force of an earthquake independently of its effects on people and civilization. (375)

right ascension A horizontal angle measured in hours, minutes, and seconds around the celestial equator. There are 24 hours, measured eastward from the place on the sky where the apparent path of the sun crosses the celestial equator. (482)

rigid The condition of materials that tend to resist twisting. (224)

rille A long narrow depression or valley on the moon's surface. (431)

ripple marks Small ridges that are formed by wave action on sediment in shallow water. (219)

rocks The basic solid material of the earth's crust (and the crusts of moons and some other planets). (145)

runoff Water that flows on the surface or through the ground into streams and lakes. (110)

S (wave) A shear type earthquake wave. (374)

salinity The number of grams of material dissolved in 1000 g of seawater. (123)

saturation A condition of balance in which an equal number of molecules leave a water surface and return to it. (60)

scale (map) The fixed ratio of the distances shown on a map to the actual distances on the earth. (31)

scale model A model that represents a real object at a smaller or larger size. (8)

sea-floor spreading The process by which material rising at the ocean ridges moves toward the ocean trenches, carrying the sea floor with it. (411)

seamount An underwater mountain. (230)

second 1/60 of a minute. (17)

sedimentary rock Rock made up of fragments of other rocks and minerals, usually deposited in water. (159)

seismic wave A wave that radiates from the point of origin of an earthquake. It moves in all directions through solid rock. (373)

seismograph A very delicate instrument that detects passing earthquake waves. (11)

seismologist A specialist in the study of earthquakes. (375)

shear wave A secondary earthquake wave that causes individual rock particles to vibrate from side to side at right angles to the direction that the wave is traveling in. A shear wave cannot pass through liquids. It is also called an S-wave. (374)

shield A large area where Precambrian rocks are exposed. (346)

sialic silicate A silicate mineral rich in silicon and aluminum, such as feldspar, quartz, or white mica. (153)

silicate A mineral containing units of oxygen and silicon atoms. (153)

simatic silicate A silicate rich in magnesium and silicon, such as pyroxene, amphibole, olivine, and black mica. (153)

sleet Ice pellets that form when raindrops or partly melted snowflakes fall through a layer of cold air and freeze. (63)

soil Loose weathered material at the surface of the earth in which plants can grow. (171)

soil profile The layers in a particular sample of soil. (175)

solar eclipse An eclipse of the sun caused by the new moon being lined up directly between the earth and the sun. (445)

species A group of organisms that can breed and produce fertile offspring. (332)

specific gravity The mass of a substance compared to that of an equal volume of water. (154)

spectrogram A picture of a spectrum. (493)

spectroscope An instrument used to observe a spectrum or measure wavelengths. (490)

stable (atmosphere) The state when the atmosphere resists either rising or sinking motion. (76)

stratus cloud A sheet or layer of cloud particles that covers a large portion of the sky. (50)

streak (mineral) The color of a powdered mineral. (156)

streamline (weather map) A line along which a major wind current is blowing at a given time. (72)

subsoil The middle layer of soil, containing small particles and minerals washed down from above. (176)

sulfate A compound containing a unit that is made up of one sulfur atom surrounded by four oxygen atoms. (154)

sulfide A compound containing one or more metals combined with sulfur. (154)

sunspot A relatively dark, cool area on the surface of the sun that is a region of violent activity. (475)

supernova The explosion of a neutron star—the brightest known object in space. (509)

temperature A measure of the average kinetic energy of molecular motion. (49)

texture (rock) The size and arrangement of mineral grains. (159)

thermistor An instrument that measures heat flow in drill holes and mines. (383)

thrust fault A fault zone along which a great block of crust slides on top of another block. (223)

till An unsorted, unlayered jumble of clay, sand, pebbles, and boulders that settles out of a glacier as the ice melts. (362)

topographic map A map that shows the landscape features of an area. (34)

topsoil The uppermost layer of soil, in which humus collects. (176)

total eclipse An eclipse in which the sun

or moon is completely blocked and appears in total darkness. (446)

trace element An element that is found in very small quantities. (125)

trace mineral A mineral usually present in very small quantities that often can be used to determine the origin of a sedimentary rock. (276)

trade wind A more or less continuous wind that blows from the east or northeast at low latitudes. (83)

transpiration The process by which moisture is carried through plants from the roots to the leaves, where it changes to vapor and escapes to the atmosphere through leaf openings. (109)

travel-time graph A graph that shows the difference in travel-time of P and S earthquake waves in relation to their distance from the epicenter. (378)

trellised drainage The ladder-like drainage pattern typical of areas where folded sedimentary rocks have eroded into parallel ridges and valleys. (263)

trough The bottom of a wave. (127)

tsunami The sea wave created by an earthquake or the eruption of a volcano. (226)

tundra soil Soil formed in regions of low yearly average temperatures, slight rainfall, and slow evaporation. (180)

turbidity current A mass of stirred up sediment that flows down a submarine canyon and spills out onto the deep-sea plain. (201)

typhoon A tropical cyclone in the Western Pacific Ocean. (79)

ultraviolet radiation A very short-wave radiation of high energy. (87)

unconformity A zone where rocks of different ages meet. The zone represents missing rocks, a gap in the rock record. (265)

unstable (air) A condition in which air continues to move and gain speed if it is given a slight push upward or downward. (76)

valley glacier A glacier that moves downslope by following an old streamcut valley. (185)

variation (organisms) The differences between individuals within a species. (332)

vertebrate An animal with a backbone. (335)

viscosity The resistance to flow in liquids. (224)

visible spectrum White light that has been passed through a prism and appears as a continuous band of colors ranging from violet to red. (489)

volcanic rock Igneous rock that formed from magma that has reached the earth's surface by volcanic eruptions and then solidified. (159)

volume The amount of space a substance occupies. (5)

vortex A whirlpool. (79)

warm front The interface at which warm air replaces cold air. (72)

water budget A water balance. It accounts for the income, storage, and loss of water over an area. (109)

water cycle The exchange of water among the ocean, the air, and the land. (48)

water table The top of a saturated soil zone. (107)

water vapor Water in the form of a gas. (46)

wave height The height of the crest of a wave above the trough. (127)

wavelength Distance between crests. (127)

weather The condition of the atmosphere at any one time and place. (11)

weathering The natural process of breaking down rocks. (170)

white dwarf A dense, iron-rich, white-hot, star. (508)

zenith The part of the celestial sphere that is directly above an observer. (478)

Acknowledgments

Technical diagrams by Vantage Art, Inc.
Investigation drawings by Les Morrill.

Cover photo © Ted Grant, Masterfile, Toronto

ii M. Borum, The Image Bank
xvi Steve McCutcheon, Alaska Pictorial Service
1 Courtesy of Judith Wray
2 Englebert/Photo Researchers, Inc.
4 (both) John S. Shelton
5 ESCP
7 Chalmer Roy
8 (top) Grant Heilman, (bottom) AGI-EBF film: "Why do we still have mountains?"
14 Georg Gerster/Photo Researchers, Inc.
24 (both) Lick Observatory Photographs
25 Philip Jon Bailey
32 (top) ESCP, (bottom) T. S. Lovering/U.S. Geological Survey
33 T. S. Lovering/U.S. Geological Survey
34 U.S. Geological Survey
38 (both) Courtesy of Donnelley Printing Company, Cartographic Services
42 © 1980 Bill Staley, After-Image, Inc.
43 Gary Ladd
44 Lynn M. Stone/Photo Researchers, Inc.
51 (a. and e.) Howard Bluestein, (b.) Tad Nichols, (c.) Jim Harrison, (d.) © Michael Philip Manheim 1975
52 (top) Grant Heilman, (bottom) Tad Nichols
63 Clyde H. Smith/Peter Arnold Photo Archives
64 NCAR
66 Tennessee Valley Authority
68 Frederick Ayer/Photo Researchers, Inc.
75 NOAA
78 Howard Bluestein
79 (left) NOAA photo by J. H. Golden, (right) NOAA
80 NASA
86 Environmental Satellite Service/NOAA
94 (left) Kim Steele, (right) Harrison Forman
96 (left) ESSA, (right) Ellis Herwig/Stock, Boston
98 Fredrik D. Bodin
101 Ted Spiegel
103 Russ Kinne/Photo Researchers, Inc.
104 Josef Muench
111 Gary Settle for the Topeka Capital-Journal
115 © Freda Leinwand, Monkmeyer Press Photos
118 Dr. Don James/Surf Photos
121 Carola Gregor
122 Courtesy of Morton Salt Company, Division of Morton-Norwich Products, Inc.
124 Trustees of the Boston Public Library
128 (top) Scripps Institution of Oceanography, (bottom) Jan Hahn
130 Scripps Institution of Oceanography
131 Werner Wolff/Black Star
139 NOAA
140 (left) Grant Heilman, (right) Shell Oil Company
142 Jay Lurie
143 Gary Ladd
144 Camera Hawaii
147 (top) NASA, (bottom) Philip Jon Bailey
148 (a.) Charles Herbert/Western Ways Photo, (b.) Courtesy of the American Museum of Natural History, (c.) Joel E. Arem
150 (a. and c.) Lee Boltin, (b.) Courtesy of the American Museum of Natural History
153 (a., b., c. and d.) M. W. Sexton, (e.) Lee Boltin
155 (a., b. and d.) M. W. Sexton; (c.) Runk/Schoenberger, Grant Heilman
158 (a.) Courtesy of the Diamond Information Center, (b., c., d., e., f. and g.) Joel E. Arem
159 Fredrik D. Bodin
160 (top) Solarfilma, (center) Werner Stoy/Camera Hawaii, (bottom left and right) Jerome Wyckoff
161 From "World Changes and Chances: Some New Perspective for Material," Radcliffe, S. V., *Science,* Vol. 191, pp. 700–707, Fig. 1, 20 February 1976. Copyright 1976 by the American Association for the Advancement of Science.
162 From "Limits to Exploitation of Nonrenewable Resources," Cook, E., *Science,* Vol. 191, pp. 677–682, Table 1, 20 February 1976. Copyright 1976 by the American Association for the Advancement of Science.
163 From "Electronic Material, Functional Substitutions," Chynoweth, A. G., *Science,* Vol. 191, pp. 725–732, Table 1, 20 February 1976. Copyright 1976 by the American Association for the Advancement of Science.
164 From "International Trade in Raw Materials, Myths and Realities," Fried, E. R., *Science,* Vol. 191, pp. 641–646, Table 2, 20 February 1976. Copyright 1976 by the American Association for the Advancement of Science.
165 U.S. Dept. of Labor
168 Cary Wolinsky, Stock, Boston
171 (left) Robert C. Frampton, (right) Henry D. Foth
173 Fredrik D. Bodin
174 David Muench
175 (both) Soil Conservation Service
179 (top left) Soil Conservation Service; (top right) Grant Heilman; (bottom left) David Muench; (bottom right) USDA.
180 © LXXIX Gary Braasch
181 (both) Jeff Albertson/Stock, Boston
182 David Muench
183 (left) J. R. Stacy/U.S. Geological Survey, (right) Robert C. Frampton
184 (top) Soil Conservation Service, (bottom) Vernon Roby/U.S. Navy

185 (top) Alan Pitcairn/Grant Heilman, (bottom) Bradford Washburn
186 (top) Geological Survey of Canada, Ottawa, (bottom left) Loic Jahan/Photo Researchers, Inc., (bottom right) Geological Survey of Canada
187 (top) M. P. Kahl, Bruce Coleman, Inc. (bottom) U.S. Geological Survey
188 Josef Muench
189 (top) Soil Conservation Service, (bottom) © M. Woodbridge Williams, After-Image, Inc.
190 Fredrik D. Bodin
191 Josef Muench
192 U.S. Dept. of the Interior, Bureau of Land Management
194 Ron Church/The Sea Library
197 NASA Photo/Grant Heilman Photography
198 NASA
201 Conrad Limbaugh, Scripps Institution of Oceanography
205 (top) George S. Sheng, (bottom) NASA, Lyndon B. Johnson Space Center
206 (top left) Lester V. Bergman Associates, (top right and bottom) Trustees of the Boston Public Library
207 (left) From "Manganese Nodules as Indicators of Long-Term Variations in Sea Floor Environment," by Ronald K. Sorem, Washington State University, (right) Trustees of the Boston Public Library
208 Laurence R. Lowry
209 Scripps Institution of Oceanography
212 (left) Yoram Kahana/Peter Arnold Photo Archives, (right) John Running/Stock, Boston
214 Grant Heilman
218 (1., 2., 3., 4., 6., 7., 8. and 9.) ESCP, (5.) Courtesy of the American Museum of Natural History, (10.) John J. Thomas
219 (left) J. R. Stacy/U.S. Geological Survey, (right) L. W. LeRoy
220 (left) Van Bucher/Photo Researchers, Inc., (right) Texas Highway Department
221 Reprinted by permission from Shepard, *Geological Oceanography,* 1977, Crane, Russak & Company, New York.
223 (left) George Whitely/Photo Researchers, Inc., (center) Jerome Wyckoff, (right) Robert C. Frampton
227 © 1980 Gary Braasch
228 Virginia Carelton/Photo Researchers, Inc.
230 Furnished courtesy of H. S. Fleming and W. L. Brundage, Naval Research Laboratory, Washington, D.C.
231 Undersea Media
234 Josef Muench
237 David Muench
243 Robert A. Olson, Slidell, Louisiana
248 Josef Muench
250 Peabody Museum of Salem, photo by M. W. Sexton
252 David Muench
255 From *Geology Illustrated* by John S. Shelton. W. H. Freeman and Company. Copyright © 1966.
256 (both) John S. Shelton
257 (both) John S. Shelton
258 (top) Original photography supplied by the Department of Energy, Mines and Resources, National Air Photo Library, (bottom) John Running/Stock, Boston
259 John S. Shelton
260 John S. Shelton
261 (both) John S. Shelton
262 (top) Editorial Photocolor Archives, (bottom left) George Montgomery, Ames, Iowa, (bottom right) John S. Shelton
264 (all) U.S. Geological Survey
265 John S. Shelton
266 John S. Shelton
267 John S. Shelton
270 U.S. Dept. of the Interior, Bureau of Reclamation
272 Richard Frear/National Park Service
273 Runk/Schoenberger, Grant Heilman Photography
274 J. H. Mercer, Institute of Polar Studies, The Ohio State University
277 (all) Adrian Vance
278 (top) Jerome Wyckoff, (center) Walter Dawn, (bottom) Gordon Anderson
280 David Muench
281 David Muench
287 David Muench
288 L. C. Huff/U.S. Geological Survey
296 Courtesy of Marathon Oil
298 Andrew Rakoczy, Bruce Coleman, Inc.
301 David Muench
303 (a., b. and c.) Courtesy of the Laboratory of Tree Ring Research, University of Arizona, (d.) Courtesy of the American Museum of Natural History, (e.) J. Hoover Mackin
304 D. I. Arnon, P. R. Stout, F. Sipos, "Radioactive phosphorus absorption of tomato fruits at various stages of development," *American Journal of Botany,* Vol. 27, pp. 791–798 (1940).
309 (top) Grant Heilman Photography (bottom) Jordan Tourist Information Office
318 Sovfoto
321 Tom Van Devender/Western Ways Features
322 Walter Dawn
324 (top) David L. Clark, University of Wisconsin, Madison, (bottom) The Pratt Museum of Geology and Natural History, Amherst College
326 (a.) Grant Heilman, (b., c. and e.) The Smithsonian Institution, (d.) Reprinted from *Trilobites: a photographic atlas* by Riccardo Levi-Setti by permission of The University of Chicago Press. © 1975 by The University

of Chicago. All rights reserved. Published 1975. Printed in the United States of America.

327 (a.) Grant Heilman, (b. and c.) Courtesy of the Peabody Museum of Natural History, Yale University, (d.) Courtesy of the Museum of Comparative Zoology, Harvard University

328 (a.) New York State Museum and Science Service, Geological Survey, (b. and d.) Courtesy of the Peabody Museum of Natural History, Yale University, (c.) Illinois State Museum

329 Grant Heilman

330 (a.) National Coal Association, (b.) Walter Dawn, (c.) F. M. Carpenter

335 (both) Elso S. Barghoorn and J. William Schopf

337 (both) The Smithsonian Institution

338 (left) Courtesy of the American Museum of Natural History, (right) Courtesy of the Peabody Museum of Natural History, Yale University

339 Painting by Charles R. Knight, American Museum of Natural History

342 Michael J. Novia, Courtesy of Clark University, Worcester, Mass.

344 Philip Jon Bailey

348 U.S. Geological Survey

349 Marshall Kay

350 Royal Canadian Air Force

351 (a. and b.) E. S. Barghoorn and S. A. Tyler, "Microorganisms from the Gunflint Chert," *Science,* Vol. 147, pp. 563–577, (February 1965). Copyright 1965 by the American Association for the Advancement of Science. (c.) Martin F. Glaessner, University of Adelaide, South Australia

352 Courtesy of Morton Salt Company, Division of Morton-Norwich Products, Inc.

353 Jerome Wyckoff

363 U.S. Navy

366 Yoram Kahana/Peter Arnold Photo Archives

368 Tom McHugh/Photo Researchers, Inc.

369 David A. Burnett, Stock, Boston

370 William A. Garnett

373 Almaden Vineyards, Cienega Winery, Hollister, Cal.

376 © George Hall 1980, Woodfin Camp and Associates

378 From "The San Andreas Fault" by Don L. Anderson. Copyright © 1971 by Scientific American, Inc. All rights reserved.

385 Adapted from p. 84, Sass, J. H., "The Earth's Heat and Internal Temperatures," in *Understanding the Earth,* edited by Gass, I. G., Smith, P. J., and Wilson, R. C. L., M. I. T. Press, Cambridge, Mass., 1971.

386 Donald Dietz/Stock, Boston

388 NASA

392 National Air Photo Library, Canada

396 Adapted from maps by Robert S. Dietz which appeared in *Sea Frontiers,* magazine of the International Oceanographic Foundation.

397 (top) From "The Confirmation of Continental Drift" by P. M. Hurley. Copyright © 1968 by Scientific American, Inc. All rights reserved.

403 Scripps Institution of Oceanography

404 Eros Data Center, Sioux Falls, S.D.

406 Kevin Galvin

408 Eros Data Center, Sioux Falls, S.D.

411 From "Geophysical Illusions of Continental Drift: A Discussion" by Robert S. Dietz and John C. Holden, Fig. 1: American Association of Petroleum Geologists *Bulletin,* Vol. 57, No. 11 (1973), p. 2291.

412 (both) From J. Coulomb, *L'Expansion Oceanique et la Derive des Continents,* Presses Universitaires de France.

413 (top) From J. Coulomb, *L'Expansion Oceanique et la Derive des Continents,* Presses Universitaires de France.

414 From *A Revolution in the Earth Sciences* by A. Hallam, published by Oxford University Press.

416 From J. Coulomb, *L'Expansion Oceanique et la Derive des Continents,* Presses Universitaires de France.

417 (both) From *A Revolution in the Earth Sciences* by A. Hallam, published by Oxford University Press.

419 From "Geosynclines, Mountains, and Continent-Building," by R. Dietz. Copyright © 1972 by Scientific American, Inc. All rights reserved.

420 From "Geosynclines, Mountains, and Continent-Building," by R. Dietz. Copyright © 1972 by Scientific American, Inc. All rights reserved.

421 From "Geosynclines, Mountains, and Continent-Building," by R. Dietz. Copyright © 1972 by Scientific American, Inc. All rights reserved.

426 NASA

427 NASA

428 NASA

431 (both) NASA

432 (both) NASA

433 (top left) NASA, (top right) Lick Observatory Photograph, (bottom) NASA

434 (all) NASA, Lyndon B. Johnson Space Center

435 NASA

436 Hale Observatories

439 (both) Clyde H. Smith/Peter Arnold Photo Archives

443 Hale Observatories

447 (top) Dennis Milon, (bottom) Hale Observatories

448 Ralph K. Dakin

450 Fredrik D. Bodin

452 © Copyright 1965 by California Institute of Technology and Carnegie Institution of

Washington.
456 NASA
460 (all) Lowell Observatory Photographs
461 NASA
462 NASA
463 (top) A P Wirephoto, (bottom) NASA
464 (both) NASA
465 NASA
466 NASA
467 Photograph by Patrick Michaud, courtesy of *Astronomy* Magazine; copyright © 1974 by Astromedia Corp.
468 Courtesy of the American Museum of Natural History
469 (all) Hale Observatories
470 Philip Jon Bailey
472 Naval Research Laboratory photograph
476 (both) Hale Observatories
478 Hale Observatories
480, 481 Adapted from Star Maps © Mark Shafer 1970
489 Bausch & Lomb, Rochester
490 Sargent-Welch Scientific Co.
491 Hale Observatories
492 Harvard College Observatory
496 Hale Observatories
500 (top) Harvard College Observatory, (bottom) Hale Observatories
501 (top) © 1959 by California Institute of Technology and Carnegie Institution of Washington, (center) Harvard College Observatory, (bottom) Hale Observatories
502 (both) Hale Observatories
503 (both) Yerkes Observatory, University of Chicago
504 Harvard College Observatory
509 © Copyright 1959 by California Institute of Technology and Carnegie Institution of Washington.
511 © Copyright 1959 by California Institute of Technology and Carnegie Institution of Washington. Quotation from Sir James Jeans, *The Universe Around Us.* 4th edition, Cambridge University Press, 1960.
512 Joshua Tree/Editorial Photocolor Archives
514 (a.) Courtesy of the Museum of Comparative Zoology, Harvard University, (b.) Culver Pictures, Inc., (c.) Bryn Mawr College Archives, (d.) Courtesy of the University of California, Los Angeles
515 (e., h., i. and j.) The Granger Collection, (f.) Culver Pictures, Inc., (g.) Courtesy of the Museum of Comparative Zoology
516 (k. and o.) Culver Pictures, Inc., (1.) Harvard College Observatory, (m.) Radio Times Hulton Picture Library, (n.) Vassar College Art Gallery, (p.) Courtesy of the Museum of Comparative Zoology
517 (q. and r.) The Granger Collection, (a.) Jerry Cordova, (b.) Courtesy of Dr. Celso S. Barrientos
518 (c.) Roger Robinson for *Science,* (d.) University of Massachusetts News Bureau, (e.) Jonathan Leiby/Woods Hole Oceanographic Institution, courtesy of Dr. Elizabeth T. Bunce, (f.) U. S. Geological Survey, (g.) Courtesy of Dr. Gabrielle H. Donnay, (h.) Frank Medeiros/Woods Hole Oceanographic Institution
519 (i.) Photo by A. H. Coleman. Courtesy of the Museum of Comparative Zoology, Harvard University, (j.) News and Publications Service, Stanford University, Stanford, California, (k.) The University of British Columbia Information Services, (1.) Courtesy of Paleontological Research Institution, (m.) Scripps Institution of Oceanography, University of California, San Diego
520 (b.) Scripps Institution of Oceanography, University of California, San Diego, (p.) Harvard University News Office, (q.) Courtesy of Alonzo Smith, (r.) Ontario Science Centre

INDEX